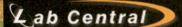 **ab Central** — Electronics Technology

Log On | **Help** | **Contact Us**

Home | **Find Labs** | **My Lab Manuals**

Quick Search: [ ] **Search**

Electronics Technology

My Account

New Account

Electronics Technology

About Elec. Tech.

Faculty Resources

POWERED BY v2

As a leader in the Electronics Technology educational market, Prentice Hall is proud to present the newest, most flexible content solution for your academic laboratory environment. LabCentral for Electronics Technology allows you to create your own custom course materials using individual laboratory experiments from many of Prentice Hall's most successful authors. Along with these handpicked experiments, you can also include your own course-specific materials, guaranteeing that your students receive a truly customized educational product.

To get started, Log On now! Or if you're new to LabCentral for Electronics Technology, create a New Account and start creating custom material today!

Or to learn more about Electronics Technology, take our Guided Tour.

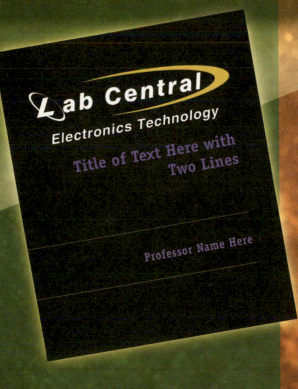

ab Central
Electronics Technology

Title of Text Here with Two Lines

Professor Name Here

As a leader in the Electronics Technology educational market, Prentice Hall is proud to present the newest, most flexible content solution for your academic laboratory environment.

LabCentral for Electronics Technology allows you to create your own custom course materials using individual laboratory experiments from many of Prentice Hall's most successful authors. Along with these handpicked experiments, you can also include your own course-specific materials, guaranteeing that your students receive a truly customized educational product.

www.prenhall.com/electronics

CONTEMPORARY ELECTRIC CIRCUITS

Insights and Analysis

Robert A. Strangeway
Milwaukee School of Engineering

Owe G. Petersen
Milwaukee School of Engineering

John D. Gassert
Milwaukee School of Engineering

Richard J. Lokken
Milwaukee Area Technical College

Prentice
Hall

Upper Saddle River, New Jersey
Columbus, Ohio

Library of Congress Cataloging-in-Publication Data

Contemporary electric circuits : insights and analysis / Robert A. Strangeway . . . [et al.].
  p.   cm.
  Includes index.
  ISBN 0-13-093426-7
  1. Electric circuits. I. Strangeway, Robert A.

TK3001 .C637 2003
621.319′24—dc21

2002021836

Editor in Chief: Stephen Helba
Acquisitions Editor: Dennis Williams
Development Editor: Kate Linsner
Production Editor: Stephen C. Robb
Design Coordinator: Karrie M. Converse-Jones
Cover Designer: Ali Mohrman
Cover art: Digital Vision
Production Manager: Pat Tonneman
Marketing Manager: Ben Leonard
Production Supervision: Lisa Garboski, bookworks

This book was set in Times and Akzidenz Grotesk BE by The Clarinda Company. It was printed and bound by Courier Kendallville, Inc. The cover was printed by Phoenix Color Corp.

Pearson Education Ltd.
Pearson Education Australia Pty. Limited
Pearson Education Singapore Pte. Ltd.
Pearson Education North Asia Ltd.
Pearson Education Canada, Ltd.
Pearson Educación de Mexico, S.A. de C.V.
Pearson Education—Japan
Pearson Education Malaysia Pte. Ltd.
Pearson Education, *Upper Saddle River, New Jersey*

Prentice Hall

10 9 8 7 6 5 4 3 2 1
ISBN 0-13-093426-7

This textbook is dedicated to our spouses,

Debby,

Lorna,

Jeanne,

and

Lisa

whose enduring love and unselfish sacrifices

made this work possible,

and to

Ray Palmer.

Preface

This book is very different from other DC/AC text in very *important* ways. While we heartily agree that the world doesn't need another 1000-plus page book on this well-established topic, we do believe that there is an urgent need for a textbook that *focuses on the primary concepts and techniques* of circuit analysis and *efficiently communicates* them. Therefore, we have streamlined our coverage of DC and AC circuit analysis and produced a book with half the usual page count. Admittedly, every special case is *not* covered. But, then, such exhaustive study diverts the students' attention and dilutes the message.

This book reflects our philosophy of always emphasizing *why* when relating the fundamental concepts. Its aim is to convey the needed core knowledge and to promote students' intellectual growth, not just cover reams of material. After the students understand the fundamentals, then they can build on that knowledge, perhaps exploring the many special cases on their own, as needed.

Basic Approaches and Important Features

The following key features of this approach to DC and AC circuit analysis in this textbook include:

- Efficient yet effective communication of the key principles
- Emphasis on what, how, and *why*
- Identification of practical aspects to keep students connected and motivated
- Elimination of redundancy (and text length) in utilization of key circuit analysis principles ("don't make a special case out of every variation"), yet use of redundancy where needed to reinforce and extend principles (for example, extending a DC concept to AC)
- Use of a "ramp-up" approach of the academic level: the level of the material and students' performance expectations generally increases as one proceeds further into the text (don't submerge the students from the first day; let the students "wade in")
- Discussion that shows students how to think about circuits problems, how to pose questions, how to try approaches, how to formulate strategies, how to solve problems requiring multistep solutions, and so on
- Use of enough but not an overwhelming number of examples to accomplish the desired learning (again, too many examples, especially with too many special cases, dilute the message)
- Effective organization for sound pedagogy: the initial chapters will be mandatory and will be set in the context of a "minimum" DC/AC circuits course, while the subsequent chapters and/or sections will be as independent as possible and selectable as needed for individual programs. Prerequisite chapters and/or sections for later chapters will be clearly established and identified at the end of this Preface.
- Chapter and section organization that is compatible with a single DC/AC course, with a DC/AC multicourse sequence, and with a DC course/AC course sequence

- Selective use of first and second person in the text to *engage* the *student* readers; although not all faculty would write in this manner, the authors have found that this writing style generally helps new college students adapt to the standard third-person writing style in engineering and engineering technology—ours is an appropriate goal for a freshman-level textbook in this field

- Gradual conceptual development of the distinction between circuits and signals

- Inclusion of learning objectives and chapter review concepts to help students realize what the fundamental concepts are, as well as to understand them

- Introduction to the concept of derivations and motivation for derivations

- Incorporation of library and/or web research exercises to expand informational topics and to promote historical awareness

- Incorporation of the role of circuit simulation where simulation is appropriate: manual circuit analysis and analysis by circuit simulation are set in the context of when it is appropriate to use each method; separate publications cover the actual circuit simulation tools (PSpice® and MultiSim®)

- A separate Laboratory Manual that reinforces circuit analysis understanding, utilizes circuit simulation, and helps students to develop electrical measurement skills and insight

- Instructor support materials, including:
  - Laboratory Manual
  - Instructor's Resource Manual with homework and laboratory manual solutions
  - Prentice Hall Test Manager (a computerized test bank)

Who Will Benefit from Using this Text?

The primary anticipated audiences include two- and four-year electrical and electronics engineering technology programs in the first DC/AC circuits courses. The text is a strong candidate for technology programs and nonelectrical/electronics engineering technology programs and for programs where a single circuits course is needed. The text can also be used in an introductory freshman survey course in an engineering program, especially a noncalculus-based course.

Organization and Coverage

The organization of the material in this textbook and the logic behind it is as follows. The first chapter is a discussion of the nature of electricity, why electricity is needed, and where electricity is used, etc. The chapter sets the tone for the readers: the electrical energy viewpoint is introduced, and this viewpoint forms the basis for our approach in subsequent chapters. Some of the motivation includes connecting the course to modern issues, such as safety, electromagnetic compliance, and CE requirements. A section covers scientific notation and unit prefixes. The importance of proper units conversion and an organized method to convert units are then examined. The computational tools used in electronics are discussed to help establish a perspective for students, especially the role of circuit simulation in circuit analysis. From both pedagogical and motivational viewpoints, the authors assert that this crucial chapter should not be "skipped."

Readers will notice the "ramp-up" approach in these first few chapters. The initial content is not introduced full-force, but rather the depth of the topical matter generally increases as the text proceeds, especially in the first few chapters. For example, the discussion of current flow is less intensive than it would be if it occurred later in the course or curriculum, but the depth is sufficient for students to continue building their conceptual foundation and understanding of electronics. Typical engineering technology students are often adapting to college itself as well as adjusting to the expectations of college coursework. Hitting students with everything at once rarely helps this process and may even reduce the students' motivation for the electronics field. A relatively short ramp-up is designed into the initial chapters for this reason.

The concepts and definitions of the fundamental electrical quantities, namely charge, current, voltage, and resistance, are built from the theme of electrical energy and power in Chapter 2. Resistance is established to represent energy conversion. Ohm's law is established as the voltage-current relationship for resistance and is connected to electric power and energy to complete the voltage, current, resistance, energy, and power relationships. Resistors are presented as an important example of resistance. Resistor types are briefly surveyed, and web information research assignments are suggested to expand this coverage. Wire resistance is addressed through the conventional calculations and tables for wires. Meters and other instrumentation coverage are *not* included; they are covered in the Laboratory Manual instead.

The motivation for different circuit configurations is introduced in Chapter 3. The circuit concepts and analysis techniques are covered for series, parallel, and series–parallel circuits. Resistor combinations, Kirchhoff's laws, and the voltage and current divider rules are explained and used. Superposition is established as one method to analyze multiple-source circuits using these same techniques. Practical sources and the current source are examined. Although this chapter is somewhat long, the circuit analysis topics are kept together to emphasize the complementary relationships between circuit analysis laws and techniques—an important concept for students to realize.

The emphasis of Chapter 4 is that an alternating current (AC) sinusoidal steady-state signal is a function of time. The topics covered include what an alternating current signal is, how the plots of sinusoids relate to the actual signal, and how the AC sinusoidal steady-state expressions relate to the plots and actual signals. Peak voltage and current, RMS voltage and current, and power in a resistance are used to continue the theme that electric circuits are used for energy transfer. The similarity of analyses between DC and AC resistive circuits is emphasized. Phasor notation is not introduced until Chapter 8.

The next two chapters utilize electric and magnetic fields to explain capacitor and inductor behavior, respectively. The authors have found that the introduction of too many electromagnetic field concepts this early in an engineering technology curriculum serves to confuse typical students. Students typically respond with fruitless memorization and frustration. Hence, the approach utilized here is to introduce *just enough* electric and magnetic field concepts to accomplish the fundamental goals of understanding how capacitors and inductors respond in DC and AC circuits. The operation of the capacitor is studied with the electric field intensity concept but without the electric flux density concept. The operation of magnets and the inductor is studied with the magnetic flux concept but without the magnetic field intensity concept. The authors are of the opinion that a fuller understanding of electric and magnetic fields should come in a subsequent physics or higher-level electromagnetic fields course, not all at once in a freshman-level electric circuits course.

The electric field concept is explained and used to understand capacitor operation in Chapter 5. Capacitance is developed from the electric field viewpoint, and the parallel plate capacitor is emphasized. Capacitor combinations are explained, and total capacitance is calculated. The action of capacitors in circuits leads to the unifying theme that a capacitor opposes an instantaneous change of voltage. DC transients and AC phase shift are developed qualitatively (quantitative treatments are given in Chapters 7 and 8, respectively). The AC discussion in section 5.5 is repeated in Chapter 8 for those who have a separate DC-then-AC course sequence. Capacitor types are briefly surveyed, and web information research assignments are suggested.

Magnetic fields are studied first in Chapter 6, both from the permanent magnet and electromagnet perspectives. The magnetic field is examined in terms of magnetic flux. The concept that moving charges create a magnetic field is thoroughly established. The magnetic induction concept (Faraday's law and Lenz's law) is explained and used to understand inductor operation. Coil inductance and inductor combinations are explained, and total inductance is calculated. The action of inductors in circuits leads to the unifying theme that an inductor opposes an instantaneous change of current. DC transients and AC phase shift are developed qualitatively (quantitative treatments are given in Chapters 7 and 8, respectively). The AC discussion in section 6.6 is repeated in Chapter 8 for those who have a separate DC-then-AC course sequence.

The existence of different signals is used as the theme to introduce the exponential signal (response) in Chapter 7. The behavior of inductors and capacitors in steady-state DC circuits is covered first. Then basic capacitive and inductive transients are examined without calculus. The unifying themes that a capacitor opposes an instantaneous change of voltage and an inductor opposes an instantaneous change of current continue to be emphasized.

Chapter 8 sets the tone for all subsequent AC sinusoidal steady-state signal circuit analysis. The role of complex numbers is related to steady-state AC concepts. Complex number math is covered in a manner to establish a strong basis for all subsequent AC circuit and power calculations. Reactance is formulated for inductors and capacitors, and the concept of impedance is naturally established with this background. The unifying themes that a capacitor opposes an instantaneous change of voltage and an inductor opposes an instantaneous change of current continue to be emphasized through the development of the $90°$ phase shift between voltage and current for each component. Ohm's law as it applies to AC circuits is established to connect conceptually the previous DC circuit analysis concepts to AC. The AC steady-state responses of RC and RL circuits are developed to underscore the need for and use of phasors.

DC series–parallel circuit analysis with real numbers is extended to AC circuit analysis with complex numbers in Chapter 9. Series, parallel, series–parallel, and multiple-source series–parallel circuits with AC sources are covered. The conceptual similarity of DC and AC series–parallel circuit analysis techniques is emphasized.

AC power is a difficult concept. The approach in Chapter 10 is to unify the understanding through one fundamental equation, namely, the complex power relationship to AC phasor voltage and current. The effectiveness in explaining and learning AC power is through reinforced, redundant use of this key equation. AC power calculation methods are examined for both RC and RL series circuits in the second section. (This organization allows for complex power coverage in a single DC/AC course without coverage of AC series–parallel circuits in Chapter 9). These methods are applied to series–parallel circuits, including those with multiple sources, and to power systems represented by block diagrams. Coverage of power factor correction illustrates an important application of the concepts examined in this chapter.

The unifying theme in Chapter 11 is the efficient analysis of multisource and/or complicated circuits via N equations in N unknowns. The underlying intention is to establish an understanding of these systematic circuit analysis techniques in order to establish a conceptual basis behind circuit simulation software. Mesh and nodal circuit analysis is explained first for DC circuits and then extended to AC circuits. The similarity of the approaches to DC and AC will be emphasized. Bridge circuits and delta/wye circuits are covered. Separate DC and AC sub-sections allow selective use of the DC sub-sections from this chapter (for use in a DC-only course, for example).

The concept of an equivalent circuit is the unifying theme of Chapter 12. The Thevenin and Norton equivalent circuits are examined first for DC signals and then extended to AC. Source models and transformations between these two equivalent circuits are established. The conditions for maximum power transfer to the load are examined in the context of the Thevenin/Norton equivalent circuit. Separate DC and AC sub-sections allow selective use of the DC sub-sections from this chapter (for use in a DC-only course, for example).

Chapter 13 opens by explaining the concept of an induced voltage in one coil due to the AC magnetic field from another coil. Ideal transformers are examined next because they are conceptually easier to understand than mutual inductors, especially in terms of the fundamental voltage and current relationships. Faraday's law is applied to two coils with complete flux linkage. The voltage, current, and impedance relations for the ideal transformer are developed and applied in analysis of circuits that contain an ideal transformer. Aspects of practical transformers are briefly addressed. Faraday's law is extended to two coils with incomplete flux linkage in the optional mutual inductors section. The phasor voltage-current relationships for the mutual inductor are developed and applied to circuits that contain a mutual inductor. A derivation of the conditions under which a mutual inductor can be modeled as an ideal transformer completes the chapter. The mutual inductors section does not contain prerequisite material for any other section of this textbook.

Chapter 14 is used to introduce three-phase circuits, why they are used, and how they are analyzed. This chapter is included for programs that need an introduction to three-phase circuits in their circuits course(s). Delta/wye circuits and conversions are utilized. Series–parallel circuit analysis techniques are utilized.

Chapter 15 is one of the most important chapters for those students who will pursue advanced electrical and electronic circuit studies. The motivation for circuit analysis with frequency as a variable is initially established through a discussion of circuit analysis for any periodic signal that is represented by a Fourier series. The concept that any periodic signal can be represented by a sum of harmonically related sinusoids is established first. The derivation of the Fourier series coefficients is *not* covered. Instead, the Fourier series of common periodic waveforms are given and intuitively justified. Previous AC circuit analysis techniques using complex number arithmetic are extended to circuit analysis with frequency as a variable, i.e., complex number algebra. The usefulness of this circuit analysis is structured by development of transfer functions and graphing Bode plots for first-order filter circuits. Previous AC circuit analysis techniques are extended to the analysis of circuits with periodic waveform inputs.

Chapter 16 begins with discussion of the bandpass filter to establish a motivation for resonant circuits. The resonance concept and resonant circuit analysis are then covered. The theme that frequency is a variable is continued from Chapter 15. The derivation and use of the resonant frequency, quality factor, and bandwidth for resonant circuits is covered. A final optional section introduces the resonant circuit frequency response in the larger context of the Bode plots for second-order circuits.

Chapters are ordered for a two-semester (three-quarter) DC/AC course sequence. Nonetheless, the textbook arrangement is designed to be suitable also for a single DC/AC circuits course and for a two-course sequence with a DC-then-AC format. The following tables are included to assist instructors in the selection of material appropriate for their curricula. The prerequisites to various chapters and/or sections that are provided in Table 1 are

TABLE 1 Chapter and Section Prerequisite Guidelines (a dash means "through")	Chapter (Section)	Prerequisite Chapters (Sections)
	1	None
	2	1
	3	1–2
	4	1–3
	5	1–3
	6	1–3
	7	1–3, 5, 6
	8	1–6
	9	1–6, 8
	10 (sect. 1, 2)	1–6, 8
	10 (sect. 1–4)	1–6, 8, 9
	11	1–6, 8, 9
	12	1–6, 8, 9, 10 (sect. 1, 2)
	13	1–6, 8, 9, 10 (sect. 1, 2)
	14	1–6, 8, 9, 10 (sect. 1, 2), 11 (sect. 4)
	15	1–9, 10 (sect. 1, 2)
	16	1–9, 10 (sect. 1, 2), 15

TABLE 2	Chapter (Section)
Sample Single DC/AC Course Chapters and Sections (a dash means "through")	1
	2
	3 (sect. 1–4)
	4
	5
	6
	8
	9 (sect. 1–3)
	10 (sect. 1, 2)
	13 (sect. 1, 2)
	14

TABLE 3		Chapter (Section)
Sample DC-then-AC Course Sequence Chapters and Sections (a dash means "through")	DC	1–3
		5–7
		11 (sect. 1, 2a, 3a, 4a)
		12 (sect. 1, 2a, 3a, 4a)
	AC	4
		8–10
		11 (sect. 2b, 3b, 4b)
		12 (sect. 2b, 3b, 4b)
		13 (sect. 1, 2)
		13 (sect. 3) is optional
		14 is optional
		15^1, 16

meant to be general guidelines. Sample selections of chapters and sections for the single DC/AC course format and the DC-then-AC courses format are provided in Tables 2 and 3 respectively.

Homework Problems

The authors believe that, beyond reading the textbook, students must also work homework problems for effective learning to occur. A corollary to this statement is that the homework problems must be concentrated on those concepts and techniques that are important and pedagogical to the student learning process. To this end the following general comments are made about the homework problem strategy in this textbook. Homework problems will address the learning objectives presented at the beginning of each chapter. A sufficient but not overwhelming number of homework problems are contained in the textbook proper. Additional test problems are included in the support materials. Expository, analysis, application, and proof homework problems are included. There usually is an even number of each type of analysis homework problems, with answers to odd-numbered problems near the end of the textbook. Circuit simulation is addressed briefly but strategically in the chapter readings, but it is incorporated significantly into the homework problems to help the students understand the role of circuit simulation tools in circuit analysis. This approach is taken to help students distinguish the concepts of circuit analysis from the tools of circuit analysis. Circuit simulation is also used in conjunction with the laboratory experiments. This practice will assist in showing the uses and limitations of circuit simulation software as well as help students develop expectations of experimental results and a better understanding of both simulation and experimental results.

We hope that the approach in and features of this textbook will result in students' understanding and ability to perform DC and AC circuit analysis but who have also developed deeper thinking skills and insight as part of their intellectual journey and professional development.

Acknowledgments

The authors would like to thank our students and colleagues. They have always challenged us to become "better teachers, better explainers, and better professors." Their inspiration was and will always be needed. We especially thank Professor Ray Palmer of MSOE who, unbeknownst to us or him, enabled this project over the last 25 years.

The authors would also like to thank the companies that gave us their permission to reproduce tables and figures in this textbook: Agilent™, Alpha Wire Company™, General Electric™, JW Miller™, KEMET™, Ohmite™, Cadence Design Systems™, Interactive Image Technologies Limited™ (Electronic Workbench) and Sprague-Goodman™.

We also wish to thank the many dedicated and patient people at Prentice-Hall whom we worked with. In particular, the ones that we are aware of are Dennis Williams, Aquisitions Editor, Steve Robb, Production Editor, Kate Linsner, Development Editor, Adam Kloza, Lara Dimmick, Editorial Assistant, Justin Schaefer, Sales Representative, and Lisa Garboski, bookworks, Production Supervision. We also thank all of the Prentice-Hall people who contributed to the success of this textbook but whom we are not aware of.

Reviewers are important to the success of any textbook. They provide a reality check without which any textbook suffers. We wish to thank the following reviewers for their most valuable commodity, their time, and their constructive comments and feedback:

Don Abernathy, DeVry Institute of Technology, Irving; Charles F. Bunting, Old Dominion University; Jeffrey R. Fancher, Western Wisconsin Technical College; George Flantinas, New Hampshire Technical Institute; George Fredericks, Northeast State Technical Community College; Mohamad S. Haj-Mohamadi, North Carolina Agricultural and Technical State University; Walter Hedges, Fox Valley Technical College; Patrick E. Hoppe, Gateway Technical College; Jean Jiang, DeVry Institute of Technology, Atlanta; Kathleen L. Kitto, Western Washington University; Dave Krispinsky, Rochester Institute of Technology; John Lunz, Milwaukee School of Engineering; Joseph C. McGowan, United States Naval Academy; Andrew Milks, Stark State College; Thomas G. Minnich, West Virginia University Institute of Technology; Robert A. Powell, Oakland Community College; and James M. Rhodes, Blue Ridge Community College.

Robert A. Strangeway, Ph.D.
Owe G. Petersen, Ph.D.
John D. Gassert, Ph.D., P.E.
Richard J. Lokken, M.S.

Contents

12   Thévenin and Norton Equivalent Circuits   287

13   Transformers and Mutual Inductors   315

14   Three-Phase Circuits and Power   339

15   Variable-Frequency Circuit Analysis   361

16 Resonant Circuits   403

THE NATURE OF ELECTRICITY

1.1 WHAT IS ELECTRICITY?

When you plug a hair dryer, a curling iron, a shaver, or another appliance into an electrical outlet, you rarely give any thought to the source of the energy that is used to operate those appliances; however, if you think about it awhile, you soon realize that electricity is something special. It is true that you could dry your hair over a campfire and shave with a sharp knife (neither is recommended), but electricity somehow makes these tasks a lot easier. Upon further thought, you might realize that electricity itself is rarely used. Instead, other forms of energy are used: heat to dry your hair, mechanical motion to move blades for shaving, light to see, and so on. In order to understand the role of electricity, one must ask the question: What is electricity?

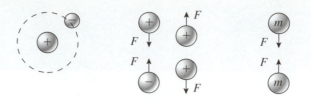

a. Simplified atom   b. Electrical forces   c. Gravitational force

Figure 1.1 Electrical and Mechanical Energy (F = force, m = mass)

Electricity is a form of energy.[1] Energy is the ability to do work (move objects, heat a liquid, etc.; energy is discussed further in Chapter 2). Electrical energy is the energy associated with electric charges. Electric charges are parts of atoms. An atom consists of three main subatomic particles: (*a*) positive charges, called protons, (*b*) negative charges, called electrons, and (*c*) neutral particles, called neutrons. The protons and neutrons are located in the nucleus of the atom; hence, the nucleus is positive, as shown in Figure 1.1a. The electrons orbit the nucleus. Figure 1.1a is a simplistic view of the atom (the atom will be examined in more detail in Chapter 2). The key point here, however, is that the attractive force between a proton and an electron is primarily electrical, not gravitational. In fact, the law of electric charges states:

Like charges repel, unlike charges attract.

Thus, two electrons will repel each other, two protons will repel each other, but an electron and a proton will be attracted to each other. The attraction of an electron and a proton and the repulsion of two protons are illustrated in Figure 1.1b. *The force between electric charges can be used to do work.* If one charge is moved while it is subject to the force of another charge, work is done because there is motion with forces present. Work is the use or conversion of energy. Hence, the electrical force between electrical charges leads to the concept of electrical energy, just as the gravitational force between masses leads to mechanical energy, as shown in Figure 1.1c. Thus, the answer to What is electricity? is that electricity is a form of energy associated with electrical charges.

If electricity is a form of energy that is associated with electrical charges and electricity is not often used directly in the form of electricity, then why is electricity used at all? The answer to this question will reveal why we live in an electronic world. To answer this question, think about how electricity is used. Normally, there is a *source* of electricity: batteries, the power generating station, solar cells, and the like. The source is where the electrical energy is produced. Then there is a *load* for the electricity: lightbulbs, motors, computers, wireless phones, and so forth. The load is where the electrical energy is used. Finally, there is the connection between the source and the load, usually electrically conducting wires (such as copper). The connection of the source to the load with wires forms an electric *circuit,* as drawn in Figure 1.2. What happens when the wires are properly connected between the source and the load?

Electrical energy transfers from the source to the load!

This realization is amazing. Now electricity can be expressed in more precise terms:

The source converts a different form of energy into electrical energy.

Examples of electrical sources are given in Table 1.1.

The load converts electrical energy into one or more other forms of energy.

Examples of electrical loads are given in Table 1.2.

The computer is an interesting example of a load. Electronics are used to perform calculations, to process data, and to complete many other tasks. Thus, electrical energy is

---

[1] Energy is used here in the physics sense and not in the spiritual or other "life force" meanings sometimes described as energy in nonscientific and fictional literature.

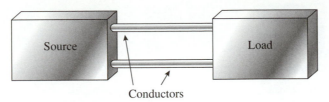

Figure 1.2 The Basic Electric Circuit

TABLE 1.1	Source	Input Energy	Output Energy
Examples of Electrical Sources	battery	chemical	electrical
	solar cell	light	electrical
	generator	mechanical	electrical
	nuclear reactor	nuclear*	electrical

*nuclear → heat → mechanical → electrical

TABLE 1.2	Load	Input Energy	Output Energy
Examples of Electrical Loads	battery charger	electrical	chemical
	lightbulb	electrical	light
	electric motor	electrical	mechanical
	electric stove	electrical	heat

used in its own energy form to perform functions. Also in a computer, electrical energy is used to form light images on a screen, to produce sounds, and to store data on a CD from the built-in laser. The computer also produces heat (a form of mechanical energy). Thus, the computer is an example where electrical energy is converted into several other forms of energy. Finally, a computer often has multiple forms of input energy—mostly electrical but also mechanical in the form of tapping of keys on the keyboard and speaking into a microphone, and in the form of light via the use of a camera.

Finally, the primary purpose of the conductors can now be surmised:

The electrical conductors guide the electrical energy from the source to the load.

Electrical conductors are often copper or other conducting metals, such as aluminum, and are often used in the form of wires. However, other forms of conductors are also used: rectangular bars for high-power applications and flexible strips (check the battery compartment of a flashlight) are examples.

1.2 THE RELEVANCE OF ELECTRICITY TO YOU

Now that you know what electricity basically is you may ask Why study electronics? The first answer that comes to these authors' minds is, *it's fun!* Electronics is a field that brings a lot of pleasure to most of those who "work" in it. We get to figure out how electronic parts, instruments, machines, and gadgets work. We figure out why they work and how to make them work better (more efficiently, faster, more safely, more reliably, etc.). We even get the chance to invent and create new parts, instruments, machines, and gadgets. Best of all, we are paid to have this fun—please do not tell our bosses. We are sure that other fields may make the same claims, but in our opinion, we are even more sure that electronics is the best field to work in. After this course, you will have the opportunity to decide for yourself.

a) blender b) box fan c) coffee maker d) curling iron

e) hair dryer f) hand blender g) hand mixer h) humidifier

i) iron j) microwave oven k) broiler oven l) food processor

m) skillet n) toaster o) vacuum cleaner p) waffle maker

Figure 1.3 Common Electronic Products Photos reprinted with permission of GE.

Figure 1.4 A Typical Specification Label on an Electronic Product Reprinted with permission of Agilent.

Almost all instruments and machines contain electronics. Think about the many appliances that you have around the home (see Figure 1.3): computers, televisions, DVD players, wired and wireless telephones, hair dryers, shavers, radios, boom boxes, stereos, automobiles, garage door openers, lights, furnaces, and the list could go on. If you have the curiosity to figure out how electronic items work, why they work, how to make them work better, and how to create new ones, then *electronics is relevant to you!* Congratulations, you have the opportunity to enjoy your primary career. For those of you who "have to take this course" as part of another discipline, the authors hope that you might learn to enjoy aspects of electronics that will help you enjoy your career field.

If you look at the back or bottom of most electronic items (do take a look!), you will see a label with symbols such as UL (Underwriter's Laboratories) or CE (in English, European Compliance), as well as numbers such as 120 V, 600 W, as illustrated in Figure 1.4. These symbols, words, and numbers mean something. The UL and CE symbols are used to indicate that the electronic item complies with certain safety and performance standards. You need to know electronics in order to understand what these standards are, how the standards are measured, how the standards are met, and so on. The numbers indicate how much electricity can be applied and how much electricity the item will use. You may be wondering why this discussion is so vague. The reason is that we need to learn what the quantities of electricity are before we can even think about rating electronic items or analyzing electric circuits. This viewpoint provides another reason why electricity is relevant to you: in order to understand standards and ratings of electronic items, you must first understand electricity.

Now that we have briefly examined the motivation for the study of electronics, how do you go about studying electronics? First, let us clarify the use of the terms *electricity* and *electronics*. Electricity is a noun that generically describes the phenomenon of electrical energy and its use in parts, equipment, and the like. The adjectives *electric* and *electrical* are used in this context; for example, this book is primarily concerned with electric circuits, which are arrangements of sources, conductors (wires), and loads. *Electronics* is a noun that generically describes equipment, instruments, and so on, that contain electrical parts and devices, especially transistors and integrated circuits, for example (do not worry if you are not familiar with these last two terms—they will come later in your education).

This discussion of electricity and electronics suggests that one should study electric circuits as a lead-in to the field of electronics. This is in fact a good method. We use it because it is the purpose of this book to help you learn, develop thinking skills, and intellectually mature in this study of electric circuits and as a part of your study of electronics in general. In the next chapter, you will learn what the key quantities in electricity are, and then you will use these quantities in electric circuits. Later, we will examine more sophisticated circuits and ways to analyze these circuits. We will introduce additional electrical components in later chapters and we will tie in some applications of electric circuits with these discussions.

The rest of this chapter contains an overview of common mathematical notation and units conversion that are used in electric circuits and comments on some of the tools used in electric circuit analysis. Please do not skip this material because this knowledge is assumed throughout the rest of this book and in future courses.

1.3 MATHEMATICAL NOTATION

The correct interpretation and expression of numbers is important in electronics. Electronic instruments, mathematical calculations, and computer simulations give numerical results. Numbers must be properly used and expressed in calculations. Thus, there are certain mathematical practices that are commonly used: rounding numbers, determining the number of significant digits, determining the range of numbers for a given accuracy specification, expressing numbers and calculations in scientific notation, and converting units, to name a few. These mathematical practices will be addressed in this and the next section.

First, a review of rounding numbers and significant digits (also called significant figures) is given. To round a number, examine the digit to the right of the digit to be rounded. If that digit is 0 to 4, keep the rounded digit the same. This is called "rounding down." If that digit is 5 to 9, increase the rounded digit by one. This is called "rounding up."

Example 1.3.1

Round the following numbers to the number of significant digits indicated:

a. 13.2 to 2 digits
b. 13.2 to 1 digit
c. 12.51 to 2 digits
d. 0.0546 to 1 digit

a. 13
b. 10
c. 13
d. 0.05

Note that in the last case the *leading zeros* are not counted in the rounding. They are not *significant digits*. Significant digits are the digits in a number that are considered when stating the accuracy, precision, or resolution of a number (these terms are discussed later in this section). The leftmost significant digit is called the *most significant digit* because it has the most impact on the size of the number. The rightmost significant digit is called the *least significant digit* because it has the least impact on the size of the number. Nonzero digits are always significant. The following rules apply to zeros in numbers when determining significant digits:

a. Leading zeros (zeros to the left of the leftmost nonzero digit) are not significant.
b. Zeros between nonzero digits are significant.
c. Zeros at the end of a number (right side) that include a decimal point are significant.
d. Zeros at the end of a number that does not include a decimal point may or may not be significant—uncertainty exists.

Example 1.3.2

For each significant digit rule, give examples and determine the number of significant digits.

a. 0.0546 has 3 significant digits
    0.5 has 1 significant digit
b. 203 has 3 significant digits
    0.0030402 has 5 significant digits
c. 20.30 has 4 significant digits
    200. has 3 significant digits
    200.0 has 4 significant digits
d. 1000 has either 1, 2, 3, or 4 significant digits

There are rules for determining the number of significant digits for the result(s) of calculations that involve numbers with a different number of significant digits:

a. When adding or subtracting numbers, the position (10's place, one's place, tenth's place, etc.) of the least significant digit in your answer is determined by the position of the *leftmost* least significant digit of the numbers being added or subtracted.
b. When multiplying or dividing numbers, your answer should have the same number of significant digits as the original number with the *fewest* number of significant digits.
c. Perform all calculations with all digits shown in all numbers, and then round the answer. Never round the numbers before performing the calculations (the reason for this will be demonstrated in the discussion on resolution later in this section).

Example 1.3.3

Perform each calculation and report the answer to the proper number of significant digits.

a. $12.3 + 14.62 = 26.92 = 26.9$

    $0.034 + 0.0528 = 0.0868 = 0.087$

    $2.4673 - 0.032 = 2.4353 = 2.435$

b. $0.437 \times 23 = 10.051 = 10.$

    $9384 \div 9383.9 = 1.000010657 = 1.000$

A separate issue, apart from the number of significant digits in a number, is the "goodness" of the number. How much trust should one have in a number? It might appear at first that the use of a given number of significant digits may imply that all the digits in the number are "good," i.e., that the number is accurate to the number of digits shown except that the least significant digit has some ambiguity due to rounding error. This statement is generally *not* true. There are three terms that are often used in describing numbers that indicate how "good" some component or instrument is: accuracy, precision, and resolution. These three terms will be described and illustrated to show how to determine the "goodness" of a number.

Accuracy is used to quantify how close the quantity is to some absolute standard or nominal value. An absolute standard is set by something or someone else. For example, the number may be provided by an organization that everyone agrees will set the standards for that industry or product. Another example of a standard is a constant of nature, such as the speed of light. Another term that is associated with accuracy is the *nominal value*. The nominal value is the value that the number should be. For example, how close is the power of a 60 watt lightbulb to the nominal value of 60 watts? ±10%? ±5%? ±1%? ±0.1%? (The plus and minus is often implied and not shown.) Whatever the number is, this would be the accuracy. How is this accuracy calculated?

$$\text{accuracy} = \frac{\text{actual value} - \text{nominal value}}{\text{nominal value}} \times 100\% \qquad (1.1)$$

Example 1.3.4

What is the accuracy if a lightbulb with a 60 watt nominal value has an actual power of 57 watts?

$$\text{accuracy} = \frac{\text{actual value} - \text{nominal value}}{\text{nominal value}} \times 100\% = \frac{57 - 60}{60} \times 100\% = -5\%$$

Note: Often, the absolute value of the numerator is taken, and the accuracy result is reported as a positive number:

$$\text{accuracy} = \frac{|\text{actual value} - \text{nominal value}|}{\text{nominal value}} \times 100\% = \frac{|57 - 60|}{60} \times 100\% = 5\%$$

Example 1.3.5

The nominal value for a set of lightbulbs is 60 watts with an accuracy of 10%. What is the range that the power of the lightbulb should be within?

Note that the accuracy is ±10%, i.e., both the plus and minus are implied.

$$\text{maximum value} = 60 + 60(10\%) = 60 + 60(0.1) = 60 + 6 = 66 \text{ watts}$$

$$\text{minimum value} = 60 - 60(10\%) = 60 - 60(0.1) = 60 - 6 = 54 \text{ watts}$$

$$\text{range} = 66 - 54 = 12 \text{ watts}$$

Precision is often confused with accuracy or with other meanings, but it usually means how repeatable a number is. For example, the factory that produces lightbulbs may always make 10,000 lightbulbs to a precision of 1% of one another, but the average power of the bulbs may be 10% from the nominal value of 60 watts. The accuracy is 10% while the precision is 1%. Often, accuracy is given as a published specification but precision is not.

Resolution is the number of digits that the number may have, especially as read from an instrument. For example, consider your calculator. After a series of calculations, it may read out eight digits. Is the number *accurate* to eight digits? Probably not. In fact, many instruments have accuracies on the order of one to one tenth percent. This implies only two or three accurate digits. Then why do these instruments have so many digits on the readout if the digits are not accurate? *Resolution "centers" the range of inaccuracy.* Hence, if the resolution is low, additional error is added to the inaccuracy because the center of the range of inaccuracy is itself inaccurate, as illustrated in Example 1.3.6.

Example 1.3.6

The measured average power of a set of lightbulbs is 60 watts. It is known that the measurement accuracy is 10%. What is the range of power if the resolution on the measured average power of the lightbulbs is (*a*) one digit and (*b*) two digits?

a.   If the resolution on the number 60 is one digit, only the 6 is significant. That means the actual value of the power is from 55 to 64 watts because 55 rounds up to 60 and 64 rounds down to 60:

upper limit due to one significant digit in the number 60:   64

lower limit due to one significant digit in the number 60:   55

Ten percent accuracy is added to the upper number, and 10% accuracy is subtracted from the lower number:

$$64 + 60(0.1) = 70 \text{ W}$$

$$55 - 60(0.1) = 49 \text{ W}$$

Hence, the range for a 10% accuracy and one digit of resolution is:

$$70 - 49 = 21 \text{ W}$$

b. If the resolution on the number 60 is two digits, both the 6 and 0 are significant. That means the actual value of the power is from 59.5 to 60.4 watts, again because of rounding. Ten percent accuracy is added to these numbers:

$$60.4 + 60(0.1) = 66.4 \text{ W}$$

$$59.5 - 60(0.1) = 53.5 \text{ W}$$

Hence, the range for a 10% accuracy and two digits of resolution is:

$$66.4 - 53.5 = 12.9 \text{ W}$$

Notice how the increased resolution reduced the range of values. For comparison, if the 60 W is a nominal value, then the range of power based on the 10% accuracy would be only:

$$60 \pm 10\% = 60 \pm 60(0.1) = 60 \pm 6 = 66 \text{ and } 54, 66 - 54 = 12 \text{ W}$$

Example 1.3.6 illustrates that high resolution is required to effectively remove the imprecision due to rounding the "center" number of the range of values. Then the accuracy specification can be easily used to determine the range of values. If the nominal value is given or known, normally the center value is assumed to equal the nominal value with a high resolution, as shown in the last calculation in Example 1.3.6.

Now that the background has been established for determining accuracy, precision, and resolution of a number, one can see that digits are not always "significant" in a physical sense even though they are significant in the mathematical sense. To complicate matters further, the full number of significant digits is often not shown or indicated. Yet, it is known from the context of the situation that there are more significant digits than shown. A common rule of thumb in these cases is to *use four or five significant digits during calculations and to round the answer to three significant digits* (the reason for this guideline is discussed in Section 1.5). This rule of thumb will be used in this text except where it is important in the particular application to explicitly show the full number of significant digits.

What happens when numbers are very large or very small, i.e., the decimal point is far to the right or far to the left? For example, consider the numbers:

$$1{,}967{,}000{,}000{,}000{,}000 \quad \text{and} \quad 0.0000000000001462$$

It is cumbersome to write all these digits out and, furthermore, error-prone. Instead, *scientific notation* is used: factor out a convenient power of 10 and multiply the resulting number by 10 to that power. For example, the two previous numbers in scientific notation are:

$$1{,}967{,}000{,}000{,}000{,}000 = 1.967 \times 10^{15}$$

$$0.0000000000001462 = 1.462 \times 10^{-13}$$

Notice that if the decimal point in 1.967 is moved 15 places to the right, then the original number results. For the second case, if the decimal point in 1.462 is moved to the left by 13 places, then the original number results. The sign of the power of 10 tells you which way to move the decimal point. A positive power of 10 is a larger number—move the decimal point to the right. A negative power of 10 is a smaller number—move the decimal point to the left.

The usefulness of scientific notation is that the number of significant digits for any number is easy to show—just show the number of significant digits. For example, how does one show that the number 340,000 has four significant digits? The trailing zeros are not significant if the decimal point is not shown. However, using scientific notation:

$$340,000 = 3.400 \times 10^5$$

and the four significant digits are apparent. In the addition and subtraction of numbers that are expressed in scientific notation, the numbers should be expressed with the same power of 10. Also, multiplication and division with numbers in scientific notation is straightforward. From the laws of exponents, add the powers of 10 if the numbers are multiplied and subtract the power of 10 of the denominator from the power of 10 of the numerator if the numbers are divided.

Example 1.3.7

a. Add and subtract 3.47×10^5 and 2.8×10^4
b. Multiply and divide 1,967,000,000,000,000 and 0.0000000000001462

a. $3.47 \times 10^5 + 2.8 \times 10^4 = 34.7 \times 10^4 + 2.8 \times 10^4 = (34.7 + 2.8) \times 10^4 = 37.5 \times 10^4$
    $= 3.75 \times 10^5$

$3.47 \times 10^5 - 2.8 \times 10^4 = 34.7 \times 10^4 - 2.8 \times 10^4 = (34.7 - 2.8) \times 10^4 = 31.9 \times 10^4$
    $= 3.19 \times 10^5$

b. $1,967,000,000,000,000 = 1.967 \times 10^{15}$

$0.0000000000001462 = 1.462 \times 10^{-13}$

product $= 1.967 \times 10^{15} \times 1.462 \times 10^{-13} = (1.967)(1.462) \times 10^{15-13} = 2.876 \times 10^2$

quotient $= 1.967 \times 10^{15} / 1.462 \times 10^{-13} = (1.967/1.462) \times 10^{15-(-13)} = 1.345 \times 10^{28}$

Note that scientific calculators perform these operations automatically; however, it is important that you know these basic mathematical operations with numbers that are expressed in scientific notation so that you can make estimates and identify obvious errors. It is your job, not the calculator's job, to know that the numbers are correct.

Powers of 10 that are multiples of 3 are commonly used in engineering and science, and they are given special symbols. These symbols are used as prefixes to units. For example, a kilometer (km) is 1000 meters (m). This practice saves a lot of writing and is easier to use in conversation. The symbols are listed in Table 1.3 and the use of these symbols is

TABLE 1.3	Symbol	Name	Power of 10
Unit Prefixes: Symbols for Powers of 10	T	tera	+12
	G	giga	+9
	M	mega	+6
	k	kilo	+3
	c	centi	−2
	m	milli	−3
	μ	micro	−6
	n	nano	−9
	p	pico	−12

illustrated in the next example. Note that letter case is important—be careful to show upper- or lowercase.

___ **Example 1.3.8** _____

Convert the following numbers and units to use the common symbols for the powers of 10.

a. 2000 m
b. 0.0001 m

Note from context that the "m" in these numbers is the unit meter, not the unit prefix "milli."

a. $2000 \text{ m} = 2 \times 10^3 \text{ m} = 2 \text{ km}$ (kilometers)
b. $0.0001 \text{ m} = 1 \times 10^{-4} \text{ m} = 1 \times 10^{-1} \times 10^{-3} \text{ m} = 0.1 \text{ mm}$ (millimeters), or
    $0.0001 \text{ m} = 1 \times 10^{-4} \text{ m} = 1 \times 10^2 \times 10^{-6} \text{ m} = 100 \text{ } \mu\text{m}$ (micrometers)

Sometimes the use of these prefixes can be confusing. For example, "min" looks like minutes, not milli-inches. In such cases, a hyphen is often used between the prefix and the units. Can you name these examples: mm, M-in., G-hr, μs, and n-in.?[2]

1.4 UNITS CONVERSION

The proper conversion of units is critical in technology. Recall the incident with the NASA Mars planet orbiter spacecraft in 1999. Teams of engineers and scientists failed to convert the units they used in the software that controlled the spacecraft. The software commanded the probe to perform actions based on the erroneous units, and the spacecraft crashed into Mars instead of orbiting the planet.[3]

Thus, an important aspect of all engineering and technology studies is converting units properly. One reliable method is to treat units like numbers in the sense that they must balance on both sides of the equation. Thus, ratios of units can be used to make units conversions. For example, to convert miles into kilometers, one could use:

$$(1 \text{ mile}) \left(\frac{5280 \text{ ft}}{\text{mile}} \right) \left(\frac{12 \text{ in.}}{\text{ft}} \right) \left(\frac{2.54 \text{ cm}}{\text{in.}} \right) \left(\frac{1 \text{ m}}{100 \text{ cm}} \right) \left(\frac{1 \text{ km}}{1000 \text{ m}} \right) = 1.609 \text{ km}$$

Notice how the miles, ft, in., cm, and m cancel in this conversion. Thus, the conversion factor is 1.609 km/mile.

Also notice that the number of significant digits in the answer does not agree with the number of significant digits in the original numbers. Often, one sees this in units conversions because the numbers involved are exact or may have more significant zeros than shown. The number of significant digits in exact numbers is as many as needed to match the other numbers. For example, there are 12.00 inches in 1.000 ft, 100.0 cm in 1.000 m, and 1000. m in 1.000 km. The number of digits in inexact unit conversions should be looked up. For example, there are 5280. ft in 1.000 mile, and 2.540 cm in 1.000 in.

The strategic arrangement of conversion factors is the key to units conversions. One must arrange the ratios to go from the initial units and end up with the desired final units. The following two examples illustrate this strategy.

---

[2] millimeters, mega-inches, giga-hours, microseconds, and nano-inches
[3] http://www/abcnews.go.com, "Metrics Error Crashes Probe," 11/10/2000.

—— **Example 1.4.1** _____

If the speed of a car is 50.0 miles per hour (MPH), and the driver travels at this speed for half of one day, how many kilometers has the car traveled?

Given: speed $= s = 50$ MPH $= 50$ miles/hr

time $= t = 0.5$ day

The number of significant digits is assumed to be three (from the 50.0 MPH figure), even though the "half day" is ambiguous.

Important Note: Often, the full number of significant digits are not written out during intermediate calculations (to save writing time and space), but then the result is expressed with the proper number of significant digits.

Desired: distance $= d$

Strategy: use the relationship between s, t, and d: $s = \dfrac{d}{t}$

Solution: _____

$$d = st = \left(\frac{50 \text{ miles}}{\text{hr}}\right)(0.5 \text{ day})$$

The units do not match and therefore do not cancel properly. Thus, the approach is to strategically convert hours into days (or vice versa) so that the time units cancel. Also, one must convert miles into kilometers.

$$d = st = \left(\frac{50 \text{ miles}}{\text{hr}}\right)\left(\frac{24 \text{ hr}}{1 \text{ day}}\right)(0.5 \text{ day})\left(\frac{1.609 \text{ km}}{\text{mile}}\right) = 965.4 \text{ km}$$

Hence, $d = 965$ km to three significant digits.

Note the problem solution layout in this example. The *given* information is clearly identified, and symbols are defined and assigned for clarity. The *desired* quantity is clearly identified and assigned a symbol. The *strategy* is determined using the relationship(s) (equations) between the given quantities and the quantity(ies) to be determined. Then the *solution* is obtained by manipulating the equations into a form that expresses the unknown quantity in terms of the known quantities. The values inserted into the equations are clearly shown, along with units, and the final result is expressed with the proper number of significant digits and units.

Why should anyone go to all of this trouble when the calculation could easily be performed on a calculator or in your head? Why show all this work? The showing of this work is *documentation*. Documentation is critical for the following reason: it allows someone else who is not familiar with this work, but who has the proper background, to follow, understand, and reproduce the work. Would you buy a product if you could not understand the operating manual? How much confidence do you have in an explanation that nobody can follow? Would you trust critical results if you could not understand the documentation and reproduce the results? The answers to these questions are self-explanatory.

There are other reasons to document your work carefully. What if the solution to the problem involves five steps and three different equations? Can you and anyone else reading the work hold all of these equations and several seven-digit numbers in you head? How long would it take to reproduce this work without documentation if you had to perform a similar problem a year from now? How could somebody check whether a mistake was made in the original calculations? The authors hope that the explanations in the last two paragraphs convince you to develop good documentation habits because they are critical to your future long-term success and career—remember the Mars probe. Also, consult with your instructor to see if any other particular documentation is required for the course.

___ Example 1.4.2 _____

If an old car gets 15 miles to the gallon, and the driver travels 10 km, how much fuel will be used on this trip?

Given: fuel usage rate $= R = 15$ MPG (miles per gallon)

$d = 10$ km

Desired: amount of fuel $= f$

Strategy: determine the relationship between R, d, and f (it is not obvious): try using the units to determine the relationship:

Solution: _____

$$\left(\frac{\text{miles}}{\text{gallon}}\right)(\text{gallons}) = \text{miles} \rightarrow Rf = d$$

$$f = \frac{d}{R} = \left(\frac{10 \text{ km}\left(\frac{1 \text{ mile}}{1.609 \text{ km}}\right)}{\frac{15 \text{ miles}}{\text{gallon}}}\right) = \left(\frac{10}{1.609} \text{ miles}\right)\left(\frac{1 \text{ gallon}}{15 \text{ miles}}\right) = 0.414 \text{ gallon} = 0.4 \text{ gallon}$$

Thus, the fuel used is 0.4 gallon. Why is only one significant digit shown? The distance is 10 km, and this number technically has only one significant digit. Sloppy! (Or, perhaps, the one significant figure in 10 km is all that is known.) However, recall that the correct number of significant digits is often not shown. Using the rule of thumb that was established earlier, the answer should be 0.414 gallon. More importantly, you should recognize that there is uncertainty in the accuracy of the numbers in this problem and should not draw any conclusion that relies on more than one significant digit of the result.

1.5   COMPUTATIONAL TOOLS IN ELECTRONICS _____

The decimal numbers that are often encountered in electronics show the necessity of computational tools. The handheld calculator is often sufficient. A scientific calculator that can handle complex number mathematical operations and can solve simultaneous equations with both real and complex numbers is recommended. Two calculators that met these requirements at the time this textbook was written are the Texas Instruments TI-86 calculator and the Hewlett-Packard HP-48 calculator. Other models exist, and new models are always being introduced. Consult with your instructor for recommended (or even required) calculators at your college.

The personal computer (PC) is also an important computational tool in electronics. Various circuit simulation and mathematical software packages are commercially available. Circuit simulation software is used to predict the performance of a circuit. The arrangement of the circuit and the values for the components are entered. The type of performance that is desired must also be entered into the software. Then the software is instructed to perform the simulation. Finally, how the results are to be displayed must be entered into the software. This brief overview of the simulation process leads to an important question: What is the role of simulation software in electronics?

Simulation software does not do everything. It does not design the circuit, nor does it say what to do with the circuit. This is where the intelligence of the software user comes in. Usually a circuit and the component values must be entered into the software as a starting point. This is why you must learn and understand how to analyze circuits—you are the one who must provide the initial circuit. Then you can harness the power of the computer. The circuit simulation software can be used to vary component values, perform thousands of repetitive calculations, make graphs, optimize values (find the best values for the desired results), and so on. One may perform these simulations to observe trends for a better insight into the behavior of the circuit. Finally, the real test is the construction and measurement of the circuit—does reality agree with predictions? A better phrased question is: How well does the actual circuit performance agree with the predictions? This is where accuracy comes back into the picture!

From the discussion in the previous paragraph, one can see that the initial values in a circuit need not have more than a few significant digits because the values will be adjusted from the initial values by the simulation runs. This reason is behind the "use four or five significant digits during calculations and round the answer to three significant digits" rule of thumb. It is not a strict rule, just a sound suggestion, given no other constraints in a particular situation.

> *In this text, four or five significant digits will be used during calculations, and the answer will be rounded to three significant digits unless other constraints apply.*

Electronics Workbench® and PSpice® are two common circuit simulation software packages available at the time that this textbook was written. Separate guides for both packages are commercially available. Some homework problems require the use of circuit simulation software. The separate guides should be consulted for background discussion, abbreviated instructions, and examples of circuit simulations.

Mathematical software packages are used to perform mathematical calculations. MATLAB® and MathCAD® are two common mathematical software packages available at the time that this textbook was written. Although the reader is welcome and encouraged to use either of these mathematical packages, most of the calculations in this textbook can be performed using a handheld scientific calculator or with a spreadsheet program, such as Microsoft Excel® or Corel Quattro Pro®. It is recommended that a spreadsheet program or one of the mathematical software programs be used for calculations that are too tedious to be performed repetitively using a calculator.

CHAPTER REVIEW

1.1   What Is Electricity?
 • Electricity is a form of energy.
 • Atoms contain neutrons and charged particles: protons, and electrons.
 • The force between charges is summarized in the law of electric charges: like charges repel, and unlike charges attract.
 • Electricity is a form of energy that is associated with electric charges.
 • Electrical sources convert a different form of energy into electrical energy.
 • Electrical loads convert electrical energy into one or more other forms of energy.
 • Conductors guide the electrical energy from the source to the load.

1.2   The Relevance of Electricity to You
 • It is fun!
 • The plethora of electronic products in today's society means that there are numerous career fields in electronics.
 • One can understand and apply electrical standards and ratings from the study of electronics.
 • *Electricity* is a noun that generically describes the phenomenon of electrical energy and the use of electrical energy in parts, equipment, etc. (adjectives *electric* and *electrical*).
 • *Electronics* is a noun that generically describes equipment, instruments, and systems that contain electrical parts and devices, especially transistors and integrated circuits.
 • The study of electric circuits is a good approach to starting an education in the field of electronics.

1.3   Mathematical Notation
 • A digit is rounded down (stays the same) if the next digit to the right is 0 to 4 and is rounded up if the next digit to the right is 5 to 9.
 • Zeros must be considered in rounding numbers:
   • Leading zeros (zeros to the left of the left-most nonzero digit) are not significant.
   • Zeros between nonzero digits are significant.
   • Zeros at the end of a number (right side) that include a decimal point are significant.
   • Zeros at the end of a number that does not include a decimal point may or may not be significant—uncertainty exists.
 • The number of significant digits in a calculated result depends on the number of signifi-

cant digits in the original numbers and in the types of mathematical operations used in the calculations (addition/subtraction and multiplication/division):

- When adding or subtracting numbers, the position (10's pace, one's place, tenth's place, etc.) of the least significant digit in your answer is determined by the position of the leftmost least significant digit of the numbers being added or subtracted.
- When multiplying or dividing numbers, your answer should have the same number of significant digits as the original number with the fewest number of significant digits.

- Perform all calculations with all digits shown in all numbers, and then round the answer. Never round the numbers before performing the calculations.
- Accuracy is used to quantify how close some quantity is to some absolute standard or nominal value.
- Precision is used to quantify how repeatable a number is.
- Resolution is the number of digits in a number.
- Numbers with high resolution are used to "center" the range of inaccuracy.
- Often, four or five significant digits are used during calculations and the results(s) are rounded to three significant digits. Usually, computational software is used when more significant digits are needed.
- Scientific notation is used to express large and small numbers for convenience. The notation also helps in showing the proper number of significant digits.
- When performing addition and subtraction of numbers that are expressed in scientific notation, the numbers should be expressed with the same power of 10.
- For the multiplication and division of numbers that are expressed in scientific notation, add

the powers of 10 if the numbers are multiplied, and subtract the power of 10 of the denominator from the power of 10 of the numerator if the numbers are divided.
- Unit prefixes are symbols for common powers of 10.

1.4 Units Conversion
- The proper conversion of units is mandatory to obtain valid results. Disastrous results may occur if units are not converted properly.
- A systematic units conversion technique similar to multiplying ratios is a reliable method for converting units.
- Documenting your work is essential so that others may follow and understand it and so that errors can be identified and corrected.
- A systematic problem solution procedure (Given: ..., Desired: ..., Strategy: ..., Solution: ...) is central to good documentation.

1.5 Computational Tools in Electronics
- Calculators with complex number capability are useful in electronics.
- Computers and software are important tools in electronics, but one needs to understand electronics in order to use the tools appropriately and properly.
- Circuit simulation software has a role in electronics:
  - It is used to perform repetitive calculations, prepare charts, optimize values, etc.
  - It is used to obtain better insight into the behavior of the circuit.
  - It is never used to replace understanding circuit analysis.
  - Circuit analysis is essential to obtaining initial values to use in circuit simulation.
- Other mathematical software packages exist to be used in the same manner as circuit simulation software.

HOMEWORK PROBLEMS

Your ability to answer the questions in this section will demonstrate your understanding of what electricity is and why electricity is used. It will also demonstrate your understanding of the function of electrical sources and loads.

1.1 State the type of energy conversion (from _____ energy to _____ energy) for the following items.
a. electric fan
b. battery charger
c. flashlight
d. CD player

1.2 State the type of energy conversion (from _____ energy to _____ energy) for the following items.
a. automotive alternator
b. malt mixer
c. neon sign
d. keyboard

1.3   Identify at least three kitchen appliances that have the UL and/or CE labels on them.

1.4   Identify at least three nonkitchen household appliances that have the UL and/or CE labels on them.

1.5   Name three items that transform electrical energy into another form of energy and describe how that transformation affects your life.

1.6   Name three nonelectrical energy sources that are used to generate electrical energy.

1.7   What is the most common form of nonelectrical energy used to produce electrical energy? What is the source of your answer?

Your correct responses to the questions in this section will demonstrate your ability to express the results of calculations with the proper number of significant digits.

1.8   Round the following numbers to the number of significant digits indicated. Use scientific notation if necessary.
  a. 1009 to 1 significant digit
  b. 1009 to 2 significant digits
  c. 1009 to 3 significant digits
  d. 1009 to 4 significant digits
  e. 0.09326 to 2 significant digits
  f. 0.09326 to 3 significant digits
  g. 9.225 to 3 significant digits
  h. 9.2249 to 3 significant digits

1.9   Round the following numbers to the number of significant digits indicated. Use scientific notation if necessary.
  a. 999 to 1 significant digit
  b. 999 to 2 significant digits
  c. 999 to 3 significant digits
  d. 999 to 4 significant digits
  e. 0.04324 to 2 significant digits
  f. 0.04324 to 3 significant digits
  g. 7.375 to 3 significant digits
  h. 7.3749 to 3 significant digits

Your correct responses to the questions in this section will demonstrate your ability to perform calculations with and express numbers in scientific notation.

1.10   Calculate the result of the following mathematical expressions. Express the answer with the proper number of significant digits. Use scientific notation if necessary.
  a. $1002 + 345.8$
  b. $0.004739 - 0.0002319$
  c. $765.2 \cdot 34$
  d. $0.0937 \div 2284$
  e. $1.002 \times 10^{-3} - 1.002 \times 10^{-4}$
  f. $3.539 \times 10^5 + 8.498 \times 10^4$
  g. $3.405 \times 10^{-2} \div 2.6700 \times 10^{-1}$
  h. $0.9492 \times 10^3 \times 3.61043 \times 10^{-4}$

1.11   Calculate the result of the following mathematical expressions. Express the answer with the proper number of significant digits. Use scientific notation if necessary.
  a. $39.4 + 345.6$
  b. $0.004739 - 0.05828$
  c. $0.09734 \cdot 34.2$
  d. $2590 \div 2289$
  e. $0.997 \times 10^{-3} - 9.997 \times 10^{-4}$
  f. $2.57133 \times 10^5 + 8.228 \times 10^4$
  g. $7.390 \times 10^{-2} \div 2.6702 \times 10^{-3}$
  h. $1.9465 \times 10^{-2} \times 3.639 \times 10^{-4}$

The following questions ask that you demonstrate that you can convert units in an organized manner and perform calculations with and express results with unit prefixes.

1.12   Convert the following numbers and units to the units indicated. Express the answer with the proper number of significant digits and units. Use scientific notation if necessary.
  a. 1000 km to miles
  b. 1.00×10^3 miles to m
  c. 34.5 in. to ft
  d. 230. lb to kg
  e. 2.00×10^4 μin. to in.
  f. 0.00200 in. to μin.
  g. 77.72 mm to m
  h. 185 hp (horsepower) to W

1.13   Convert the following numbers and units to the units indicated. Express the answer with the proper number of significant digits and units. Use scientific notation if necessary.
  a. 1000. mm to yd
  b. 9.009×10^3 km to miles
  c. 34.5 ft to in.
  d. 80 kg to lb
  e. 23.000×10^{-5} M-in. to in.
  f. 0.00040 in. to m-in.
  g. 34.1×10^4 mm to m
  h. 3900 W to hp (horsepower)

The problems in the following section will demonstrate your ability to express the range of values for a given quantity based on accuracy and resolution.

1.14   Determine the range, the maximum value, and the minimum value for each nominal value and accuracy.
a. 12 (3% accuracy)
b. 0.45 (0.1% accuracy)
c. 1.392   (8% accuracy)

1.15   Determine the range, the maximum value, and the minimum value for each nominal value and accuracy.
a. 0.935 (2% accuracy)
b. 265 (0.1% accuracy)
c. 1.44 (5% accuracy)

1.16   A stopwatch has an accuracy of 1%. The maximum reading that can occur on the stopwatch is 99 s. Over what range of time does a readout of 34 s fall if the stopwatch readout has
a. 2 digits?
b. 3 digits?

1.17   A stopwatch has an accuracy of 2%. The maximum reading that can occur on the stopwatch is 99 s. Over what range of time does a readout of 72 s fall if the stopwatch readout has
a. 2 digits?
b. 3 digits?

Challenge problems and research that further demonstrate, by your proper completion, that you have achieved the expected outcomes of this chapter.

1.18   You are in the market for a digital multimeter (a meter that is used to measure electrical quantities) and you have found one that does everything that you want. The meter has an accuracy of 1.0 % in all ranges and displays three digits as a standard feature. The salesman offers to add two additional digits to the display for a meager incremental cost of $150. In other words, the display would read 12.000 instead of 12.0 if you were measuring a 12 V battery.
a. Should you spend the additional money? Justify your answer.
b. If the meter range was selectable, would it make a difference? Justify your answer.

1.19   You decide to take a friend to the drag races over the weekend, and remembering your lecture in your electronics, you begin to question the race times. You know that there is going to be delay between the time the car actually crosses the ¼-mile mark and the time the timing circuit actually stops. Upon investigation, you find that the variation in the detection time can be as much as ± 0.15 s. What is the range and accuracy of the measurement if the elapsed time is:
a. 5.015 s?
b. 22.5 s?
c. 7.552 s?

1.20   Assume that the drag racer in problem 1.19 was able to reach the ¼-mile mark in 6.325 s. Also assume that the race car generated 455 hp for the time of the race in order to achieve that time. Being the exceptional student that you are, you decide to figure out how much energy was expended in kilowatt-hours (kWh). Express your answer in scientific notation.

1.21   You are budgeting for a fishing trip to Canada and you need to figure in the cost of fuel. Your SUV gets about 18 MPG, and all the maps suggest that you will be traveling about 1250 km. Based on today's currency exchange rate (Internet research) and an expected gasoline cost of $0.519 per liter:
a. How much money, in U.S. dollars and cents, do you need to budget for gasoline?
b. How many significant figures is this?
c. Develop a spreadsheet that can be used to answer the above questions.

1.22   Today, you and a group of fellow students are traveling to watch your friends participate in the SAE mini-Baja competition. For fun, you all decide to travel across Canada rather than through the United States. While traveling, you discover that the fuel gauge is not working. Not to be daunted, you continue your travels knowing that you can estimate how far you can go between fill-ups based on the fact that your SUV has been averaging about 22 MPG on the highway and has a gas tank that holds 18 gallons. A while back, about 115 miles when your tank was almost empty, you purchased $10 American of fuel at a cost of $0.51 per liter Canadian. You just passed a fuel station about 10 km back and the next one is 50 km down the road. Should you go back and purchase more fuel or continue on to the next station? Justify your answer.

1.23   You have learned that the definition of a farad is one coulomb of charge for every volt of potential energy (Q/V). Assume that charge is flowing at a constant rate of 35 microcoulombs per millisecond for 0.006667 hour from a charge source (also known as a current source). After the specified time, you measure the voltage and find it to be 9.67 volts. How many farads of charge do you have? Be sure to use the appropriate number of significant figures and express your answer in engineering notation.

1.24   Your job is to count the number of electrons that accumulated in problem 1.23. Assume that you can count at a rate of 367 electrons per minute.

How old will you be when you finish your counting? Express your answer in years using engineering notation and use four significant figures.

ELECTRICAL QUANTITIES

As a result of successfully completing this chapter, you should be able to:

1. Describe the difference between energy and power.
2. Calculate the energy, power, or time given any two of the quantities.
3. Describe voltage, current, and the role of voltage and current in power and the transfer of electrical energy from the source to the load.
4. Calculate electrical power, resistance, voltage, or current given any two of the quantities.
5. Describe the difference between electron current flow and conventional current flow.
6. Describe the concepts that the defining equations for current and voltage represent.
7. Describe what resistance represents and the difference between resistance and a resistor.
8. Describe some reasons that resistors are used in electric circuits.
9. Calculate the resistance, resistivity, or dimensions of wires given the other quantities.

2.1 ENERGY, POWER, VOLTAGE, AND CURRENT IN ELECTRIC CIRCUITS

In Chapter 1, the energy viewpoint of electricity was introduced. Electrical sources convert a different form of energy into electrical energy. Electrical loads convert electrical energy into another form of energy, hopefully to perform useful work (turn motors, heat air, etc.). Electrical conductors connect the source to the load and provide the path for the electrical energy flow. Notice that the units given in the lightbulb examples in Chapter 1 were watts, not joules. The watt (W) is the SI unit of power, and the joule (J) is the SI unit of energy. What is the difference between energy and power?

Energy is often defined as the ability to do work, whereas power is how fast energy is used or generated. This definition of energy is somewhat a play with words, although work has a precise mechanical definition (force times distance). Energy can be stored, transported, and converted from one form into another. The conversion of energy from one form into another is "doing work," i.e., objects are moved, chemicals react, heat is generated or removed, light is generated or absorbed, and so on. The quantity of energy represents how much work can be done.

To obtain a feel for the unit of energy, i.e., the joule, let's examine how much energy is required to heat one cup of water (approximately 237 grams (g)) from room temperature, approximately 22 °C, to boiling (100 °C). It takes 4.184 J of energy to heat 1 gram of water 1 °C (this will be covered in a physics or chemistry course). A typical cup of water might hold 237 g of water. Thus, it would take:

$$\left[\frac{\left(\frac{4.184 \text{ J}}{°C} \right)}{g} \right] (237 \text{ g}) = 992 \text{ J/°C}$$

i.e., it would take 992 J of energy to heat 237 g of water 1 °C. To heat the 237 g of water from room temperature to boiling, the increase in temperature is 100 °C − 22 °C = 78 °C. Hence, the total amount of energy required to heat 237 g of water from room temperature to boiling is

$$\left(\frac{99.16 \text{ J}}{°C} \right) (78 \text{ °C}) = 7740 \text{ J}$$

Thus, the joule is a "small" amount of energy based on everyday experience, but it has precise physical definitions that allow one to calculate and predict amounts of energy.

As stated previously, power is how fast energy is used or generated. When a 60 W lightbulb is on, it is converting 60 J of electrical energy into light energy *every second*. Energy is an amount, and power is the rate of energy conversion. Electrical power is the rate of energy conversion to or from electrical energy. How much total energy does a 60 W light bulb use? You cannot say. The total amount of energy used depends on how long the lightbulb is on. In fact, you pay for electrical energy, not power.

How does one predict energy usage? A relationship between energy and power is needed, i.e., an equation. Power is how fast energy is converted (used), which suggests that time is involved. For example, speed is how fast distance is covered, and the units are m/s. The equation for speed is distance divided by time. By analogy, power is energy divided by time:

$$\boxed{P = \frac{E}{t}} \qquad \qquad \textbf{(2.1)}$$

where: P is the power in watts (W),

   E is the energy in joules (J),

and t is the time in seconds (s).

This fundamental relationship (when it is "fundamental," it is important! Fundamental relations are boxed in this textbook) between energy and power also relates to the units: the watt is one joule per second. This equation can be algebraically manipulated to determine energy: $E = Pt$. Hence, the amount of energy used is directly proportional to the time that the device is on.

___ Example 2.1.1 _____

A 60 W lightbulb is left on for 10 hours every night. How much energy is used in one month (assume 31 days)?

Given: $P = 60 \text{ W}$

   $t = (10 \text{ hr/day})(31 \text{ days})$

Desired: E

Strategy: Use the relationship between energy, power, and time: $P = \dfrac{E}{t}$

Solution:

$$E = Pt = 60\ \text{W (31 days)} \left(\frac{1\frac{\text{J}}{\text{s}}}{\text{W}}\right)\left(\frac{10\ \text{hr}}{\text{day}}\right)\left(\frac{60\ \text{min}}{\text{hr}}\right)\left(\frac{60\ \text{s}}{\text{min}}\right) = 66.96\ \text{MJ} = 67.0\ \text{MJ}$$

Hence, 67.0 MJ of energy are used every month by a nominal 60 W lightbulb that is on 10 hours every night. Notice that three significant digits were assumed, but technically there is only one significant digit based on the original numbers given. However, no accuracy specifications were given, so the true number of significant digits is uncertain. Again, to obtain a feel for the numbers, 67 MJ of energy would be equivalent to heating about 865 cups (about 54 gallons) of water from room temperature to boiling, as discussed earlier in this section.

When people pay their electric utility bills, they are really paying for the amount of electrical energy that they used, not how fast that they used the energy. The same fundamental relationship between electrical energy and power applies, but the units are different. The unit of time is hours, and the unit of energy is kWh, which is power in kilowatts multiplied by time in hours.

Example 2.1.2

A 60 W lightbulb is left on for 10 hours every night in one month (assume 31 days). How much is the cost per month if electricity costs $0.062 per kWh?

Given: price = $0.062 / kWh

time = (10 h/day)(31 days)

power = 60 W

Desired: cost

Strategy:

Relationship between cost, price, and time: cost = (price)(energy)

Relationship between energy, power, and time: $P = E/t$

Determine energy in kWh first from power and time, then determine cost from price and energy.

Solution:

$$E = Pt = (60\ \text{W})\left(\frac{1\ \text{kW}}{1000\ \text{W}}\right)(31\ \text{days})\left(\frac{10\ \text{h}}{\text{day}}\right) = 18.60\ \text{kWh}$$

$$\text{cost} = (\text{price})(E) = \left(\frac{\$0.062}{\text{kWh}}\right)(18.60\ \text{kWh}) = \$1.15$$

Thus, the monthly cost of running a 60 W lightbulb for 10 hours per day is $1.15 when the price of electricity is $0.062/kWh.

It has been established how to determine energy from power, and that the amount of time is needed to determine the energy used. Then how is the power of an electrical device determined? One method is straightforward: the power rating is given. This is a superficial answer because someone else had to determine the power rating (probably someone like you in the future: an electrical engineer, an engineering technologist, or a technician). One could measure the power, but this method is inefficient. The power should be known and/or specified during the design phase, not after the product is manufactured. There must be another way to *predict* electrical power. This is the case, and the prediction involves two new terms that you may have heard of before: voltage and current. *Both* voltage and current

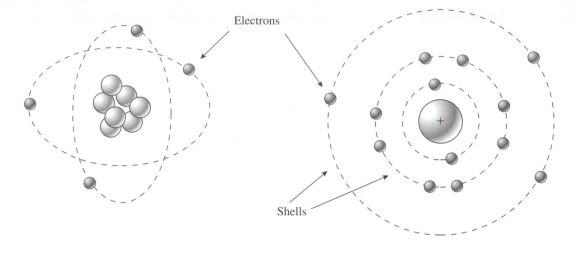

a. The atom   b. Electron shells in aluminum

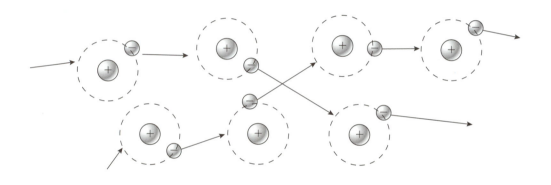

c. Electron flow in a conductor

Figure 2.1 The atom, shells, and current flow in a conductor

must exist in order for electrical energy to transfer from the source to the load. In the following paragraphs, qualitative discussions of voltage and current are given first. Then the
fundamental relationships of voltage and current are given and discussed.

Recall the basic electric circuit consisting of a source, a load, and the electrical conductors that connect them (Figure 1.2 in Chapter 1). Electric *current* consists of flowing
charges. Also recall from Chapter 1 that an atom has positive charges (protons) in the
nucleus and negative charges (electrons) in orbit around the nucleus, as illustrated in
Figure 2.1a. The electrons stay in orbit because of the electrical attraction between the protons and the electrons. The orbits of the electrons are called *shells*. There can be several
shells in an atom, as illustrated in Figure 2.1b. Each shell has a maximum number of electrons that it can possibly hold. If a shell is filled and there are more electrons in the atom,
the extra electrons go into the next shell away from the nucleus.[1] Only the outermost shell
is usually important to electricity, and it is called the *valence* shell.

The number of electrons in the valence shell is one of the central factors that determines the electrical properties of a material. Insulators often have valence shells that are
filled or nearly filled. The electrons are bound to their respective nuclei because there is
no room for an electron to jump to another atom. On the other hand, conductors, such

---

[1] The details of shells are usually covered in chemistry courses. A detailed understanding of the
atom is essential in the study of semiconductor materials that are used in electronic devices.

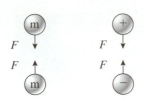

Figure 2.2 Potential energy

a. Gravitational   b. Electrical

as copper and aluminum, have one, two, or three electrons in their valence shells. The valence shells are not filled in these conducting materials. There is plenty of room for the valence shell electron of one atom to jump to the valence shell of another atom. Hence the valence electrons are free to move from atom to atom, as sketched in Figure 2.1c. The primary characteristic of a conductor is the property that the outer shell electrons are not bound to only one atom and can easily move from atom to atom within the conductor. Thus, conductors are used to make electrical connections between the sources and the loads in electric circuits.

If current is the flow of these charges in the conductor, what makes the charges flow? An electrical "force" is needed to push/pull the electrons. This electrical force is called *voltage.* The force arises from a potential energy difference. But the potential energy is electrical, not gravitational. (If gravitational forces and mechanical energy concepts are not familiar to you at this time, skip from here to the next paragraph.) Refer to Figure 2.2a. If a mass is moved away from another mass, such as lifting a rock up from the surface of Earth, the mass gains potential energy, which is the potential to do work. If the rock is let go, it gains velocity as it approaches Earth, i.e., the rock gains kinetic energy as it loses potential energy.

In a similar manner, if charges are separated (as they are in a source), then there is an *electrical potential energy difference* between the separated charges, as shown in Figure 2.2b. Recall from the law of charges that like charges repel and unlike charges attract. If two unlike charges are separated farther apart, then someone or something must expend energy to counter the attraction of the unlike charges. Hence, the system of separated charges has gained potential energy, namely, electrical potential energy because the forces are electrical in nature. This potential energy can be used to do work, as will be seen later in the voltage discussion. For now, the current discussion will be continued.

If a conductor is connected between the separated charges, electrons are attracted toward the positive side. Hence, electrons will flow from the negative side to the positive side of the separated charges in the source, as shown in Figure 2.3. If a load is connected to the source, electrical energy will transfer from the source to the load. An electrical potential difference develops across the load, as indicated by the presence of the signs across the load in Figure 2.3. In the source, a nonelectrical form of energy is converted into electrical potential energy difference. In the load, the electrical potential energy difference is converted into another form of energy. The transfer of electrical energy from the source to the load can occur only when the electrical potential energy difference exists *and* when the electrons flow in the conductors.

How an electric circuit operates really is quite involved—reread the previous paragraph a few times as you continue reading this section (it will make more sense with each rereading).

Symbols and conventions (a method of expressing something on which everyone agrees) can be used to show circuit operation instead of writing all the words in the previ-

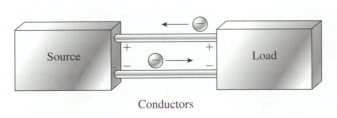

Conductors

Figure 2.3 Electron flow and electrical potential energy difference in an electric circuit

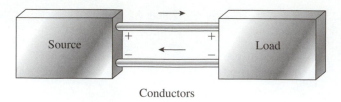

Figure 2.4 Voltage polarity signs and current arrows in an electric circuit

Conductors

ous paragraph. Electrical potential energy difference (voltage) is shown by a plus sign and a minus sign, representing the positive and negative sides of the separated charge, as shown in Figure 2.4. Current is shown by an arrow drawn parallel to the conductor and is shown in the direction from *positive to negative* of the source through the conductor path.

Why is current shown going from positive to negative when electrons would flow from negative to positive? The answer is historical: Benjamin Franklin realized that there were two types of charges and that one type moved, but he did not have atomic theory to help him. So he guessed, and he guessed wrong. He said that positive charges flow from the positive to the negative side of the electrical potential energy difference. From his time on, literature was written based on his incorrect guess. Thus, to this day almost all electrical engineers and technologists use the convention that current consists of positive charges flowing from positive to negative in the conductors. However, they also know that based on atomic theory, current really consists of negative charges (electrons) flowing from negative to positive in the conductors. This positive-to-negative current flow of assumed positive charges is named *conventional current flow*. Occasionally one does see literature where electron flow is assumed, and this negative-to-positive flow of electrons is called *electron current flow*.

Information Research Exercise 2.1.1   (library and/or web)

Investigate the contributions that Benjamin Franklin made (*a*) to the understanding of electricity and (*b*) to science in general.

A careful observation of the voltage and current in Figure 2.4 reveals that conventional current leaves the positive terminal of the source, flows through the conductor, enters the positive terminal of the load, leaves the negative terminal of the load, flows through the other conductor, and enters the negative terminal of the source. The current directions into the load and into the source are opposite. This convention may make sense if one considers that the source is generating (converting to) electrical energy and the load is using (converting from) electrical energy. In fact, this convention even has a name: the passive sign convention. "Passive" means that the load is not generating electrical energy. More will be said about this convention in the next section.

The qualitative descriptions of current and voltage can be improved with formal definitions of them. Current was described as consisting of flowing charges. A charge is an atomic particle with an electrical attraction, namely protons and electrons. Hence, a charge in electric current would be the electron. Charge is given the symbol Q and could be measured in the number of electrons. However, the raw number of charges flowing in typical circuits is huge and unwieldy, especially in conversation. A typical current is often in the range of 10^{15} to 10^{21} charges per second (the time units are to be clarified in the next paragraph). A different unit of charge is used to make the numbers manageable. Charge is expressed in SI units of coulombs, which is abbreviated C. What is the coulomb unit? One coulomb is equal to 6.242×10^{18} charges (whether protons or electrons, but electrons are usually referred to). Thus, the current numbers would be approximately in the range of 0.001 to 1000 if coulombs were utilized. Units often make the accompanying numbers convenient. A dozen eggs is convenient: 12 eggs. Instead of saying 6.242×10^{18} e^- (the abbreviation for electrons) flow in a circuit, one says more conveniently that 1 C of electrons flow in the circuit.

Current is precisely defined as the *rate* of charge flow, i.e., how much charge, in coulombs, passes a point in the circuit per second. The unit of current is the ampere, abbreviated A. Thus, the equation for current is:

$$I = \frac{Q}{t}$$ (2.2)

where: I is the current in amperes (A),

   Q is the charge in coulombs (C),

and t is the time in seconds (s).

Thus, one ampere of current equals one coulomb of electric charge passing a point in a circuit per second. Notice how this equation "firms up" the definition of current. If twice as many charges flow past a point in a circuit per second, the current is double. Thus, more charge flowing past a point in the circuit per second means that the current is larger.

The symbol for current is I, not C, and the symbol for charge is Q, not C. One reason for the use of different symbols is that the symbol C is already commonly used for something else. Another reason is historical: the symbol has been used for the quantity for so many years that it has become a standard. However, beware of the reuse of symbols (C will be reused in Chapter 5) Also, do not confuse units C with symbol C.

Information Research Exercise 2.1.2   (library and/or web)

The unit of charge, the coulomb, and the unit of current, the ampere, are attributed to individuals with the same last names. What were their contributions to electricity?

Example 2.1.3

If the current in a flashlight is 20 mA, how many coulombs of charge flow through the miniature lamp per second?

Given: $I = 20$ mA

   $t = 1$ s

Desired: Q

Strategy: Use the relationship between current, charge, and time: $I = Q/t$

Solution:

$$Q = It = 20 \text{ mA} \left(\frac{1 \text{ A}}{1000 \text{ mA}} \right)(1 \text{ s}) = 0.0200 \text{ C}$$

Is this result considered much charge? Convert the units of charge from coulombs to the number of actual charges:

$$Q = 0.0200 \text{ C} \left(\frac{6.242 \times 10^{18} \text{ charges}}{1 \text{ C}} \right) = 1.25 \times 10^{17} \text{ charges}$$

Yes, this is a lot of charge. Again notice the convenience of the coulomb units.

Recall that voltage was described as electrical potential energy difference. Voltage is precisely defined as the electrical potential energy difference per unit of charge. Voltage is given the symbol V, and the units of voltage are volts, abbreviated with the same symbol V. Thus,

$$V = \frac{E}{Q}$$ (2.3)

where: V is the voltage in volts (V),

E is the energy of the electrical potential energy difference in joules (J), and

Q is the charge in coulombs (C).

Thus, one volt of voltage is one joule of electrical potential energy difference for each coulomb of charge.

A caution on notation is important here. In this text, the symbol E has been used for energy. In some literature, E is used as the symbol for electromotive force, an older name for voltage. Furthermore, E is sometimes used as the symbol for the voltage of a source and V is used as the symbol for the voltage of a load. In both cases, V is used as the abbreviation for the units of voltage, i.e., volts. You must determine from context which symbol stands for which quantity (this situation is another example of the pitfalls of memorization without understanding).

Information Research Exercise 2.1.3   (library and/or web)

The quantity voltage and the unit of voltage are both named after Alessandro Volta. (*a*) What were his contributions to electricity? (*b*) What modern electrical component did he develop?

Example 2.1.4

A D-size battery is rated at 1.5 V. How much electrical energy can the battery provide to the load for each coulomb of charge that flows?

Given: $V = 1.5$ V

$Q = 1$ C

Desired: E

Strategy: Use the relationship between voltage, energy, and charge: $V = E/Q$

Solution:

$$E = VQ = (1.5 \text{ V})(1 \text{ C}) = 1.50 \text{ J}$$

Thus, a D-size battery can deliver 1.50 J of electrical energy for each coulomb of charge that flows in the circuit.

Why is this additional stipulation of "per unit charge" present in the definition of voltage? Consider a typical battery. It is labeled in volts, not joules. For example, standard AAA, AA, C, and D batteries are rated 1.5 V, not 1.5 J. In other words, batteries convert chemical energy into an electrical potential energy difference of 1.5 J for every coulomb of charge (until the chemical energy in the battery is depleted). If there is less current, there is less charge and less electrical energy transferred per unit time, but always in the ratio of 1.5 J/C, i.e., 1.5 V.

Can there be voltage without current? Yes. A C-size battery may be able to create an electrical potential energy difference between the battery terminals of 1.5 J for every coulomb of charge. However, no electrical energy will transfer from the battery to a load unless the current flows. This is why the energy difference is *potential*. The energy has the potential to do work, i.e., to be converted to another form of energy. Current must also flow for this electrical potential energy to do work in the load. This fact can be deduced from a manipulation of the fundamental power equation. Multiply both the numerator and denominator by Q, which is actually multiplying by one (one times any quantity does not change the value of that quantity), regroup, and identify the subresults:

$$P = \frac{E}{t} = \left(\frac{E}{t}\right)\left(\frac{Q}{Q}\right) = \left(\frac{E}{Q}\right)\left(\frac{Q}{t}\right) = VI \tag{2.4}$$

$$\boxed{P = \frac{E}{t} = VI} \tag{2.5}$$

Thus, both voltage and current must be present for energy to transfer from the source to the load, and power is the rate of the energy transfer. Equation (2.5) is a fundamental equation: electrical power is determined by the voltage and the current and, consequently, the electrical energy is determined by voltage, current, and time:

$$E = Pt = VIt \tag{2.6}$$

Equations (2.5) and (2.6) have finally answered the question of at least one way to predict power and energy from electrical information, i.e., the voltage and the current. Now power and energy usage in a product design can be predicted before it is fabricated.

___ Example 2.1.5 _____

A lamp with a 60 W lightbulb is plugged into (connected) to a 120 V source. How much current flows in the circuit?

Given: $P = 60$ W

$V = 120$ V

Desired: I

Strategy: Use the relationship between voltage, current, and power: $P = VI$

Solution: _____

$$I = \frac{P}{V} = \frac{60 \text{ W}}{120 \text{ V}} = 0.500 \text{ A}$$

Thus, ½ A flows. *Note:* This voltage produced a current level that is dangerous (potentially fatal) for humans!

_____

This has been an involved discussion, but from it we have learned that current is the amount of charge flowing past a point in a circuit per unit time and that voltage is the electrical potential energy difference per unit charge. Both voltage and current must be present for electrical energy to transfer from the source to the load, and electrical energy and power can be determined from voltage and current: $P = VI$, and $E = Pt$. Current consists of flowing electrons, but everyone considers current to be flowing positive charges (conventional current flow).

2.2 ELECTRICAL LOADS, RESISTANCE, AND OHM'S LAW

There is a missing piece to this electricity "puzzle" as presented thus far. Consider the following reasoning. Power can be determined from energy and time, but the energy used or the time over which the energy is used are usually not known in advance. Alternatively, power can be determined from voltage and current. Voltage is often known in advance: the number of batteries used, the standard voltage supplied by the electric power utility, and so

Figure 2.5 Schematic circuit diagram of a DC circuit

on. However, current is not often known in advance, and consequently power cannot be predicted. In the case where power is known, for example the 60 W lightbulb, how was the power of 60 W determined? Again, a piece of information is missing. The missing information is the concept of *resistance*. Resistance relates voltage and current for given load by a famous law called Ohm's law:

$$R = \frac{V}{I}$$ (2.7)

where: R is the resistance in ohms (Ω, the uppercase Greek letter omega),

V is the voltage in volts (V), and

I is the current in amperes (A).

Resistance is a property that represents the ratio of voltage to current as electrical energy is converted into another form of energy. Thus, resistance is a property of the load. One can speak of the voltage-to-current ratio of the source, but this is not often done. Why not? Most electrical sources have a constant voltage (or current), and the current (or voltage) can vary, depending on the resistance of the load.

Even the conductors have resistance. A small portion of the electrical energy is converted into heat when the current flows through the conductors because flowing electrons "bump into" atoms in the conductors. Often, the resistance of the conductors is negligible compared with the resistance of the load, and the conductor resistance is ignored; however, the resistance of conductors is significant in the design of electrical power distribution systems. The resistance of conductors is determined in Section 2.4.

Before an example with resistance is given, the drawing of circuit diagrams will be explained. Circuits can be drawn using symbols instead of boxes and words. A circuit diagram is shown in Figure 2.5. The symbol with a short and a long set of parallel lines is an electrical source that always has a positive terminal, a negative terminal, and a constant voltage between the terminals. The long bar is the positive side of the voltage source and the short bar is the negative side. It is called a *direct current* source because the current normally goes in only one direction. The standard abbreviation for direct current is either "dc" or "DC." Voltage sources that supply direct current are called DC voltage sources.

The straight lines between other symbols are conductors, as utilized previously in Figures 2.3 and 2.4. The symbol with the zigzag line represents resistance. The actual load can be a motor, a lightbulb, a resistor, etc. The symbol is generic for resistance—it represents a component that converts electrical energy into another form of energy. The entire diagram is called a *schematic* circuit diagram, often abbreviated in conversation to "the schematic."

Ohm's law is quite useful because voltage and current are related by a property of the load: its resistance. Often, resistance is directly predictable (this will become apparent in the next two sections). Once voltage and current are related, power and energy calculations are straightforward.

Example 2.2.1

A 200 ohm load is connected to a 20 V source per the circuit diagram in Figure 2.5. Determine (*a*) the current and (*b*) the power delivered to the load.

a. Given: $R = 200 \ \Omega$

$V = 20$ V

Desired: I

Strategy: Use the relationship between R, V, and I: $R = V / I$

Solution: _____

$$I = \frac{V}{R} = \frac{20 \text{ V}}{200 \text{ }\Omega} = 0.100 \text{ A}$$

b. Given: $R = 200 \text{ }\Omega$

$V = 20 \text{ V}$

$I = 0.1 \text{ A}$

Desired: P

Strategy: Use the relationship between P, V, and I: $P = VI$

Solution: _____

$$P = VI = (20 \text{ V})(0.1 \text{ A}) = 2.00 \text{ W}$$

Thus, for a 200 ohm load connected to a 20 V source, the current is 0.100 A and the power delivered from the source to the load is 2.00 W.

It is easy to combine the electrical power equation $P = VI$ and Ohm's law $V = IR$ to calculate power directly using resistance:

$$P = VI = (V)\left(\frac{V}{R}\right) = \frac{V^2}{R} \tag{2.8}$$

or

$$P = VI = (IR)(I) = I^2R \tag{2.9}$$

Thus, there are three convenient expressions for power:

$$P = VI = I^2R = \frac{V^2}{R} \tag{2.10}$$

but they all really come from $P = VI$ and $V = IR$, two fundamental laws of electricity. The authors emphasize that *you do not need to memorize every equation.* Know the fundamentals, and you can figure out the other equations, often with one or two steps of simple algebra. This strategy leaves you more time to think about and enjoy the new ideas about electricity instead of worrying about how to memorize a huge number of equations that you will probably forget soon after the next exam anyway. Example 2.2.2 illustrates how Equation (2.10) may save steps in the solution process.

___ **Example 2.2.2** _____

A 200 ohm load is connected to a 20 V source. Determine the power delivered to the load.

Given: $R = 200 \text{ }\Omega$

$V = 20 \text{ V}$

Desired: P

Strategy: use the relationship between P, V, and R: $P = \frac{V^2}{R}$

Solution: _____

$$P = \frac{V^2}{R} = \frac{(20\ \text{V})^2}{200\ \Omega} = 2.00\ \text{W}$$

which matches the answer in Example 2.2.1. Note that a step was saved using this approach.

It is now time to readdress the *passive sign convention*. In a circuit the load is usually a passive device—it does not generate electrical energy; instead it converts electrical energy into another form. The conventional current enters the load on the positive voltage side of the load and leaves the load on the negative voltage side of the load, as shown in Figure 2.6. The power in the load is agreed to be positive when the conventional current enters the component on the positive voltage side of the load. This agreement is the passive sign convention. The source also conforms to this convention. The current leaves the source on the positive voltage side—just the opposite of the load. Hence, the power of a source is negative. The passive sign convention will be investigated further in later chapters of this book.

Some examples of resistance are given in Table 2.1. In each load example, an energy conversion is taking place, and the resistance is the voltage-to-current ratio of that load and represents that energy conversion. Resistors are covered in more detail in Section 2.3, and wire resistance is covered in detail in Section 2.4.

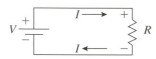

Figure 2.6 Passive sign convention in a DC circuit

TABLE 2.1	Load Example	Input Energy	Output Energy	Consequence of Resistance
Examples of Resistances	lightbulb	electrical	light	heating a filament until it glows
	electric motor	electrical	mechanical	creating magnetic fields that turns a shaft
	electric stove	electrical	heat	heating an element to generate heat to cook food
	resistor	electrical	heat	heating a material; used in electric circuits
	wire	electrical	heat	heating the metal when current flows

2.3 RESISTORS

Resistors are an important specific example of *resistance*. A material is selected such that when a current flows through the material a significant voltage develops across the terminals that connect the material to the rest of the circuit. Consequently, the material heats up. Electrical energy is converted into heat energy. Thus, are resistors generally used as "heaters"? No. Resistors are used in electric circuits to control voltage and current levels. What happens to the current in Ohm's law if the voltage source is constant and the resistance increases? The current decreases. Hence, the current level can be controlled.

___ Example 2.3.1 ___

A 6 V battery is connected to a resistor. What is the current if the resistance of the resistor is (*a*)1.0 kΩ, (*b*) 2.2 kΩ, (*c*) 3.3 kΩ, (*d*) 4.7 kΩ, (*e*) 8.2 kΩ, and (*f*) 10. kΩ? (*g*) What is the trend?

Given: $V = 6$ V

      a. $R = 1.0$ kΩ

      b. $R = 2.2$ kΩ

      c. $R = 3.3$ kΩ

      d. $R = 4.7$ kΩ

      e. $R = 8.2$ kΩ

      f. $R = 10.$ kΩ

Desired: I in each case; the trend

Strategy: Ohm's law

Solution:

a.

$$I = \frac{V}{R} = \frac{6}{1 \text{ k}\Omega} = 6 \text{ mA}$$

The exact same procedure is used for parts b and c, so just report the results:

    b.   2.73 mA

    c.   1.82 mA

    d.   1.28 mA

    e.   0.73 mA

    f.   0.60 mA

    g.   Trend: create a graph (a spreadsheet is useful here); see Table 2.2 and Figure 2.7

TABLE 2.2	R (kΩ)	I (mA)
Resistance and Currents for Example 2.3.1	1.0	6.000000
	2.2	2.727273
	3.3	1.818182
	4.7	1.276596
	8.2	0.731707
	10.0	0.600000

Current–Resistance Trend

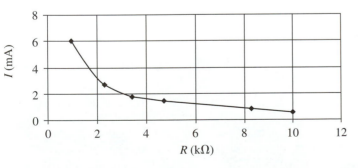

Figure 2.7 Current–resistance plot of data in Table 2.2

The trend between current and resistance is an inverse relationship: as resistance increases, current decreases, and vice versa. Hence, controlling the resistance in a circuit is one way to control current through the circuit.

Another example of the use of resistors in circuits is to divide voltages. If the load consists of two resistors, as shown in Figure 2.8, part of the source voltage is across the upper

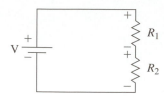

Figure 2.8 Dividing the source voltage using two resistors

resistor and the remainder of the source voltage is across the lower resistor. For example, if the source was 15 V, and R_1 and R_2 were both 2 kΩ, there would be 7.5 V across each resistor. The source voltage has been divided across each of the two resistors and a lower voltage is now available. The use of resistors for applications such as this one will be examined in detail in the next chapter.

There are several types of resistors. They differ in many ways, such as shape, size, and lead attachment method, but three important considerations are:

■ composition: materials used

■ accuracy: how close is the actual value to the nominal value

■ power rating: how much power can be dissipated (converted to heat) before it fails

Each consideration is briefly examined next. Information research exercises are suggested for those who need more in-depth information at this time.

Common resistor types include carbon composition, metal film, wire wound, and surface mount, as pictured in Figure 2.9. Carbon composition resistors (Figure 2.9a) basically consist of a mixture of carbon and an electrically inert material, as sketched in Figure 2.10. The ratio of the carbon to the inert material in the mixture determines the resistance. The

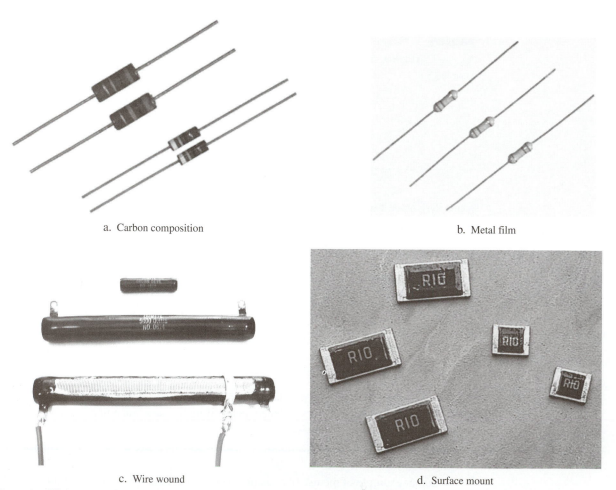

a. Carbon composition

b. Metal film

c. Wire wound

d. Surface mount

Figure 2.9 Pictures of several resistor types Reprinted with permission of Ohmite Manufacturing.

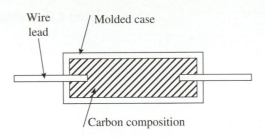

Wire lead
Molded case
Carbon composition

Figure 2.10 Carbon composition resistor construction

composition is sealed inside a cylindrical case, with wire leads inserted into each end of the cylinder. The resistance value of the resistor is often shown by color bands on the surface of the case (discussed later in this section).

Metal-film resistors (Figure 2.9b) and wire-wound resistors (Figure 2.9c) both depend on the resistance of a conductor (discussed in detail in the next section). A thin layer of metal is deposited on an insulating surface to form a metal film resistor. The film may be in a pattern, such as a spiral around a cylinder. In wire-wound resistors, a wire with the desired resistance is bonded to a ceramic material. The ceramic material draws heat away from the wire. Hence, wire-wound resistors are often used when high-power resistors are needed. Surface-mount resistors (Figure 2.9d), also called chip resistors, are just resistors constructed in a planar (flat) manner with the leads being conductors on two opposite ends of the chip. The resistor is soldered to the circuit on these ends.

Information Research Exercise 2.3.3   (web and/or vendor catalogs)

Investigate the materials used in and the construction of carbon, metal-film, wire-wound, and surface-mount resistors.

The nominal resistance value of a resistor may be printed on its surface. However, it is often represented by several bands of different colors, called the *color code.* The color code is shown in Figure 2.11. Each color in the first three bands stands for a number:

Black	0	Green	5
Brown	1	Blue	6
Red	2	Violet	7
Orange	3	Gray	8
Yellow	4	White	9

Notice the rainbowlike trend in the colors arrangement. Although there are plenty of charts and tables in which to look up these color codes, one of many mnemonics is "Big Bucks Run Over Your Garden But Vultures Go Wild!"

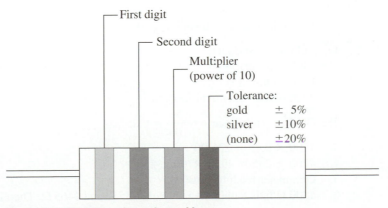

First digit
Second digit
Multiplier (power of 10)
Tolerance:
gold   ± 5%
silver ±10%
(none) ±20%

Figure 2.11 Resistor color code markings

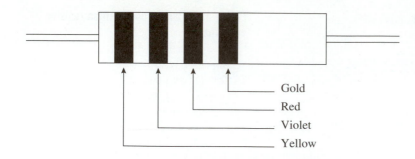

Figure 2.12 Resistor color code example

Gold
Red
Violet
Yellow

Notice that the colored bands are closer to one end of the resistor than the other. This is the end at which to start interpreting the color code, as drawn in Figure 2.11. The first band is the most significant digit of the nominal value, and the second band is the next most significant digit of the nominal value. The third band represents the *multiplier*. The multiplier is the power of 10 and is multiplied with the first two digits. For example, in Figure 2.12, the color bands are read as yellow–violet–red, and the nominal resistance value would be:

$$47 \times 10^2 = 4700 \ \Omega = 4.7 \ k\Omega$$

The fourth band is the tolerance, i.e., the accuracy. The colors used in the tolerance band are shown with their corresponding accuracies in Figure 2.10. The range of acceptable in-tolerance values for the resistor can be determined. A generic strategy follows:

Tolerance Test of a Resistor

1. Read the color code off the resistor; determine the nominal resistance value.
2. Convert the tolerance from a percent to a decimal (divide by 100).
3. Multiply the decimal and the ideal resistance together.
4. a. Add the result of step 3 to the nominal resistance. This is the upper limit on the range of resistance.
   b. Subtract the result of step 3 from the nominal resistance. This is the lower limit on the range of resistance.
5. Measure the resistance of the resistor on an ohmmeter (this instrument that is used to measure resistance is covered in a separate laboratory manual).
6. If the ohmmeter reading falls within the range of step 4, the resistor is "in-spec" (within the specification). Otherwise, the resistor is "out-of-spec."

Example 2.3.2

Determine the acceptable range of values for a resistor with a color code of yellow–violet–red–gold. The ohmmeter reading is 4800 Ω.

Using the tolerance test procedure:

1. Yellow 4 (1st digit)
   Violet 7 (2nd digit)
   Red 2 (multiplier)
   4700 ohms = nominal resistance
2. Gold 5% (tolerance) $\rightarrow \dfrac{5}{100} = 0.05$
3. $4700 \times 0.05 = 235.00$
4. $4700 + 235 = 4935; 4700 - 235 = 4465$
5. Ohmmeter reading: 4800 Ω
6. 4800 Ω falls within the range of 4935 Ω and 4465 Ω. Therefore, the resistor is in-spec.

Standard resistor values are available. The first two digits of the resistance value of these resistors are 10, 12, . . . (see Information Research Exercise 2.3.2). Most of the multipliers are available. Consider these standard values when selecting resistors for circuit designs.

The power rating of a resistor must be considered before placing it in a circuit. If the power dissipated in a resistor exceeds the power rating, the resistor will fail and will be permanently destroyed. Usually, it burns up and releases smoke. If any electronic part burns, do *not* breathe the fumes—they may be toxic! Power ratings are often standardized: ⅛ W, ¼ W, ½ W, . . . (see Information Research Exercise 2.3.2). The size and cost of the resistor generally increases as the power rating is increased.

Information Research Exercise 2.3.2   (web and/or vendor catalogs)

a. Some resistors have five color bands. Identify what each band represents in this labeling scheme.
b. Identify the first two digits of standard resistor values.
c. Identify the power rating values of standard resistors.

2.4 WIRE SIZES AND RESISTANCE

It was mentioned earlier in this chapter that as charges flow in a conductor the flowing charges sometimes "bump" into atoms in the conductor, and this imparts energy to those atoms. The atoms will vibrate even more because of these collisions. Thus, some of the electrical energy is converted into heat energy. This energy conversion is a resistance called *wire resistance*. Often, the resistance of the wires is ignored in circuits because the other resistances in the circuit are much larger than the resistance of the conductors. However, in some cases, such as power distribution systems, where large currents are present, even a small wire resistance results in a significant voltage drop (think about this in terms of Ohm's law). Thus, the prediction of the resistance of a conductor is needed in order to determine the voltage drop in the conductors.

What would the resistance of a conductor depend on? If the wire were longer, there would be more charge collisions with atoms in the conductor. Hence, wire resistance is directly proportional to the length (ℓ) of the conductor:

$$R \propto \ell \tag{2.11}$$

If the conductor had a larger diameter, i.e., cross-sectional area, then the flowing charges could spread out further across the cross-section as they flowed, and the collisions with atoms in the conductor would decrease. Hence, wire resistance is inversely proportional to the cross-sectional area (A) of the conductor:

$$R \propto \frac{1}{A} \tag{2.12}$$

Finally, the particular metal used as the conductor makes a difference. For example, copper is a better conductor than iron. A proportionality constant, called the *resistivity*, is defined to represent the effect of the material on the resistance of the conductor. The lowercase Greek letter rho (ρ) is used as the symbol for resistivity. The larger the resistance due to the material, the larger resistivity is. Hence, wire resistance is directly proportional to resistivity:

$$R \propto \rho \tag{2.13}$$

If Equations (2.11), (2.12), and (2.13) are combined into one equation, then the equation for wire resistance results:

$$R = \frac{\rho \ell}{A} \tag{2.14}$$

where: R is the resistance of the wire in ohms,

ρ is the resistivity of the conductor material in ohm-meters,

ℓ is the length of the wire in meters, and

A is the cross-sectional area of the wire in meters squared.

These quantities are labeled on the sketch in Figure 2.13 for a straight wire with a circular cross section.

SI units are shown. However, many resistivity tables use units of μΩ-cm, as shown in Table 2.3. The resistivity values reported in this table are approximate. The exact value of resistivity depends on many factors such as the manufacturing process, the purity of the conductor material, and the temperature. In any event, the units in Equation (2.14) must be converted properly so that the distance units of ρ, ℓ, and A are the same, as illustrated in Example 2.4.1.

Figure 2.13 Wire resistance quantities

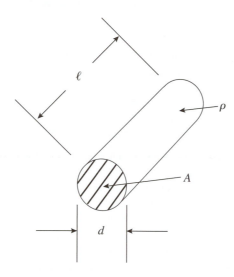

TABLE 2.3	Material	Resistivity (μΩ-cm) at 25 °C
Resistivities of Common Conductor Materials (SI units)	aluminum	2.83
	carbon	3500
	copper	1.72
	gold	2.45
	iron	12.3
	Nichrome	100
	tungsten	5.5
	silver	1.65

_____ Example 2.4.1 _____

Determine the resistance of a square bar of copper that is 0.5 in. × 0.5 in. and 3 ft long.

Given: $\ell = 2$ ft

$A = 0.5$ in. × 0.5 in.

material: copper: ρ = 1.72 μΩ-cm

Desired: R

Strategy: Use the relationship between R, ρ, ℓ, and A: $R = \dfrac{\rho \ell}{A}$

Solution: _____

$$R = \frac{\rho\ell}{A} = \frac{(1.72 \times 10^{-6}\,\Omega\text{-cm})(2\,\text{ft})\left(\dfrac{12\,\text{in.}}{\text{ft}}\right)\left(\dfrac{2.54\,\text{cm}}{\text{in.}}\right)}{(0.5\,\text{in.})(0.5\,\text{in.})\left(\dfrac{2.54\,\text{cm}}{\text{in.}}\right)^2} = 65.0\,\mu\Omega$$

Note how the distance units were converted to be the same, centimeters in this case.

_____

In many handbooks, *conductivity* is given instead of resistivity. The relationship between conductivity σ (lowercase Greek letter sigma) and resistivity is straightforward:

$$\sigma = \frac{1}{\rho} \qquad\qquad (2.15)$$

where: ρ is the resistivity in Ω-m, and

σ is the conductivity in siemens/m (S/m).[2]

When the conductor is a wire of circular cross section, the cross-sectional area (A) is the area of a circle, as illustrated in Example 2.4.2.

Example 2.4.2 _____

Determine the resistance of a copper wire that has a diameter of 5.22 mm and a length of 67.5 m.

Given: $\ell = 67.5$ m

$d = 5.22$ mm

material: copper: $\rho = 1.72\ \mu\Omega$-cm

Desired: R

Strategy: Use the relationship between R, ρ, ℓ, and A: $R = \dfrac{\rho\ell}{A}$, where $A = \pi r^2 = \dfrac{\pi d^2}{4}$

Solution: _____

$$A = \frac{\pi d^2}{4} = \frac{\pi(5.22\,\text{mm})^2}{4} = 21.401\,\text{mm}^2$$

$$R = \frac{\rho\ell}{A} = \frac{(1.72 \times 10^{26}\,\Omega\text{-cm})\left(\dfrac{1\,\text{m}}{100\,\text{cm}}\right)(67.5\,\text{m})}{(21.401\,\text{mm}^2)\left(\dfrac{1\,\text{m}}{1000\,\text{mm}}\right)^2} = 54.3\,\text{m}\Omega$$

_____

There is another system of units used for wire resistance calculations in the United States. The length of wires are specified in feet in this system. There are standard wire sizes in this system that are specified as an American Wire Gage (AWG) size, as listed in Table 2.4 Each AWG wire size has a corresponding cross-sectional area in peculiar units

_____

[2] The unit siemens will be discussed in Chapter 3.

TABLE 2.4

American Wire Gage Sizes
for Circular Wires*

TECHNICAL DATA
AWG/METRIC CONDUCTOR CHART

AWG	STRANDING	APPROX. O.D. INCHES	APPROX. O.D. MM	CIRCULAR MIL AREA	SQUARE INCHES	SQUARE MM	WEIGHT LBS./ 1000 FT.	WEIGHT KG/KM	D.C. RESISTANCE OHMS/ 1000 FT.	D.C. RESISTANCE OHMS/ K/M
36	Solid	0.0050	0,127	25.0	—	0,013	0.076	0,113	445.0	1460,0
36	7/44	0.006	0,152	28.0	—	0,014	0.085	0,126	371.0	1271,0
34	Solid	0.0063	0,160	39.7	—	0,020	0.120	0,179	280.0	918,0
34	7/42	0.0075	0,192	43.8	—	0,022	0.132	0,196	237.0	777,0
32	Solid	0.008	0,203	67.3	0.0001	0,032	0.194	0,289	174.0	571,0
32	7/40	0.008	0,203	67.3	0.0001	0,034	0.203	0,302	164.0	538,0
32	19/44	0.009	0,229	76.0	0.0001	0,039	0.230	0,342	136.0	448,0
30	Solid	0.010	0,254	100.0	0.0001	0,051	0.30	0,45	113.0	365,0
30	7/38	0.012	0,305	112.0	0.0001	0,057	0.339	0,504	103.0	339,0
30	19/42	0.012	0,305	118.8	0.0001	0,061	0.359	0,534	87.3	286,7
28	Solid	0.013	0,330	159.0	0.0001	0,080	0.48	0,72	70.8	232,0
28	7/36	0.015	0,381	175.0	0.0001	0,072	0.529	0,787	64.9	213,0
28	19/40	0.016	0,406	182.6	0.0001	0,093	0.553	0,823	56.7	186,0
27	7/35	0.018	0,457	219.5	0.0002	0,112	0.664	0,988	54.5	179,0
26	Solid	0.016	0,409	256.0	0.0002	0,128	0.770	1,14	43.6	143,0
26	10/36	0.021	0,533	250.0	0.0002	0,128	0.757	1,13	41.5	137,0
26	19/38	0.020	0,508	304.0	0.0002	0,155	0.920	1,37	34.4	113,0
26	7/34	0.019	0,483	277.8	0.0002	0,142	0.841	1,25	37.3	122,0
24	Solid	0.020	0,511	404.0	0.0003	0,205	1.22	1,82	27.3	89,4
24	7/32	0.024	0,610	448.0	0.0004	0,229	1.36	2,02	23.3	76,4
24	10/34	0.023	0,582	396.9	0.0003	0,202	1.20	1,79	26.1	85,6
24	19/36	0.024	0,610	475.0	0.0004	0,242	1.43	2,13	21.1	69,2
24	41/40	0.023	0,582	384.4	0.0003	0,196	1.16	1,73	25.6	84,0
22	Solid	0.025	0,643	640.0	0.0005	0,324	1.95	2,91	16.8	55,3
22	7/30	0.030	0,762	700.0	0.0006	0,357	2.12	3,16	14.7	48,4
22	19/34	0.031	0,787	754.1	0.0006	0,385	2.28	3,39	13.7	45,1
22	26/36	0.030	0,762	650.0	0.0005	0,332	1.97	2,93	15.9	52,3
20	Solid	0.032	0,813	1020.0	0.0008	0,519	3.10	4,61	10.5	34,6
20	7/28	0.038	0,965	1111.0	0.0009	0,562	3.49	5,19	10.3	33,8
20	10/30	0.035	0,889	1000.0	0.0008	0,510	3.03	4,05	10.3	33,9
20	19/32	0.037	0,940	1216.0	0.0010	0,620	3.70	5,48	8.6	28,3
20	26/34	0.036	0,914	1031.9	0.0008	0,526	3.12	4,64	10.0	33,0
20	41/36	0.036	0,914	1025.0	0.0008	0,523	3.10	4,61	10.0	32,9
18	Solid	0.040	1,020	1620.0	0.0013	0,823	4.92	7,32	6.6	21,8
18	7/26	0.048	1,219	1769.6	0.0014	0,902	5.36	7,98	5.9	19,2
18	16/30	0.047	1,194	1600.0	0.0013	0,816	4.84	7,20	8.5	21,3
18	19/30	0.049	1,245	1900.0	0.0015	0,969	5.75	8,56	5.5	17,9
18	41/34	0.047	1,194	1627.3	0.0013	0,830	4.92	7,32	6.4	20,9
18	65/36	0.047	1,194	1625.0	0.0013	0,829	4.91	7,31	6.4	21,0
16	Solid	0.051	1,290	2580.0	0.0020	1,310	7.81	11,60	4.2	13,7
16	7/24	0.060	1,524	2828.0	0.0022	1,442	8.56	12,74	3.7	12,0
16	65/34	0.059	1,499	2579.9	0.0020	1,316	7.81	11,62	4.0	13,2
16	26/30	0.059	1,499	2600.0	0.0021	1,326	7.87	11,71	4.0	13,1
16	19/29	0.058	1,473	2426.3	0.0019	1,327	7.35	10,94	4.3	14,0
16	105/36	0.059	1,499	2625.0	0.0021	1,339	7.95	11,83	4.0	13,1
14	Solid	0.064	1,630	4110.0	0.0032	2,080	12.40	18,50	2.6	8,6
14	7/22	0.073	1,854	4480.0	0.0035	2,285	13.56	20,18	2.3	7,6
14	19/27	0.073	1,854	3830.4	0.0030	1,954	11.59	17,25	2.7	8,9
14	41/30	0.073	1,854	4100.0	0.0032	2,091	12.40	18,45	2.5	8,3
14	105/34	0.073	1,854	4167.5	0.0033	2,125	12.61	18,77	2.5	8,2

379

Web Site: www.alphawire.com
Email: info@alphawire.com

Toll Free: 1-800-52 ALPHA • Telephone: 908-925-8000 • Fax: 908-925-6923
Europe/UK Telephone: +44 (0) 1932 772422 • Europe/UK Fax: +44 (0) 1932 772433

*Adapted with permission from Master Catalog, Alpha Wire Company, 2001, pp. 379–80.

TECHNICAL DATA
AWG/METRIC CONDUCTOR CHART

380

AWG	STRANDING	APPROX. O.D. INCHES	MM	CIRCULAR MIL AREA	SQUARE INCHES	MM	WEIGHT LBS./ 1000 FT.	WEIGHT KG/KM	D.C. RESISTANCE OHMS/ 1000 FT.	D.C. RESISTANCE OHMS/ K/M
12	Solid	0.081	2,05	6,530.0	0.0052	3,31	19.80	29,50	1.7	5,4
12	7/20	0.096	2,438	7,168.0	0.0057	3,66	21.69	32,28	1.5	4,8
12	19/25	0.093	2,369	6,087.6	0.0048	3,105	18.43	27,43	1.7	5,6
12	65/30	0.095	2,413	6,500.0	0.0051	3,315	19.66	29,26	1.8	5,7
12	165/34	0.095	2,413	6,548.9	0.0052	3,340	19.82	29,49	1.6	5,2
10	Solid	0.102	2,59	1,038.0	0.0083	5,26	31.4	46,80	1.0	3,4
10	37/26	0.115	2,921	9,353.6	0.0074	4,770	28.31	41,13	1.1	3,6
10	49/27	0.116	2,946	9,878.4	0.0078	5,038	29.89	44,48	1.1	3,6
10	105/30	0.116	2,946	10,530.0	0.0083	5,370	31.76	47,26	0.98	3,2
8	49/25	0.188	4,775	16,589.0	0.0130	8,403	47.53	70,73	0.67	2,2
8	133/29	0.166	4,216	16,851.0	0.0132	8,536	51.42	76,52	0.61	2,0
8	655/36	0.166	4,216	16,625.0	0.0131	8,422	49.58	73,78	0.62	2,0
6	133/27	0.210	5,334	26,799.0	0.0210	13,575	81.14	120,74	0.47	1,5
6	259/30	0.210	5,334	26,029.0	0.0204	13,185	78.35	116,59	0.40	1,3
6	1050/36	0.204	5,181	26,250.0	0.0206	13,297	79.47	118,26	0.39	1,3
4	133/25	0.257	6,527	42,613.0	0.0334	21,587	129.01	191,98	0.24	0,80
4	259/28	0.261	6,629	41,388.0	0.0324	20,966	158.02	235,15	0.20	0,66
4	1666/36	0.290	7,366	41,650.0	0.0327	21,099	126.10	187,65	0.25	0,82
2	133/23	0.328	8,331	67,763.0	0.0532	34,327	205.62	305,98	0.15	0,50
2	259/26	0.325	8,255	65,811.0	0.0516	33,338	198.14	294,85	0.16	0,52
2	665/30	0.335	8,509	66,832.0	0.0524	33,856	201.16	299,35	0.16	0,52
2	2646/36	0.379	9,626	66,150.0	0.0519	33,510	200.28	298,04	0.16	0,52
1	133/22	0.365	9,271	85,439.0	0.0671	43,282	257.60	383,34	0.12	0,40
1	259/25	0.375	9,525	82,982.0	0.0651	42,037	251.20	373,81	0.13	0,41
1	836/30	0.377	9,575	84,015.0	0.0659	42,562	247.10	367,71	0.13	0,42
1	2107/34	0.375	9,525	83,753.0	0.0657	42,428	253.29	376,92	0.12	0,41
1/0	133/21	0.464	11,786	107,743.0	0.0846	54,581	327.05	486,68	0.096	0,31
1/0	259/24	0.422	10,668	104,636.0	0.0821	53,007	316.76	471,37	0.099	0,32
2/0	133/20	0.500	12,700	135,926.0	0.1067	68,858	412.17	613,35	0.077	0,25
2/0	259/23	0.473	12,014	131,960.0	0.1036	66,849	400.41	595,85	0.077	0,25
3/0	259/22	0.509	12,928	166,381.0	0.1306	84,286	501.70	746,58	0.062	0,20
3/0	427/24	0.538	13,665	167,401.0	0.1314	87,802	522.20	777,09	0.059	0,19
4/0	259/21	0.606	15,392	209,815.0	0.1647	106,289	638.88	950,72	0.049	0,16
4/0	427/23	0.605	15,367	212,342.0	0.1667	107,569	660.01	982,16	0.047	0,15

alpha

Toll Free: 1-800-52 ALPHA • Telephone: 908-925-8000 • Fax: 908-925-6923
Europe/UK Telephone: +44 (0) 1932 772422 • Europe/UK Fax: +44 (0) 1932 772433

Web Site: www.alphawire.com
Email: info@alphawire.com

TABLE 2.5	Material	Resistivity (CM-Ω/ft) at 25 °C
Resistivities for Common Conductor Materials (U.S. units)	aluminum	17.0
	carbon	21,000
	copper	10.4
	gold	14.7
	iron	74.0
	Nichrome	600
	tungsten	33
	silver	9.9

called circular mils (CM). Do not confuse CM with cm, i.e., circular mils with centimeters. The CM is peculiar because the cross-sectional area in CM is *defined* to be the diameter of the circular wire, in mils, squared:

$$A \text{ (CM)} = [d(\text{mil})]^2 \tag{2.16}$$

where: A is the cross-sectional area in circular mils (CM),

d is the circular wire diameter in mils, and

1 mil = 1/1000 in.

Notice that the *mil* is not an abbreviation for millimeter. The mil is shorthand for one-thousandth of one inch. Also notice that the area in CM is not a true area—the factor $\pi/4$ is missing for the area to be the true area of a circle (recall that the area of a circle is $\pi r^2 = \pi d^2/4$). However, if the factor $\pi/4$ is incorporated into the resistivity constant, the calculation of area is simplified—simply square the diameter of the wire in mils. The resistivities of common conductor materials, corresponding to those in Table 2.3, are listed in Table 2.5 in these U.S. units. How were the units for resistivity in this system obtained?

$$R(\Omega) = \frac{\rho L(\text{ft})}{A(\text{CM})} \tag{2.17}$$

Solve for the resistivity to determine the units:

$$\rho = \frac{A(\text{CM})R(\Omega)}{L(\text{ft})} = \rho \left(\frac{\text{CM} - \Omega}{\text{ft}} \right) \tag{2.18}$$

Example 2.4.3

Determine the resistance of a 220-ft-long circular cross-section aluminum wire. The diameter of the wire is 0.05 in.

Given: $\ell = 220$ ft

$d = 0.05$ in.

material: aluminum: $\rho = 17.0$ CM-Ω/ft

Desired: R

Strategy: Use the relationship between R, ρ, ℓ, and A: $R = \dfrac{\rho \ell}{A}$, where $A(\text{CM}) = [d(\text{mils})]^2$

Solution:

$$A = d^2 = \left[(0.05 \text{ in.})\left(\frac{1000 \text{ mils}}{\text{in.}} \right) \right]^2 = 2500 \text{ mils}^2 = 2500 \text{ CM}$$

$$R = \frac{\rho\ell}{A} = \frac{\left(17\dfrac{\text{CM}-\Omega}{\text{ft}} \right)(220 \text{ ft})}{2500 \text{ CM}} = 1.496 \ \Omega = 1.50 \ \Omega$$

If the wire is a standard AWG size, the cross-sectional area is read directly from Table 2.3, as illustrated in Example 2.4.4.

Example 2.4.4

Determine the resistance of a 220-ft-long 16 AWG solid aluminum wire.

Given: $\ell = 220$ ft

16 AWG: $A = 2580.0$ CM

material: aluminum: $\rho = 17.0$ CM-Ω/ft

Desired: R

Strategy: Use the relationship between R, ρ, ℓ, and A: $R = \dfrac{\rho\ell}{A}$

Solution:

$$R = \frac{\rho\ell}{A} = \frac{\left(17\dfrac{\text{CM}-\Omega}{\text{ft}} \right)(220 \text{ ft})}{2580.0 \text{ CM}} = 1.4496 \ \Omega = 1.45 \ \Omega$$

Notice that the difference between the resistances in Examples 2.4.3 and 2.4.4 is less than 4%. Why?[3]

The resistances in the examples of this section were calculated for wires at 25 °C. If the temperature of the wire were different, the resistance would change. If temperature increases, should wire resistance increase or decrease? What happens to the vibrations of the atoms in the conductor as temperature increases? The vibrations should increase. If the atom vibrations increase, what happens to the wire resistance? The wire resistance should increase, too, because there are more collisions between the flowing charges and the vibrating atoms. Hence, in general,

Wire resistance increases as temperature increases.

Although these calculations are not examined here, if you need to calculate the effect of temperature on conductor resistance, investigate the topic of temperature coefficients of conducting metals.

---

[3] Compare the diameter of a 16 AWG solid wire with 0.05 in.

CHAPTER REVIEW

2.1 Energy, Power, Voltage, and Current in Electric Circuits
  - Electrical sources convert a different form of energy into electrical energy.
  - Electrical loads convert electrical energy into another form of energy.
  - Conductors guide the electrical energy from the source to the load.
  - Energy is the ability to do work (convert energy from one form into another).
  - Power is the rate at which energy is used or generated (converted).
  - Users pay for electrical energy, not power.
  - Electrical power can be determined from voltage and current.
  - Current is the amount of charge passing a point in a circuit per second.
  - Voltage is the electrical potential energy difference per unit charge.
  - Both voltage and current must be present for electrical energy to transfer from the source to the load.

2.2 Electrical Loads, Resistance, and Ohm's Law
  - Resistance is the voltage-to-current ratio for a load and is expressed as Ohm's law.
  - Resistance represents the conversion of electrical energy into another form of energy.
  - Power is related to voltage, current, and resistance.
  - There are many types of resistance, including resistors and resistance in conductors.

2.3 Resistors
  - Resistors are an important specific example of resistance.
  - Resistors are used to control voltage and current levels in circuits.
  - There are several types of resistors—carbon composition, metal-film, wire-wound—that differ by composition, power rating, and accuracy.
  - Resistors are manufactured with common nominal values. The color code is used to determine the resistance value.
  - The power rating of a resistor must be considered in circuits, or resistor failure may occur.

2.4 Wire Sizes and Resistance
  - Conductors have resistance that can be predicted from length, cross-sectional area, and material resistivity.
  - Two common systems of units are used in wire resistance calculations: SI and U.S. (CM-Ω/ft). The U.S. system is based on the American Wire Gage system of wire sizes.

HOMEWORK PROBLEMS

Energy and Power: Correctly solving the problems in the following section will demonstrate that you have an understanding of energy and power and that you understand the difference between the two. Note: One horsepower (hp) = 746 W.[4]

2.1 A lightbulb has a nominal value of 75 W. Determine the energy used if the bulb is on for 3 days.

2.2 A lightbulb has a nominal value of 100 W. Determine the energy used if the bulb is on for 26 hours.

2.3 If 20 kJ are dissipated in a resistance over a 17.6-day interval, what is the power in mW? Use the proper number of significant digits in your answer.

2.4 If 3 kJ are converted to mechanical energy by a motor in 2 minutes, what is the power in kW?

2.5 An electric heater is rated at 1500 watts and is run for 18.0 hours per day. What is the cost per day if the cost of electricity is $0.0625 per kWh?

2.6 An air conditioner is rated at 1200 watts and is run for 16.0 hours per day. What is the cost per day if the cost of electricity is $0.0625 per kWh?

2.7 If a battery can keep a flashlight on for 3 hours, and the bulb in the flashlight is rated at 1.2 W, how much energy is stored in the battery in J?

2.8 If a battery can keep a penlight on for 1 hour, and the lamp in the penlight is rated at 0.5 W, how much energy is stored in the battery in J?

2.9 If an industrial battery can turn a motor/machinery fixture for 8 hours, and the motor is rated at 2 hp, how much energy is stored in the battery in J? In kWh? (1 hp = 746 W)

2.10 If a portable battery can operate a motor in a piece of equipment for 2 hours, and the motor is

---

[4] The horsepower (hp) is another unit of power often used with electric motors and the like. The conversion factor to watts is: 1 hp = 746 W.

rated at 0.3 hp, how much energy is stored in the battery in J? In kWh? (1 hp = 746 W)

2.11 A house contains five 60 W, two 75 W, and three 40 W lightbulbs. If electricity costs $0.12/kWh, and the lights are on for 9 hours per night, determine (*a*) the amount of energy used per night in kWh and (*b*) the total cost per night.

2.12 A factory has eight 50 hp motors and 25 industrial lights rated at 250 W each. If electricity costs $0.085/kWh, and the factory is in operation for 16 hours per day, determine (*a*) the amount of energy used per day in kWh and (*b*) the total cost per day. (1 hp = 746 W)

2.13 Briefly state the difference between power and energy.

Voltage and Current: Correctly solving the problems in the following section will demonstrate that you have an understanding of voltage and current and that you understand the difference between the two.

2.14 What is conventional current flow?

2.15 If 22.6×10^{20} electrons pass a point in an electric circuit each 45.4 minutes, what is the current in amperes?

2.16 What is the current, in mA, if 57.6×10^{20} electrons flow past a point in a circuit in 2.394 minutes?

2.17 Given that a current of 22.0 mA flowed for 10 minutes, how many coulombs of charge were moved in the circuit?

2.18 What is the difference between voltage and energy?

2.19 How much energy can a voltage source deliver for every 18 C of charge if the voltage rating is 8 kV?

2.20 How much energy can a 1.5 V battery deliver for every 20 mC of charge?

2.21 Assuming the source voltage is 120 V, determine the current in a 25 W lightbulb.

2.22 Assuming a source voltage of 0.8 V, determine the current in a 50 mW LED (light-emitting diode).

2.23 If a 5 hp electric motor, an electric heater that draws 7 A with a voltage drop of 120 V, and ten

60 W lightbulbs are all on the same circuit, what is the total power dissipation? (1 hp = 746 W)

2.24 Assume that a 200 W appliance is on five days per week for 9.5 hours per day, and the cost for electric power is $0.125/kWh. What is the annual cost, to the nearest dollar, to operate this circuit?

2.25 278×10^{23} e^- flow past a point in a circuit. 45 MJ of energy is transferred to the load at a power of 16 W. What is the current in the circuit?

2.26 35×10^{22} e^- flow past a point in a circuit. 92 kJ of energy is transferred to the load at a power of 43 W. What is the current in the circuit?

2.27 If a 7.50 hp electric motor, an electric heater that draws 9.00 A with a voltage drop of 240. V, and five 75.0 W lightbulbs are all on the same circuit, what is the total power dissipation?

2.28 Assume that a 7.50 hp electric motor, an electric heater that draws 9.00 A with a voltage drop of 240. V, and five 75.0 W lightbulbs are all on the same circuit and operated for 16 hours per day. How much energy in J is consumed per day? How much in kWh?

Resistors: Correctly solving the problems in the following section will demonstrate that you have an understanding of what a resistor is and of the difference between resistance and a resistor.

2.29 Your are evaluating a resistor in the laboratory. In doing so, you connect a voltage source and set it to 10.0 V. You then measure the current in the circuit and find it to be 10.3 mA.
   a. What is the standard value of the resistor if the tolerance is 5%?
   b. What are the color bands that you would expect to be on the resistor?

2.30 A 12 V battery is connected to a series of resistors, one at a time. The current measured in each is as follows: (*a*) 11.6 mA, (*b*) 5.7 mA, (*c*) 1.25 mA, (*d*) 1.41 mA, (*e*) 3.71 mA, and (*f*) 2.47 mA.
   a. What is the resistance value for each of the resistors?
   b. Assuming a tolerance of 5%, what are the standard values of these resistors?
   c. What are the color bands that you would expect to be on the resistors?

2.31 You have received a package filled with carbon composition resistors and you need to sort them by resistance value. Based on the color code, create a table listing the nominal value, the minimum value, the maximum value, and the range:
   a. red, red, red, silver
   b. brown, black, orange, gold
   c. black, red, black, gold
   d. orange, violet, orange, silver
   e. blue, gray, yellow, gold
   f. yellow, orange, red
   g. black, brown, brown, silver
   h. brown, yellow, gold, gold

2.32 Research the "standard values" of carbon composition resistors and the resistor color code. Is there any trend in the resistance values?

Correctly solving the problems in the following section will demonstrate that you have an understanding of electrical loads, resistance, and Ohm's law.

2.33   What is the voltage drop across a 14.5 kΩ resistor if the current flow is 3.00 mA?

2.34   What is the rate of energy consumption by a 14.5 kΩ resistor if the current flow is 3.00 mA?

2.35   What is the current flow through a 12.5 kΩ resistor if 3.00 MV are applied?

2.36   A 3.5 hp motor operates at 120 V. What is the resistance of the motor?

2.37   If the current in a resistive circuit is 3.97 A,
   a. what is the power dissipated if the resistance is 1.75 Ω?
   b. what is the hp rating?

2.38   If you have a 5.00 hp compressor in your workshop and it is connected to a 220 V circuit, what is the operating current?

2.39   If the voltage drop in a resistive circuit is 35.8 V, what is the power dissipated if the resistance is a 6.75 Ω resistor?

2.40   If the voltage drop across a motor is 35.8 V, what is the power dissipated if the motor has an operating resistance of 6.75 Ω?

2.41   A resistor is rated at 125 mW. If 2.185 kJ are dissipated when 47.8 C of charge pass through the resistor, and the current is 34.2 μA, what is the actual power dissipated in the resistance?

2.42   Assume that a resistive device can be operated continuously at 50% of its maximum rating. If the total energy used is 7.375 kJ after 75.5 C of charge have been moved at a rate of 1.000 mA, what is the maximum power rating of the device?

2.43   If the voltage drop across a lighting circuit is 120 V, what is the power dissipated if the lights have an operating resistance of 144 Ω?

2.44   In a short statement:
   a. State what a resistance (not a resistor) represents in general.
   b. Give an example.

Correctly solving the problems in the following section will demonstrate that you have an understanding of the importance of wire size and the resulting resistance of that wire.

2.45   If a copper wire has a diameter of 5.22 mm and a length of 67.5 m, what is its resistance at room temperature?

2.46   If a copper wire has a diameter of 4.54 mm and a length of 38.5 m, what is its resistance at room temperature?

2.47   A coil is wound from 22 AWG copper wire (0.0253 in. diameter). Determine the length of wire required if the DC resistance of the coil is 0.1 Ω.

2.48   If 2.50 miles of copper wire are required to connect a circuit to a source (1.25 miles in each of the two conductors), and the maximum resistance allowed is 1.5 Ω,
   a. what is the area in cm^2 of the wire for this resistance limitation?
   b. is this a maximum or a minimum area? Justify your answer.

RESEARCH AND SELF-STUDY:

2.49   After a bit of research, you find that the specific energy of low-volatile bituminous coal is 36.4 MJ/kg that bituminous coal has a carbon content ranging from 45 to 86% carbon and a heat value of 10,500 to 15,500 BTUs per pound, and that a BTU is equivalent to approximately 0.293 watt-hour. Is all the information provided necessary to answer the following question: How many tons of coal would be necessary to operate a 10.0 hp electric motor for 30 days? Can you verify this research? What are your sources?

2.50   As a bright young future engineering technologist, you have come up with an idea that you believe will motivate people to exercise. Your idea consists of a flywheel, an electric generator, an exercise bicycle, and an entertainment center. In order to operate the entertainment center, a person must exercise to transfer energy into the flywheel that in turn will operate the electric generator. The entertainment center requires 100 W and the exercise regimen calls for a half hour of continuous exercise in order to operate the entertainment center for 45 minutes. This will allow for a cool-down period while the flywheel keeps the generator running. How many food calories will be necessary to keep the music playing? (one food calorie = 1000 heat calories; 1 calorie = 4.186 J)

SERIES-PARALLEL ANALYSIS OF DC CIRCUITS

As a result of successfully completing this chapter, you should be able to:

1. Explain why there are series, parallel, and series-parallel circuits.
2. Describe the fundamental properties of DC series and parallel circuits and subcircuits.
3. Calculate all voltages, currents, and powers in single source DC series, parallel, and series-parallel circuits.
4. Describe why and how superposition is used in multiple source DC series-parallel circuits.
5. Calculate all voltages, currents, and powers in multiple source DC series, parallel, and series-parallel circuits.
6. Describe ideal and practical DC voltage and current sources.
7. Analyze DC series-parallel circuits that contain a current source.
8. Explain the role of circuit simulation in the analysis of DC series-parallel circuits.

3.1 WHY ARE THERE DIFFERENT TYPES OF CIRCUITS?

In the last chapter we learned that resistance is the voltage-to-current ratio of a load, as expressed by Ohm's law. Resistance represents the conversion of electrical energy into some other form of energy. It was also suggested that a circuit may consist of more than one load. Two circuit configurations are shown in Figure 3.1. In the configuration in Figure 3.1a, the resistances are connected end to end in a single path, i.e., the resistances are connected in *series*. In the configuration in Figure 3.1b, one side of all the resistances are connected and the other side of all the resistances are also connected, i.e., the resistances are connected in *parallel*. The first question is: Why would anyone

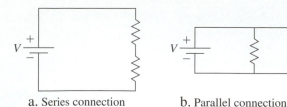

Figure 3.1 Two circuit configurations for two resistances

a. Series connection   b. Parallel connection

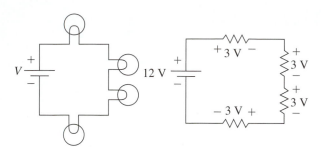

Figure 3.2 Example of a series circuit

a. Four lamps in series   b. Schematic circuit

want to connect loads in either manner? Some reasons are given in this introductory section.

In the series connection, one immediately sees that the current has only one path.

The current is the same through any element, source or load, in a series circuit.

However, the voltage that is supplied by the source must be shared by the two resistances. The source voltage *divides* between the loads. One could look at this from an alternative viewpoint: if a load requires a smaller voltage than the voltage supplied by the source, then two or more loads of the proper resistances could be connected in series to *divide* the voltage. This is the voltage divider mentioned at the end of Chapter 2.

An example of loads that require a lower voltage than that supplied by the source is a string of lights, as shown in Figure 3.2. Each lamp in this example requires 3 V. The source is 12 V. If four lamps are connected in series, then all four lamps together will require 12 V, which matches the voltage of the source.

There is a significant disadvantage of a series connection of resistances. What happens if one of the resistances fails in a manner that breaks the continuity of the current path? Then it becomes an *open circuit* (often just called an *open*), which has infinite resistance, and there will be no current flow through any component. All energy transfer from the source to the loads ceases. Although this is usually safe, the circuit operation is halted. If one of the resistances fails in a manner such that the component conducts current with little or no voltage drop, then the resistance becomes a *short circuit* (again, often just called a *short*). A short has zero resistance ideally, no voltage is across that load, and the other loads have a larger portion of the source voltage. This may be all right unless the loads are not rated for a higher voltage. Then they may fail too, and this is generally not safe!

In the parallel connection in Figure 3.1b, one sees that the electrical potential energy difference across each load is the same if the resistance of the conductors is negligible compared with the resistance of the loads.

The voltage is the same across any element, source or load, in a parallel circuit.

But the current that is supplied by the source must be shared among the two resistances. The source current *divides* among the loads.

An example of loads that require a parallel connection is shown in Figure 3.3. Each lamp is rated for the entire source voltage. However, the source must be able to supply enough current for all three lamps, which is three times the current that is required by one lamp if the lamps are identical.

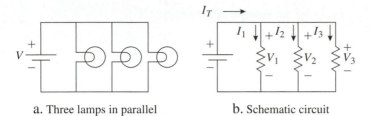

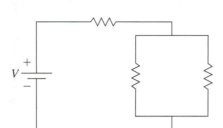

a. Three lamps in parallel

b. Schematic circuit

Figure 3.3 Example of a parallel circuit

Figure 3.4 An example of a series-parallel circuit

There is a significant advantage and a significant disadvantage to a parallel connection of resistances. If one load were to become an open, the other loads would still have the same voltage across them and would still continue to operate. This is usually good. For example, if a light burns out (the filament opens) in a house, the others continue to work. On the other hand, if one of the loads becomes a short, all the current goes through the short, and the current is very high. If the resistance approaches zero and the voltage is constant, then the current approaches infinity. Practically, the current capability of the source is met, and fuses blow, circuit breakers trip, wires melt and become an open, or the source fails. The last two cases are generally very unsafe.

One might easily extrapolate the series and parallel connections to a combination of both, as shown in Figure 3.4. This circuit configuration is one example of a *series-parallel* circuit. The analysis of series-parallel circuits combines the analysis of series and parallel circuits. This circuit is generally very useful in the analysis and design of electronic circuits.

Now that the basic motivation for connecting resistances in series and in parallel has been established, we may proceed to predicting voltages, currents, resistances, powers, and energies. In the rest of this chapter, voltage, current, and resistance laws and circuit analysis techniques are established to help make predictions about any circuit with series and/or parallel connections. Power and energy are easily calculated if voltage, current, and resistance are already determined. A subsequent section shows a straightforward technique for determining voltages and currents in a series-parallel circuit that contains multiple sources. A section addresses practical sources and introduces another type of source called the current source. A final section covers the role of circuit simulation software in the analysis of DC series-parallel circuits.

3.2 DC SERIES CIRCUITS

How does one start to establish rules and laws for series circuits? We have to intellectually explore what we know and see if laws, rules, and analysis techniques result. What do we know about electricity so far? Electricity is a form of energy; hence, it obeys the law of conservation of energy:

> *Energy can neither be created nor destroyed but may be transferred and may be converted from one form into another.*

Do you recall any previous discussion in which the electrical energy was split up (divided)? (If not, reread the previous section.) Yes, it was the series circuit. The energy of

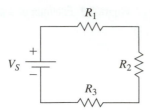

Figure 3.5 An example of a series circuit

the source must be shared by all the loads. For example, in Figure 3.5, the sum of the energies in the three resistances must equal the energy supplied by the source to satisfy the law of conservation of energy:

$$E_S = E_1 + E_2 + E_3 \tag{3.1}$$

Furthermore, we know that electrical energy is related to voltage, and voltage is the electrical potential energy difference per unit of charge. Then we realize that the current, i.e., charge per unit time, is the same in a series circuit because there is only one path for the charge to flow. Thus, each energy term in Equation (3.1) can be divided by the same charge Q:

$$\frac{E_S}{Q} = \frac{E_1}{Q} + \frac{E_2}{Q} + \frac{E_3}{Q} \tag{3.2}$$

Because voltage is electrical energy divided by charge (per unit charge), *the voltage supplied by the source must equal the sum of the voltages of the loads*. This conclusion is one of the fundamental laws of circuits and is named after its discoverer: *Kirchhoff's voltage law (KVL)*:

$$V_S = V_1 + V_2 + V_3 \tag{3.3}$$

Note that these series of mathematical steps allowed us to obtain a useful result from previously known results. This series of steps is called a *derivation*. A derivation is a series of mathematical steps and reasons for developing a result from other known results. Do you need to memorize all the steps in the derivation? Absolutely not! The purpose of the derivation is to obtain and justify another result using previously known results. Hence, your obligation is to make sure you understand and can explain each step in the derivation and then to explain the significance of and the use of the new result. There are several derivations in this text, again not for memorization but to be sure we understand where results come from and to be sure we are justified in using those results. See problem 3.41 in the homework at the end of the chapter for an alternative derivation of KVL.

Kirchhoff's voltage law is actually even more general than the statement given in the previous paragraph. There may be more than one source, for example. The sum of the voltages "supplied" must equal the sum of the voltages "used." This statement needs to be firmed up if the law is to be always applicable (not all circuits are series circuits). This clarification is made with the definition of two new terms: *voltage drop* and *voltage rise*. In Figure 3.6, the voltage polarities are labeled. Recall that the positive side of the voltage across a load is on the side of the load that the conventional current enters. Start at the negative terminal of the source and move clockwise (CW) around the circuit. As you move around the circuit, the voltage of the source *rises* from negative to positive. As you move through each load, the voltage *drops* from positive to negative. Hence, in this circuit, going CW around the circuit gives one voltage rise, V_S (voltage of the source), and three voltage drops, V_1, V_2, and V_3. The general form of KVL states that the sum of the voltage rises must equal the sum of the voltage drops in any complete (closed) path:

$$V_S = V_1 + V_2 + V_3 \tag{3.4}$$

which is identical to Equation (3.3).

In complicated circuits, it becomes difficult to "jump around" and find all the rises and then jump around again to find all the drops. Hence, KVL is stated in an even more general form:

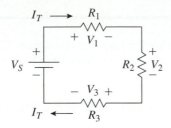

Figure 3.6 Voltage polarities in a series circuit

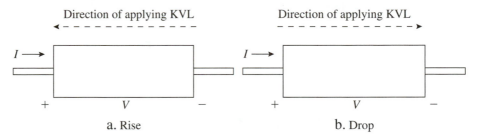

Direction of applying KVL

a. Rise

Direction of applying KVL

b. Drop

Figure 3.7 Voltage rises and voltage drops in general

In any closed path, the sum of the voltage rises and drops must equal zero.

For the circuit in Figure 3.6, the application of KVL is:

$$+V_S - V_1 - V_2 - V_3 = 0 \tag{3.5}$$

Note that voltage rises were assigned a positive sign, and voltage drops were assigned a negative sign. Some books and literature use just the opposite signs for rises and drops. It does not really matter which convention you use as long as you stick with the same assignment and that you document your assignment. In this textbook the following assignment is used:

Voltage rises are assigned a positive sign, and voltage drops are assigned a negative sign.

A summary of voltage rises and drops is given in Figure 3.7. Solid arrows indicate conventional current directions, and dashed arrows indicate the direction that KVL is applied. The type of component and the current direction do not matter. The voltage polarity rise (negative to positive) or drop (positive to negative) in the direction that KVL is applied is the key.

What if one moved around the circuit in Figure 3.6 counterclockwise (CCW)? Try it. Assign signs to the voltage polarities using the aforementioned convention:

$$-V_S + V_1 + V_2 + V_3 = 0 \tag{3.6}$$

This is identical to Equation (3.5). Why is this so? Multiply through both sides of Equation (3.5) by −1. Equation (3.6) results. Hence, we conclude that in the application of KVL, the direction around the closed path does not matter—just pick a direction and stick with it. A general mathematical statement of KVL is:

$$\sum_{n=1}^{N} V_n = +V_{\text{rise1}} + V_{\text{rise2}} + V_{\text{rise3}} + \ldots - V_{\text{drop1}} - V_{\text{drop2}} - V_{\text{drop3}} - \ldots = 0 \tag{3.7}$$

The summation sign (capital Greek letter sigma) tells you to add up N voltages (N is an integer). The index, n, is incremented for each voltage encountered around the closed path: $n = 1, 2, 3, ..., N$. Do not become hung up in mathematical notation. Just go around the closed circuit path, adding each voltage rise and subtracting each voltage drop. The sum must be zero.

Example 3.2.1

The source voltage for a series circuit with three resistances is 15 V. The voltage drops across each resistance are shown in Figure 3.8. Show that KVL is satisfied.

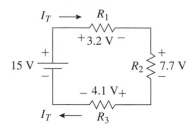

Figure 3.8 Example of KVL in a series circuit

Given: $V_S = 15$ V

$V_1 = 3.2$ V

$V_2 = 7.7$ V

$V_3 = 4.1$ V

Desired: Show that KVL is satisfied.

Strategy: Use KVL and evaluate the sum.

Solution:

The direction to apply KVL is *arbitrarily* selected to be CCW.
Apply KVL:

$$-V_S + V_3 + V_2 + V_1 = -15 + 4.1 + 7.7 + 3.2 = 0$$

KVL is satisfied.

It is good that we can check our work using KVL (you should always check your work), but is KVL useful for finding unknown quantities? Yes. What if a voltage is unknown? Let's rework Example 3.2.1 assuming V_2 is unknown.

Example 3.2.2

The source voltage for a series circuit with three resistances is 15 V. The voltage drops across resistances one and three are 3.2 V and 4.1 V, respectively. What is the voltage drop across the resistance R_2?

Given: $V_S = 15$ V

$V_1 = 3.2$ V

$V_3 = 4.1$ V

Desired: V_2

Strategy: Set up KVL and solve for V_2

Solution:

The direction to apply KVL is arbitrarily selected to be CCW. Apply KVL:

$$-V_S + V_3 + V_2 + V_1 = -15 + 4.1 + V_2 + 3.2 = 0$$

The solution of this equation gives $V_2 = 7.7$ V.

Hence, KVL is a useful circuit analysis tool. In addition, KVL is useful in the derivation of an important result: the total resistance of a series circuit. A series circuit with three resistances will be used for convenience, but the derivation could be done for any number of series resistances. Start with KVL.

$$+V_S - V_1 - V_2 - V_3 = 0 \tag{3.8}$$

Move the negative terms to the other side of the equation to make them positive.

$$V_S = V_1 + V_2 + V_3 \tag{3.9}$$

Use Ohm's law to express the voltage across each resistance: $V = IR$:

$$V_S = I_1 R_1 + I_2 R_2 + I_3 R_3 \tag{3.10}$$

But we realize that the current is the same anywhere in a series circuit:

$$I_S = I_1 = I_2 = I_3 = I \tag{3.11}$$

The source voltage is across all the loads, i.e., the total resistance of the circuit. Use Ohm's law again.

$$IR_T = IR_1 + IR_2 + IR_3 \tag{3.12}$$

Divide through both sides of the equation by I:

$$R_T = R_1 + R_2 + R_3 \tag{3.13}$$

The conclusion from this derivation is:

The total resistance of a series circuit is the sum of the load resistances.

For N resistances in series,

$$\boxed{R_T = R_1 + R_2 + R_3 + \ldots + R_N = \sum_{n=1}^{N} R_n} \tag{3.14}$$

Thus, the application of KVL to a series circuit produces a very useful result. The usefulness is illustrated in the next example.

Example 3.2.3

For the circuit shown in Figure 3.9, determine

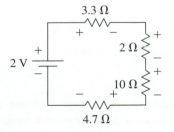

Figure 3.9 Series circuit for Example 3.2.3

a.   the total resistance,
b.   the total current,
c.   the current through each resistance,
d.   the voltage across each resistance,
e.   if the voltages satisfy KVL
f.   the total power delivered from the source to the loads
g.   the power used in each load
h.   if the powers satisfy *conservation of power*

Given: $V_S = 2\ V$

$R_1 = 3.3\ \Omega$

$R_2 = 2\ \Omega$

$R_3 = 10\ \Omega$

$R_4 = 4.7\ \Omega$

Desired: a. R_T

b. I_T

c. $I_1, I_2,$ and I_3

d. $V_1, V_2,$ and V_3

e. Check using KVL.

f. $P_T = P_S$ (P_S is the power delivered by the source)

g. $P_1 = V_1 I_1$, etc.

h. Check conservation of power.

Strategy: Use the series circuit total resistance equation.

Use Ohm's law to find I_T.

The current through each resistor is the same in a series circuit.

Use Ohm's law to find the voltage across each R.

Check using KVL.

Use $P_T = V_S I_T$ to determine the total power.

Use $P = VI$ with individual component voltages and currents to determine the power of each component.

Add component powers to see if it equals the power delivered by the source.

Solution: _____

a. $R_T = R_1 + R_2 + R_3 + R_4 = 3.3 + 2 + 10 + 4.7 = 20.0\ \Omega$

Note that all answers are reported to three significant digits per the rule of thumb from Chapter 1. However, during the following calculations, 4 or 5 significant digits are used (held in the calculator memory). Often these digits are not written out, when the digit is zero, to save writing time.

b. $I_T = \dfrac{V_S}{R_T} = \dfrac{2\ V}{20\ \Omega} = 0.100\ A$

c.   series circuit $\Rightarrow I_T = I_1 = I_2 = I_3 = I_4 = 0.100\ A$

d. $V_1 = IR_1 = (0.1\ A)(3.3\ \Omega) = 0.33\ V = 0.330\ V$

$V_2 = 0.200\ V$

Note that if the same calculation is repeated with different numbers, only the first calculation needs to be shown and then the other results are just reported.

$$V_3 = 1.00 \text{ V}$$

$$V_4 = 0.47 \text{ V}$$

e. KVL: ■ Select a direction to go around the closed path: CW

■ Label all voltage polarities on resistances based on the current direction—see Figure 3.10.

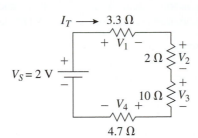

Figure 3.10 Voltage polarities for KVL in Example 3.2.3

■ Apply KVL:

$$+V_S - V_1 - V_2 - V_3 - V_4 = 0$$

$$+2 - 0.33 - 0.2 - 1 - 0.47 = 0 \qquad \text{KVL is satisfied.}$$

f. $P_T = V_T I = (2 \text{ V}) (0.1 \text{ A}) = 0.200 \text{ W}$

g. $P_1 = V_1 I = (0.33 \text{ V}) (0.1 \text{ A}) = 0.0330 \text{ W}$

$P_2 = 0.0200 \text{ W}$

$P_3 = 0.100 \text{ W}$

$P_4 = 0.0470 \text{ W}$

h. Conservation of power: The power supplied by the source is equal to the power used in all loads.

$$P_S = P_1 + P_2 + P_3 + P_4 = 0.033 + 0.02 + 0.1 + 0.047 = 0.200 \text{ W}$$

This answer matches P_T in part (f), so conservation of power is satisfied.

The extensiveness of the previous example demonstrates the usefulness of Ohm's law, KVL, and the electrical power equation. However, what if all of these quantities were not needed? What if only one voltage was needed? There is a shortcut that is often used with series circuits: the *voltage divider rule,* abbreviated VDR. A short derivation of the voltage divider rule is given next. See if you can explain what was done in each step. Then see if you can explain the result.

1. Assume a series circuit with three resistances.
2. Let R_x be any one of the resistances in series; use resistance 2, for example.
3. $I_T = I_2$ (3.15)

4. $\dfrac{V_T}{R_T} = \dfrac{V_2}{R_2}$ (3.16)

Note: V_T is the total voltage, which is identical to the voltage source in a series circuit.

5. $V_2 = R_2 \dfrac{V_T}{R_T}$ (3.17)

This is the VDR for resistance 2. What is the significance of this rule? First, only one step is necessary to determine the voltage across that resistance (two steps if the total resistance needs to be calculated). Second, this rule states that the total voltage divides proportionate to resistance. If R_2 is larger with respect to the other resistances, then more of the total voltage drops across R_2. Notice that the R_2/R_T is the ratio (proportion) of R_2 with respect to the total resistance. An alternative viewpoint is that V_T/R_T is just the current, which is the same anywhere in a series circuit, and this is multiplied by R_2 to give V_2. Hence, the voltage divider rule is just a convenient form of Ohm's law. These two reasons demonstrate why the voltage divider rule is important: faster calculations and more insight into series circuits. By the way, the explanations for the steps in the VDR derivation are:

1. problem setup assumption
2. definition of quantities
3. fundamental property of current in a series circuit
4. substitution of variables using Ohm's law
5. algebra to obtain the final result

Observe that two or more series resistances could be grouped together and treated as one resistance. This is why V_x is used in the general form of the VDR:

$$V_x = R_x \frac{V_T}{R_T}$$

(3.18)

Example 3.2.4

In the circuit of Example 3.2.3, determine (*a*) the voltage across R_4 and (*b*) the voltage across the series combination of R_2 and R_3 using the voltage divider rule.

Given: $R_T = 20\ \Omega$

$R_2 = 2\ \Omega$

$R_3 = 10\ \Omega$

$R_4 = 4.7\ \Omega$

$V_S = 2\ V$

Desired:

a. V_4

b. V_{2+3} (the voltage across R_2 and R_3 collectively)—see Figure 3.11

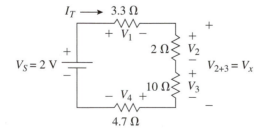

Figure 3.11 V_x in Example 3.2.4

Strategy: Define R_x and V_x and use in the VDR; $V_T = V_S$

Solution:

a. $R_x = R_4 = 4.7\ \Omega$

$R_T = 20\ \Omega$

$V_x = V_4$

$$V_x = R_x \frac{V_T}{R_T}$$

$$V_4 = R_4 \frac{V_T}{R_T} = (4.7 \ \Omega)\left(\frac{2 \ V}{20 \ \Omega}\right) = 0.470 \ V$$

This matches the result in Example 3.2.3d.

b. $R_x = R_2 + R_3 = 2 + 10 = 12 \ \Omega$

   $V_x = V_2 + V_3$

$$V_x = R_x \frac{V_T}{R_T} = (12 \ \Omega)\left(\frac{2 \ V}{20 \ \Omega}\right) = 1.20 \ V$$

Before leaving this example, let us check our work. Does $V_x = V_2 + V_3$ using the values obtained in Example 3.2.3?

$$V_x = V_2 + V_3 = 0.2 + 1 = 1.20 \ V$$

Thus, the VDR result matches this result.

Ask this question: Are equalities such as $V_x = V_2 + V_3$ in Example 3.2.4 valid in general? How could you answer this question? Try KVL! Set up a path for KVL through R_2 and R_3 and then through the air to complete the path. You might question the "through the air" part, but imagine placing an ideal voltmeter across R_2 and R_3, as sketched in Figure 3.12. The ideal voltmeter measures voltage but has infinite resistance, so that no current flows through the voltmeter. Label the polarity for V_x, either way, but it is obvious in this circuit that the positive side of V_x is on the positive side of R_2 and the negative side of V_x is on the negative side of R_3. It is also customary to label this voltage V_{ab}, where terminals a and b are labeled on the circuit diagram, and the polarity is assigned (here, terminal a is positive and terminal b is negative). Add the voltages up as you go around the closed path:

$$+V_3 + V_2 - V_x = 0$$

$$V_x = +V_2 + V_3$$

Thus, $V_x = +V_2 + V_3$ is valid. In KVL, the sign in front of each variable is often shown, even if positive, as a reminder that the sign has been considered. More importantly, the concept to recognize is:

KVL works around any closed path, including "through the air."

As a final exercise, explain each step in the derivation of the voltage divider rule:

1.   Assume a series circuit of N resistances.
2.   Let $R_x \equiv$ any one or more resistances in series ($\equiv$ means "defined as").
3. $I_T = I_x$ (3.19)
4. $\dfrac{V_T}{R_T} = \dfrac{V_x}{R_x}$ (3.20)

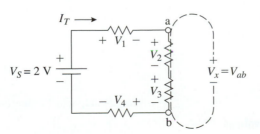

Figure 3.12 Path for KVL through the air

$$5. \quad \boxed{V_x = R_x \frac{V_T}{R_T}} \qquad\qquad (3.21)$$

3.3 DC PARALLEL CIRCUITS

In Section 3.1, the electrical potential energy difference across each component in the parallel circuit was the same. The assumption in this statement is that the resistance of the conductors is negligible. If the conductor resistance were not negligible, one would have to draw additional resistances in the schematic diagram to represent the resistances of the wires. Under the negligible resistance assumption, all of the electrical potential energy difference is directly across each load. Hence,

The voltage across each component in a parallel circuit is the same.

The current, on the other hand, has multiple paths. Every time the current reaches a junction of three or more wires, the current can split or recombine. Can the direction of the current be determined in each load? Yes. Recall that if the voltage polarity is known across a load, the current direction is also known: current enters the load on the side where the voltage is positive and leaves the load on the side where the voltage is negative. But how much current splits off into each load? This is the key question. First, two new terms are introduced. Each leg of a parallel circuit is called a *branch*. Each connection of conductors that has no voltage difference across itself is called a *node*. A parallel circuit with three parallel resistances, as shown in Figure 3.13, will be used to illustrate nodes. There are two nodes: one on the top, and one on the bottom. If you placed a voltmeter lead at the positive terminal of the source and the other lead at the top of R_3, the voltage would be zero. The connection is *common* (the same).

The total current from the source splits into three branches. If 1 billion charges leave the positive terminal of the source every second and split into the three load branches as they leave the upper node, then 1 billion charges will recombine at the lower node and enter the negative side of the source every second. Charge cannot build up in a wire or resistance because the voltage keeps forcing the charges to move. Thus, for every charge that enters a node, a charge must leave that node. This can be expressed in terms of current:

The sum of the currents entering a node must equal the sum of the currents leaving that node.

This statement is fundamental, and it is called Kirchhoff's current law (KCL). As with voltage rises and voltage drops, one can arbitrarily assign a sign to the current entering a

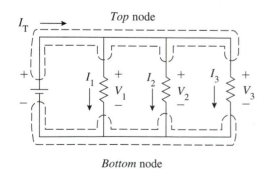

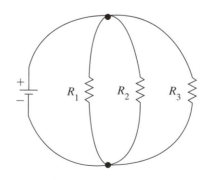

a) Schematic

b) One possible actual connection

Figure 3.13 Parallel circuit with all nodes identified

node as positive and the current leaving a node as negative. Some books use just the opposite convention—just pick one and be consistent. In this book, we will assign signs to current in KCL as stated. Hence, KCL can be restated as

The sum of the currents entering and leaving a node must equal zero.

In mathematical form, for the upper node of the circuit in Figure 3.13,

$$+I_S - I_1 - I_2 - I_3 = 0 \tag{3.22}$$

For the lower node of the circuit in Figure 3.13,

$$+I_1 + I_2 + I_3 - I_S = 0 \tag{3.23}$$

This equation is identical to Equation (3.22). Hence, just as KVL was a fundamental relation in series circuits, KCL is a fundamental relation in parallel circuits. The general mathematical statement of KCL for a node with N branches connected to it is

$$\sum_{n=1}^{N} I_n = +I_1 + I_2 + I_3 + \ldots + I_N = 0 \tag{3.24}$$

where all currents are shown positive, but if the current is leaving the node, the current must be assigned a negative sign.

How might you prove KCL? Assume three parallel resistances and start with the conservation of power relation:

$$P_S = P_1 + P_2 + P_3 \tag{3.25}$$

$$V_S I_S = V_1 I_1 + V_2 I_2 + V_3 I_3 \tag{3.26}$$

But the voltage is constant in a parallel circuit:

$$V_S = V_1 = V_2 = V_3 = V \tag{3.27}$$

Equation (3.26) becomes:

$$V I_S = V I_1 + V I_2 + V I_3 \tag{3.28}$$

Divide through both sides of the equation by V, and KCL results:

$$+I_S = +I_1 + I_2 + I_3 \tag{3.29}$$

which is mathematically identical to Equation (3.22).

Information Research Exercise   (web and/or library)

Kirchhoff must have been important—KVL and KCL were named in his honor. When did he arrive at these laws? Relate this time to the historical context and to the progress in the discovery and utilization of electricity to that point.

Important results can be derived from KCL: the total resistance of a parallel circuit and the current divider rule. The derivation of the total resistance of a parallel circuit is given first. A parallel circuit with three parallel branches is used for convenience. See if you can explain each step in the derivation.

$$+I_S - I_1 - I_2 - I_3 = 0 \tag{3.30}$$

$$+I_S = I_1 + I_2 + I_3 \tag{3.31}$$

$$\frac{V_T}{R_T} = \frac{V_1}{R_1} + \frac{V_2}{R_2} + \frac{V_3}{R_3} \tag{3.32}$$

$$V_T = V_1 = V_2 = V_3 = V \tag{3.33}$$

$$\frac{V}{R_T} = \frac{V}{R_1} + \frac{V}{R_2} + \frac{V}{R_3} = V\left(\frac{1}{R_1} + \frac{1}{R_2} + \frac{1}{R_3}\right) \tag{3.34}$$

$$\frac{1}{R_T} = \frac{1}{R_1} + \frac{1}{R_2} + \frac{1}{R_3} \tag{3.35}$$

$$R_T = \frac{1}{\dfrac{1}{R_1} + \dfrac{1}{R_2} + \dfrac{1}{R_3}} \tag{3.36}$$

Thus, the total resistance of a parallel circuit is not just a simple addition of resistances. It involves two reciprocals: one for each individual parallel resistance, and one for the sum of those reciprocals. You can easily expand the number of parallel branches to N branches:

$$R_T = \frac{1}{\dfrac{1}{R_1} + \dfrac{1}{R_2} + \dots + \dfrac{1}{R_N}} \tag{3.37}$$

Example 3.3.1

Determine the total resistance of the circuit shown in Figure 3.14.

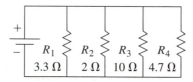

Figure 3.14 Parallel circuit for Example 3.3.1

R_1	R_2	R_3	R_4
3.3 Ω	2 Ω	10 Ω	4.7 Ω

Given: $R_1 = 3.3\ \Omega$
$R_2 = 2\ \Omega$
$R_3 = 10\ \Omega$
$R_4 = 4.7\ \Omega$

Desired: R_T

Strategy: Use the parallel circuit total resistance equation.

Solution:

$$R_T = \frac{1}{\dfrac{1}{R_1} + \dfrac{1}{R_2} + \dfrac{1}{R_3} + \dfrac{1}{R_4}} = \frac{1}{\dfrac{1}{3.3} + \dfrac{1}{2} + \dfrac{1}{10} + \dfrac{1}{4.7}} = \frac{1}{1.1158} = 0.89622\ \Omega = 0.896\ \Omega$$

There is something striking about this total resistance—try to identify it. The total resistance of the parallel circuit is *less than* the resistance of the branch with the lowest resistance. This observation is always true for parallel circuits. One can think of this result in another way. In a parallel circuit, the total current divides into several branches. The same voltage is across all branch resistances and across the total resistance. However, the total resistance has more current than any individual branch. The same voltage and higher current in the total resistance as opposed to any individual branch resistance translates into the total resistance being smaller than any individual branch resistance.

Another consequence in parallel circuits is that the branches with lower resistances will have a larger current than those branches with higher resistances. Hence, the smallest resistance dominates the total resistance of a parallel circuit.

Circuits with two parallel resistances are encountered often in practice. The total parallel resistance of two parallel resistances can be found using Equation (3.37), but a simpler form is usually used. Here is a short development:

$$R_T = \frac{1}{\dfrac{1}{R_1} + \dfrac{1}{R_2}} \tag{3.38}$$

One can multiply the numerator and denominator by the same quantity without changing the value of the expression:

$$R_T = \left(\frac{1}{\dfrac{1}{R_1} + \dfrac{1}{R_2}}\right)\left(\frac{R_1 R_2}{R_1 R_2}\right) = \frac{R_1 R_2}{\dfrac{R_1 R_2}{R_1} + \dfrac{R_1 R_2}{R_2}} = \frac{R_1 R_2}{R_2 + R_1} \tag{3.39}$$

$$R_T = R_1 \parallel R_2 = \frac{R_1 R_2}{R_1 + R_2} \tag{3.40}$$

The two parallel bars mean "in parallel with." This is a handy formula that you will use often.

Another special parallel circuit case is N parallel resistances, where all the resistances have the same value R. Try to derive that $R_T = R / N$. As an example, if five $10\ k\Omega$ resistors are in parallel, the total resistance is $2\ k\Omega$.

Before leaving the total resistance topic, let us examine another common approach: using *conductance*. Conductance is one over resistance and has the symbol G:

$$G = \frac{1}{R} \tag{3.41}$$

The SI unit for conductance is the siemens (S). If Ohm's law is used for R, then

$$G = \frac{I}{V} \tag{3.42}$$

Conductance is often used when most of the circuits are parallel circuits. Why? Look at what happens to the total resistance equation for a parallel circuit when conductance is used. Start with Equation (3.35):

$$\frac{1}{R_T} = \frac{1}{R_1} + \frac{1}{R_2} + \frac{1}{R_3} \tag{3.43}$$

$$G_T = G_1 + G_2 + G_3 \tag{3.44}$$

In general, for N parallel branches,

$$G_T = G_1 + G_2 + G_3 + \ldots + G_N = \sum_{n=1}^{N} G_n \tag{3.45}$$

In series circuits, resistances in series add. In parallel circuits, *conductances in parallel add*.

Example 3.3.2

Determine the total conductance of the circuit in Example 3.3.1. Check the result against the total resistance found in Example 3.3.1.

Given: $R_1 = 3.3\ \Omega$
$R_2 = 2\ \Omega$
$R_3 = 10\ \Omega$
$R_4 = 4.7\ \Omega$

Desired: G_T

Strategy: Convert all individual resistances into conductances and sum.

Solution: _____

$$G_1 = \frac{1}{R_1} = \frac{1}{3.3 \ \Omega} = 0.30303 \ S$$

$$G_2 = 0.5 \ S$$

$$G_3 = 0.1 \ S$$

$$G_4 = 0.21277 \ S$$

Notice that some of the numbers are not "nice," i.e., we needed to record the four or five nonzero significant digits during these intermediate calculations.

$$G_T = G_1 + G_2 + G_3 + G_4 = 0.30303 + 0.5 + 0.1 + 0.21277 = 1.1158 \ S$$

Thus, $G_T = 1.12$ S. As a check, determine R_T from G_T. Be sure to use more significant digits during the calculation before rounding to three significant digits.

$$R_T = \frac{1}{G_T} = \frac{1}{1.1158 \ S} = 0.896 \ \Omega$$

This total resistance matches the result obtained in Example 3.3.1.

The current divider rule (CDR) is useful for finding the current in one branch without having to determine all branch currents. This is especially useful if the voltage across the parallel circuit is not known but the total current is known. The derivation of the CDR is now given. Again, try to explain each step.

Assume a parallel circuit with N parallel branches and the current in branch x (1 or 2 or 3 or whatever branch we select) is desired. Let V_T be the voltage across the entire circuit and V_x be the voltage across one branch in the parallel circuit. Then:

$$V_T = V_x \tag{3.46}$$

$$I_T R_T = I_x R_x \tag{3.47}$$

$$\boxed{I_x = \frac{I_T R_T}{R_x}} \tag{3.48}$$

That was easy! This is the general expression for the current divider rule. Note that $I_T R_T$ is the voltage across the parallel circuit, and when this voltage is divided by R_X, which is Ohm's law again, the current through that branch is obtained. Again, the CDR is especially useful when the total current is known and the total voltage is not known—one step is saved. However, this rule also gives insight into how the current divides in a parallel circuit. What happens to the branch current as the resistance of that branch increases? The current decreases because R_X is in the denominator of Equation (3.48). This is expected because the current decreases when the resistance increases and the voltage is constant.

Example 3.3.3 _____

If the total current in Figure 3.14 (see Example 3.3.1) is 1.72 A, determine the current through R_4.

Given: $R_4 = 4.7 \ \Omega$

$$R_T = 0.89622 \ \Omega \text{ (from Example 3.3.1)}$$

$$I_T = 1.72 \text{ A}$$

Desired: I_4

Strategy: CDR

Solution:

$$I_x = \frac{I_T R_T}{R_x}$$

$$I_4 = \frac{I_T R_T}{R_4} = \frac{(1.72 \text{ A})(0.89622 \text{ }\Omega)}{4.7 \text{ }\Omega} = 0.32798 \text{ A} = 0.328 \text{ A}$$

As with the VDR, there is a convenient form of the CDR when there are only two parallel resistances. A short development of this form is:

$$I_1 = \frac{I_T R_T}{R_1} \tag{3.49}$$

$$R_T = \frac{R_1 R_2}{R_1 + R_2} \tag{3.50}$$

$$I_1 = \frac{I_T \left(\dfrac{R_1 R_2}{R_1 + R_2} \right)}{R_1} = I_T \left(\frac{R_2}{R_1 + R_2} \right) = \frac{I_T R_2}{R_1 + R_2} \tag{3.51}$$

The current in branch 2 could be developed in the same way (try it). Note that the resistance in the numerator is in the opposite branch of the current to be found. This suggests that if the resistance of one branch increases, and the total current is the same, then the other branch will have more current. The CDR for two parallel resistances is summarized in Equation (3.52).

$$\boxed{I_1 = \frac{I_T R_2}{R_1 + R_2} \qquad\qquad I_2 = \frac{I_T R_1}{R_1 + R_2}} \tag{3.52}$$

These parallel circuit results will be illustrated in the next three examples.

___ Example 3.3.4

For the parallel circuit shown in Figure 3.15, determine (a) I_2 and (b) I_1 using the CDR. (c) Check the results using KCL. Determine (d) R_T and (e) V_T.

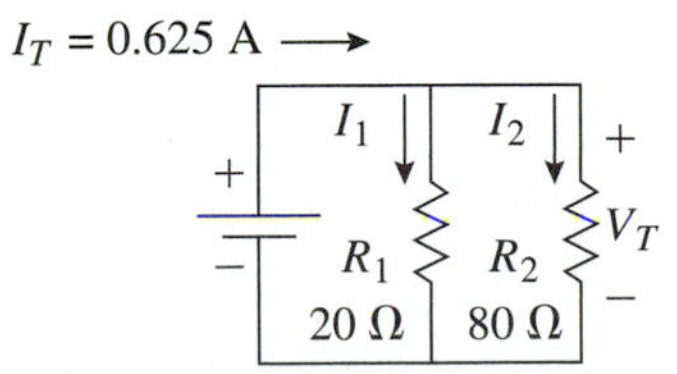

$I_T = 0.625 \text{ A} \longrightarrow$

Figure 3.15 Parallel circuit for Example 3.3.4

Given: $R_1 = 20 \text{ }\Omega$

$R_2 = 80 \text{ }\Omega$

$I_T = 0.625 \text{ A}$

Desired: a. I_2

b. I_1

c. KCL check

d. R_T

e. V_T

Strategy: a. and b. use the CDR for two R's

c. KCL

d. R_T for two parallel R's

e. Ohm's law

Solution:

$$I_2 = \frac{I_T R_1}{R_1 + R_2} = \frac{(0.625 \text{ A})(20 \text{ }\Omega)}{20 \text{ }\Omega + 80 \text{ }\Omega} = 0.125 \text{ A}$$

$$I_1 = \frac{I_T R_2}{R_1 + R_2} = \frac{(0.625 \text{ A})(80 \text{ }\Omega)}{20 \text{ }\Omega + 80 \text{ }\Omega} = 0.500 \text{ A}$$

At the upper node:

$$+I_T = +I_1 + I_2 = 0.500 + 0.125 = 0.625 \text{ A}$$

KCL is satisfied.

$$R_T = \frac{R_1 R_2}{R_1 + R_2} = \frac{(20)(80)}{20 + 80} = 16.0 \text{ }\Omega$$

$$V_T = I_T R_T = (0.625 \text{ A})(16 \text{ }\Omega) = 10.0 \text{ V}$$

Example 3.3.5

For the circuit shown in Figure 3.16, determine (*a*) G_T, (*b*) R_T, (*c*) I_2, and (*d*) P_2.

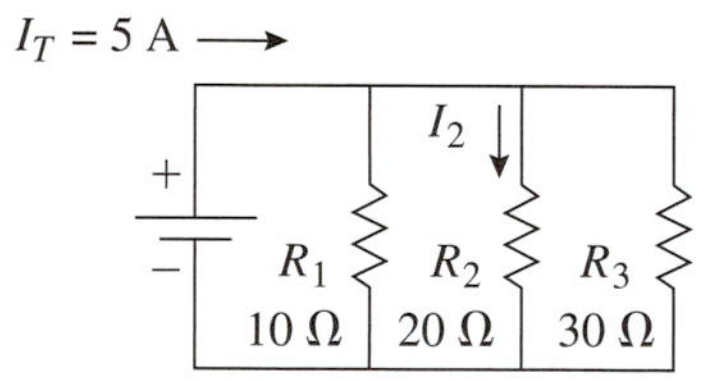

$I_T = 5 \text{ A} \longrightarrow$

Figure 3.16 Parallel circuit for Example 3.3.5

Given: $R_1 = 10 \text{ }\Omega$

$R_2 = 20 \text{ }\Omega$

$R_3 = 30 \text{ }\Omega$

$I_T = 5 \text{ A}$

Desired: a. G_T

b. R_T

c. I_2

d. P_2

Strategy: Use standard parallel circuit equations for G_T, $R_T = 1 / G_T$
CDR for I_2
$P_2 = (I_2)^2 R_2$

Solution:

$$G_T = G_1 + G_2 + G_3 = \frac{1}{R_1} + \frac{1}{R_2} + \frac{1}{R_3} = \frac{1}{10 \ \Omega} + \frac{1}{20 \ \Omega} + \frac{1}{30 \ \Omega} = 0.18333 \ \text{S} = 0.183 \ \text{S}$$

$$R_T = \frac{1}{G_T} = \frac{1}{0.18333 \ \text{S}} = 5.4545 \ \Omega = 5.45 \ \Omega$$

$$I_2 = \frac{I_T R_T}{R_2} = \frac{(5 \ \text{A})(5.4545 \ \Omega)}{20 \ \Omega} = 1.364 \ \text{A} = 1.36 \ \text{A}$$

$$P_2 = I_2^2 R_2 = (1.364 \ \text{A})^2 \ (20 \ \Omega) = 37.2 \ \text{W}$$

Example 3.3.6

For the circuit shown in Figure 3.17, determine R_3.

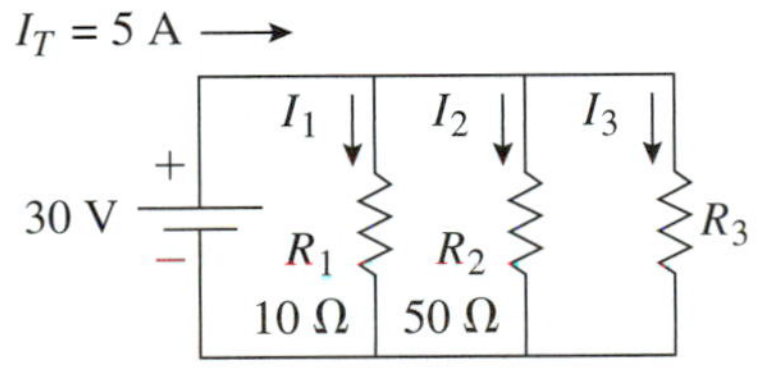

$I_T = 5 \ \text{A} \longrightarrow$

Figure 3.17 Parallel circuit for Example 3.3.6

Given: $R_1 = 10 \ \Omega$
$R_2 = 50 \ \Omega$
$V_T = 30 \ \text{V}$
$I_T = 5 \ \text{A}$

Desired: R_3

Strategy: Find I_1 and I_2 using Ohm's law.
Determine I_3 using KCL.
Determine R_3 using Ohm's law.

Solution:

$$I_1 = \frac{V_S}{R_1} = \frac{30 \ \text{V}}{10 \ \Omega} = 3 \ \text{A}$$

$$I_2 = \frac{V_S}{R_2} = \frac{30 \ \text{V}}{50 \ \Omega} = 0.6 \ \text{A}$$

KCL:

$$+I_T - I_1 - I_2 - I_3 = 0$$

$$I_3 = +I_T - I_1 - I_2 = 5 - 3 - 0.6 = 1.4 \ \text{A}$$

$$R_3 = \frac{V_S}{I_2} = \frac{30 \text{ V}}{1.4 \text{ A}} = 21.429 = 21.4 \, \Omega$$

Check:

$$R_T = \frac{1}{\dfrac{1}{R_1} + \dfrac{1}{R_2} + \dfrac{1}{R_3}} = 6 \, \Omega, \quad I_T = \frac{V_T}{R_T} = \frac{30 \text{ V}}{6 \, \Omega} = 5 \text{ A} \quad \text{checks}$$

3.4 DC SERIES-PARALLEL CIRCUITS

For series circuits, KVL, R_T, the VDR, and the fact that the current is the same throughout a series circuit have been established. For parallel circuits, KCL, R_T, the CDR, and the fact that the voltage is the same throughout a parallel circuit have been established. The next mission is even more interesting: analyzing circuits that are combinations of series and parallel circuits, which are called *series-parallel circuits*. The circuit designer can utilize series-parallel circuits in numerous ways. For example, what if there are three electronic devices that are connected in parallel and the source is 12 V^2? If one of the devices fails and becomes a short (shorts out), the source could be protected if an additional resistor is in series with the source and load, as shown in Figure 3.18. The load in this case is the combination of the three parallel resistances. If one electronic device shorts, all the source voltage will drop across the series resistor. The resistance value of the resistor can be calculated, using Ohm's law, to limit the current to a safe value. However, in normal operation, the source voltage will divide across this resistor and the load. This is just a simple example of a series-parallel circuit. There are many more interesting examples that will come in later electronic circuits courses (transistors, operational amplifiers, etc.).

The general approach to analyzing a given series-parallel circuit will be investigated next. First, identify the groups of resistances that are in series and those that are in parallel. Then, apply the appropriate series and parallel circuit laws to the resistances in that group. You must be able to identify groups of resistances as being in series or parallel with other groups of resistances, and apply the appropriate series and parallel circuit laws to those groups.

To illustrate this strategy, consider the circuit in Figure 3.18. The three parallel resistances are one group of resistances. Parallel circuit laws would be applied to the three resistances in that group. The series resistor is in series with the *group* of three parallel resistances. In fact, one could visualize the circuit as shown in Figure 3.19, where R_5 is the equivalent resistance of R_1, R_2, and R_3 in parallel, which is shown as $R_1 \parallel R_2 \parallel R_3$. This entire load is in series with the source, so this is fundamentally a series circuit.

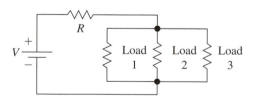

Figure 3.18 Example of a series-parallel circuit

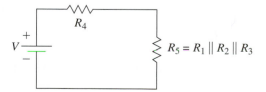

Figure 3.19 Visualization of the series-parallel circuit in Figure 3.18

Example 3.4.1

Determine V_x in the circuit shown in Figure 3.20.

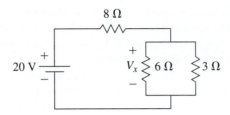

Figure 3.20 Series-parallel circuit for Example 3.4.1

Given: $R_1 = 8\ \Omega$

   $R_2 = 6\ \Omega$

   $R_3 = 3\ \Omega$

   $V_S = 20\ V$

Note: From now on, you do not need to rewrite all these resistance and voltage source values from a given circuit—just be sure they are carefully labeled on the schematic, as shown in Figure 3.21.

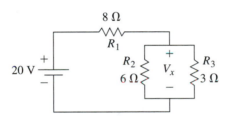

Figure 3.21 Labeled series-parallel circuit for Example 3.4.1

Desired: $V_x = V$ across $R_2 \| R_3$

Strategy: $R_T = R_1 + R_2 \| R_3$ (R_1 is in series with the parallel group R_2 and R_3)
   VDR: V_x is V across $R_2 \| R_3$

Solution:

$$R_x = R_2 \| R_3 = \frac{R_2 R_3}{R_2 + R_3} = \frac{(6)(3)}{6 + 3} = 2\ \Omega$$

$$R_T = R_1 + R_x = 8 + 2 = 10\ \Omega$$

$$V_x = R_x \frac{V_T}{R_T} = (2)\frac{20\ V}{10} = 4\ V = 4.00\ V \qquad (V_T = V_S \text{ for this VDR setup})$$

Note: You could also have found V_x by a longer route:

$$R_x = R_2 \| R_3 = \frac{R_2 R_3}{R_2 + R_3} = \frac{(6)(3)}{6 + 3} = 2\ \Omega$$

$$R_T = R_1 + R_x = 8 + 2 = 10\ \Omega$$

$$I_T = \frac{V_S}{R_T} = \frac{20\ V}{10\ \Omega} = 2\ A$$

$$V_X = I_T R_X = (2\ A)(2\ \Omega) = 4\ V = 4.00\ V$$

Thus, there are usually multiple approaches to analyzing a series-parallel circuit, but if you take some time to *think* about your strategy, you will usually come up with an *efficient* approach.

Was the circuit of Example 3.4.1 fundamentally a series circuit or fundamentally a parallel circuit? The source is the key. If a single resistance is in series with the source on either side (or both sides) of the source, then the circuit is fundamentally a series circuit. If the current path from the source divides on both sides of the source, then the circuit is fundamentally a parallel circuit. In Example 3.4.1, R_1 is a single resistance in series with the source; hence, it is fundamentally a series circuit.

As a reminder, a group of series resistances is called a *branch*. In parallel circuits, each branch consists of one resistance. In series-parallel circuits, a branch may have several series resistances. The use of the term *branch* in series-parallel circuit analysis is helpful for clarity, as illustrated in the next example.

Example 3.4.2

Determine I_2 and V_4 in the circuit shown in Figure 3.22.

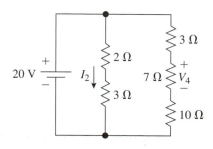

Figure 3.22 Series-parallel circuit for Example 3.4.2

Given: Label resistances as shown in Figure 3.23.

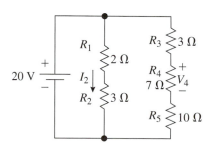

Figure 3.23 Labeled series-parallel circuit for Example 3.4.2

Desired: I_2

V_4

Strategy: fundamentally a parallel circuit

Identify resistance groups.

branch A: $R_1 + R_2$ (branch A consists of R_1 in series with R_2)

branch B: $R_3 + R_4 + R_5$

$R_T = R_A \parallel R_B$, but R_T is not needed

$I_2 = V_S / R_A$

V_4: use VDR within branch B.

Solution:

$$R_A = R_1 + R_2 = 2 + 3 = 5 \, \Omega$$

$$I_2 = \frac{V_S}{R_A} = \frac{20 \text{ V}}{5 \, \Omega} = 4 \text{ A} = 4.00 \text{ A}$$

$$R_B = R_3 + R_4 + R_5 = 3 + 7 + 10 = 20 \, \Omega$$

$$V_4 = R_4 \frac{V_T}{R_B} = (7)\frac{20 \text{ V}}{20} = 7 \text{ V} = 7.00 \text{ V} \qquad (V_T = V_S \text{ for this branch})$$

In the VDR and the CDR, the total voltage or current becomes the total voltage or current for that group of resistances, not necessarily the entire circuit. Also, KVL and KCL are valid and can be used where appropriate.

___ **Example 3.4.3** _____

Determine (a) I_T, (b) I_5, and (c) V_{xy} in the circuit shown in Figure 3.24 (already labeled).

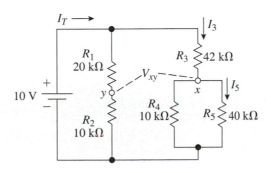

Figure 3.24 Series-parallel circuit for Example 3.4.3

Desired: a. I_T
          b. I_5
          c. V_{xy}

Strategy: fundamentally a parallel circuit
          Identify resistance groups.
          $R_A = R_1 + R_2$
          $R_B = R_3 + R_4 \| R_5$
          $R_T = R_A \| R_B$
          $I_T = V_S / R_T$
          Use Ohm's law for I_3, then CDR for I_5.
          Use KVL for V_{xy}.

Solution: _____

a. $R_A = R_1 + R_2 = 20 \text{ k}\Omega + 10 \text{ k}\Omega = 30 \text{ k}\Omega$

$$R_C = R_4 \| R_5 = \frac{R_4 R_5}{R_4 + R_5} = \frac{(10 \text{ k}\Omega)(40 \text{ k}\Omega)}{10 \text{ k}\Omega + 40 \text{ k}\Omega} = 8 \text{ k}\Omega$$

$R_B = R_3 + R_C = 42 \text{ k}\Omega + 8 \text{ k}\Omega = 50 \text{ k}\Omega$

$$R_T = R_A \parallel R_B = \frac{(30\ k\Omega)(50\ k\Omega)}{30\ k\Omega + 50\ k\Omega} = 18.75\ k\Omega$$

$$I_T = \frac{V_S}{R_T} = \frac{10\ V}{18.75\ k\Omega} = 0.53333\ mA = 0.533\ mA$$

b. $\quad I_3 = \dfrac{V_S}{R_B} = \dfrac{10\ V}{50\ k\Omega} = 0.2\ mA$

CDR: ("total" current for use in CDR) $R_4 \parallel R_5 = R_C$ ("total"resistance for use in CDR)

$$I_5 = \frac{I_3 R_C}{R_5} = \frac{(0.2\ mA)(8\ k\Omega)}{40\ k\Omega} = 0.04\ A = 40.0\ \mu A$$

c.   KVL:—select a closed path through resistances with known voltages and through the
air between terminals x and y; see Figure 3.25

■   Arbitrarily assign a polarity to the terminals (arbitrary; x is +; y is −).

■   Label voltage polarities on resistances in the path if not done so already.

■   Select a direction (arbitrary: CW or CCW) around the closed path and apply
KVL.

■   Solve for V_{xy}.

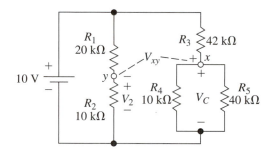

Figure 3.25 KVL setup in Figure 3.24

KVL in CCW direction:

$$-V_{xy} - V_2 + V_C = 0$$
$$V_{xy} = -V_2 + V_C$$

Need V_2; use the VDR:

$$V_2 = R_2 \frac{V_T}{R_A} = (10\ k\Omega)\frac{10\ V}{30\ k\Omega} = 3.3333\ V$$

Need V_C; use Ohm's law:

$$V_C = I_3\, R_C = (0.2\ mA)(8\ k\Omega) = 1.6\ V$$

$$V_{xy} = -V_2 + V_C = -3.3333 + 1.6 = -1.7333\ V = -1.73\ V$$

V_{xy} is negative—is this permitted? Yes. It means that the polarity of the actual voltage is the
opposite of the polarity that was arbitrarily selected.

___ **Example 3.4.4** _____

Determine (a) I_4, (b) V_{ab}, and (c) V_{cd} in the circuit shown in Figure 3.26 (already labeled).

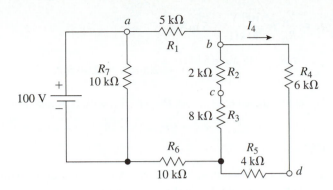

Figure 3.26 Series-parallel circuit for Example 3.4.4

Desired: a. I_4

b. V_{ab}

c. V_{cd}

Strategy: Redraw the circuit to clearly show what is in series and what is in parallel (see Figure 3.27).

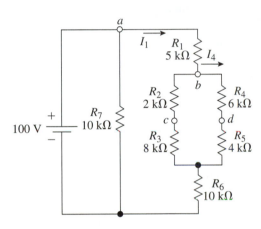

Figure 3.27 Redrawn series-parallel circuit for Example 3.4.4

$R_A = R_2 + R_3$

$R_B = R_4 + R_5$

$R_C = R_A \parallel R_B$

$R_D = R_1 + R_C + R_6$

(do not need R_T)

Find I_1 using Ohm's law and then CDR for I_4.

Use Ohm's law for V_{ab}.

Set up and solve KVL for V_{cd}.

Solution: _____

a. $R_A = R_2 + R_3 = 2\,k\Omega + 8\,k\Omega = 10\,k\Omega$

$R_B = R_4 + R_5 = 6\,k\Omega + 4\,k\Omega = 10\,k\Omega$

$R_C = R_A \parallel R_B = \dfrac{10\,k\Omega}{2} = 5\,k\Omega$

$R_D = R_1 + R_C + R_6 = 5\,k\Omega + 5\,k\Omega + 10\,k\Omega = 20\,k\Omega$

$$I_1 = \frac{V_S}{R_D} = \frac{100\ \text{V}}{20\ \text{k}\Omega} = 5\ \text{mA}$$

$$I_4 = \frac{I_1 R_C}{R_B} = \frac{(5\ \text{mA})(5\ \text{k}\Omega)}{10\ \text{k}\Omega} = 2.5\ \text{mA} = 2.50\ \text{mA}$$

b. V_{ab} is just the voltage across R_1 , i.e., V_1, with terminal a +, and terminal b −

$$V_{ab} = I_1 R_1 = (5\ \text{mA})(5\ \text{k}\Omega) = 25\ \text{V} = 25.0\ \text{V}$$

c.   KVL:

■ Select a closed path through resistances with known voltages and through the air between terminals c and d; see Figure 3.28.

■ Assign a polarity to the terminals (arbitrary: c is +, d is −).

■ Label voltage polarities on resistances in the path if not done so already.

■ Select a direction (arbitrary: CW) around the closed path and apply KVL.

■ Solve for V_{cd}.

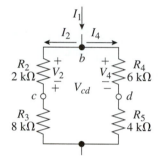

Figure 3.28 KVL setup in Figure 3.27 (only part of the circuit is shown)

KVL in CW direction:

$$+V_{cd} + V_2 - V_4 = 0$$

$$V_{cd} = +V_4 - V_2$$

Need V_4; use Ohm's law:

$$V_4 = I_4 R_4 = (2.5\ \text{mA})(6\ \text{k}\Omega) = 15\ \text{V}$$

Need V_2; use KCL at node b and then Ohm's law (see Figure 3.28)

$$+I_1 - I_2 - I_4 = 0$$

$$I_2 = +I_1 - I_4 = +5\ \text{mA} - 2.5\ \text{mA} = 2.5\ \text{mA}$$

$$V_2 = I_2 R_2 = (2.5\ \text{mA})(2\ \text{k}\Omega) = 5\ \text{V}$$

KVL: $$V_{cd} = +V_4 - V_2 = +15 - 5 = +10\ \text{V} = +10.0\ \text{V}$$

The positive result means that the voltage polarity for V_{cd} is the same as the initial arbitrarily selected polarity.

3.5 ANALYSIS OF MULTIPLE-SOURCE DC CIRCUITS USING SUPERPOSITION

So far we have learned how to analyze complex series-parallel circuits with several resistances. Can you think of any types of circuits that we have not analyzed yet? One type is the circuit with *several sources* as well as several resistances. Thus, our next step in the mission to analyze any DC circuit is to analyze circuits with more than one source.

How might we arrange multiple sources in a single circuit? Think of resistance combinations. One way is to combine sources in series, as shown in Figure 3.29. Why would anyone want to put sources in series? Think of KVL. Yes, to obtain a larger source voltage! In fact, think of various battery-powered electronic products that you have: flashlight, TV remote control, handheld calculator, portable electronic games, and so on. The batteries are electrically placed in series (note the conductor routing inside the battery compartment).

What is the total voltage of all the sources together considered as one source? Again, KVL is the tool. Select a closed path for KVL through the sources and through the air to set up the total voltage V_T. Arbitrarily select a direction, say CCW, apply KVL, and solve for V_T:

$$+V_T - V_{S1} - V_{S2} - V_{S3} = 0 \tag{3.53}$$

$$V_T = V_{S1} + V_{S2} + V_{S3} \tag{3.54}$$

Thus, the voltages of the sources in series add, assuming that the polarities are arranged as in Figure 3.29. What would happen if one of the sources was reversed? It would subtract instead of add to the total voltage—this is one reason why batteries should not be inserted backward!

Is there anything else that can be said about sources in series? Again, think of resistances in series. The current is the same through all the sources. Thus, sources are placed in series to obtain a larger voltage, but the current is the same through all the sources.

The previous discussion suggests that there may be applications where more current, not voltage, is needed. Is there a way to increase current at the same voltage level? Think of KCL and combination of sources. Yes, place the sources in parallel, as shown in Figure 3.30.

What is the total current of all of the sources together considered as one source? Again, KCL is the tool. Select a node for KCL to which all the sources are connected, say the bottom node, apply KCL, and solve for I_T:

$$+I_T - I_{S1} - I_{S2} - I_{S3} = 0 \tag{3.55}$$

$$I_T = I_{S1} + I_{S2} + I_{S3} \tag{3.56}$$

Thus, the currents of the sources in parallel add, assuming that the polarities are arranged as in Figure 3.30. What would happen if one of the sources were reversed? It would subtract instead of add to the total current. Let us take this reasoning one step further—what if

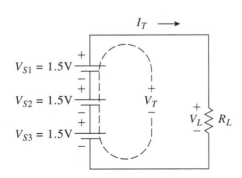

Figure 3.29 Series connection of multiple sources

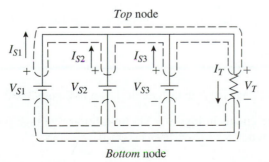

Figure 3.30 Parallel connection of multiple sources

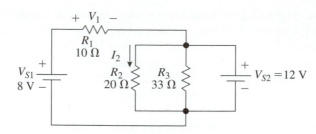

Figure 3.31 Example of a circuit where the sources are neither in series nor in parallel

the voltages of the sources were not equal? Then the voltage in a parallel circuit would not be the same, and this would produce a KVL violation. Try it. If the path for KVL were through one source and completed back through another source, and the voltages of the sources were not equal, there would be a net voltage around a closed path, and that would violate KVL. A reversed source leads to the same result. Practically, there is always a small resistance in the source and in the conductors connected to the sources, so huge currents would flow from one source to the other to make up the voltage drop necessary to satisfy KVL. This current does not reach the load, so current is being wasted. All of this is obviously undesirable and can be dangerous; hence,

Parallel voltage sources should always have the same voltage and matching polarities.

Another series-parallel scenario that can occur is one in which the sources are in different parts of the circuit and, hence, are not in series nor in parallel. An example is shown in Figure 3.31. How is this circuit analyzed? A very powerful yet simple technique exists: *superposition.* What is superposition? Superposition is a circuit analysis technique to handle multiple source circuits. In superposition, one analyzes the circuit one source at a time using series-parallel circuit analysis techniques. The other sources are "deactivated." How does one deactivate a voltage source? Ask this question: How can a voltage source be set to zero volts? *Replace the voltage source with a short.* A short guarantees that the voltage will be zero across that short. Finally, the results are added (with proper signs) to determine the desired voltage(s) and current(s) in the circuit.

Here is a strategy for using superposition in the analysis of multiple source circuits:

1. Label the desired quantities to be found, assign an arbitrary polarity to each voltage to be found, and assign an arbitrary direction to each current to be found.
2. Deactivate all sources except one. Analyze the circuit for the desired quantities using series-parallel circuit analysis. Be sure to note voltage polarities and current directions carefully.
3. Deactivate that first source and reactivate another source. Analyze the circuit for the desired quantities using series-parallel circuit analysis. Again, be sure to note voltage polarities and current directions carefully. Continue with this process until all sources have been accounted for.
4. Add the corresponding resultant voltages and currents from the analysis of each source, but be sure to assign the proper sign to each quantity: positive if it agrees with the voltage polarity or current direction that was initially assigned, and negative if opposite.

This last step is actually the superposition step: the voltages and currents are *superposed.* Let us illustrate this technique with the circuit in Figure 3.31.

Example 3.5.1

Determine (*a*) V_1 and (*b*) I_2 in the circuit shown in Figure 3.31.

Given: The circuit in Figure 3.31. An arbitrary voltage polarity for V_1 and current direction for I_2 have been assigned.

Desired: a. V_1

b. I_2

Strategy: Superposition:

deactivate source 2 and determine V_1' and I_2' due to source 1;

deactivate source 1 and determine V_1'' and I_2'' due to source 2;

$V_1 = V_1' + V_1''$; $I_2 = I_2' + I_2''$

Solution:

Replace V_{S2} with a short. What happens to the circuit in Figure 3.31? R_2 and R_3 are shorted out! See Figure 3.32. Thus, $I_2' = 0$. All of V_{S1} drops across R_1. Hence, $V_1' = +8$ V (positive because the polarity of V_1' is the same as assigned in Figure 3.31).

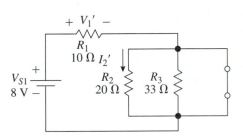

Figure 3.32 Circuit in Figure 3.31 with V_{S2} replaced by a short

$$I_2' = 0$$

$$V_1' = +8 \text{ V}$$

Replace V_{S1} with a short. What happens to the circuit in Figure 3.31? R_1 is now in parallel with R_2 and R_3, as shown in Figure 3.33 and Figure 3.34. Determine V_1'' and I_2'':

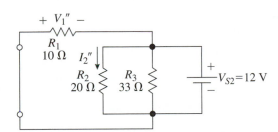

Figure 3.33 Circuit in Figure 3.31 with V_{S1} replaced by a short

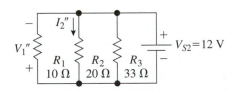

Figure 3.34 Circuit in Figure 3.33 redrawn

$$I_2'' = \frac{V_{S2}}{R_2} = \frac{12 \text{ V}}{20 \text{ }\Omega} = +0.6 \text{ A}$$

(positive because I_2'' is in the same direction as I_2)

$$V_1'' = -V_{S2} = -12 \text{ V}$$

(voltage is the same across a parallel circuit; negative because V_1'' has the opposite polarity of V_1)

The final step is to superpose the voltages and the currents:

$$V_1 = V_1' + V_1'' = +8 - 12 = -4 = -4.00 \text{ V}$$

$$I_2 = I_2' + I_2'' = 0 + 0.6 = +0.600 \text{ A}$$

The total voltage across R_1 is the opposite of the arbitrarily selected voltage polarity, and the total current through R_2 is in the same direction as the arbitrarily selected current direction.

This example, although simple (the series-parallel circuit analysis was simple), illustrates the procedure used in superposition.

3.6 Practical Voltage Sources and the DC Current Source

As previously mentioned in the context of voltage sources in parallel, practical voltage sources have a series resistance internal to them. In other words, a practical voltage source does have some internal loss mechanisms that depend on the type of source it is (battery, generator, solar cell, etc.). Also, some of this internal resistance is conductor resistance. Hence, a practical voltage source has a *model* of an ideal voltage source in series with a resistance R_{int}, as shown in Figure 3.35. A model is a representation of an actual device in terms of ideal elements, and it must be useful in making predictions of the performance of that device. Again, it needs to be emphasized that R_{int} is not necessarily a resistor but is a resistance in general.

The little circles in Figure 3.35 are used to indicate the terminals of the component. The outside world (the rest of the circuit) is connected to these terminals. Often, the internal resistance of a practical voltage source is ignored. When can you make this assumption? When the resistance of the source is negligible with respect to the resistances in the circuit, where "negligible" is defined in the context of the application. For example, many laboratory sources have an internal resistance of 50 Ω. If the resistances in the circuit are on the order of kiloohms, one can ignore R_{int} if the error in the total circuit resistance, 5% or less is acceptable (50 Ω is $\frac{1}{20}$th of 1000 Ω). On the other hand, if the resistances in the circuit are on the order of tens of ohms, one cannot ignore R_{int}. R_{int} must be placed into the schematic. However, the source voltage is across the terminals of the voltage source, not the ideal voltage source. Hence, what happens to terminal voltage V_t as the current from the source increases? Apply KVL to Figure 3.35:

$$+V_S - V_{Rint} - V_t = 0 \tag{3.57}$$

$$V_t = V_S - V_{Rint} = V_S - I_{\pm}R_{int} \tag{3.58}$$

Thus, as the current from the source increases, the voltage at the terminals of the source decreases. What happens to the sound from the radio in a car when the car is started? The sound goes off while the car is being started because the huge current from the battery (100s of amperes) lowers the terminal voltage of the battery too much to operate the radio. Once the car is started, the huge current drain is not present anymore, and the car radio has enough voltage to operate once again.

The *current source* is another type of source. It is the "opposite" of the voltage source. What are the properties of an ideal voltage source?

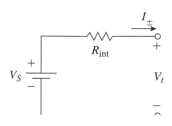

Figure 3.35 Model of a practical voltage source

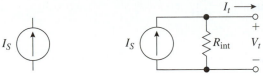

a. Ideal current source   b. Model of a practical current source

$I_t \longrightarrow$

Figure 3.36 Schematic symbols of current sources

An ideal voltage source has a constant voltage and can deliver any current value.

What would be the expected properties of an ideal *current source?*

An ideal current source delivers a constant current at any voltage value.

Current sources are usually electronic circuits. They are useful in certain applications where a constant current, not voltage, is required. Current source applications will be covered in subsequent electronics courses. For now, assume that the current source exists and concentrate on how to analyze circuits that contain them. The schematic symbol for a current source is shown in Figure 3.36a.

Like the practical voltage source, the practical current source has internal resistance. In the model of a practical current source, however, R_{int} is in parallel, not in series, with the ideal current source. Why is this true? The answer can be found by posing another question. Recall in superposition that sources must be deactivated. How is a current source deactivated? The current source must be replaced by an *open* to make the current equal zero. However, a resistance in series with an open is meaningless—it is still infinite resistance. Hence, the internal resistance of a practical current source is in parallel with the ideal current source in the model, as shown in Figure 3.36b.

Often, the internal resistance of a practical current source is ignored. When can one make this assumption? When the resistance of the circuit is negligible with respect to the resistance in the source. Why? Because almost all of the current will go into the circuit and not R_{int} when the circuit resistance is much less than the source resistance (recall the reasoning behind parallel circuits and the CDR). However, when this assumption is not true, the current at the terminals will be less than the ideal current, as shown by KCL at the node where the ideal current source, R_{int}, and the external circuit connect:

$$+I_S - I_{Rint} - I_t = 0 \tag{3.59}$$

$$I_t = I_S - I_{Rint} = I_S - \frac{V_T}{R_{int}} \tag{3.60}$$

Thus, the current of a practical current source will drop as the terminal voltage increases.

How are circuits that contain current sources analyzed? Just use the basic property of the current source: the current through it is constant and the voltage across it can be anything—that is, the voltage is determined by the load.

___ Example 3.6.1 _____

For the circuit shown in Figure 3.37, determine (*a*) G_T, (*b*) R_T, (*c*) I_2, (*d*) P_2 and (*e*) the voltage across the parallel circuit.

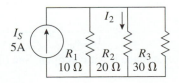

$I_2 \downarrow$

I_S
5A

R_1 R_2 R_3
10 Ω   20 Ω   30 Ω

Figure 3.37 Parallel circuit for Example 3.6.1

Given: $I_T = I_S = 5$ A; all R values

Desired: a. G_T

b. R_T

c. I_2

d. P_2

e. V_T

Strategy: Use standard parallel circuit equations for G_T, $R_T = 1/G_T$

CDR for I_2

$P_2 = (I_2)^2 R_2$

$V_T = I_T R_T$

Solution:

$$G_T = G_1 + G_2 + G_3 = \frac{1}{R_1} + \frac{1}{R_2} + \frac{1}{R_3} = \frac{1}{10\ \Omega} + \frac{1}{20\ \Omega} + \frac{1}{30\ \Omega} = 0.18333\ \text{S} = 0.183\ \text{S}$$

$$R_T = \frac{1}{G_T} = \frac{1}{0.18333\ \text{S}} = 5.4545\ \Omega = 5.45\ \Omega$$

$$I_2 = \frac{I_T R_T}{R_2} = \frac{(5\ \text{A})(5.4545\ \Omega)}{20\ \Omega} = 1.364\ \text{A} = 1.36\ \text{A}$$

$$P_2 = I_2^2 R_2 = (1.364\ \text{A})^2\,(20\ \Omega) = 37.2\ \text{W}$$

$$V_T = I_T R_T = (5\ \text{A})\,(5.4545\ \Omega) = 27.272\ \text{V} = 27.3\ \text{V}$$

Recall that in the analysis of multiple-source circuits using superposition, a current source is deactivated by replacing it with an open, as illustrated in the next example.

Example 3.6.2

Determine the voltage across the 10 Ω resistance in the circuit shown in Figure 3.38.

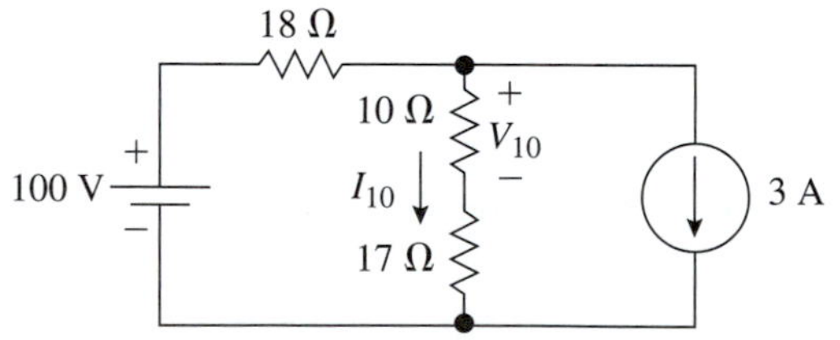

Figure 3.38 Circuit for Example 3.6.2

Given: Circuit in Figure 3.38

Desired: V_{10}

Strategy: Superposition: formulate substrategies for each activated source.

Solution:

Deactivate the 3 A current source: replace it with an open. See Figure 3.39.

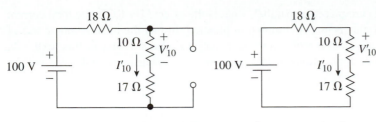

Figure 3.39 Circuit in Figure 3.38 with the 3 A current source deactivated

a. Open in place of the current source

b. Resultant circuit

Substrategy: VDR

$$V_x = R_x \frac{V_T}{R_T}$$

$$V'_{10} = R_{10} \frac{V_T}{R_T} = (10) \frac{100 \text{ V}}{(18 + 10 + 17)} = 22.222 \text{ V}$$

Deactivate the 100 V source; replace it with a short. See Figure 3.40.

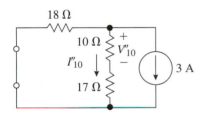   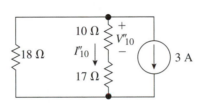

Figure 3.40 Circuit in Figure 3.38 with the 100 V voltage source deactivated

a. Short in place of the voltage source

b. Resultant circuit

Substrategy: CDR → I_{10}
Ohm's law → V_{10}

$$I''_{10} = \frac{-I_T R_B}{R_A + R_B} = -\frac{(3 \text{ A}) (18 \text{ Ω})}{(18 + 27)\Omega} = -1.2 \text{ A}$$

Note: I''_{10} and V''_{10} were shown in the same direction and polarity as originally assigned; hence, both are *negative* due to the direction of the current source.

$$V''_{10} = I''_{10} \, 10 \text{ Ω} = (-1.2 \text{ A})(10 \text{ Ω}) = -12 \text{ V}$$

Superposition:

$$V_{10} = V'_{10} + V''_{10} = +22.222 + (-12) = +10.222 = +10.2 \text{ V}$$

Hence, the polarity of the total voltage across the 10 Ω resistance is the same as in Figure 3.38.

If current sources are placed in series, what is the effect? The current must be the same in a series circuit, so series current sources must have the same current rating. However, a larger voltage could be obtained, especially if a practical current source has a voltage output limitation. *Current sources in series are rarely used in practice.*

If current sources are placed in parallel, what is the effect? By KCL, the total current will be larger! Again, this is not so common in practice. In fact the most common use of current sources is equivalent circuits, a subject that is examined later in this textbook. There are several other applications of current sources in electronics, and these will be covered in a subsequent electronics course.

3.7 THE ROLE OF CIRCUIT SIMULATION SOFTWARE IN DC SERIES-PARALLEL CIRCUIT ANALYSIS

The use of computer software programs to analyze systems, circuits, and even components is very extensive in electronics. Hence, it is important to understand both how computer simulation programs extend your capability to perform tasks and what their limitations are. Fundamental to this understanding is to begin by asking questions about the computer. Does it "think" for you? Does it provide useful "intelligence"? The answer to the first question is emphatically no. But the answer to the second question is an emphatic yes if the simulations are done in a deliberate and careful manner.

The results of the simulation process is a lot of calculations being performed rapidly. The simulation programs provide the instructions on how the calculations are to be performed and the rules to be followed. A resistor exists within the software in the form of the equation given by Ohm's law, relating the quantities of voltage and current. The information in a schematic exists in the form of how components are connected to each other and the application of KVL and KCL to the circuit. Hence, the computer solves equations and does the bookkeeping of tracking the data that is generated. You could do these same calculations yourself. But there are tremendous differences. The computer can do the calculations much faster and more accurately and provide results and displays of data that are truly astounding.

How can the speed and accuracy of the computer be utilized to aid you in performing your job? Even small circuits very quickly become difficult to solve using manual calculations. Imagine a circuit with hundreds or thousands of components. The effort would be too great and the time taken too long if done with hand calculations. Simulations do not just make it practical to tackle complicated circuits, but they also provide a wealth of data regarding any voltage, current, or other variable that pertain to the circuit.

There are at least three key reasons for using computer simulations. One reason has already been addressed, that of being able to efficiently solve complicated circuits. The second reason is the ability to design or answer performance questions *without building the actual circuit.* When you look at the performance specifications for various electronic items, notice that the specifications are generally given as a performance range. These are the allowed tolerances for the various parameters. Why not a single exact number? The reason is that every component that is used in a circuit has its value specified in terms of tolerances.[1] Hence, it is essential in designing a circuit to check out the impact that all the tolerances will have on circuit performance. This is also sometimes referred to as a worst-case analysis. Could you do the same thing by actually building the circuit with all the worst-case component values? Where would you get the components with these particular values? It would be a very imperfect, slow process. Computer simulations allow you to specify the range of the values of a component and usually within seconds or minutes you will have all the necessary results. A common situation, which is similar to the component tolerance case, is the "what if" question. What if a different component with modified specifications is used? Will the circuit operate satisfactorily? Again, computer simulations can more readily answer the question.

The third reason for using computer simulations is to calculate the performance as a function of some variable. An example of this would be to "sweep," meaning to allow the value to change, such as the source voltage of a DC circuit. Of interest is to see how the current in the circuit varies as a function of the source voltage. Such an output is generally

---

[1] Recall the resistor tolerance discussion in Chapter 2.

displayed on a graph. The calculations involved are quite extensive and yet are done extremely rapidly.

To understand circuit simulation software, you should keep in mind that specific procedures must be performed, regardless of which software package is used. Here are some common procedures that all computer simulation programs will require:

1. Choose and place components
2. Move and rotate components
3. Connect the components together
4. Edit the values of the components
5. Assign variable names and specify variables of interest
6. Simulate the circuit
7. View the results of the simulation
8. Save the results

These tasks are similar to the familiar word processing steps of writing and editing. These common procedures, necessary for all simulations, allow you to look for specific features in the menu bars.

Figure. 3.41 shows a sample schematic generated using the software simulation package PSpice® from Cadence Design Systems. Also shown is the associated DC solution, which details the voltage at each node. The components were chosen from a components library, interconnected, edited, and the circuit was then simulated. The simulation results shown are one of the many available options.

Another popular software package is MultiSim® from Interactive Image Technologies. Figure 3.42 shows the same schematic and the DC analysis results using MultiSim® for generating the circuit schematic and the circuit simulation. Note the major difference in how the results are displayed. MultiSim® attaches meters to the circuit nodes in a manner duplicating the setup you would have in the lab. Separate windows display the readings of each "meter." It is left to the reader to decide which software package would be preferred. While there are strong similarities between the software programs, you can see there are also considerable differences in how details are handled. One must refer to the software program's user manual for these details until a certain amount of expertise is acquired.

The usefulness of simulations directly correlates to the accuracy of the information entered into the computer. This is where the old adage "garbage in, garbage out" comes from. It needs to be clearly understood that the computer does not improve upon poor input data. It is with practice that you will learn when to rely on computer simulations and when to perform measurements in the laboratory. Please note, however, there is a steady shift to using the computer as much as possible in the design and analysis of electronic circuits.

Does this mean you do not really have to know any of the material in this book? Will the computer and simulations answer all questions? No, nothing could be further from the truth. Who will ask the questions in the first place? Who will specify the circuit and com-

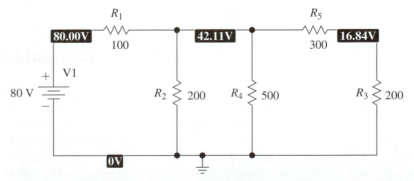

Figure 3.41 PSpice® circuit schematic with DC solution (PSpice® simulation output, used with permission of Cadence Design Systems)

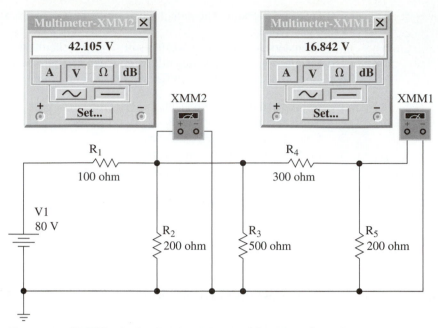

Figure 3.42 MultiSim® circuit schematic with DC solution (MultiSim® simulation output, used with permission of Interactive Image Technologies)

ponent values for the simulation to be done? Who will fill the blank piece of paper that is the starting point of all designs? You must do the thinking, and the computer simulations will be one of your central tools to assist you in performing your tasks.

CHAPTER REVIEW

3.1 **Why Are There Different Types of Circuits?**
- Loads are placed in series, in parallel, or in series-parallel to control voltage and current levels in circuits.

3.2 **DC Series Circuits**
- The current is the same throughout a series circuit.
- Kirchhoff's voltage law (KVL): the sum of the voltages around any closed path, including through the air, must equal zero.
- The total resistance in a series circuit is the sum of the individual resistances.
- Voltage division occurs in series circuits. The voltage divider rule (VDR) is a direct expression of this voltage division.

3.3 **DC Parallel Circuits**
- The voltage is the same throughout a parallel circuit.
- Kirchhoff's current law (KCL): the sum of the currents entering and leaving a node must equal zero.
- Conductance is the reciprocal of resistance (1 over resistance). Conductance is useful in parallel circuits.

- The total conductance of a parallel circuit is the sum of the individual conductances.
- Current division occurs in parallel circuits. The current divider rule (CDR) is a direct expression of this current division.

3.4 **DC Series-Parallel Circuits**
- Series-parallel circuits are analyzed by applying the analysis techniques for series and parallel circuits to series groups (branches) and parallel groups of resistances within the series-parallel circuit.

3.5 **Analysis of Multiple-Source DC Circuits Using Superposition**
- Series voltage sources add (in voltage) to obtain a larger source voltage. The current is the same through all of the series sources.
- Parallel voltage sources add (in current) to obtain a larger source current capability. The voltage and polarities of parallel voltage sources must be the same.
- When multiple sources are neither in series nor in parallel, superposition can be used to determine voltages and currents in the circuit.
- Superposition is the addition (with proper sign) of the corresponding voltages and cur-

rents due to each source in the circuit. Series-parallel circuit analysis is used to analyze the circuit for each source.

- In superposition, a voltage source is deactivated by replacing it with a short to guarantee zero voltage.

3.6 **Practical Voltage Sources and the DC Current Source**
- An ideal voltage source maintains the same voltage and can deliver any current level.
- Practical sources have internal resistance.
- The model of a practical voltage source is an ideal voltage source in series with the internal resistance.
- The terminal voltage of a practical voltage source drops as the current draw increases.
- An ideal current source delivers the same current at any voltage level.

- The model of a practical current source is an ideal current source in parallel with the internal resistance.
- Series-parallel circuits with current sources can be analyzed using series-parallel circuit analysis techniques, including superposition.
- In superposition a current source is deactivated by replacing it with an open to guarantee zero current.

3.7 **The Role of Circuit Simulation Software in DC Series-Parallel Circuit Analysis**
- Circuit simulation software is used to perform circuit simulations with speed and accuracy.
- Circuit simulations allow for the generation of results before the circuit is built.
- Circuit simulations allow for "what if" scenarios and tolerance investigations.
- A knowledge of circuit analysis is critical for utilizing circuit simulation software properly.

HOMEWORK PROBLEMS

It is necessary that you practice analyzing basic circuits in order to understand analysis principles as they apply to practical applications. The following problems will exercise your analysis skills.

Proper completion of the circuit analysis in the following section will demonstrate your ability to perform series circuit analysis.

3.1 For the circuit in Figure P3.1, determine (*a*) R_T, (*b*) V_{ab}, and (*c*) P_2.

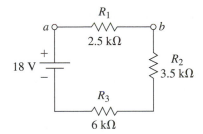

Figure P3.1 Schematic drawing for problem 3.1

3.2 For the circuit in Figure P3.2, determine (*a*) V_T, (*b*) I_2, and (*c*) P_T.

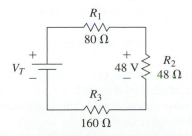

Figure P3.2 Schematic drawing for problem 3.2

3.3 For the circuit in Figure P3.3, determine (*a*) I_T, (*b*) R_T, (*c*) V_{ab} using the VDR, and (*d*) V_{ab} using KVL.

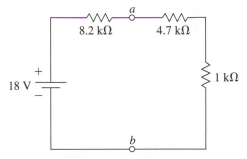

Figure P3.3 Schematic drawing for problem 3.3

3.4 For the circuit in Figure P3.4, determine (*a*) R_T, and (*b*) I_T, (*c*) P_T, (*d*) V_{ab} using VDR, and (*e*) V_{ab} using KVL.

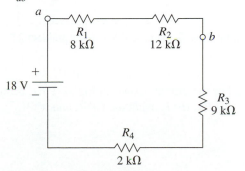

Figure P3.4 Schematic drawing for problem 3.4

3.5   A circuit has three resistors in series with a 36 V source. The current through the first resistor is 8 A. The power dissipated in the second resistor is 90 W. The voltage across the third resistor is 20 V. Determine the three resistor values.

3.6   In problems 3.1 through 3.5, you were asked to analyze four series circuits. Provide short answers to the following questions:
   a. Why do you need to understand series circuits?
   b. Can you name three practical uses for or examples of series circuits?
   c. What are the problems if a component is short-circuited?

   d. What are the problems if a component is open-circuited?

3.7   You are asked to wire a doorbell at the guard station in a factory. The door is 1000 m away from the guard station. The bell draws 100 mA of current and will operate as long as the voltage drop across the bell is at least 9.5 V. Your power supply is 12 V. What size wire would you use? Justify your answer. Draw the schematic and represent the wire and the bell as a resistance. Remember, there is a wire to and from the voltage source.

Proper completion of the circuit analysis in the following section will demonstrate your ability to perform parallel circuit analysis.

3.8   In the circuit shown in Figure P3.5, determine (a) V_T, (b) I_2 using the CDR, and (c) I_2 using Ohm's law.

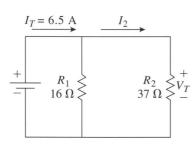

Figure P3.5 Schematic drawing for problem 3.8

3.9   For the circuit shown in Figure P3.6, determine (a) G_T, (b) R_T, (c) I_3, and (d) P_1.

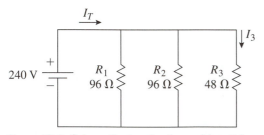

Figure P3.6 Schematic drawing for problem 3.9

3.10   For the circuit shown in Figure P3.7, determine (a) R_T, (b) I_2 using the CDR, and (c) V_{ab}.

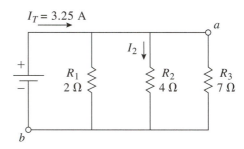

Figure P3.7 Schematic drawing for problem 3.10

3.11   For the circuit shown in Figure P3.8, determine (a) G_T, (b) R_T, (c) I_T, and (d) I_3.

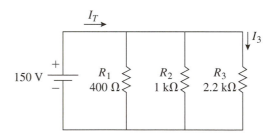

Figure P3.8 Schematic drawing for problem 3.11

3.12   In problems 3.8 through 3.11, you were asked to analyze four basic parallel circuits. Provide short answers to the following questions:
   a. Why do you need to understand parallel circuits?
   b. Can you name three practical uses for or examples of parallel circuits?
   c. What are the potential problems if a component is short-circuited?
   d. What are the potential problems if a component is open-circuited?

3.13 You want to add some additional circuits in the kitchen of your residence because there are not enough outlets. You know that there are going to be a 1000 W microwave, two lamps that will take 100 W light bulbs, and a ceiling fan that has a ¼ hp motor and five 40 W lamps connected in parallel to the 120 V supply. Your job is to estimate a circuit breaker and the wire size for a power distribution circuit in your residence. (746 W = 1 hp)

a. Draw the schematic.
b. What is the current for the individual items?
c. What is the total current?
d. Using the *National Electric Code* make a recommendation on the wire size and the circuit breaker rating.

Proper completion of the circuit analysis in the following section will demonstrate your ability to perform series-parallel circuits analysis.

3.14 For the circuit shown in Figure P3.9, determine (a) V_3 using the VDR, and (b) I_1 using the CDR.

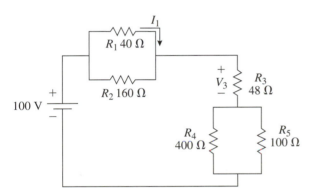

Figure P3.9 Schematic drawing for problem 3.14

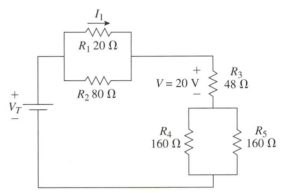

Figure P3.11 Schematic drawing for problem 3.16

3.15 For the circuit shown in Figure P3.10, determine (a) R_T, (b) I_2, and (c) V_{ab}.

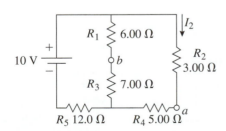

Figure P3.10 Schematic drawing for problem 3.15

3.17 For the circuit shown in Figure P3.12, determine (a) R_T, (b) I_T, (c) I_3, and (d) V_{ab}.

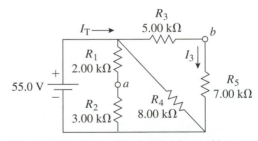

Figure P3.12 Schematic drawing for problem 3.17

3.16 For the circuit shown in Figure P3.11, determine (a) V_T and (b) I_1.

3.18 In the circuit shown in Figure P3.13, determine (a) I_2 using the CDR, and (b) I_1 using KCL.

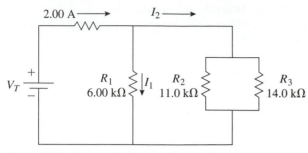

Figure P3.13 Schematic drawing for problem 3.18

3.19   Determine V_3 and V_S in the circuit shown in Figure P3.14.

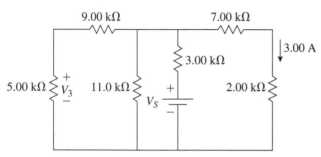

Figure P3.14 Schematic drawing for problem 3.19

3.20   For the circuit shown in Figure P3.15, determine (a) R_T, (b) I_2, and (c) V_{ab}.

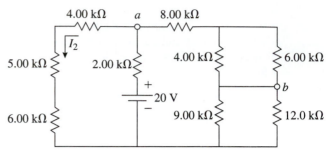

Figure P3.15 Schematic drawing for problem 3.20

Proper completion of the circuit analysis in the following section will demonstrate your ability to use superposition to analyze multiple-source DC series-parallel circuits.

3.21   Determine V_A in the circuit shown in Figure P3.16.

3.22   Determine I_2 in the circuit shown in Figure P3.17.

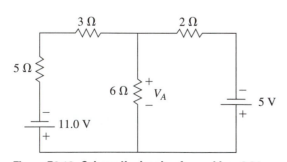

Figure P3.16 Schematic drawing for problem 3.21

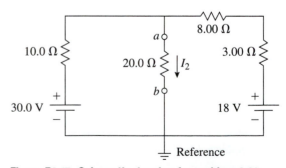

Figure P3.17 Schematic drawing for problem 3.22

3.23   Determine I_x in the circuit shown in Figure P3.18.

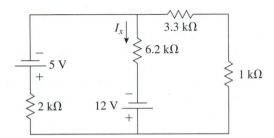

Figure P3.18 Schematic drawing for problem 3.23

3.24   Determine I_2 in the circuit shown in Figure P3.19.

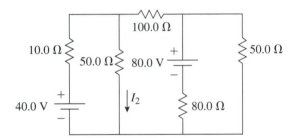

Figure P3.19 Schematic drawing for problem 3.24

Properly answering the following questions and completion of the circuit analysis in the following section will demonstrate your understanding of what a DC current source is and how to analyze DC series-parallel circuits that contain a current source.

3.26   What are the properties of an ideal current source?

3.27   How does an ideal current source differ from an ideal voltage source?

3.28   In the circuit shown in Figure P3.21, determine:
   a. G_T
   b. R_T
   c. I_1 using KCL
   d. I_2 using the CDR
   e. I_3 using the CDR
   f. Demonstrate that KCL holds.

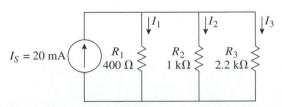

Figure P3.21 Schematic drawing for problem 3.28

3.25   For the circuit shown in Figure P3.20, determine (a) I_T, and (b) V_{ab}.

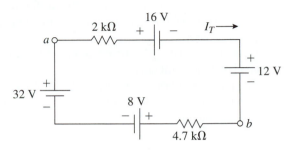

Figure P3.20 Schematic drawing for problem 3.25

3.29   In the circuit shown in Figure P3.22, determine:
   a. G_T
   b. R_T
   c. I_1 using KCL
   d. I_2 using the CDR
   e. I_3 using the CDR
   f. Demonstrate that KCL holds.

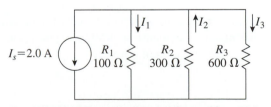

Figure P3.22 Schematic drawing for problem 3.29

Properly answering the following questions and completion of the circuit analysis in the following section will demonstrate your understanding of what practical DC voltage and current sources are and how to analyze DC circuits that contain a practical current source.

3.30 In your own words, describe the difference between an ideal current source and a practical current source.

3.31 Draw a schematic of a practical current source and label the components.

3.32 In your own words, describe the difference between an ideal voltage source and a practical voltage source.

3.33 Draw a schematic of a practical voltage source and label the components.

3.34 Rework problems 3.9, and 3.11 replacing the ideal voltage source with a practical voltage source. Assume that the source resistance is 20 Ω. Compare your results to the prior solutions.

The following problems ask that you use a computer software circuit simulation package to verify the results of prior calculations. The problems will also ask you to demonstrate that you understand the power of using a computer once you understand the analysis process.

3.35 Verify the results in the odd-numbered problems 3.15 through 3.25 using Electronic Workbench® or PSpice®.

3.36 Verify the results in the even-numbered problems 3.14 through 3.24 using Electronic Workbench® or PSpice®.

3.37 For the circuit shown in Figure P3.16, assume that the 6 Ω resistor has a 5.0% tolerance. If all other resistor values are nominal, what is the greatest change in V_A from its nominal value? Is the variance in the resistance positive or negative for this greatest change?

3.38 For the circuit shown in Figure P3.19, assume that the 50 Ω resistor through which I_2 flows has a 5.0% tolerance. If all other resistor values are nominal, what is the greatest change in I_2 from its nominal value? Is the variance in the resis-

tance positive or negative for this greatest change?

3.39 For the circuit shown in Figure P3.17, assume that all resistors are at their nominal resistance values. What is the change in V_{ab} if the 8 Ω resistor is out of tolerance by
  a. +10%
  b. −10%
  c. +20%
  d. −20%

3.40 For the circuit shown in Figure P3.18, assume that all resistors are at their nominal resistance values. What is the change in I_x if the 2 kΩ resistor is out of tolerance by
  a. +10%
  b. −10%
  c. +20%
  d. −20%

CHALLENGE PROBLEMS:

Your ability to develop the derivations in the following will further demonstrate your knowledge of material presented in Chapters 1, 2, and 3.

3.41 In Section 3.2 of the text, Kirchhoff's voltage law was developed using the principle of conservation of charge. Your challenge, should you accept it, is to demonstrate KVL using the principle of conservation of power.

3.42 In Section 3.3 of the text, the equation for total resistance in a parallel circuit was developed using Kirchhoff's current law. Your challenge, should you accept it, is to develop that equation using the principle of conservation of power or some other method.

AC Sinusoidal Steady-State Signals and Resistive Circuits

As a result of successfully completing this chapter, you should be able to:

1. Describe an AC sinusoidal signal.
2. Describe why AC is important.
3. Determine peak and effective amplitudes, radian and cyclic frequency, and phase from a given AC voltage or current versus time waveform.
4. Sketch an AC waveform with properly scaled and labeled peak amplitude and period from a mathematical time-domain expression of an AC voltage or current waveform.
5. Describe the concept of average power for an AC signal.
6. Determine effective (RMS) values of AC voltages and currents.
7. Determine AC voltages, currents, and average powers in series-parallel resistive circuits using series-parallel analysis techniques.

4.1 What Is AC? Why Is AC Important?

4.2 Mathematical Expression of AC Signals

4.3 Power in AC Resistive Circuits and Effective Values

4.4 Analysis of Resistive Circuits with AC Signals

Chapter Review

Homework Problems

4.1 What Is AC? Why Is AC Important?

So far we have learned what electricity is, what types of quantities are used in electric circuits, and how to analyze any series-parallel circuit with DC voltage and/or current sources. There are other aspects of electricity to explore. A second type of source, the current source, was introduced near the end of the previous chapter. There are also other electrical components, namely inductors and capacitors, that will be introduced after this chapter. Here we will consider the type of electrical signal. A DC source generates a steady-state DC signal. What does *steady-state* mean?

Consider the graph of a DC steady-state signal voltage (or current) versus time, as shown in Figure 4.1. The plot is $v(t)$ versus t. The reason that a lowercase v is used for voltage is because the voltage *might* vary with respect to time. The (t) part is used to indicate the independent variable, i.e., the voltage might vary as a function of time [recall $y(x)$ in algebra: y is the dependent variable, x is the independent variable, and y varies as a function of x]. The DC voltage stays the same in this case, i.e., the voltage is constant. The assumption is that the voltage has been constant for all time and will remain constant for all time. The three dots in the graphs indicate that the same pattern continues, a constant level in this case. This assumption is literally absurd because nobody can have a source on forever, but the assumption is valid in the region of time after the voltage has been turned on and has steadied out, up to the time the voltage is turned off, as sketched in Figure 4.2. Thus, steady state means, in the DC sense, that the voltage and current do not vary over time.

The plot of voltage or current versus time, i.e., the plot of an electronic quantity versus time, is called a *waveform*. The plot shown in Figure 4.2 also suggests that there are many other waveforms, and each of these possible waveforms is the plot of an electronic signal. An infinite number of signals exists, but some are very common in electronics because the signal performs useful electronic tasks. Probably the most common and useful electronic signal is the alternating current sinusoidal steady-state signal, which is abbreviated as AC. AC is applicable to voltage too, not just current, and is an industry standard abbreviation. What is AC?

Imagine a battery that is mounted on an axial rod, as shown in Figure 4.3. As the battery rotates, the battery terminals will touch the spring contacts for part of the rotation, and then they will touch the spring contacts in the opposite polarity during part of the other half of the rotation. When the battery terminals are not touching the spring contacts, the voltage will be zero. This pattern repeats over and over as the battery rotates and is plotted in Figure 4.4. The voltage polarity across the load and current direction through the load

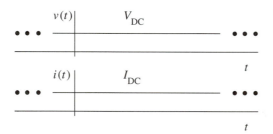

Figure 4.1 Graphs of Steady-State DC Voltage and Current

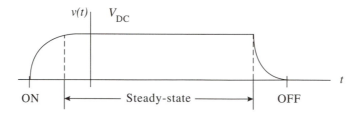

Figure 4.2 Graph of a DC Voltage with the Steady-State Region Labeled

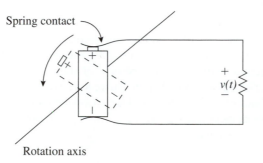

Figure 4.3 A Rotating Battery Circuit Setup

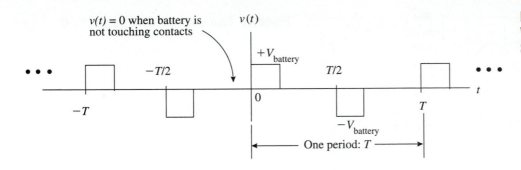

Figure 4.4 AC Voltage Waveform for the Circuit Setup in Figure 4.3

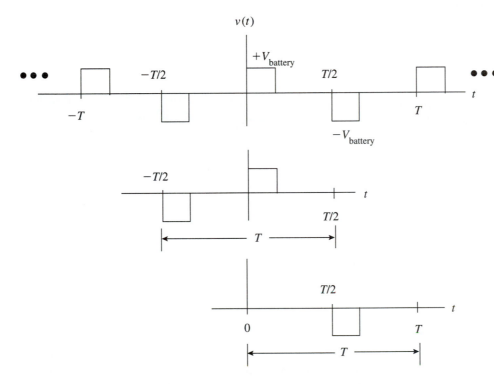

Figure 4.5 Two Different Cycles from the Waveform in Figure 4.4

alternates; hence, this waveform is one example of alternating current (AC). One could just as easily say "alternating voltage," but industry-standard terminology is AC. The T in Figure 4.4 stands for *period,* which is the length of time over which the pattern repeats. The part of the waveform that repeats is called a *cycle.* The period can begin and end at different spots as long as the length of time is T seconds, as shown in Figure 4.5. The key is that the period corresponds to the time for one complete cycle. Thus, the cycle keeps repeating every period. One cycle of the waveform is completed every T seconds.

Although this method of generating AC is easy to visualize, it is not very efficient because the voltage is zero for too long, and rotating a battery is not practical. Instead, there is a periodic waveform, i.e., a waveform that has a cycle that repeats periodically (with period T), that is easy to generate and has remarkable properties. In fact you will spend a significant portion of your electronics career using this waveform and learning more about it each time you study it. What is it? It is the sinusoidal waveform: $y = \sin(\theta)$. A sinusoidal waveform is a sine or cosine wave, just as in trigonometry—see Figure 4.6. In electronics, however, the independent variable is time (t)—instead of angle—see the example waveform in Figure 4.7.

What does this graph mean? If a voltmeter was placed across a source that generates a sinusoidal signal, and if both humans and the voltmeter could respond fast enough to observe the voltage as it changed, one would see the voltage rise to a peak like a sine function rises from 0 to 1 as the angle changes from 0° to 90°. Then the voltage would decrease back to 0, again like a sine function between 90° and 180°, the polarity would reverse, and

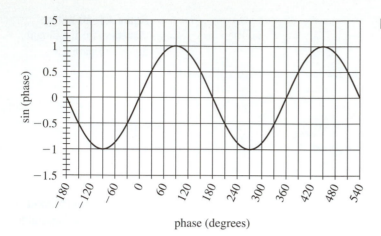

Figure 4.6 Sinusoidal Waveform: sin(θ) vs. θ

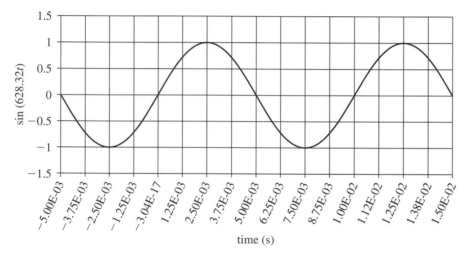

Figure 4.7 Sinusoidal Waveform: sin(628.32*t*) vs. *t*

the voltage would increase in this negative polarity to -1. Then this voltage would return to 0 as the angle reached 360°, and the cycle would start over again. The signal with this sinusoidal waveform is called the alternating current sinusoidal steady-state (AC) signal (again, AC applies to both alternating voltage and current).

Why is this sinusoidal waveform called "steady state"? It certainly does not stay at a constant value as time goes on. Steady state for AC waveforms is used in the sense that the waveform is periodic (T never changes) and that each cycle has the same peak magnitude as all the other cycles. A sine function always peaks at $+1$ and -1, so the sinusoid is a steady-state signal.

The sinusoid may seem like a strange waveform for electronics, but it is actually very easy to generate compared with most other waveforms. This waveform will be useful for performing tasks for which DC and other waveforms are not useful. A few interesting applications of AC are mentioned here, but the explanations behind the applications will have to wait until we learn more about AC.

■ AC is useful in circuits because the peak voltage and current levels can easily be changed without losing power, as would be the case in a voltage divider circuit, by using a component called the transformer.

■ AC along with some new components that will be introduced soon, the capacitor and the inductor, can be used to select a channel, such as on a radio or a television.

■ Many electromechanical devices (motors and generators, for example) use or produce AC to operate properly and efficiently.

■ Electrical power transmission can be very efficient with AC.

■ Several sinusoids can be combined to create new signals for specific applications (this is an advanced topic—look at some of the figures in Chapter 15).

These applications span electronics, electrical power, and electronic communications. The AC signal is very useful; hence, one needs to be able to make predictions of circuit performance, just as was done with DC circuits. In order to make circuit predictions, AC waveforms must be expressed mathematically, which is the topic of the next section.

4.2 MATHEMATICAL EXPRESSION OF AC SIGNALS

How is the equation for sine expressed as a function of t? The standard expression for a sinusoidal voltage waveform is

$$v(t) = \sin(\omega t) \tag{4.1}$$

where ω is a constant and is assigned the lowercase Greek letter omega as its symbol. The significance of ω will be examined soon. For now, treat it as a constant. As an example, let the value of ω be 628.32 rad/s. What is "rad/s"? The s is seconds. There are two common types of units for angles: radians (abbreviation: rad) and degrees (symbol: °). Recall that $180° = \pi$ rad. Check the units in Equation 4.1:

$$\text{unit of } (\omega t) = \left(\frac{\text{rad}}{\text{s}}\right)(\text{s}) = \text{rad} \tag{4.2}$$

The quantity in parentheses, which one takes the sine of, is called the *argument*. The argument of a trigonometric function should always be in radians or degrees. In electricity, the argument of a sine (or cosine) is normally in radians. Hence, even though what ω is has not been established, it is now apparent that it serves the function of converting time into an angle.

Return to the waveform in Figure 4.7. The value of ω is 628.32 rad/s. The horizontal axis is time. Notice that the waveform looks like Figure 4.6 except for the horizontal axis labeling. Is there a correspondence between θ and ωt? Indeed: $\theta = \omega t$. Try Equation 4.1 at a few time values from Figure 4.7:

$$\sin[(628.32 \text{ rad/s})(-0.005 \text{ s})] = \sin(-3.1416 \text{ rad}) = 0$$

$$\sin[(628.32 \text{ rad/s})(+0.0025 \text{ s})] = \sin(1.5708 \text{ rad}) = +1$$

$$\sin[(628.32 \text{ rad/s})(+0.015 \text{ s})] = \sin(9.4248 \text{ rad}) = 0$$

Thus, multiplication of a time value by ω converts the product into an angle. But what is ω? It has a name: the *radian frequency* (sometimes called angular velocity, due to a mechanical analogy). The radian frequency indicates how many radians of the sinusoidal waveform occur each second. There is also *cyclic frequency*, often just called frequency, which indicates how many cycles of the sinusoidal waveform occur each second. The symbol for cyclic frequency is f, and the unit is hertz (Hz). (A former name for f is "cycles per second," abbreviated cps.) "Cycles" is not really a unit, so the official unit hertz equals 1/s. Are there relationships between ω, T, and f, where T is the period, or the time for one cycle. Invert T, and the result is the number of cycles per second, which is frequency:

$$f = \frac{1}{T} \tag{4.3}$$

Let us examine Equation (4.3) to see if we can relate it to ω. There are 2π rad in one cycle of the sinusoid, so let us multiply $1/T$ by this ratio and see if anything useful results:

$$\left(\frac{1 \text{ (cycle)}}{T \text{ (S)}}\right)\left(\frac{2\pi \text{ (rad)}}{1 \text{ (cycle)}}\right) = \left(\frac{2\pi \text{ (rad)}}{T \text{ (S)}}\right) \tag{4.4}$$

But rad/s is radian frequency! Hence, f and T are related to ω:

$$\omega = \frac{2\pi}{T} = 2\pi f \qquad (4.5)$$

where: ω = radian frequency (rad/s),

T = period (s), and

f = cyclic frequency (Hz = 1/s)

Although simple, this result is fundamental. In practice both ω and f are called frequency, and you should know which one is being used from the units or from the context.

Example 4.2.1

Determine the (*a*) period, (*b*) cyclic frequency, and (*c*) radian frequency for the waveform shown in Figure 4.7.

Given: waveform plot in Figure 4.7

Desired: a. T

b. f

c. ω

Strategy: Determine T, the length of time for one cycle, from the plot.

$f = 1/T$

$\omega = 2\pi f$

Solution:

a.   Pick a convenient starting time (often a zero crossing): 5×10^{-3} s.
Move along the waveform until it starts over: 1.5×10^{-2} s.

$$T = 1.5 \times 10^{-2} - 5 \times 10^{-3} = 1 \times 10^{-2} = 0.0100 \text{ s}$$

The same result would be obtained if the period was started at $t = 0$ (which is 3.04×10^{-17} on the graph—often, zero will be seen as an extremely small number in computer-generated plots, because a computer does not have infinite resolution nor infinite accuracy). Follow the waveform until it starts repeating (the start of the next cycle), and read that time value: 0.01 s.

$$T = 0.01 \text{ s} - 0 \text{ s} = 0.0100 \text{ s}$$

b. $f = \dfrac{1}{T} = \dfrac{1}{0.01\ s} = 100 \text{ Hz} = 100. \text{ Hz}$

c. $\omega = 2\pi f = 2\pi(100) = 628.32 = 628$ rad/s

Now that the frequency concept has been established, the next step is to determine the mathematical expression for an AC signal. Start with Equation (4.1), repeated here as Equation (4.6):

$$v(t) = \sin(\omega t) \qquad (4.6)$$

The peak values of the sine function are $+1$ and -1. To obtain a different peak value, just multiply the sine by that peak value (one times "anything" equals that "anything"):

$$v(t) = V_P \sin(\omega t) \qquad (4.7)$$

Finally, the waveform may not always cross over the horizontal axis at $t = 0$ like a sine wave does. This is called phase shift, and in later chapters you will see how important it is (phase shift is one of the most interesting aspects of AC). For now, assume that this phase shift may occur. The symbol θ is often used for phase shift. This is a *different* θ from the θ in Figure 4.6. The θ in Figure 4.6 was the independent variable in the plot. Because time is the independent variable in AC voltage and current plots, one is free to use θ to mean phase shift, which is a constant. To shift the waveform in phase, add to or subtract from the angle in the argument:

$$v(t) = V_P \sin(\omega t + \theta) \tag{4.8}$$

How is the phase shift determined? Recall that ω is a time-to-angle converter. The time shift that the sine wave zero crossing is from $t = 0$ on the horizontal axis could be converted into phase shift. The zero crossing of interest is when the wave crosses 0 V from below the horizontal axis (negative values) to above (positive values), not positive to negative. How is the zero crossing determined? Look at the plot of a sine wave and locate the time where it crosses the horizontal axis closest to $t = 0$. This is the time shift Δt. Once the time shift is read from the plot, it can be converted to an angle:

$$\theta = \omega \, \Delta t \tag{4.9}$$

where: θ is the phase shift (degrees or radians)

ω is the radian frequency (rad/s)

Δt is the time shift of the upward zero crossing of the sine wave from $t = 0$ (s)

Notice that θ is expressed in either radians or degrees. In fact, it is more often expressed in degrees than in radians! However, one must convert θ into radians because $\omega \Delta t$ is in radians (or convert $\omega \Delta t$ into degrees) and then use the proper mode on a calculator or computer software: radians or degrees. What are the units of Equation (4.9)? The angle is in radians because ω is in rad/s and Δt is in s. How is the angle converted from radians into degrees? Multiply by 1 with the proper units conversion:

$$\theta[\text{rad}] = \omega \left[\frac{\text{rad}}{\text{s}} \right] \Delta t[\text{s}]$$

$$\theta[°] = \omega \left[\frac{\text{rad}}{\text{s}} \right] \Delta t[\text{s}] \left(\frac{180°}{\pi \, \text{rad}} \right) \tag{4.10}$$

Some people like to determine θ in degrees first. They determine the phase shift from the fraction that Δt is of the whole period T. This is especially convenient when taking measurements from an oscilloscope (an instrument used to display and measure waveforms—see the laboratory manual for a discussion of oscilloscopes). Then θ in radians can be determined:

$$\theta[°] = \Delta t \left(\frac{360°}{T} \right)$$

$$\theta[\text{rad}] = \theta[°] \left(\frac{\pi \, \text{rad}}{180°} \right) \tag{4.11}$$

The last issue is the sign of the phase shift. If θ is positive, then at $t = 0$, the argument of the sine function already has a value greater than zero. The total phase of the argument at $t = 0$ is already positioned beyond the beginning of the sine wave cycle. Hence, the wave is advanced in time, which means the sine wave is shifted to the left. The opposite argument holds for negative phase shift. Hence,

A positive phase shift advances the signal in time—the waveform is shifted to the left.
A negative phase shift delays the signal in time—the waveform is shifted to the right.

Example 4.2.2

For the signal in Figure 4.8, determine (*a*) $v(t)$ expression and (*b*) $v(150 \, \mu s)$.

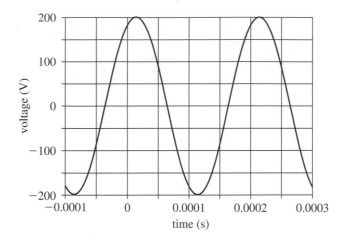

Figure 4.8 Waveform for Example 4.2.2

Given: the waveform in Figure 4.8

Desired:

a. $v(t)$ expression

b. $v(150 \, \mu s)$ [value of $v(t)$ at $t = 150 \, \mu s$]

Strategy:

a.   determine from the plot: V_P, T, Δt, whether advanced or delayed

  $\omega = 2\pi/T$

  $\theta = \omega \, \Delta t$

  $v(t) = V_P \sin(\omega t + \theta)$

b.   insert $t = 150 \, \mu s$ into $v(t)$ result

Solution:

a. $V_P = 200$ V

  $T = 200 \, \mu s$

  $\Delta t = 37 \, \mu s$ (estimated from the graph in Figure 4.8; the beginning of the sine wave cycle, i.e., where the sine wave crosses from negative to positive, occurs 37 μs before $t = 0$)

$$\omega = \frac{2\pi}{T} = \frac{2\pi}{200 \, \mu s} = 31{,}416 \text{ rad/s}$$

$$\theta = \omega \, \Delta t = \left(31{,}416 \frac{\text{rad}}{\text{s}}\right)(37 \, \mu s) = 1.1628 \text{ rad}\left(\frac{180°}{\pi \, \text{rad}}\right) = 66.6°$$

The closest upward zero crossing to $t = 0$ is before $t = 0$, i.e., it is advanced, so the phase shift is positive.

$$v(t) = V_P \sin(\omega t + \theta) = 200 \sin(31{,}416t + 66.6°)$$

b.   Choose degrees or radians for the argument: radians

$$v(150 \, \mu s) = 200 \sin\left[31{,}416(150 \times 10^{-6}) + 66.6°\left(\frac{\pi \, \text{rad}}{180°}\right)\right]$$

$$= 200 \sin(5.8748 \text{ rad}) = -79.4 \text{ V}$$

This result is confirmed on the graph (Figure 4.8).

___ **Example 4.2.3** ___

For the waveform in Figure 4.9, determine (*a*) *v*(*t*) and (*b*) *v*(1.6 ms)

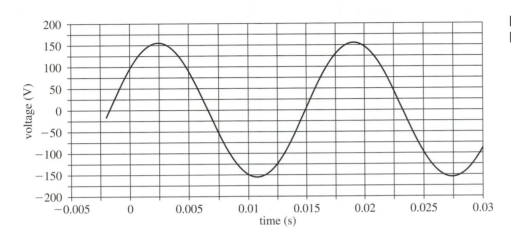

Figure 4.9 Waveform for Example 4.2.3

Given: the waveform in Figure 4.9

Desired: a. $v(t)$

          b. $v(1.6 \text{ ms})$

Strategy: a. from the plot: V_P, T, Δt, whether advanced or delayed

$$f = 1/T$$

$$\omega = 2\pi f$$

$$\theta = 360° \; \Delta t/T$$

$$v(t) = V_P \sin(\omega t + \theta)$$

          b. Plug in $t = 1.6$ ms

Solution: _____

a.   estimated from Figure 4.9: $V_P = 155$ V, $T = 16.67$ ms, $\Delta t = 1.62$ ms

$$f = \frac{1}{T} = \frac{1}{16.67 \text{ ms}} = 60 \text{ Hz}$$

$$\omega = 2\pi f = 2\pi 60 = 377 \text{ rad/s}$$

$$\theta = \frac{360°}{T} \Delta t = \left(\frac{360°}{16.67 \text{ ms}}\right)(1.62 \text{ ms}) = 35° \text{ (advanced)}$$

$$v(t) = V_P \sin(\omega t + \theta) = 155 \sin(377t + 35°)$$

b.   Choose degrees or radians for the argument: degrees

$$v(1.6 \text{ ms}) = 155 \sin\left[(377)(1.6 \text{ ms})\left(\frac{180°}{\pi \text{ rad}}\right) + 35°\right] = 145 \text{ V}$$

— **Example 4.2.4**

Given the following expression for $i(t)$, sketch and label the plot of $i(t)$.

$$i(t) = 18.5 \sin(245t - 27°)$$

Given: expression for $i(t)$: $I_P = 18.5$ A

$\omega = 245$ rad/s

$\theta = -27°$

Desired: plot of $i(t)$: T

Δt

Strategy: $\omega = 2\pi/T$; solve for T.

$\theta = \omega \Delta t$; solve for θ.

Sketch and label plot.

Solution: _____

$$T = \frac{2\pi}{\omega} = \frac{2\pi}{245} = 25.65 \text{ ms}$$

$$\Delta t = \frac{\theta}{\omega} = \frac{27°\left(\dfrac{\pi \text{ rad}}{180°}\right)}{245 \text{ rad/s}} = 1.92 \text{ ms (delayed due to } -27°)$$

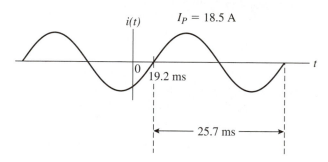

Figure 4.10 AC Waveform for Example 4.2.4

4.3 POWER IN AC RESISTIVE CIRCUITS AND EFFECTIVE VALUES

As with any electric circuit, the electrical power must transfer from the source to the load. This fact does not change just because we have a different type of electronic signal. Recall that power was easily determined for DC: $P = VI$. Both V and I are constant in steady-state DC, so this calculation is straightforward. In AC, however, both $v(t)$ and $i(t)$ are sinusoidal; hence, to determine power, the sinusoids must be multiplied. Let us try this graphically first for the steady-state DC case—see Figure 4.11.

In the graphical multiplication, the value of $v(t)$ and $i(t)$ at each time instance are multiplied to give $p(t)$, which is instantaneous power, i.e., the power at that time instance. In DC, this product is easy because $v(t) = V_{DC}$ and $i(t) = I_{DC}$, where the subscript DC is used here for clarity:

$$p(t) = v(t)\, i(t) = V_{DC} I_{DC} \tag{4.12}$$

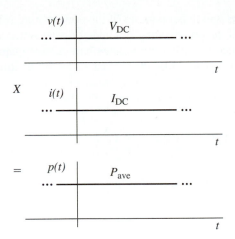

Figure 4.11 Graphical Multiplication of Steady-State DC $v(t)$ and $i(t)$ to Give $p(t)$ and P_{ave}

which is a constant power. The average power P_{ave} is the same as this constant power in the DC case, so

$$p(t) = v(t)\, i(t) = V_{DC}I_{DC} = P_{ave} \qquad (4.13)$$

Now consider the AC case. Note that $v(t)$ and $i(t)$ are not constant versus time like in the DC case, so the instantaneous power is not constant. The graphical multiplication of $v(t)$ and $i(t)$ is shown in Figure 4.12. Again, the value of $v(t)$ and $i(t)$ at each time instance are multiplied to give $p(t)$. This is trickier than the DC case but can be performed easily with software, such as a spreadsheet.

Notice the striking result that occurs for $p(t)$ in Figure 4.12: $p(t)$ "pulsates" but it is all positive (technically, nonnegative). When the negative voltage is multiplied by the negative current, positive power results. The average power is halfway between the power peaks and zero in the $p(t)$ plot. The power peak is V_PI_P; thus, the average power is

$$P_{ave} = \frac{V_PI_P}{2} \qquad (4.14)$$

The average power result for AC is different from the DC case. Do not forget this division by two; otherwise, all of your AC power calculations will be off by a factor of two and the

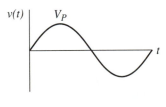

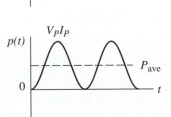

Figure 4.12 Graphical Multiplication of AC $v(t)$ and $i(t)$ to Give $p(t)$ and P_{ave}

electronic components will either burn up (if the power rating is half of what it should be) or be too expensive (if the power rating is double of what is needed).

Wouldn't it be nice if there were voltage and current quantities for AC that would give the proper average power like the DC case? Such quantities do exist and are routinely used. They are called the *effective value* or, equivalently, the *RMS (root-mean-square) value* of the voltage (and current). To obtain the effective value, manipulate Equation (4–14):

$$P_{ave} = \frac{V_P I_P}{2} = \frac{V_P I_P}{\sqrt{2}\sqrt{2}} = \left(\frac{V_P}{\sqrt{2}}\right)\left(\frac{I_P}{\sqrt{2}}\right) = V_{RMS}I_{RMS} = V_{eff}I_{eff} \qquad (4.15)$$

Thus, the effective (or RMS) values of voltage and current are defined to be the peak values divided by the square root of two:

$$\boxed{\begin{aligned} V_{RMS} = V_{rms} = V_{eff} = \frac{V_p}{\sqrt{2}} \\ I_{RMS} = I_{rms} = I_{eff} = \frac{I_p}{\sqrt{2}} \end{aligned}} \qquad (4.16)$$

The subscripts eff and rms (RMS) have identical meanings; however, the RMS subscript is used most commonly and will be used in this textbook. The term *effective* is used as often as RMS in sentences. Average power is then calculated with these effective values as it is in the DC case and can be written a number of ways for resistive loads (*not* valid for other components):

$$P = V_{RMS}I_{RMS} \qquad (4.17)$$

$$P = \frac{V_{RMS}^2}{R} = I_{RMS}^2 R = \frac{V_P^2}{2R} = \frac{I_P^2}{2}R \qquad (4.18)$$

Example 4.3.1

If $v(t) = 34\sin(377t)$ and $i(t) = 0.27\sin(377t)$, determine (*a*) the effective values and (*b*) the average power.

Given: $v(t) = 34\sin(377t)$
$i(t) = 0.27\sin(377t)$

Desired: a. V_{RMS}
I_{RMS}
b. P_{ave}

Strategy: a. $V_{RMS} = \frac{V_P}{\sqrt{2}}$ $I_{RMS} = \frac{I_P}{\sqrt{2}}$

b. $P = V_{RMS}I_{RMS}$

Solution:

a. $V_{RMS} = \frac{V_P}{\sqrt{2}} = \frac{34}{\sqrt{2}} = 24.042 = 24.0\ V\ (RMS)$

$I_{RMS} = \frac{I_P}{\sqrt{2}} = \frac{0.27}{\sqrt{2}} = 0.19092 = 0.191\ A\ (RMS)$

Note: Sometimes the (RMS) is stated after the units to indicate that the value is effective, not peak. Similarly, (peak) is often stated after the units to indicate peak values.

b. $P = V_{RMS}I_{RMS} = (24.042)(0.19092) = 4.5900 = 4.59 \text{ W}$

Thus, the RMS (or effective, whichever you like) values make AC average power calculations easy; however, you must restrict the use of these power equations to resistive circuits. Although other types of circuits have not been covered yet, we will see later that different power equations exist for other components.

How did the names "effective value" and "RMS value" come about? The effective value name means that the voltage and current values are *effectively* the same as the DC voltage and current values that give the same average power. The RMS (root-mean-square) name comes from the derivation of average power for a periodic waveform. Unlike the graphical development of the average power equations for AC that was shown, the full derivation requires calculus. It will not be presented here, but we hope that insight like this will motivate you to study mathematics for understanding as well as techniques and to progress into the calculus courses with the same intent.

4.4 ANALYSIS OF RESISTIVE CIRCUITS WITH AC SIGNALS

Now that the importance of the AC signal electronic signal has been established, how does one analyze circuits that contain an AC source? First, the symbols for AC voltage sources and current sources should be documented. They are shown in Figure 4.13. The tilde serves as a reminder of an AC sinusoidal (steady-state) source as well as distinguishing the symbol from the corresponding DC sources.

Now, if an AC voltage is across a resistance, what is the AC current? As the voltage increases in the sine wave, there is more electric potential energy difference across the resistance and more electrical energy should convert into another form. But the implication from Ohm's law is that more current should flow. In fact the current should "track" the voltage: as the voltage increases, so does the current. As the voltage decreases, so does the current. When the voltage changes polarity, the current direction should turn around (alternate). Now is a good time to reinspect Figure 4.12. Notice how the voltage and current track, i.e., they are *in-phase*. Then average power results, as shown in the same figure, and electrical energy is being converted into another form of energy in the resistance.

How does the knowledge that voltage and current are in-phase in a resistive circuit with an AC source help us analyze the circuit for desired voltages and currents? Because the voltage and the current are directly proportional to each other, then the analysis situation is similar to circuits with DC sources. In fact, *DC series-parallel circuit analysis techniques are applicable to AC series-parallel circuits with resistances only.* One can use either peak or effective (RMS) values for voltages and currents (pick one or the other—do not switch around between the two in the same problem).

In multiple source circuits, two additional stipulations are necessary to use superposition: the frequencies of the sources must be the same and the sources must be in-phase. In Chapter 9, the reasons for these restrictions will be examined, the first restriction will be justified, and the second restriction will be removed after the introduction of additional concepts.

a. AC voltage source

b. AC current source

Figure 4.13 Schematic Symbols for AC Voltage and Current Sources

____ Example 4.4.1 _____

For the circuit in Figure 4.14, determine: (a) R_T, (b) $i_T(t)$, (c) P_3, and (d) $v_{ab}(t)$.

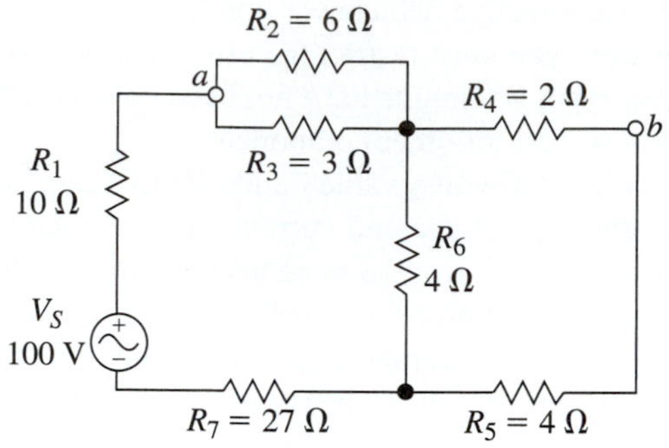

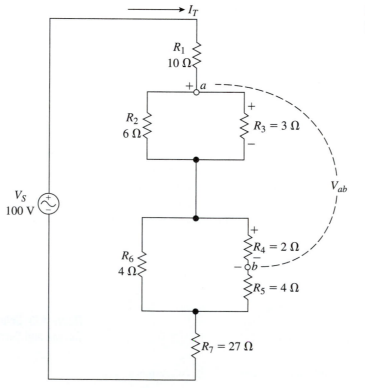

Figure 4.14 Circuit for Example 4.4.1

Given: circuit in Figure 4.14

Desired: a. R_T

         b. $i_T(t)$

         c. P_3

         d. $v_{ab}(t)$

Strategy: Redraw circuit (Figure 4.15); fundamentally a series circuit.

$$R_T = 10 + (6 \| 3) + [4 \| (2 + 4)] + 27$$

Use peak values (arbitrary).

$$I_T = V_S/R_T \rightarrow i_T(t)$$
$$I_3 \text{ by CDR, } P_3 = I_3^2/(2R_3)$$
$$V_{ab} \text{ by KVL, } \rightarrow v_{ab}(t)$$

Figure 4.15 Circuit in Figure 4.13 Redrawn

Solution: _____

a. $R_T = 10 + \dfrac{(6)(3)}{6+3} + \dfrac{(4)(2+4)}{4+(2+4)} + 27$

$$= 10 + 2 + 2.4 + 27 = 41.400\ \Omega = 41.4\ \Omega$$

b. $I_T = \dfrac{V_S}{R_T} = \dfrac{100}{41.4} = 2.4155$ A (peak)

$i(t) = 2.42 \sin(\omega t)$ [no specific frequency information given; hence, ω is left as a variable in the time-domain current expression]

c. $I_3 = \dfrac{I_T R_2}{R_2 + R_3} = \dfrac{(2.4155)(6)}{6+3} = 1.6103$ A

$P_3 = \dfrac{I_3^2 R_3}{2} = \dfrac{(1.6103)^2 (3)}{2} = 3.8896 = 3.89$ W

d.   KVL—going CW: $+V_3 - V_{ab} + V_4 = 0$
Need V_4:   VDR $\to V_6$, VDR $\to V_4$

$$V_6 = \dfrac{V_S[4\,\|\,(2+4)]}{R_T} = \dfrac{(100)(2.4)}{41.4} = 5.7971\ V$$

$$V_4 = \dfrac{V_6(R_4)}{R_4 + R_5} = \dfrac{(5.7971)(2)}{2+4} = 1.9324\ V$$

Now, apply KVL to determine V_{ab}:

$$V_{ab} = +V_3 + V_4 = I_3 R_3 + V_4 = (1.6103)(3) + 1.9324 = 6.7633\ V$$

$$v_{ab}(t) = 6.76 \sin(\omega t)$$ [no especific frequency information is given]

   The revelation of this section dramatically enlarges the scope of the previous chapter: all of the DC circuit analysis techniques in Chapter 3 are applicable to AC circuits with resistances. Either peak or RMS voltage and current values can be used. In addition, now that methods to determine the average power with AC signals have been established, then energy could be calculated if needed. In short, our previous learning has been *leveraged* to cover an entire new area. This scenario is one of the pleasing facts about the study of electronics: the more you understand, the more that can be understood, and the more interesting topics that you are in a position to understand and enjoy. In fact you are now in a position to study, understand, and apply two electronic components that are different from resistances: the capacitor and the inductor. These components are the subjects of the next two chapters.

CHAPTER REVIEW

4.1   What Is AC? Why Is AC important?
- An AC signal periodically alternates voltage polarity and current direction.
- The period is the length of time for one cycle of an AC signal.
- An AC signal varies sinusoidally in voltage and current and is in steady state: the period, frequency, and peak amplitude are constant.

4.2   Mathematical Expression of AC Signals
- The mathematical expression for an AC signal includes peak amplitude, radian frequency, and phase shift.
- Radian frequency is how many radians of the signal occur per second.
- Cyclic frequency is the number of cycles of the signal per second.
- $\omega = 2\pi f = 2\pi/T$
- Phase is expressed in radians or degrees. The conversion factor is $180° = \pi$ r.
- Positive phase shift occurs when the zero crossing (negative to positive) of the AC signal is before $t = 0$. The signal is shifted toward the left. Negative phase shift occurs when the zero crossing (negative to positive) of the AC signal is after $t = 0$. The signal is shifted toward the right.
- Phase shift is related to time shift by radian frequency or period.

4.3   Power in AC Resistive Circuits and Effective Values
- An AC signal has an average power:
$$P_{ave} = \frac{V_P I_P}{2} = V_{RMS} I_{RMS} = V_{eff} I_{eff}$$
- The effective value (which is identical to the RMS value) of a voltage or current is effectively the same, from a power viewpoint, as a DC voltage or current of the same value.
- The effective value of an AC voltage or current is determined by dividing the peak value by $\sqrt{2}$.

4.4   Analysis of Resistive Circuits with AC Signals
- DC series-parallel circuit analysis techniques are applicable to AC series-parallel circuits with resistances only. Peak or effective values for voltages and currents are used.

HOMEWORK PROBLEMS

Understanding alternating current steady state (AC): provide one or two short sentences for the following questions:

4.1   What is a waveform?

4.2   What is the difference between DC and AC signals?

4.3   What is the most common form of AC signal?

4.4   What does steady-state mean for an AC signal?

4.5   Name a few reasons for the common use of AC.

4.6   Name some applications for AC.

The following questions will demonstrate your understanding of trigonometric functions as they are used in the study of AC analysis.

4.7   Given the following expression: $v(t) = \sin(\omega t)$,
   a. What does t represent?
   b. What does ω represent?

4.8   Given the following expression:
$v(t) = 120 \sin(377t)$,
   a. What is the peak value of the AC signal?
   b. What is the frequency, f, of the AC signal?
   c. What is the period, T, of the AC signal?
   d. What is the frequency, ω, of the AC signal?

4.9   Given the following expression:
$v(t) = 10 \sin(2\pi 1000t)$,
   a. What is the peak value of the AC signal?
   b. What is the frequency, f, of the AC signal?
   c. What is the period, T, of the AC signal?
   d. What is the frequency, ω, of the AC signal?

4.10   Given the following expression:
$i(t) = 0.025 \sin(1577t)$,
   a. At what time does the peak value of the AC signal occur?
   b. What is the frequency, f, of the AC signal?
   c. What is the period, T, of the AC signal?
   d. At what time does the negative peak value of the AC signal occur?

4.11   Given the following expression:
$v(t) = 24 \cos(2\pi 500t)$,
   a. What is the peak value of the AC signal?
   b. What is the frequency, f, of the AC signal?
   c. What is the period, T, of the AC signal?
   d. What is the frequency, ω, of the AC signal?

4.12 Given the following expression:
$v(t) = 24 \cos(2\pi 2500t)$,
a. At what time does the peak value of the AC signal occur?
b. What is the frequency, f, of the AC signal?
c. What is the period, T, of the AC signal?
d. At what time does the negative peak value of the AC signal occur?

4.13 Given the following expression:
$i(t) = 0.50 \sin(2\pi 250t)$,
a. What is the value of $i(t)$ at $t = 1.00$ ms?
b. What is the value of $i(t)$ at $t = 0.50$ ms?
c. What is the value of $i(t)$ at $t = 1.50$ ms?
d. What is the value of $i(t)$ at $t = 2.00$ ms?

4.14 Given the following expression,
$i(t) = 0.750 \cos(2\pi 100t)$:
a. What is the value of $i(t)$ at $t = 10.00$ ms?
b. What is the value of $i(t)$ at $t = 5.00$ ms?
c. What is the value of $i(t)$ at $t = 15.0$ ms?
d. What is the value of $i(t)$ at $t = 20.0$ ms?

4.15 Sketch the waveform for the following expressions. Be sure to properly label the axes.
a. $i(t) = 75.0 \sin(2\pi 100t)$ mA
b. $v(t) = 24 \sin(2\pi 500t)$ V
c. $v(t) = 24 \sin(2\pi 500t + 90°)$ V
d. $i(t) = 150 \cos(2\pi 100t - 90°)$ mA
e. $v(t) = 120 \sin(2\pi 500t - 30°)$ V
f. $i(t) = 150 \cos(2\pi 100t + 210°)$ mA

4.16 Given the voltage waveform in Figure P4.1:
a. What is the frequency of the signal?
b. What is the phase shift of the waveform?
c. What is the maximum amplitude of the AC signal?
d. What is the sine function that describes the waveform?
e. What is the cosine function that describes the waveform?

4.17 Given the voltage waveform in Figure P4.2:
a. What is the maximum voltage V_P?
b. What is the frequency, ω, in radians per second?

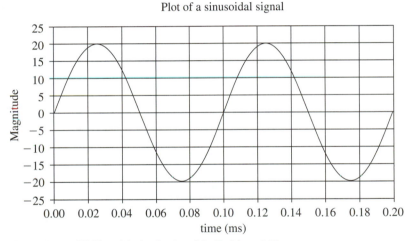

Figure P4.1 AC Signal to be Analyzed in Problem 4.16

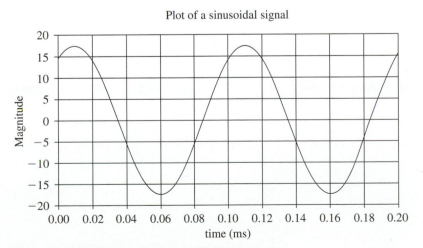

Figure P4.2 AC Signal to be Analyzed in Problem 4.17

c. What is the phase shift in degrees?
d. Write the equation for the voltage $v(t)$ as a sine function.

e. Write the equation for the voltage $v(t)$ as a cosine function.

Power with AC Signals and Effective Values: Completion of the following questions will demonstrate that you have an understanding of the concept of effective value of a steady-state sinusoidal signal and the relationship to power.

4.18 Assume that each of the following signals has been measured across or through a 100 Ω resistor. Determine the RMS value of the signal and the power that would be dissipated:

a. $i(t) = 75.0 \sin(2\pi 100t)$ mA
b. $v(t) = 24 \sin(2\pi 500t)$ V
c. $v(t) = 24 \sin(2\pi 500t + 90°)$ V
d. $i(t) = 150 \cos(2\pi 100t - 90°)$ mA
e. $v(t) = 120 \sin(2\pi 500t - 30°)$ V
f. $i(t) = 150 \cos(2\pi 100t + 210°)$ mA

4.19 Assume that you have measured the following voltages across a 330 Ω resistor. Determine the RMS value of the current through the resistor:

a. $v(t) = 75.0 \sin(2\pi 100t)$ V

b. $v(t) = 24 \sin(2\pi 500t)$ mV
c. $v(t) = 24 \sin(2\pi 500t + 90°)$ V
d. $v(t) = 150 \cos(2\pi 100t - 90°)$ mV
e. $v(t) = 120 \sin(2\pi 500t - 30°)$ V
f. $v(t) = 150 \cos(2\pi 100t + 210°)$ μV

4.20 Assume that you have measured the following currents through a 670 Ω resistor. Determine the RMS value of the voltage across the resistor:

a. $i(t) = 75.0 \sin(2\pi 100t)$ mA
b. $i(t) = 2.4 \sin(2\pi 500t)$ A
c. $i(t) = 24 \sin(2\pi 500t + 90°)$ μA
d. $i(t) = 150 \cos(2\pi 100t - 90°)$ mA
e. $i(t) = 120 \sin(2\pi 500t - 30°)$ A
f. $i(t) = 150 \cos(2\pi 100t + 210°)$ μA

Often it is necessary that you practice analyzing basic circuits in order to understand analysis principles as they apply to practical applications. The following problems will exercise your analysis skills.
AC Signals in Resistive Circuits: Completing analysis of the following problems will demonstrate that you have an understanding of a steady-state sinusoidal signal in resistive series, parallel, and series-parallel circuits.

Series Circuits:

4.21 For the circuit in Figure P4.3, determine (*a*) R_T, (*b*) V_{ab}, and (*c*) P_2.

4.22 For the circuit in Figure P4.4, determine (*a*) V_S, (*b*) I_2, and (*c*) P_T.

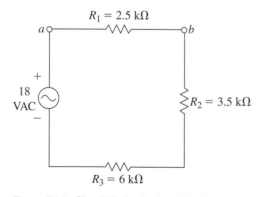

Figure P4.3 Circuit to be Analyzed in Problem 4.21

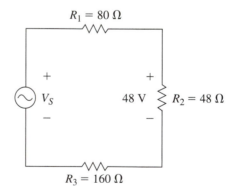

Figure P4.4 Circuit to be Analyzed in Problem 4.22

4.23 For the circuit in Figure P4.5, determine (a) I_T, (b) R_T, (c) V_{ab} using the VDR, and (d) V_{ab} using KVL.

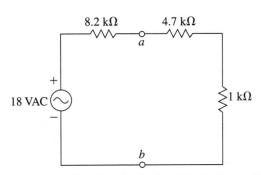

Figure P4.5 Circuit to be Analyzed in Problem 4.23

4.24 For the circuit in Figure P4.6, determine (a) R_T, (b) I_2, (c) P_T, (d) V_{ab} using VDR, and (e) V_{ab} using KVL.

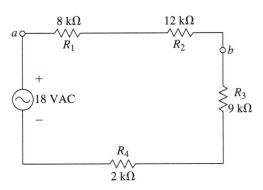

Figure P4.6 Circuit to be Analyzed in Problem 4.24

Parallel Circuits:

4.25 In the circuit shown in Figure P4.7, determine (a) V_T, (b) I_2 using the CDR, and (c) I_2 using Ohm's law.

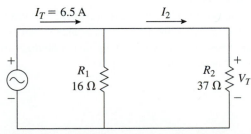

Figure P4.7 Circuit to be Analyzed in Problem 4.25

4.26 For the circuit shown in Figure P4.8, determine (a) G_T, (b) R_T, (c) I_3, and (d) P_T.

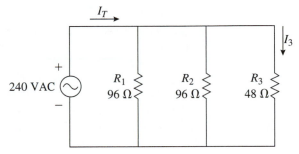

Figure P4.8 Circuit to be Analyzed in Problem 4.26

4.27 For the circuit shown in Figure P4.9, determine (a) R_T, (b) I_2 using the CDR, and (c) V_{ab}.

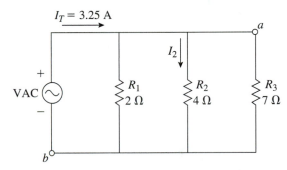

Figure P4.9 Schematic Drawing for Problem 4.27

4.28 For the circuit shown in Figure P4.10, determine (a) G_T, (b) R_T, (c) I_T, and (d) I_3.

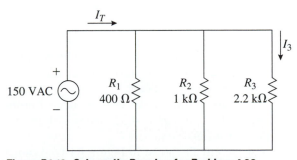

Figure P4.10 Schematic Drawing for Problem 4.28

Series-Parallel Circuits:

4.29 For the circuit shown in Figure P4.11, determine (a) V_3 using the VDR, and (b) I_1 using the CDR.

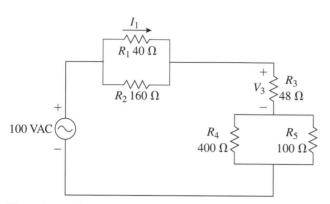

Figure P4.11 Schematic Drawing for Problem 4.29

4.30 For the circuit shown in Figure P4.12, determine (a) R_T, (b) I_2, and (c) V_{ab}.

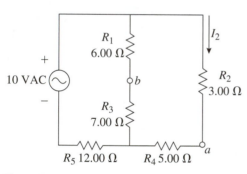

Figure P4.12 Schematic Drawing for Problem 4.30

4.31 For the circuit shown in Figure P4.13, determine (a) V_T and (b) I_1.

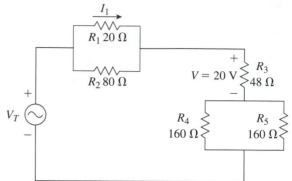

Figure P4.13 Schematic Drawing for Problem 4.31

4.32 For the circuit shown in Figure P4–14, determine (a) R_T, (b) I_T, (c) I_3, and (d) V_{ab}.

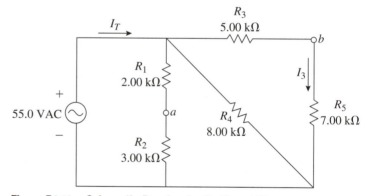

Figure P4.14 Schematic Drawing for Problem 4.32

The next two following problems ask that you use a computer software circuit simulation package to verify the results of prior calculations. The last two problems ask that you demonstrate the use of simulation software in tolerance analysis.

4.33 Verify the results in the odd-numbered problems 4.21 through 4.31 using Electronic Workbench® or PSpice®.

4.34 Verify the results in the even-numbered problems 4.22 through 4.32 using Electronic Workbench® or PSpice®.

4.35 For the circuit shown in Figure P4.14, assume that the resistors are all 5.0% tolerance resistors.

Which resistor will cause the greatest change in I_T and I_3, and is the variance in the resistance positive or negative?

4.36 For the circuit shown in Figure P4.12, assume that the resistors are all 5.0% tolerance resistors. Which resistor will cause the greatest change in V_{ab} if it is out of tolerance by 5%

ELECTRIC FIELDS AND CAPACITORS

As a result of successfully completing this chapter, you should be able to:

1. Describe an electric field and the relationship of electric field to voltage.
2. Define and describe capacitance from charge and energy viewpoints.
3. Calculate the capacitance of ideal parallel-plate capacitors.
4. Determine the total capacitance of series and parallel combinations of capacitors.
5. Describe the fundamental action of capacitors in a DC circuit.
6. Describe the fundamental action of capacitors in an AC circuit.
7. State the capacitor specifications of capacitance and breakdown voltage.
8. Describe the basic construction of a capacitor.

5.1 WHAT IS AN ELECTRIC FIELD? WHAT IS CAPACITANCE?

Visualize the setup in Figure 5.1. Two parallel metal plates are separated with nothing in between. When the switch is closed at an arbitrary time, which will be called $t = 0$, what should the current be immediately after the switch is closed? From the circuit analysis in previous chapters, one would conclude that the current is zero because there is an open circuit in series; If KVL is applied around the series circuit, it appears KVL is not satisfied. All voltages except the source are zero; hence, the sum of the voltages around the closed path is not zero. Somehow this violation must be resolved. The operation of the circuit immediately after the switch is closed must be examined more closely.

Recall the basic law of electric charges: *like charges repel, unlike charges attract.* When the switch is closed, the positive charges (using conventional current) from the positive side of the source will be attracted to the negative side of the source. Hence,

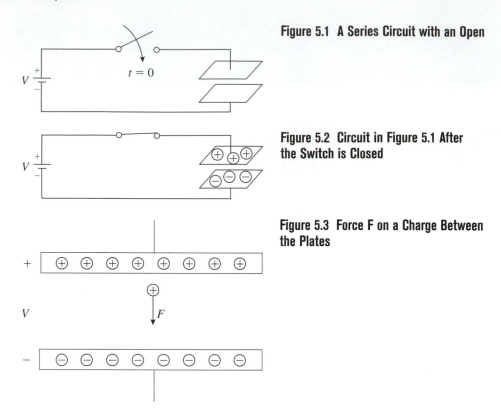

Figure 5.1 A Series Circuit with an Open

Figure 5.2 Circuit in Figure 5.1 After the Switch is Closed

Figure 5.3 Force F on a Charge Between the Plates

some positive charges will flow out of the positive terminal of the battery and will collect on the upper plate. These positive charges on the upper plate will repel positive charges on the lower plate. The positive charges that are repelled from the lower plate are attracted to the negative side of the source. A deficiency of positive charge, which can be viewed as negative charge, remains on the lower plate. The situation shown in Figure 5.2 results.

Something amazing has just happened. What has been created between the plates? A voltage! Recall from Chapter 2 that separated electric charges create an electric potential energy difference. How large will this voltage be? Again, think of KVL: charge must flow until the voltage across the metal plates equals the voltage of the source. This is how KVL is satisfied. Hence, some current had to flow for at least some time to build up the charge on the plates. Then the current went to zero. KVL is always satisfied. The details of how the voltage builds up and the current decreases toward zero will be investigated later. We need to concentrate on the separated metal plates for now.

An expanded side view of the metal plates is shown in Figure 5.3. What if a positive charge were placed between the plates and let go? The positive plate would repel the free charge and the negative plate would attract it. The free charge would accelerate toward the negative plate, i.e., it would gain kinetic energy. Where did this energy come from? It came from the repulsion by the upper plate and the attraction by the lower plate. There is a "force field" between the plates that exerts forces on electric charges. The field as a whole is called an *electric field,* and the quantity that represents the electric field at each point in space is called the *electric field intensity.* The symbol for electric field intensity that is used in this text is $\mathcal{E}$ (script E; do not confuse $\mathcal{E}$ with E, energy). It is drawn as arrows from the positive to the negative plates, as shown in Figure 5.4. However, the field exists everywhere between the plates, not just where the arrows are shown.

This electric field is somehow directly associated with the voltage between the plates: the electric field exists everywhere between the plates, but the voltage is *between* two points, namely the plates. Since the electric field is "spread out" between the plates, one can examine how the electric field intensity might change if the spacing between the plates is changed. If the distance between the plates is increased, but the voltage between the plates is kept constant, then there is less voltage per unit distance (fewer volts/meter) from one plate to the other—see Figure 5.5. This is a clue to the connection between voltage and

Figure 5.4 Representation of an Electric Field $\mathcal{E}$

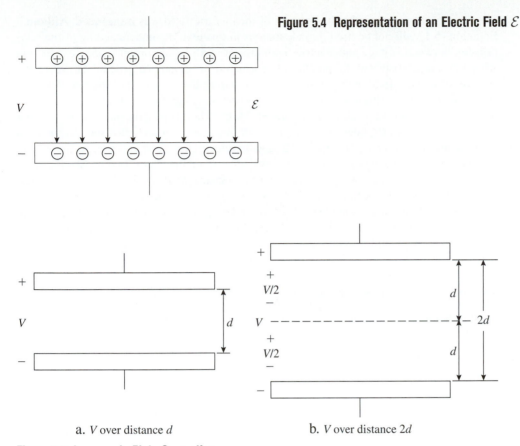

a. V over distance d b. V over distance $2d$

Figure 5.5 Increase in Plate Separation

electric field intensity. Voltage is the electric potential energy difference between two points—distance must be a factor in the relationship between voltage and electric field intensity.

Go through the following thought experiment. If the voltage in both cases in Figure 5.5 is the same, then the energy required to move a free charge from one plate to the other plate must be the same if the same charge is used. But the charge must be moved twice the distance in Figure 5.5b relative to Figure 5.5a. Hence, the energy difference over distance d in Figure 5.5b must be half of that in Figure 5.5a. The force exerted on a free charge that is placed between the plates in Figure 5.5b would be half of that in Figure 5.5a. Then the electric field intensity in Figure 5.5b must be half of that in Figure 5.5a because the force exerted on the charge is proportional to the electric field intensity (from the original statement when $\mathcal{E}$ was introduced: $\mathcal{E}$ is a "force field"). See the trend? The electric field intensity is directly proportional to voltage (higher voltage, more electric potential energy difference, more force on a free charge, hence a larger electric field intensity) and inversely proportional to distance (larger distance, fewer volts per meter, less electric potential energy difference per meter, less force on a free charge, hence a smaller electric field intensity).

The preceding thought experiment suggests the following relationship between electric field intensity and voltage:

$$\mathcal{E} = \frac{V}{d} \tag{5.1}$$

where: $\mathcal{E}$ is the electric field intensity in V/m,

V is the voltage in V, and

d is the plate separation distance in m.

Notice how the units of $\mathcal{E}$ incorporate voltage and distance.

One might wonder why the concept of the electric field was introduced. Although Equation (5.1) will not be used for calculations in this text, the understanding of the relationship between voltage and electric field intensity will be important in the next section when the characteristics of this parallel-plate structure are examined in detail. However, the concept of electric field is important to *understanding* the operation of the parallel-plate structure. Recall how this discussion started: KVL was not satisfied in Figure 5.1 unless current flowed for some time, allowing separated charge to build up on the plates, and a voltage was created across the plates. A consequence of this investigation is that the electric field between the plates is related to the voltage, and this electric field contains energy, since it can do work by moving any free charge that is placed between the plates. In other words, separated charge on the two plates *stores* energy in the electric field that exists between the plates.

Do you realize the significance of the two-plate structure? It is another type of electrical component, and it is called the *capacitor.* Unlike resistance, which *converts* electrical energy into another form of energy, the capacitor *stores* electrical energy in the electric field. This energy could be withdrawn from the capacitor at a later time. The capacitor is a fundamentally different component from resistance, and it will have different but interesting properties that will be examined in this chapter. However, the formal definition of *capacitance* should be established first.

Capacitance is the property of storing electrical energy in an electric field.

A capacitor is a configuration of conductors, usually two conductors that are separated with an insulating material, that stores electrical energy in the electric field.

The capacitor that consists of two parallel plates separated by a distance is only one example of a capacitance, but it is by far the most important example in electronics. So what is the equation that defines capacitance? More charge on the plates creates a stronger electric field that stores more energy, so capacitance is proportional to the amount of charge Q. In fact, many people view capacitance as the capacity to store separated charge, but it is really the capacity to store electrical energy in the electric field. A larger voltage creates more separated charge, so to remove the effect of different voltages, capacitance is the capacity to store separated charge *per volt*:

$$C = \frac{Q}{V} \qquad (5.2)$$

where: C is the capacitance in farads (F),

   Q is the charge in C, and

   V is the voltage in V.

Thus, one farad of capacitance is equal to one coulomb of separated charge stored in the capacitor with one volt applied. Note that uppercase C stands for two different things: the symbol for capacitance and the symbol for the unit coulomb. How does one know which C is being used? As in many situations, the context makes the difference. You must be the one who determines which quantity is being discussed at any given time.

Information Research Exercise 5.1.1   (library and/or web)

The unit of capacitance, the farad, is no accident—check out Michael Faraday. What is Faraday's contribution in connection with capacitance? What are some of Faraday's other contributions?

The equation for capacitance in Equation (5.2) is a fundamental equation. The more charge (and, hence, energy in the electric field) that a structure can store for each volt applied, the larger the capacitance. This equation allows us to reason out what capacitance is. It is also used to formally derive the capacitance of different capacitor structures (such as one tube inside another tube), but this task requires calculus.[1] However, Equation (5.2) is not directly usable for circuit calculations in this form. The equation for the capacitance of a parallel-plate capacitor will be developed in the next section of this chapter.

How much electric energy does a capacitance store? The formal derivation of the result also requires calculus.[2] The result will be stated and intuitively justified. The electrical energy stored in the electric field of a capacitance is:

$$E = \tfrac{1}{2}CV^2 \qquad (5.3)$$

where E is the stored energy in the electric field in J,

       C is the capacitance in F, and

       V is the voltage across the capacitance in V.

The following reasoning can be used to intuitively justify Equation (5.3). If the capacitance of the capacitor is larger, it can store more charge on the plates and can store more energy in the electric field. Hence, the electric energy is proportional to C. The proportionality of electric energy to the voltage squared for a capacitor can be thought of in a manner similar to the electric power proportionality to voltage squared for a resistance. The difference is that energy is stored in a capacitor whereas energy is converted in a resistance. Another energy viewpoint is to substitute Q/V for the C in Equation (5.3):

$$E = \frac{1}{2}CV^2 = \frac{1}{2}\left(\frac{Q}{V}V\right)V = \frac{1}{2}QV \qquad (5.4)$$

Capacitance was described as the property of storing electric energy in an electric field. It can be seen in Equation (5.4) that the energy stored is proportional to the product of charge stored and voltage. Thus, the connection between the charge stored on the plates in Equation (5.2) and the property of storing electric energy is now established.

As with any new component that is introduced, a schematic symbol for the capacitor is needed. There are three symbols for capacitors, as shown in Figure 5.6. The symbol shown in Figure 5.6a is reserved for capacitors that can have the voltage applied with either polarity. Some capacitors, by their construction, must have the proper voltage polarity applied, and this is shown with a "+" on the terminal that must have the positive side of the voltage, as shown in Figure 5.6b. These capacitors are called polarized capacitors.

> **Warning:** *Application of the wrong voltage polarity to a polarized capacitor is dangerous.*

If the improper voltage polarity is applied to a polarized capacitor, it will usually fail and may even explode! This is dangerous, so be especially careful when using polarized capacitors. More will be said about polarized capacitors in Section 5.5. The schematic symbol in Figure 5.6c is used for capacitors that may be adjusted (a variable capacitor). Samples of adjustable capacitors are shown in Figure 5.6d.

The importance of capacitance resides fundamentally in the property that capacitors store electrical energy in the electric field. This property will cause the capacitor to behave differently than a resistance in an electric circuit (this is covered in Section 5.4). This different circuit behavior means that different electronic effects can be obtained. For example, capacitors can be used to make a channel selector in a television or can be used to reduce AC current levels in a circuit. Be assured that there are several interesting effects that will unfold in this text in connection with capacitors.

---

[1] See Hayt and Buck, *Engineering Electromagnetics,* 6th ed., Ch. 5, for example.
[2] See Nilsson and Riedel, *Electric Circuits,* 6th ed., Ch. 6, for example.

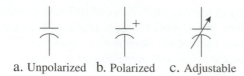

a. Unpolarized   b. Polarized   c. Adjustable

Figure 5.6 Schematic Symbols for Capacitors (a–c) and samples (d)
Part d Photo Courtesy of Sprague-Goodman Electronics, Inc.

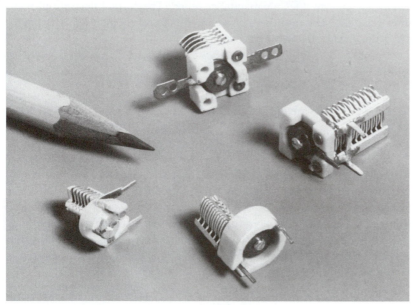

d. Samples of adjustable capacitors

5.2 CAPACITANCE OF A PARALLEL-PLATE CAPACITOR

The capacitor is a fundamental electric component. It stores electric charge, which really means that it stores a difference of positive charges on one plate with respect to the other plate. This separation of charge creates an electric field between the capacitor plates, which is measured as a voltage. Hence, the capacitor stores electrical energy in the electric field between the plates. So where does one go from here? Hopefully you are thinking about questions like "How much energy?" or "How much charge?" The capacitance of the capacitor must be determined in order to calculate energy and charge. In this section, an intuitive reasoning approach will be used to arrive at the capacitance of a parallel-plate capacitor. The equation for this capacitance will be derived in a future course, such as physics and/or electromagnetics.

The geometry of the capacitor must be designated first. The assumption of two parallel plates of equal area is made (also, a derivation would require that the plate area is much greater than the plate spacing). The area of one plate is labeled *A*, and the distance between the plates is labeled *d*, as shown in Figure 5.7.

Figure 5.7 Parallel-Plate Capacitor Geometry

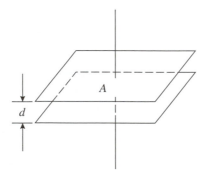

What if the plate area is larger? More charge could "fit" onto the plates, or there is more area for charges to repel charges on the other plate. Alternatively, there is more volume between the plates for the electric field, so more energy can be stored in that electric field. In short, the capacitance is directly proportional to the plate area:

$$C \propto A \qquad (5.5)$$

where the "$\propto$" means "is proportional to." On the other hand, if the distance increases between the plates, the force by the charges on one plate from the charges on the other plate decreases (recall the electric field intensity discussion related to Figure 5.5, and that the force on a charge is proportional to $\mathcal{E}$). Hence, capacitance is inversely proportional to the plate separation distance:

$$C \propto \frac{1}{d} \qquad (5.6)$$

These two proportions can be expressed as one combined proportion:

$$C \propto \frac{A}{d} \qquad (5.7)$$

Alternatively, Equation (5.7) can be expressed as an equation if a proportionality constant, k, is used:

$$C = k\frac{A}{d} \qquad (5.8)$$

What should be the next question at this point? How is k determined? (That's the question.) What has not been considered yet? The material that is between the capacitor plates has not been addressed. In the previous section, air (or vacuum) was assumed to be in between the plates. For now, assume that is the case. Then if the capacitor dimensions (A and d) and the capacitance are carefully measured, then the experimental value of the proportionality constant can be solved for in Equation (5.8). This task has been performed by past scientists, and the proportionality constant for air (and vacuum) is approximately:

$$k = 8.854 \times 10^{-12} \qquad (5.9)$$

What are the units of k? One should arrive at farads/meter (F/m) in SI units. This number turns out to be a special constant in nature, and it is given a special symbol and two names. The symbol is the lowercase Greek letter epsilon (ϵ; do not confuse ϵ with the symbol for electric field intensity: $\mathcal{E}$), and the names are *permittivity* or *dielectric constant*. Technically, this constant is assigned to no material, i.e., vacuum. The value of permittivity for air is approximately equal to that for vacuum. To designate this universal constant for the permittivity of vacuum, a subscript o is attached to the epsilon:

$$\epsilon_o = 8.854 \times 10^{-12} \text{ F/m} \qquad (5.10)$$

Then the equation for the capacitance of a parallel-plate capacitor with vacuum between the plates is:

$$C = \epsilon_o \frac{A}{d} \qquad (5.11)$$

Is there something disturbing about the capacitor setup so far? Think about the lack of material between the plates. What holds the plates in place? If the capacitor is bumped, could the capacitor plates move and short out? This situation is obviously not desirable from a practical standpoint. An insulating material is often placed between the plates to hold the plates in position with a constant spacing. The insulating material also has electrical properties that affect the dielectric constant. How does that happen? Is this beneficial or harmful? An investigation is in order.

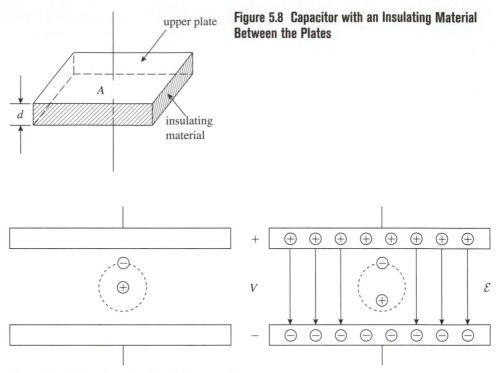

Figure 5.8 Capacitor with an Insulating Material Between the Plates

Figure 5.9 Effect of an Electric Field on an Atom

If an insulating material is placed between the capacitor plates, as shown in Figure 5.8, then the material is immersed in the electric field because of the charges on the capacitor plates. The insulating material is made up of atoms and molecules. How does the electric field affect them? A simple atom, situated between two plates, is sketched in Figure 5.9a. If the atom is exposed to an electric field, as in Figure 5.9b, the positive nucleus is attracted toward the negative plate and the electron orbit is shifted toward the positive plate. Although the atom is still electrically neutral as a whole, one end of the atom is, on the average, more positive than the other end. Equivalently, one end of the atom is more negative than the other end. The atom in this state is described to be polarized and is now called a *dipole*. The insulating material between the capacitor plates is called a *dielectric* (*di* meaning "two", and *electric* for "electric poles"—the atom has two electric poles: one positive and one negative).

Several dipoles are sketched in Figure 5.10. The positive ends of the dipoles "cancel" the negative ends of adjacent dipoles for atoms in the middle of the dielectric. However, for atoms next to the plates, the ends are not cancelled, and a net polarity exists for the dielectric when viewed as a whole, as shown in Figure 5.11. What does this net polarization of the dielectric create? It looks like separated charge, which creates its own electric field. What is the direction of the electric field because of the dielectric? It is *opposite* of the

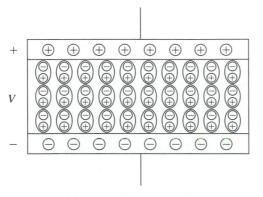

Figure 5.10 Dipoles in the Dielectric—All Dipoles Shown

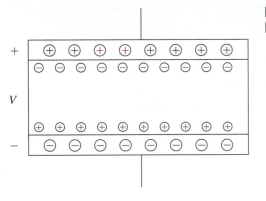

Figure 5.11 Dipoles in the Dielectric—Net Dipoles after Internal Cancellation

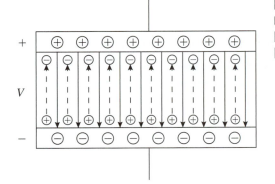

Figure 5.12 Electric Fields Due to the Charges on the Plates (Solid Lines) and Due to the Net Polarization of the Dielectric (Dashed Lines)

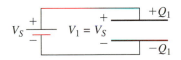

a. Air dielectric

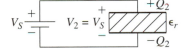

b. With dielectric material

Figure 5.13 Capacitor Comparison With and Without Dielectic Material

electric field because of the charges on the plates—see Figure 5.12. The electric field due to the net polarization of the dielectric reduces the total electric field between the capacitor plates (assuming no more charge flows onto the capacitor plates). Thus, what should happen to the voltage across the plates? Should it decrease? Does this decrease in voltage mean that a dielectric lowers capacitance? Don't answer too quickly. This situation also needs to be investigated further.

Visualize a voltage source V_S connected across a parallel plate capacitor with an air dielectric, as shown in Figure 5.13a. Label $+Q_1$ on one plate, $-Q_1$ on the other plate, and V_1 between the plates. Now, redraw the previous case with a dielectric between the plates, as shown in Figure 5.13b. What is the voltage across the capacitor? It still must be V_1 because of the source connection directly across the capacitor (KVL must hold true at all times). What is the charge on the plates? The electric field due to the dipoles opposes the electric field due to the charge on the capacitor plates, so *more* charge must flow onto the capacitor plates to overcome the dipole opposition. Thus, $Q_2 > Q_1$. More charge on the capacitor plates means that more electric energy is stored in the capacitor, but it is not seen as an increased voltage across the capacitor. Part of the energy is stored in the electric field due to charges on the capacitor plates and part of the energy is stored in the dipoles due to the dielectric. Thus, a dielectric between capacitor plates results in more energy stored in the capacitor and more charge stored on the capacitor plates. Has the capacitance increased or decreased? It has increased. Note that this explanation is consistent with Equation (5.3): more energy is stored with a larger capacitance (all else being the same).

How is the effect of the dipoles in the dielectric incorporated into the capacitance expression for a parallel-plate capacitor? The dielectric constant is multiplied by another constant,

TABLE 5.1	dielectric	ϵ_r (F/m)
Dielectrics and Approximate Relative Permittivity Values	vacuum	1
	air	1.0006
	ceramic	10's to 1000's
	mica	5
	Mylar	3
	paper, paraffined	2.2–2.5
	Teflon	2.1

called the *relative permittivity* or the *relative dielectric constant,* that is the factor by which the permittivity of the material is greater than the permittivity of free space. In symbol form:

$$\epsilon = \epsilon_o \epsilon_r \tag{5.12}$$

where ϵ = dielectric constant = permittivity (F/m)

ϵ_o = free space (vacuum) permittivity = 8.854×10^{-12} F/m

ϵ_r = relative permittivity = relative dielectric (dimensionless)

There is a caution to be noted: some people say "dielectric constant" when they mean "relative dielectric constant." Some common dielectrics that are used in capacitors and the corresponding approximate relative dielectric constant values are listed in Table 5.1. The permittivity value for any manufactured material depends on the manufacturing process, the level of impurities, etc. Hence, these values are not exact, and approximate values are listed in the table.

The complete expression for the capacitance of a parallel-plate capacitor can now be stated by inserting the complete expression for permittivity into Equation (5.11):

$$\boxed{C = \frac{\epsilon A}{d} = \epsilon_o \epsilon_r \frac{A}{d}} \tag{5.13}$$

Notice that Equation (5.2) [$C = Q/V$] is useful for reasoning and understanding capacitor behavior, whereas Equation (5.13) is useful for actual capacitance predictions, as shown in the next example.

Example 5.2.1

Determine the capacitance of a parallel-plate capacitor with a Teflon dielectric, a plate spacing of 0.1 mm, and a plate surface area of 3 in.^{2}

Given: ϵ_r = 2.1 (Teflon)

d = 0.1 mm

A = 3 in.^{2}

Desired: C

Strategy: $C = \epsilon_o \epsilon_r \dfrac{A}{d}$

Solution:

$$C = \epsilon_o \epsilon_r \frac{A}{d} = \left(8.854 \times 10^{-12} \frac{F}{m}\right)(2.1)\frac{(3 \text{ in.}^2)\left(\dfrac{25.4 \text{ mm}}{\text{in.}}\right)^2 \left(\dfrac{1 \text{ m}}{1000 \text{ mm}}\right)}{0.1 \text{ mm}}$$

$$= 3.5987 \times 10^{-10} = 360. \text{ pF}$$

Notice that even 0.1 mm of plate spacing results in a small capacitance. Thus, plate spacing is often smaller than 0.1 mm!

5.3 COMBINATIONS OF CAPACITORS

As with resistances, it is possible to combine capacitors in series, parallel, and series-parallel combinations. For example, this is done to obtain a nonstandard value of capacitance, or to model a complex electronic device with several capacitances. The equations for the total capacitance of series and parallel combinations of capacitors will be developed not only for the equations themselves but for understanding the concepts that underlie them.

The first case is capacitors in parallel. Three capacitors are shown in parallel in Figure 5.14, but N parallel capacitors could be shown in general. What can be said about the voltage across each capacitor? They are in parallel, so the voltage across each capacitor is the same. What can be said about the total charge stored on the capacitors as a group? The total stored charge should be the sum of the stored charges of the individual capacitors (likewise with the stored electrical energies). Hence, with the same voltage, more stored charge and electrical energy means that the total capacitance increases when capacitors are in parallel.

How can the expression for total capacitance of parallel capacitors be derived? Three parallel capacitors will be assumed, but there can be N parallel capacitors in general. Start with the known conditions:

$$V_T = V_1 = V_2 = V_3 \tag{5.14}$$

$$Q_T = Q_1 + Q_2 + Q_3 \tag{5.15}$$

But the stored charge is related to capacitance and voltage by $C = Q/V$:

$$C_T V_T = C_1 V_1 + C_2 V_2 + C_3 V_3 \tag{5.16}$$

But the voltage is the same across parallel capacitor [Equation(5.14)]:

$$C_T V = C_1 V + C_2 V + C_3 V \tag{5.17}$$

Divide both sides of Equation (5.17) by V:

$$C_T = C_1 + C_2 + C_3 \tag{5.18}$$

which is the equation for the total capacitance of three capacitors in parallel. In general, for N parallel capacitors:

$$C_T = \sum_{n=1}^{N} C_n \tag{5.19}$$

Thus, add the capacitances of capacitors in parallel. An alternate viewpoint is to visualize the total plate area increasing as capacitors are added in parallel. Increased plate area increases capacitance.

The second case is capacitors in series. Three capacitors are shown in series in Figure 5.15, but N series capacitors could be shown in general. What can be said about the charge on each plate as compared to the charge on the adjacent plate of the next capacitor? If one

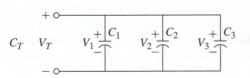

Figure 5.14 Three Capacitors in Parallel

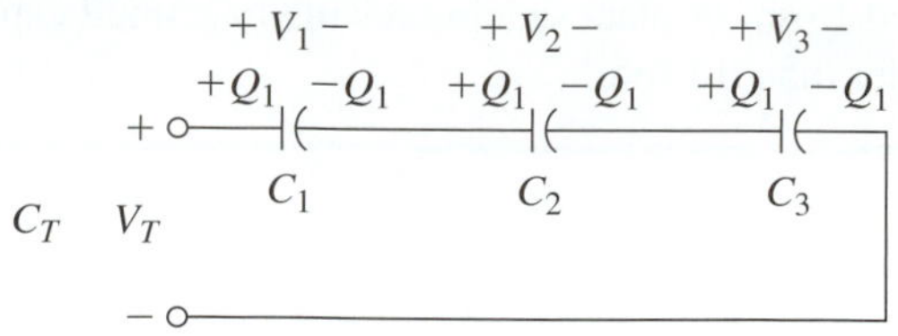

$$+ V_1 -\qquad + V_2 -\qquad + V_3 -$$

Figure 5.15 Three Capacitors in Series

charge is added to the leftmost plate, one charge is repelled from the opposite plate of the same capacitor onto the leftmost plate of the next capacitor, and so on. Thus, the total charge on one plate must be the same but of the opposite sign of the total charge on the adjacent plate of the next capacitor. The total charge on the adjacent plates of each pair of capacitors must be zero—there is no path for a net charge to enter the nodes between capacitors. The charge on the plates of a given capacitor are equal (but opposite in sign); hence, the charge on each plate of the total capacitance is the same as the charge on each plate of each capacitor, as sketched in Figure 5.15.

What can be said about the voltage across each capacitor as compared to the source voltage (think of series circuits)? By KVL, the total voltage must be the sum of the individual capacitor voltages. With a higher voltage across the total capacitance and the same charge on the total capacitance as on each of the capacitors, then the total capacitance must be less than any of the individual capacitances. Start with the known conditions to derive the total capacitance:

$$Q_T = Q_1 = Q_2 = Q_3 \tag{5.20}$$

$$V_T = V_1 + V_2 + V_3 \tag{5.21}$$

But each voltage is related to capacitance and stored charge by $C = Q/V$:

$$\frac{Q_T}{C_T} = \frac{Q_1}{C_1} + \frac{Q_2}{C_2} + \frac{Q_3}{C_3} \tag{5.22}$$

But the charge is the same for each series capacitor [Equation (5.20)]:

$$\frac{Q}{C_T} = \frac{Q}{C_1} + \frac{Q}{C_2} + \frac{Q}{C_3} \tag{5.23}$$

Divide both sides of Equation (5.23) by Q:

$$\frac{1}{C_T} = \frac{1}{C_1} + \frac{1}{C_2} + \frac{1}{C_3} \tag{5.24}$$

Solve for C_T. The equation for the total capacitance of three capacitors in series is:

$$C_T = \frac{1}{\dfrac{1}{C_1} + \dfrac{1}{C_2} + \dfrac{1}{C_3}} \tag{5.25}$$

or, in general, for N series capacitors:

$$C_T = \frac{1}{\displaystyle\sum_{n=1}^{N} \frac{1}{C_n}} \tag{5.26}$$

Thus, for series capacitor, the total capacitance is less than the smallest individual capacitance. An alternative viewpoint is to view capacitors in series like a capacitor with a larger plate separation. A larger plate separation decreases capacitance.

Finally, consider the case of a series-parallel combination of capacitors. How would the total capacitance be determined? Use series-parallel analysis with the total capacitance relations in Equations (5.19) and (5.26), as illustrated by the next example.

___ **Example 5.3.1** _____

Determine the total capacitance of the capacitor combination shown in Figure 5.16.

Given: circuit in Figure 5.16

Desired: C_T

Strategy: $C_T = C_1$ in series with $C_2 \parallel C_3$

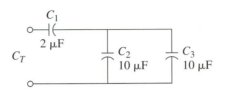

Figure 5.16 Capacitor Combination for Example 5.3.1

Solution: _____

$$C_2 + C_3 = 10\ \mu F + 10\ \mu F = 20\ \mu F$$

$$C_T = \cfrac{1}{\cfrac{1}{C_1} + \cfrac{1}{(C_2 + C_3)}} = \cfrac{1}{\cfrac{1}{2} + \cfrac{1}{20}} = 1.82\ \mu F$$

Notice that the smaller series capacitance dominates the total capacitance in this combination.

5.4 THE FUNDAMENTAL ACTION OF CAPACITORS IN DC CIRCUITS

Now that this new component, the capacitor, has been introduced, how does it behave in circuits? Recall that the behavior of resistive circuits was straightforward: more voltage meant more current, and vice versa, as expressed by Ohm's law. The DC voltage and current are directly proportional. Consider the case when the DC voltage is first turned on. The switch is closed at $t = 0$, which is an arbitrarily set reference time. In a resistive circuit, the current and the voltage track each other, as sketched in Figure 5.17.

Now consider the case of a circuit with capacitance, as shown in Figure. 5.18. The resistance is shown because it will be significant, and every circuit has some resistance (perhaps just wire resistance). The capacitor looks like an open to a steady-state DC signal. Again, consider the case when the DC voltage is first turned on. When the switch is first closed, current flows and charge begins to build on the capacitor plates. Thus, the voltage across the capacitor builds up from zero volts—it does not jump up to a voltage value as it did in the resistive circuit case. Note that just after the switch is closed, all of the voltage is

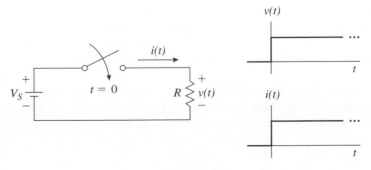

Figure 5.17 Resistive Circuit When DC is Turned On

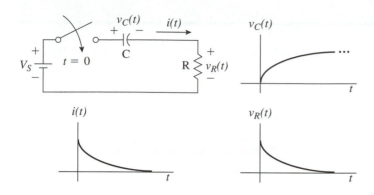

Figure 5.18 Circuit with Capacitance When the DC is Turned On

across the resistance and, by Ohm's law, the current will be at its maximum value at that instance in time. Despite the "open," the current flows in the circuit for some amount of time.

The capacitor voltage polarity opposes the source voltage polarity, so as the charge builds up on the capacitor plates and the capacitor voltage increases, then by KVL, there is less voltage across the resistor. But less voltage across the resistor means less current, as shown in Figure. 5.18. The time span over which the voltages and current are changing is called the *transient*. Eventually the capacitor voltage equals the source voltage, there is no voltage across the resistor, and the circuit current is zero. The circuit has reached *steady state*, where the capacitor now looks like an open circuit. An interesting observation is that at any given time, the voltage across the capacitor plus the voltage across the resistor equals the source voltage, as expected by KVL.

What is learned from this reasoning? The capacitor voltage did not "jump up," but instead took time to change. Why? Because it took time to move charge onto the plates, and the capacitor voltage is directly related to the separated charge on the plates ($V = Q/C$). Hence, the fundamental action of capacitors in circuits is:

The voltage across a capacitor cannot change instantaneously.

An alternative, equivalent statement is:

Capacitors oppose an instantaneous change in voltage.

where the "voltage" is assumed to be the voltage across the capacitor. The resistance in the circuit plays a role in this charging of the capacitor because the resistance will limit the current. Predictions of DC transients in circuits with capacitance (capacitive circuits) will be made in Chapter 7. For now, the new situation that voltages and currents are not constant, at least for a while (during the transient), is emphasized when there is a capacitor in the circuit.

5.5 THE FUNDAMENTAL ACTION OF CAPACITORS IN AC CIRCUITS

Consider the case of an AC signal.[3] The voltage and current are always changing. Hence, the capacitor will always be reacting. This case is very special. Recall that phase shift was stated and determined in Chapter 4, but the reason that a sine wave might have a phase shift was not explained. We are now ready to understand why phase shift happens. Start with the AC circuit shown in Figure 5.19. The AC source delivers a sinusoidal voltage and current. The voltage across and the current through the resistance always are in-phase, as established in Chapter 4.

For the capacitor in Figure 5.19, however, the voltage can only change if the charge on the capacitor plates changes. Current is the rate of charge flow, not just the number of charges. Hence, current must be "on" first before any charges can build up (or leave) the

_____

[3] The discussion in this section is repeated in Chapter 8 for those who will study AC later.

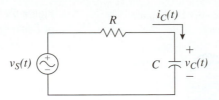

Figure 5.19 An AC Circuit with a Capacitor

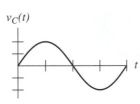

Figure 5.20 AC Voltage and Current Waveforms for a Capacitor

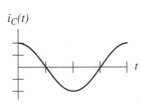

capacitor plates. Thus, one would expect the current to *lead* the voltage, i.e., the current is advanced in phase relative to the voltage waveform, as shown in Figure 5.20. The current leads in order to deposit charge on (or remove charge from) the capacitor plates, and the voltage changes in response to the change of charge on the plates. Notice that when the current is zero, the capacitor voltage has leveled out at a maximum point (in either polarity)—zero current means no charge movement on to or off of the plates, and this means the voltage should not be changing (at that instant). Conversely, when the current is maximum, the voltage is changing the fastest—note the slope of the voltage waveform at the times when current is a maximum.

How far (in terms of angle) should the AC current lead the voltage for a capacitor? The answer is remarkable: 90°. Why would this angle be such a "nice" value? The answer comes from power and energy for the capacitor, which is covered in Chapter 10. An ideal capacitor does not dissipate energy, it stores energy. For this to happen, the mathematics of AC voltage, current, and power require the voltage and current to be 90° out-of-phase.

In summary, the capacitor opposes an instantaneous change of voltage. This effect is seen in a capacitive circuit when a DC signal is applied. The capacitor voltage will build up or decay over time. This effect is also seen in a capacitive circuit with an AC signal. The capacitor current will lead the capacitor voltage by 90°.

5.6 CAPACITOR TYPES

Capacitors may be literally constructed with two parallel plates separated by a dielectric, but this is often impractical because of the large plate area requirements for the desired value of capacitance. One way to increase the plate area but keep the capacitor size manageable is to "roll" the plates together with the dielectric in between, attach leads, and cover it with an insulating layer, as sketched in Figure 5.21. Another method is to sandwich several layers, as sketched in Figure 5.22. The capacitance ranges and polarity restrictions differ in these various constructions (see Information Research Exercise 5.6.1).

Another manner in which these various capacitor constructions differ is the *breakdown voltage* (or maximum voltage rating). The breakdown voltage is the maximum voltage that can be applied to the capacitor before the dielectric "breaks down," i.e., is no longer an insulator. The breakdown voltage mostly depends on the electrical properties of the dielectric material and the thickness of the dielectric, which is normally the plate spacing. The relevant electrical property of the material is called the dielectric breakdown

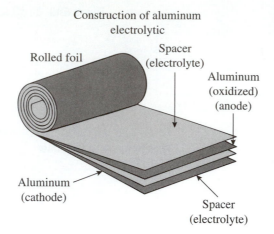

Construction of aluminum electrolytic

Figure 5.21 "Rolled-Up" Type of Capacitor Construction
Photo courtesy of KEMET Corporation

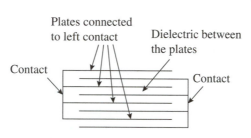

Figure 5.22 Multilayer Type of Capacitor Construction

strength. It is the electric field intensity at which the dielectric starts conducting. Recall that the units of electric field intensity are V/m. The ideal breakdown voltage is determined by both the dielectric breakdown strength and the thickness of the dielectric:

breakdown voltage = (dielectric breakdown strength) (dielectric thickness)

However, the dielectric breakdown strength of capacitor dielectric strongly depends on the manufacturing process, impurity levels, and so on. The dielectric breakdown strength and the thickness of the dielectric are already determined in any given capacitor, so the user relies on the breakdown voltage specification that is given by the manufacturer when selecting capacitors.

There is another practical consideration in capacitors: leakage resistance. Dielectrics and manufacturing processes are not perfect. There are impurities in materials, for example. Most capacitors have negligible amount of current that flows between the plates, but sometimes it is not negligible, especially in circuits that contain high resistance values. Then the capacitor must be represented in circuit analysis by an ideal capacitor in parallel with a large resistance. This resistance is called the leakage resistance. It is the purpose here to make you aware that leakage resistance exists and that it could be significant in some applications.

Information Research Exercise 5.6.1   (web, manufacturers' catalogs, and/or library) _____

For each capacitor type, determine

a.   the construction and materials,
b.   typical capacitance value ranges,
c.   typical voltage breakdown ratings, and
d.   polarity restrictions.

1. paper
2. Mylar
3. tantalum
4. mica
5. ceramic

6. electrolytic
7. air adjustable
8. trimmer
9. chip (surface mount)

How are capacitors used in circuits? One use has been suggested in Section 5.4. The capacitor can be used to control how fast a voltage rises in a circuit. The transient in RC circuits is examined in more detail in Chapter 7. Capacitors can be used to block steady-state DC signals. Capacitors are used in AC circuits to perform several functions, such as filtering (Chapters 15 and 16) and power factor correction (Chapter 10). In short, the capacitor is another very useful component in electronics and will be used throughout the remainder of this textbook as well as in many of your subsequent courses in electronics.

CHAPTER REVIEW

5.1 **What Is an Electric Field? What Is Capacitance?**
- Like charges repel and unlike charges attract: this phenomenon is explained using the electric field, which is a force field that affects electric charges.
- An electric field exists between the separated charges on the capacitor plates.
- Capacitance is the property of storing electrical energy in an electric field.
- Capacitance is defined as the separated charge that is stored on the capacitor plates per volt across the capacitor plates ($C = Q/V$).
- The energy stored in the electric field of a capacitor is $E = \frac{1}{2}CV^2$ and is proportional to the product of charge stored and voltage.

5.2 **Capacitance of a Parallel-Plate Capacitor**
- The capacitance of a parallel-plate capacitor is directly proportional to the plate area and the dielectric constant, and it is inversely proportional to the plate separation:

$$C = \frac{\epsilon A}{d} = \epsilon_o \epsilon_r \frac{A}{d}$$

- The dielectric constant (permittivity) expresses the effect of the dielectric material that is between the plates on the capacitance.
- Dipoles in the dielectric set up an electric field that opposes the applied electric field. This effect causes more charge to be stored on a capacitor. Hence, capacitance is directly proportional to the dielectric constant.
- The relative dielectric constant (relative permittivity) is the factor by which the permittivity of the dielectric material is greater than the permittivity of air.

5.3 **Combinations of Capacitors**
- The capacitance of parallel capacitors is the sum of the individual capacitances because more charge is stored for a given applied voltage.
- The capacitance of series capacitors is less than the capacitor with the least capacitance because the same charge is on each capacitor plate, but the voltage is divided among the capacitors.

5.4 **The Fundamental Action of Capacitors in DC Circuits**
- Capacitors oppose an instantaneous change of voltage.
- When DC is switched on in a circuit with capacitance, the voltages and currents in the circuit go through a transition from the initial values to the DC steady-state values. This transition is called the (DC) transient.

5.5 **The Fundamental Action of Capacitors in AC Circuits**
- Capacitors oppose an instantaneous change of voltage.
- When AC is applied to a circuit with capacitance, there is a 90° phase shift between the AC voltage and the current: current leads voltage by 90°. The current must "come first" to change the charge on the plates, which changes the voltage across the capacitor.

5.6 **Capacitor Types**
- There are several capacitor types and constructions that are used in different applications. There are several dielectric materials commonly used.
- Capacitance and breakdown voltage are important specifications for a capacitor.
- The leakage resistance of a capacitor may be important in certain applications.

HOMEWORK PROBLEMS

Proper completion of the following questions will demonstrate your ability to describe an electric field and the relationship of electric field to voltage and to define and describe capacitance from charge and energy viewpoints.

5.1   What is formed when two parallel-plate conductors are connected to a voltage source?
   a. an electric field, a force field, or a gravitational field?
   b. Briefly explain your answer to *a*.
   c. What creates this field?
   d. What effect does the distance of separation have on this field?

5.2   Based on the material presented in this chapter, would you consider a switch in the open position to be a capacitor? Justify your response.

5.3   What is significant about the "open" in a capacitor?

5.4   Both a capacitor and a resistor are considered to be electrical components. What is the difference between the two?

5.5   What does a capacitor store? How?

5.6   The energy stored in a capacitor is stored in what form?

Proper completion of the following questions will demonstrate your understanding of the physical factors that affect the amount of capacitance in a capacitor. It will also demonstrate your ability to calculate the capacitance of ideal and practical parallel-plate capacitors.

5.7   What physical parameters affect the value of a capacitor?

5.8   How does the material that is between the plates of a capacitor affect the amount of capacitance?

5.9   A capacitor has a plate area of 10 in.^{2}, a plate separation of 0.1 mm, and a Teflon dielectric. Determine the capacitance of this capacitor.

5.10   A 10 μF capacitor has a plate separation of 0.05 mm and a Mylar dielectric. What is the plate surface area in square feet?

5.11   A sheet of mica is available in a thickness of 0.02 in. Design a capacitor with a 100 pF capacitance. Express the plate area in square centimeters.

5.12   A parallel-plate capacitor has a plate separation of 0.025 mm and a ceramic dielectric with $\epsilon_r = 250$. Determine the plate surface area in square meters if $C = 10$ μF.

5.13   A 10 μF capacitor has 12 V applied. What is the separated charge stored on the plates?

5.14   A 0.33 μF capacitor has 200 μC of separated charge on the plates. What is the voltage across the capacitor?

5.15   There are 48 V applied to a 47 μF capacitor. What is the separated charge stored on the plates?

5.16   There are 5.00×10^{-8} C of charge stored on a 1.00 μF capacitor. What is the voltage across the capacitor?

5.17   A 100 nF capacitor has 250 μC of separated charge on the plates. What is the voltage across the capacitor?

5.18   Are the calculations for problems 5.13 through 5.17 practical? Why or why not? (Hint: Which quantities can or cannot be readily measured.)

Proper completion of the following questions will demonstrate your ability to determine the total capacitance of series, parallel, and series-parallel combinations of capacitors.

5.19   Determine the total equivalent capacitance of the capacitor combination shown in Figure P5.1.

5.20   Determine the total equivalent capacitance of the capacitor combination shown in Figure P5.2.

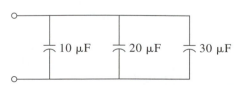

Figure P5.1 Capacitor Combination to be Evaluated in Problem 5.19

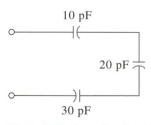

Figure P5.2 Capacitor Combination to be Evaluated in Problem 5.20

5.21 Determine the total equivalent capacitance of the capacitor combination shown in Figure P5.3.

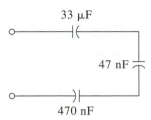

Figure P5.3 Capacitor Combination to be Evaluated in Problem 5.21

5.22 Determine the total equivalent capacitance of the capacitor combination shown in Figure P5.4.

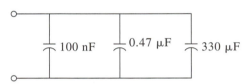

Figure P5.4 Capacitor Combination to be Evaluated in Problem 5.22

5.23 Determine the total equivalent capacitance of the capacitor combination shown in Figure P5.5.

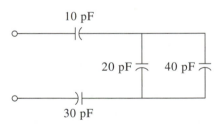

Figure P5.5 Capacitor Combination to be Evaluated in Problem 5.23

5.24 Determine the total equivalent capacitance of the capacitor combination shown in Figure P5.6.

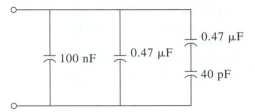

Figure P5.6 Capacitor Combination to be Evaluated in Problem 5.24

5.25 Determine the total equivalent capacitance of the capacitor combination shown in Figure P5.7.

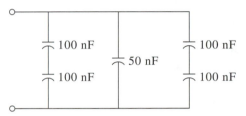

Figure P5.7 Capacitor Combination to be Evaluated in Problem 5.25

5.26 Determine the total equivalent capacitance of the capacitor combination shown in Figure P5.8.

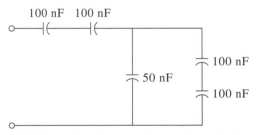

Figure P5.8 Capacitor Combination to be Evaluated in Problem 5.26

Proper answers to the following questions will demonstrate your ability to describe the fundamental action of capacitors in DC and AC circuits.

5.27 What does a capacitor oppose an instantaneous change of? Why?

5.28 What is the difference between the DC transient response and the DC steady-state response of a capacitor?

5.29 What role does a resistor play in a capacitive circuit?

5.30 Describe the change in voltage across a capacitor, i.e., does it change instantaneously?

5.31 What is the phase relationship between the AC voltage and the current for a capacitor?

5.32 Why is there a difference in the phase relationship between the AC voltage and the current for a capacitor?

Proper answers to the following questions will demonstrate your understanding of the specifications of a capacitor, including the breakdown voltage and its basic construction.

5.33   What is the meaning of breakdown voltage as it applies to a capacitor?

5.34   Why is the breakdown voltage an important specification for a capacitor?

5.35   What are the basic physical configurations of a capacitor?

5.36   Why is it important to know if a capacitor is polarized?

5.37   What is the leakage resistance of a capacitor?

5.38   When is the leakage resistance of a capacitor significant? When can it be ignored?

MAGNETIC FIELDS AND INDUCTORS

6.1 WHAT IS MAGNETISM? WHAT IS ELECTROMAGNETISM?

Imagine two bar magnets. Each end of the bar magnet has a different *pole:* one end is a north (N) pole and the other end is a south (S) pole. If two north poles are moved toward each other, what would be the force on the poles: attraction or repulsion? Recall the basic law of magnetic poles: *like poles repel, unlike poles attract.* Hence, the poles would repel each other.

The electric field in capacitors was represented with solid lines and arrows. The lines started on a positive charge and ended on a negative charge, as shown in Figure 6.1a. In a similar manner, the magnetic field is represented with dashed lines and arrows, starting on an N pole, and ending on an S pole, as shown in Figure 6.1b. The magnetic poles in this figure originate from the permanent magnet, often called a bar magnet because of its shape. As with the electric field, the magnetic field fills the volume, not just where the arrows are drawn. Two symbols are used for the magnetic field. One symbol is B (B does not stand for anything obvious, but it is the accepted symbol), called *magnetic flux density,* and it represents the strength of the magnetic field, just as

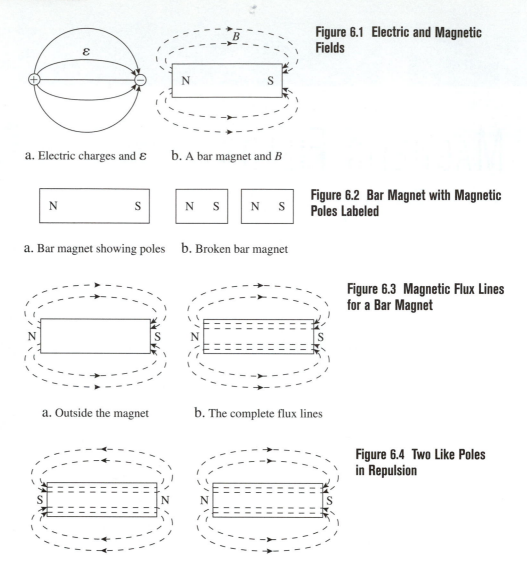

Figure 6.1 Electric and Magnetic Fields

a. Electric charges and $\mathcal{E}$ b. A bar magnet and B

Figure 6.2 Bar Magnet with Magnetic Poles Labeled

a. Bar magnet showing poles   b. Broken bar magnet

Figure 6.3 Magnetic Flux Lines for a Bar Magnet

a. Outside the magnet   b. The complete flux lines

Figure 6.4 Two Like Poles in Repulsion

$\mathcal{E}$ did for the electric field intensity. The other symbol is the uppercase Greek letter phi Φ, called the *magnetic flux,* and it is the entire set of the lines used to represent the magnetic field. Each line in a magnetic field sketch is called a magnetic flux line.

There is an important difference in the two sketches of Figure 6.1: *there are no individual, distinct magnetic charges* as there are electric charges. This fact raises two questions: How is the magnetic field created, and Where do the magnetic flux lines go? Consider the following known fact about bar magnets. A bar magnet has an N pole and an S pole, as shown in Figure 6.2a. However, if the bar magnet is broken in half, one of the halves does not have only a north pole and the other half does not have only a south pole. Instead, two smaller bar magnets result, each with an N pole and an S pole, as shown in Figure 6.2b.

The magnetic field of a bar magnet is shown in Figure 6.3a. If one could "look inside" the magnet, each magnetic flux line would continue *through the material of the bar magnet itself* and end on itself, as shown in Figure 6.3b. The magnetic flux lines must form closed loops, and this property of magnetic fields is stated as follows:

Magnetic flux lines are solenoidal.

With this picture of magnetic fields, one can qualitatively visualize the repulsion of two like poles, as shown in Figure 6.4, and the attraction of two unlike poles, as shown in Figure 6.5. In Figure 6.4, the magnetic flux lines must "push" against those of the other magnet because there is no place for the solenoidal flux from one magnet to form a complete path through the other magnet. However, in Figure 6.5, the poles attract because there

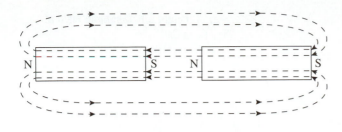

Figure 6.5 Two Unlike Poles in Attraction

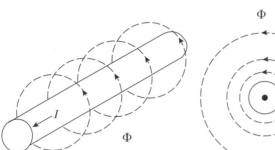

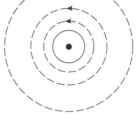

Figure 6.6 Magnetic Flux Around a Current Flowing Through a Straight Wire

a. Three-dimensional view   b. One end view   c. The other end view

is a natural path for the magnetic flux lines to close in on themselves through both magnets. The magnets attempt to become one larger magnet, i.e., attract each other.

The question, How is the magnetic field created? still lingers. We resort to an experiment that you may have performed many years ago. Wrap an iron nail with wire. Connect the wire to a battery. The iron nail assembly will attract other iron objects—it is magnetic. This method of creating a magnetic field is called *electromagnetism*. What is the source of the electromagnetism? If the iron nail was removed but the wire coil was left intact, the coils would still attract iron objects, but much more weakly. Thus, the iron nail is not responsible for the magnetism. The copper wire, by itself, does not attract iron objects. That only leaves one other thing going on—what is it? It is the current, i.e., flowing charges. This is the fundamental concept of *electromagnetism*:

Moving *electric charges create a magnetic field.*

Two common examples illustrate this concept. The first is current flowing through a straight wire, as illustrated in Figure 6.6a. The magnetic flux lines go *around* the current and form complete paths, i.e., are solenoidal. The direction of the flux lines can be found by the *right-hand rule (RHR) for straight wires:*

Point your right-hand thumb in the direction of the conventional current flow. Your right-hand fingers point in the direction of the magnetic flux lines around the wire.

In the end view of the wire, shown in Figure 6.6b, the dot in the middle of the wire means that the current is coming out of the page toward you (as if viewing the tip of an arrow coming toward you). If the current is going away from you, a cross is put inside the end view of the wire, as illustrated in Figure 6.6c (as if viewing the tail fins of an arrow going away from you). Does the RHR predict the magnetic flux direction in this case? Try it—it does. Note that Φ has been used to label the magnetic flux lines.

The other common case of electromagnetism is the coil, as shown in Figure 6.7. The wire is wrapped around some core, such as iron or plastic. Only a few *turns* of wire are shown. A turn is one complete wrap of the wire around the core. Some coils have thousands of turns. The current directions are shown at the ends of the wire and along the wire of the coil. The ends of the core are labeled N and S. How were those poles determined? Look at the cross-sectional view, as shown in Figure 6.8. The coil is cut in half along its

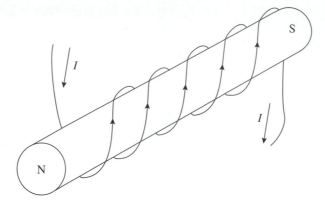

Figure 6.7 Magnetic Field of a Coil

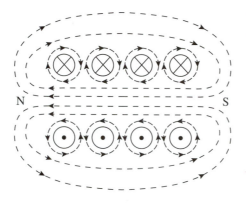

Figure 6.8 Cross-sectional View of the Coil in Figure 6.7

 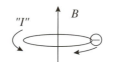

a. Simple atom model   b. Magnetic field

Figure 6.9 Atomic Magnetic Fields

length, resulting in a lengthwise cross-sectional view. The current is coming out of the page on the bottom side and into the page on the top side. The RHR is used to determine the magnetic field of each wire end view. Note that between the wires the magnetic fields due to two adjacent turns are equal and opposite—they cancel between the wires; however, on the inside (and outside) the direction of the magnetic fields due to each turn are in the same direction—they add to create the total magnetic flux. Then note which end that the flux comes out of, i.e., the N pole end. This reasoning is how the poles were labeled, and it is summarized in the *RHR for coils:*

> Point the fingers of your right hand in the direction that the current flows around the coil. Your right-hand thumb points in the direction of the north pole end of the coil.

Now, the question of how a bar magnet can have a magnetic field can be answered. All atoms have electrons that orbit the nucleus. A simple atom is shown in Figure 6.9a. The electron, a charged particle, orbiting the nucleus *looks like a current,* and a magnetic field is produced, as shown in Figure 6.9b (use the RHR for coils, and remember that the electron current direction is opposite of the conventional current direction). All of the atoms in the material contribute to the total magnetic field. If the atoms are randomly oriented, then the total magnetic field is zero. But if most of the atoms are aligned (the magnet is magnetized), then a strong magnetic field results. If the atoms stay aligned, then the result is called a *permanent magnet.*

By the previous reasoning, all atoms should produce a magnetic field and all materials should be magnetic; however, only certain metals, such as iron, cobalt, and nickel, are magnetic. The reasons must be explained by the field of science known as quantum mechanics; however, a "hand-waving" explanation may suffice for now. Usually the electrons in the atoms of most materials balance each other out in a magnetic sense. For every electron, there is another electron or other atomic effect that compensates and cancels the magnetic field. In magnetic materials, this cancellation does not occur, and the material is magnetic. (Again, this reasoning is not foolproof and is offered only as a temporary explanation. There are many textbooks on quantum mechanics and magnetism. You might start with a chemistry or physics textbook[1]—ask one of your science instructors.)

Thus, whether you consider permanent magnets or coils, the magnetic field arises from the motion of charge. Now that the concepts of magnetic fields and electromagnetism have been established, one may ask how they are used in electricity. The answer will start with a discussion of inductance and inductors.

6.2 WHAT IS INDUCTANCE?

We have already established that a coil that has a current flowing through it creates a magnetic field. There is energy stored in the magnetic field. For example, the magnetic field of a coil can potentially do work: attract and move magnetic materials. Hence, the inductor is another component that *stores* electrical energy. Unlike the capacitor, which stores electrical energy in an electric field, an inductor stores electrical energy in a magnetic field. *Inductance* is the name given to this property:

> *Inductance is the property of storing electrical energy in the magnetic field.*

> *An inductor is a wire configuration or other current-carrying configuration that stores electrical energy in the magnetic field.*

A coil is probably the most common type of inductor because it concentrates a magnetic field into a small volume. Inductance is given the symbol L (I was already used for current) and the units are henries (H). Can a defining equation be established for inductance as it was for capacitance (recall $C = Q/V$)? A current that flows through a conductor configuration, such as a coil, creates magnetic flux Φ. The magnetic flux represents the magnetic field, and the magnetic field stores the electrical energy. Hence, inductance is directly proportional to magnetic flux:

$$L \propto \Phi \qquad (6.1)$$

However, if there is a larger current, there is more magnetic flux. Thus, the magnetic flux created *per unit current* is appropriate, just as it was the stored charge per unit volt for capacitance. Hence, the fundamental defining equation for inductance is:

$$\boxed{L = \frac{\Phi}{I}} \qquad (6.2)$$

where: L is the inductance in henries (H),

Φ is the magnetic flux of the inductor in webers (Wb), and

I is the current in amperes (A).

Thus, one henry of inductance is equal to one weber of magnetic flux that originated from one ampere of current in the inductor. The equation for inductance in Equation (6.2) is fundamental. The more magnetic flux that a current-carrying structure can create, the larger the inductance. This equation allows one to reason out what inductance is. It is also used to formally derive the inductance of different current-carrying structures, but this task requires calculus, so it will be performed later in your curriculum. However, Equation (6.2)

---

[1] See the *Feynmann Lectures on Physics,* for example.

is not directly usable for circuit calculations in this form. The equation for the inductance of a coil will be developed in the next section of this chapter.

Information Research Exercise 6.2.1   (library and/or web)

    a.   The unit of inductance is named after Joseph Henry. What were his main contributions to the understanding of electricity?

    b.   The unit of magnetic flux is named after one of two brothers with the last name Weber. What is the full name of the Weber who made contributions to the understanding of electricity? What were his main contributions?

    c.   There are many other famous contributors to the understanding of electricity and magnetism: Gauss, Gilbert, and Oersted, for example. Name one main contribution for each individual.

How much electrical energy does an inductance store? The formal derivation of the result requires calculus.[2] The result will be stated and intuitively justified. The electrical energy stored in the magnetic field of an inductance is:

$$E = \frac{1}{2} L I^2$$

(6.3)

where: E is the stored energy in the magnetic field in joules,

           L is the inductance in henries, and

           I is the current through the inductance in amperes.

To intuitively justify Equation (6.3), consider an inductance L. If the inductance is increased, more magnetic flux per unit current exists and more energy is stored in the magnetic field. Hence, the electrical energy is directly proportional to L. The electrical energy proportionality to the current squared for an inductor can be thought of in a manner similar to electric power proportionality to current squared for a resistance. The difference is that energy is stored in an inductor, and energy is converted in a resistance.

As with any new component that is introduced, a schematic symbol for the inductor is needed. There are three symbols for inductors shown in Figure 6.10. The symbol shown in Figure 6.10a is the standard schematic symbol used for any inductor. It is also the symbol used for inductors that have a core of nonmagnetic material, often called "air-core" inductors. The schematic symbol shown in Figure 6.10b is used for an inductor that has a magnetic core material such as iron. The solid straight lines usually indicate iron, steel, or an iron alloy. The dotted lines usually indicate another class of magnetic materials, called *ferrites,* that give a stronger magnetic field than air but a weaker magnetic field than iron. Each of the symbols may have an arrow drawn through it to indicate that the inductance is adjustable, as shown in Figure 6.10c.

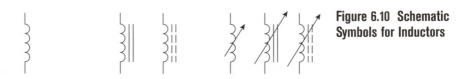

Figure 6.10 Schematic Symbols for Inductors

a. Standard   b. With magnetic core material   c. Adjustable

---

[2] See Nilsson and Riedel, *Electric Circuits,* 6th ed., Ch. 6, for example.

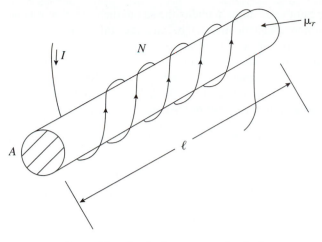

a. Quantities for inductance calculation   b. Sample coil

Figure 6.11 Coil Inductor Quantities and Sample   Part b Photo courtesy of J.W. Miller

6.3 INDUCTANCE OF A COIL

The inductor is a fundamental electric component. A current flows through the wire and creates a magnetic field. Electrical energy is stored in the magnetic field. The strength of the magnetic field is directly proportional to the current. The inductance of the inductor must be determined in order to calculate the energy that is stored in the magnetic field. In this section an intuitive reasoning approach will be used to arrive at the inductance of a coil. The inductance of two other types of inductors will be stated. Be assured that in a future course, in physics and/or electromagnetics, the equations for these inductances will be derived.

The geometry of the inductor must be designated first and is shown in Figure 6.11a. The assumption of a magnetic core material is made in this intuitive development. The cross-sectional area (A) of the core, the length of the core of the coil ℓ, and the number of wire turns (N) are labeled on the figure. A wire turn is one 360° wrap of a wire around the core (sometimes a turn is called a "wrap"). These quantities suggest asking what happens to the magnetic field and inductance:

■ if the cross-sectional area of the coil is increased?

■ if the number of turns is increased?

■ if the length of the coil and the magnetic core is increased?

■ if a material that is more magnetic is placed inside the coil?

If the cross-sectional area A of the coil increases, the circumference around the core increases, and there is a longer length of wire per turn. Thus, more magnetic flux is created per turn because there is more moving charge per turn in a longer length of wire. More magnetic flux means more energy is stored in the magnetic field. Thus, the inductance of a coil is proportional to the cross-sectional area:

$$L \propto A \qquad (6.4)$$

If there are more turns N, again, there is more wire length and more magnetic flux. However, the effect of the number of wire turns is more involved than this. The current through each turn affects all the other turns—this statement will be justified in Section 6.5. If each turn affects N other turns, and there are N turns, then the total effect is proportional to $N \times N$. Thus, the inductance of a coil is proportional to the square of the number of wire turns:

$$L \propto N^2 \qquad (6.5)$$

If the coil is longer but the number of turns remains the same, then the path for the magnetic flux is longer. There is more magnetic material because the coil is longer, but the path for the magnetic flux through the air outside of the coil also increases. It is much harder to set up magnetic flux through air than it is through a magnetic material (recall the iron nail experiment). Consequently, the magnetic field is not as intense, there is less magnetic flux, and there is less energy stored in the magnetic field. Thus, the inductance of a coil is inversely proportional to the length of the core of the coil:

$$L \propto \frac{1}{\ell} \qquad \qquad \textbf{(6.6)}$$

Finally, the magnetic material of the core needs to be considered. Given the same wire wrap arrangement, some materials have more magnetic flux than others. The ability of the magnetic material to establish magnetic flux is characterized by the parameter called *permeability* and is symbolized by the lowercase Greek letter mu (μ). Like the dielectric constant (permittivity), the permeability of vacuum, which is very close to the permeability of air, is a constant called the free-space permeability. The effect of the magnetic material is given by a factor relative to vacuum, and it is called the relative permeability. In symbol form:

$$\mu = \mu_o \mu_r \qquad \qquad \textbf{(6.7)}$$

where: μ = permeability in H/m,

μ_o = free space (vacuum) permeability = $4\pi \times 10^{-7}$ H/m, and

μ_r = relative permeability.

What does μ depend on? The material itself. More specifically, permeability depends on the magnetic properties of the material at the atomic and molecular level—again, quantum mechanics would be required for a detailed explanation. There are some major classes of magnetic materials:

■ nonmagnetic: air, wood, plastics, and nonferrous metals

■ ferrous: iron, cobalt, nickel—strongly magnetic and used in lower-frequency inductors and transformers

■ ferrites: iron oxide compounds—moderately magnetic, work at high frequencies, and are often used in inductors

Typical permeability values are listed in Table 6.1. Thus, a magnetic material increases inductance and inductance is directly proportional to permeability:

$$L \propto \mu \qquad \qquad \textbf{(6.8)}$$

The inductance of a coil is found by combining Equations (6.4) through (6.8):

$$L = \frac{\mu N^2 A}{\ell} = \frac{\mu_o \mu_r N^2 A}{\ell} \qquad \qquad \textbf{(6.9)}$$

TABLE 6.1	magnetic material	μ_r (H/m)
Magnetic Materials and Approximate Relative Permeability Values	vacuum	1
	air	1.00
	ferrites*	10's to 100's
	iron, iron alloys, and steels*	100's to 100,000's
	nickel*	400
	nonmagnetic metals	1.00

*These high-permeability magnetic materials exhibit complicated effects such as saturation, hysteresis, and anisotropy. Permeability values also depend on the manufacturing process. Consult with electromagnetic fields literature, vendor catalogs, and specialized handbooks.

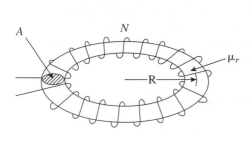

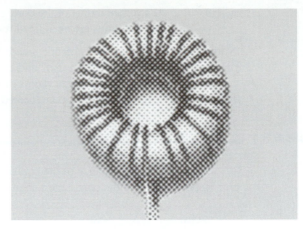

a. Quantities for inductance calculation   b. Sample toroid

Figure 6.12 Toroid Inductor Quantities and Sample   Part b Photo courtesy of J.W. Miller

where: L is the inductance in henries,

μ is the permeability in H/m,

N is the number of wire turns,

A is the cross-sectional area of the core in meters squared, and

ℓ is the length of the coil in meters

under the assumptions (to be justified in courses later in your curriculum):

■   the length of the coil is much larger than the diameter (d) of the coil ($\ell >> d$),

■   the core has a magnetic material ($\mu_r >> 1$), and

■   the windings are wrapped closely together and along the entire length of the core.

The inductance expression in Equation (6.9) is for a coil with a magnetic core. Do the units balance in this equation? What if the coil has an air core? There is an empirical equation, called *Wheeler's equation,* that predicts the inductance to within a few percent:

$$L = \frac{a^2 N^2}{9a + 10\ell}$$
(6.10)

where: L is the inductance in microhenries,

a is the radius of the core in inches,

N is the number of turns, and

ℓ is the coil length in inches.

Strictly speaking, Equation (6.10) is valid only for coil with a single layer of wire turns and should only be used for coils that have a length greater than 0.8 of the radius. Nevertheless, the results for $\ell < 0.8a$ are still often acceptable but with decreasing accuracy as the assumption is further violated.

Another common inductor is the *toroid.* A toroid has a core that forms a complete path, usually circular, as sketched in Figure 6.12. Usually, the core magnetic material is a ferrite. The advantage of a toroid is that the magnetic field stays inside the core because the solenoidal magnetic flux lines can complete their path within the core—imagine a coil that was bent around to form a toroid. The toroid does not interfere with nearby components because there is negligible magnetic flux outside the core. The inductance of a toroid is:

$$L = \frac{\mu N^2 A}{2\pi R}$$
(6.11)

where: R is the average radius of the core, as indicated in Figure 6.12, and all other quantities are as defined for a straight coil.

Notice that Equation (6.11) is identical with Equation (6.9) if the "length" is recognized as the circumference of the toroid $2\pi R$ and is valid if the radius of the cross section of the core is much less than the average radius of the core (R). If this condition is not met, then more accurate expressions should be located in technical handbooks.

Example 6.3.1

a. Determine the inductance of the coil shown in Figure 6.13.
b. Repeat part (*a*) if the core is nonmagnetic.

a. Given: $N = 1000$

$\ell = 5$ cm

$d = 1$ cm ($a = 0.5$ cm)

$\mu_r = 1200$

Desired: L

Strategy: $L = \dfrac{\mu_o \mu_r N^2 A}{\ell}, A = \dfrac{\pi}{4}d^2$

Solution:

$$L = \frac{\mu_o \mu_r N^2 A}{\ell} = \frac{\left(4\pi \times 10^{-7}\,\dfrac{H}{m}\right)(1200)(1000)^2 \left\{\dfrac{\pi}{4}\left[(1\text{ cm})\left(\dfrac{1\text{ m}}{100\text{ cm}}\right)\right]^2\right\}}{(5\text{ cm})\left(\dfrac{1\text{ m}}{100\text{ cm}}\right)} = 2.37\text{ H}$$

b. Given: $\mu_r = 1$, other quantities as per part (*a*)

Desired: L

Strategy: air core → use Wheeler's equation: $L = \dfrac{a^2 N^2}{9a + 10\ell}$, a and ℓ in inches, L in μH

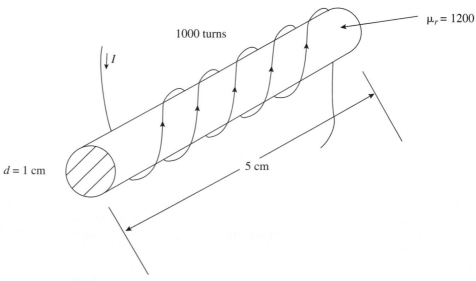

$\mu_r = 1200$

1000 turns

I

$d = 1$ cm

5 cm

Figure 6.13 Coil for Example 6.3.1

$$L = \frac{a^2 N^2}{9a + 10\ell} = \frac{\left[(0.5 \text{ cm})\left(\dfrac{1 \text{ in.}}{2.54 \text{ cm}}\right)\right]^2 (1000)^2}{9(0.5 \text{ cm})\left(\dfrac{1 \text{ in.}}{2.54 \text{ cm}}\right) + 10\,(5 \text{ cm})\left(\dfrac{1 \text{ in.}}{2.54 \text{ cm}}\right)} = 1.81 \times 10^3 \ \mu\text{H} = 1.81 \text{ mH}$$

The magnetic core increased the inductance of the coil from approximately 2 mH to 2 H.

6.4 COMBINATIONS OF INDUCTORS

As with capacitors, there are situations where inductors are combined. It may be windings within a motor or heater, or multiple inductors within an electronic circuit. The equations for the total inductance of series and parallel combinations of inductors will be developed not only for the equations themselves but for understanding the concepts that underlie them.

The first case is inductors in series. Three inductors are shown in series in Figure 6.14, but N series inductors could be shown in general. Assume that the inductors are physically spaced far enough apart so as not to interact with each other. What can be said about the current through each inductor? By KCL, the current must be the same through each inductor. What can be said about the total magnetic flux of the three inductors? Each inductor has a magnetic field, so the total magnetic flux is the sum of the magnetic fluxes in the three magnetic fields. Thus, the same current has created more total magnetic flux. Hence, the inductance of inductors in series has increased.

The development of the equation for the total inductance of inductors in series will begin with the known conditions: the current is the same through all of the series inductors and the total energy in the magnetic fields is the sum of the energies in the magnetic fields of the individual inductors:

$$I_T = I_1 = I_2 = I_3 \tag{6.12}$$

$$E_T = E_1 + E_2 + E_3 \tag{6.13}$$

The energy in the magnetic fields can be expressed in terms of the inductances and currents:

$$\tfrac{1}{2}L_T I_T^2 = \tfrac{1}{2}L_1 I_1^2 + \tfrac{1}{2}L_2 I_2^2 + \tfrac{1}{2}L_3 I_3^2 \tag{6.14}$$

Multiply through Equation (6.14) by 2. The current is the same through series inductors, so substitute Equation (6.12) into Equation (6.14):

$$L_T I_T^2 = L_1 I_T^2 + L_2 I_T^2 + L_3 I_T^2 \tag{6.15}$$

Divide through by I_T^2 to obtain the total inductance of three inductors in series:

$$L_T = L_1 + L_2 + L_3 \tag{6.16}$$

This development could be extended to N inductors in series. The result is:

$$\boxed{L_T = \sum_{n=1}^{N} L_n} \tag{6.17}$$

Figure 6.14 Three Inductors in Series

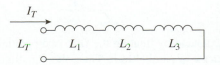

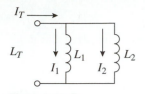

Figure 6.15 Two Inductors in Parallel

Thus, for inductors in series, the total inductance is the sum of the series inductances.

The second case is inductors in parallel. Two inductors are shown in parallel in Figure 6.15, but N parallel inductors could be shown in general. What can be said about the current through each inductor? They are in parallel, so the total current divides among the two inductors. What can be said about the magnetic flux in each inductor? The magnetic flux of each inductor is less than it would be if the total current flows through both inductors because the current through each inductor is less. Hence, for a given total current, the magnetic flux of each inductor is less than it would be in the series case. The total current, however, is the sum of the individual inductor currents. For a given total current, there is less total magnetic flux (at least relative to the series inductors case). Thus, the total inductance of inductors in parallel is less than the total inductance of the series case. It will be shown next that the total inductance of the parallel inductors case is less than any one of the individual inductances.

The development of the equation for the total inductance of inductors in parallel will begin with the inductance of two inductors in parallel. Start with the known conditions: the sum of the individual branch currents equals the total current and the total energy in the magnetic fields is the sum of the energies in the magnetic fields of the individual inductors:

$$I_T = I_1 + I_2 \tag{6.18}$$

$$E_T = E_1 + E_2 \tag{6.19}$$

Substitute the energy in the magnetic field of an inductor ($\frac{1}{2}LI^2$) into Equation (6.19):

$$\tfrac{1}{2}L_T I_T^2 = \tfrac{1}{2}L_1 I_1^2 + \tfrac{1}{2}L_2 I_2^2 \tag{6.20}$$

Multiply through by 2 and substitute $LI = \Phi$ in Equation (6.20):

$$\Phi_T I_T = \Phi_1 I_1 + \Phi_2 I_2 \tag{6.21}$$

Incorporate KCL [Equation (6.18)] into Equation (6.21):

$$\Phi_T (I_1 + I_2) = \Phi_1 I_1 + \Phi_2 I_2 \tag{6.22}$$

$$\Phi_T I_1 + \Phi_T I_2 = \Phi_1 I_1 + \Phi_2 I_2 \tag{6.23}$$

A rigorous derivation using calculus and Faraday's law shows that all the magnetic fluxes are equal:

$$\Phi_T = \Phi_1 = \Phi_2 \tag{6.24}$$

You can check this result by substituting it into Equation (6.23). The equality of the magnetic fluxes is a surprising result that is not intuitively obvious, but it is valid because it is based on KCL, conservation of energy, and the definition of inductance. With the revelation in Equation (6.24), the total inductance of parallel inductors is straightforward:

$$I_T = I_1 + I_2 \tag{6.25}$$

$$\frac{\Phi_T}{L_T} = \frac{\Phi_1}{L_1} + \frac{\Phi_2}{L_2} \tag{6.26}$$

$$\frac{\Phi}{L_T} = \frac{\Phi}{L_1} + \frac{\Phi}{L_2} \tag{6.27}$$

$$\frac{1}{L_T} = \frac{1}{L_1} + \frac{1}{L_2} \tag{6.28}$$

$$L_T = \frac{1}{\dfrac{1}{L_1} + \dfrac{1}{L_2}} \tag{6.29}$$

This process could be extended by treating the first two parallel inductors as a new "L_1" and adding another inductor in parallel to it. Thus, the equation for total inductance of N inductors in parallel is:

$$L_T = \cfrac{1}{\displaystyle\sum_{n=1}^{N} \cfrac{1}{L_n}}$$ (6.30)

As previously reasoned, the total inductance of inductors in parallel is less than the inductor with the lowest inductance.

The development for the total inductance of N inductors in parallel had a critical assumption: the resistances of the conductors in the inductors are negligible. The key to whether or not this assumption is satisfied is *"negligible in comparison to. . . "* For example, at DC, wire resistance is significant, the assumption is violated, and the equation for total parallel inductance is generally invalid. However, inductors are most often used for their AC behavior. The criteria then will be whether or not the conductor resistance is negligible compared with what the inductor looks like at some frequency (this "what" is called *reactance,* which will be introduced in Chapter 8). The purpose of this discussion is to remind you that the validity of the results of a derivation is only as good as the validity of the assumptions behind the derivation.

Example 6.4.1

Determine the total inductance of the inductor combination shown in Figure 6.16.

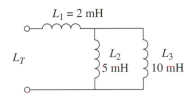

$L_1 = 2$ mH

L_T

L_2 L_3
5 mH   10 mH

Figure 6.16 Inductor Combination for Example 6.4.1

Given: circuit in Figure 6.16

Desired: L_T

Strategy: $L_T = L_1$ in series with $L_2 \parallel L_3$

Solution:

$$L_2 \parallel L_3 = \cfrac{1}{\cfrac{1}{L_2} + \cfrac{1}{L_3}} = \cfrac{1}{\cfrac{1}{5\text{ m}} + \cfrac{1}{10\text{ m}}} = 3.3333\text{ mH}$$

$$L_T = L_1 + L_2 \parallel L_3 = 2\text{ m} + 3.3333\text{ m} = 5.3333\text{ mH}$$

6.5 THE FUNDAMENTAL ACTION OF INDUCTORS IN DC CIRCUITS

Recall two key concepts about electric fields: (*1*) electric charges create an electric field, and (*2*) an electric field will exert force on another charge that is placed into the electric field. Are there similar statements that can be made about magnetic fields? There are no magnetic "charges"; however, *moving* electric charges do create a magnetic field—the phenomenon of electromagnetism. This statement corresponds to the first concept about electric fields. Is there a "force" statement that corresponds to the second statement for electric fields?

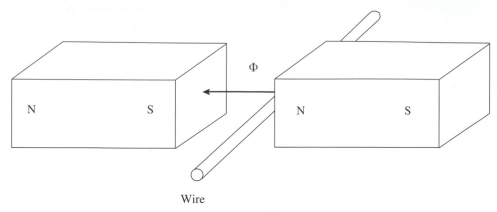

Wire

Figure 6.17 Illustration of Force on Electric Charges by a Magnetic Field

There is indeed a force concept associated with magnetic fields: if a charge is moving through a magnetic field, or a magnetic field is moving or changing strength around a stationary (relatively) electric charge, then there is a force exerted on the electric charge by the magnetic field. The condition of relative motion between the charge and the magnetic field should not be a surprise because charges have to move to create the magnetic field. A stationary, nonchanging magnetic field will not exert a force on an electric charge.[3]

To investigate this force on a charge by a magnetic field, consider the setup in Figure 6.17. Two bar magnets are aligned with a gap between them. Opposite poles of the magnets face each other. Hence, magnetic flux goes from the north pole, across the gap, to the south pole (only one magnetic flux line is shown for clarity, but flux lines fill the entire gap region). A wire is placed perpendicular to the magnet arrangement in the gap, as shown in the figure. With everything standing still, there are no forces on the charges in the wire. However, if the wire is moved upward through the magnetic field and crosses through magnetic flux lines, there is a force on the charges in the wire, as shown in Figure 6.18. There is a consequence to this action—try to identify it from Figure 6.18.

The force on the charges causes the positive charges to move toward one end of the wire and the lack of positive charges to be at the other end of the wire (the negative end).[4] Electric charges have been separated. The consequence is that *a voltage has been developed.* There can be relative motion between the conductor and the magnetic field by moving a conductor through a stationary magnetic field, by moving the magnetic field relative to the conductor (move the magnet set while holding the conductor stationary), or by changing the strength of the magnetic field. "Moving" means that the conductor is crossing through magnetic flux lines. If the conductor is moved parallel to the magnetic flux lines, there is no force on the charges. The cases of the "relative motion between the conductor and the magnetic field" are the principles behind electric generators: motion (mechanical energy) is converted into voltage (electrical potential energy difference, i.e., electric energy).

The case of "changing the strength of the magnetic field" is analogous to moving the conductor closer to or farther from a magnet. The conductor must be oriented such that the magnetic flux lines cross the conductor as the magnetic field strength changes. Changing the magnetic field strength is most often accomplished by changing the current in an electromagnet that created the magnetic field (as opposed to a permanent magnet). The change in current increases or decreases the strength of the magnetic field; hence, the magnetic field is not static but is changing over time. For example, changing current would occur automatically with an AC signal. (AC is introduced in Chapter 4.)

---

[3] The interested reader should investigate the topic of the *Lorentz force law* in a physics textbook for the mathematical relationship between the force of a charge and the strengths of electric and magnetic fields.

[4] The Lorentz force law is also used to predict the direction of the force on the charges.

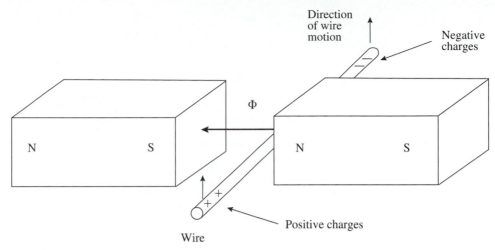

Figure 6.18 Motion of a Conductor in a Magnetic Field

These results are incorporated into a statement called Faraday's law:

Relative movement between a magnetic field and a conductor induces *a voltage in the conductor if the conductor crosses magnetic flux lines.*

or, for the case of a changing magnetic field strength, Faraday's law is often stated as:

A time-changing magnetic field induces a voltage in the conductor.

A new word was used in the Faraday's law statements. To *induce* is to cause to happen without actual physical contact between bodies, usually through the interaction with an electric field or a magnetic field (the latter in this case). If the relative motion between the conductor and the magnetic field is faster or the magnetic field strength changes more quickly, then the induced voltage is larger. Both of these cases can be expressed as the rate of change of the magnetic field:

The higher the rate of change of the magnetic field, the larger the induced voltage.

If a magnetic field is changing strength, then the charges in a conductor will experience a force, the charges will separate into positive and negative regions, and, hence, a voltage will be induced in the wire itself! This is probably irritating, because a perfect conductor is not supposed to have a voltage across it. However, this condition was established before changing magnetic fields were considered. There is now a case where a conductor can have a voltage within it—when there is a changing magnetic field. This new (to us) phenomenon will lead to an interesting concept for inductors in circuits. Before addressing the inductor operation, one more question needs to be asked: What is the polarity of the induced voltage (which end of the wire is positive)?

As previously mentioned in a footnote, there is a law from physics that can be used to make this prediction. However, in electric circuits there is even an easier method that will be used here. Complete the circuit in Figure 6.18 through a load resistance, as shown in Figure 6.19. Current will flow, and energy will be delivered to the load. A lowercase *i* is used because the current is, in general, time dependent. The current flows from positive to negative through the resistance because of the voltage induced in the wire. The charges are then forced from the negative to the positive end of the portion of the wire in the magnetic field because of the force that the magnetic field exerts on them as the wire is moved through the magnetic field. However, the current that flows in the conductor will create its own magnetic field.

The interaction of the magnetic field due to the current flowing through the wire and the magnetic field due to the magnets is revealing. Take an end view of Figure 6.19, as

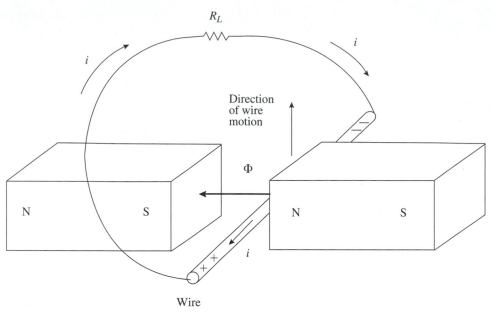

Figure 6.19 Figure 6.18 with a Complete Circuit: *i*(t) Flows

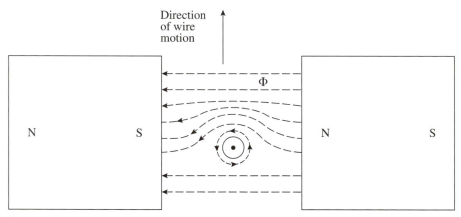

Figure 6.20 End View of Figure 6.19

shown in Figure 6.20. Recall that the dot in the middle of the wire indicates that the current is coming out of the page. Notice the distortion of the main magnetic field by the magnetic field from the current in the wire. The moving charge of the current that flows in the wire is also moving with respect to the magnetic field of the permanent magnets. Hence, this charge and the wire itself (since it contains the charge) will also be subject to a force because there is relative motion between the magnetic field and the charge. The force will be in the direction opposite to the original motion of the wire through the magnetic field of the permanent magnets.[5] Hence, energy must be used to move the wire through the magnetic field (recall the mechanical energy input). From the electrical viewpoint, the current that flows in the wire, due to the induced voltage, is providing opposition to the movement of the wire in the magnetic field of the bar magnets. In other words, the induced effect opposes the cause of that effect. This phenomenon is known as *Lenz's law:*

> *The effect of the induced voltage always opposes the original cause of the induced voltage.*

---

[5] This analysis is, again, based on the Lorentz force law.

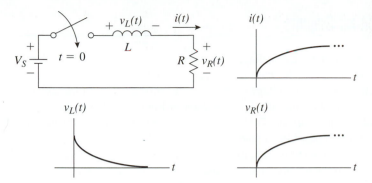

**Figure 6.21 Circuit with Inductance
When the DC is Turned on**

There is another straightforward argument for explaining Lenz's law. If the magnetic field of the current flowing through the wire and the magnetic field of the permanent magnets were not in opposition, but instead were supportive, then the voltage induced would increase, the current would increase, the magnetic field of the wire would increase, the speed of the wire moving through the field would increase (due to the supportive magnetic fields), which would mean a larger induced voltage, and the cycle would continue upward until infinite energy resulted ("blow up the universe"). This situation violates the law of conservation of energy; hence, Lenz's law must hold true.

How do Faraday's law and Lenz's law relate to inductor operation? This will be demonstrated with the circuit shown in Figure 6.21a, where an inductor, a resistance, a switch, and a voltage source are connected in series. The switch is closed at some arbitrary time called $t = 0$. What happens as the current begins to flow and the magnetic field begins to build up? The magnetic field of the coil is changing. Each turn of the coil is immersed in the magnetic fields of all of the other turns; hence, by Faraday's law, the changing magnetic field of each turn is inducing a voltage into all of the other turns in the coil (recall the N^2 in L calculations). What is the effect of the induced voltage? From Lenz's law, the induced voltage should oppose the cause of the effect. What is the cause of the effect? The current and the magnetic field are increasing. Hence, the polarity of the voltage induced into the inductor should *oppose* the increase in current—see the polarity across the inductor in Figure 6.21. In fact this opposition is largest at $t = 0$ because the current is changing the fastest at $t = 0$: from nothing to some value. The induced voltage across the inductor opposes the current change, so the current will not jump up immediately but will instead grow gradually, as shown in Figure 6.21. This is the *inductive transient,* analogous to capacitive transient in Chapter 5, when the DC source is first turned on. The voltage across the resistance will track the current by Ohm's law. The voltage across the inductor will be largest at $t = 0$, when the current is changing the fastest, and then will gradually decrease as the current changes more slowly as it approaches the DC steady-state value. Eventually the current levels out to the value of the source voltage divided by the circuit resistance, as predicted by Ohm's law. The circuit has reached DC steady state. The current is no longer changing, the magnetic field of the inductor is steady, and there is no induced voltage across the inductor when the current is constant. The inductor looks like the resistance of the wire in the inductor when the circuit has reached steady state.

Hence, what is the fundamental action of inductors in circuits?

The current cannot change instantaneously through an inductor.

Again, this is often shortened to:

Inductors oppose an instantaneous change of current.

where the current is understood to be through the inductor. Predictions of DC transients in circuits with inductance (inductive circuits) will be made in Chapter 7. For now the new situation to be emphasized is that voltages and currents are not constant, at least for awhile (during the transient), when there is an inductor in the circuit.

6.6 THE FUNDAMENTAL ACTION OF INDUCTORS IN AC CIRCUITS

Consider the case of an AC signal.[6] The voltage and current are always changing. Hence, the inductor will always be reacting. This case is also very special. Just as the case with the capacitor, the phase shift between the voltage across and the current through the inductor may now be understood. Start with the AC circuit shown in Figure 6.22. The AC source delivers a sinusoidal voltage and current. The voltage across and the current through the resistance are in-phase, as established in Chapter 4.

When the current is at zero and increasing, its rate of change is the fastest (note the slope of the sine wave at $t = 0$). The magnetic field is consequently changing at its fastest rate. Thus, the induced voltage across the inductor should be at the maximum (peak) value—see $v_L(t)$ in Figure 6.23. As the current increases, its rate of change lessens. Hence, the induced voltage across the inductor should decrease. When the current reaches the maximum of the sine wave, the rate of change and the slope are zero at that time instant. Therefore, the induced voltage should be zero, as shown in Figure 6.23. The remainder of the cycle could be similarly argued. Hence, the inductor voltage is *leading* the inductor current because the inductor voltage is proportional to the rate of change of the magnetic field or, equivalently, is directly proportional to the rate of change of the current (not the actual value of the current). The voltage is induced "ahead" of the current change and thus opposes the current change.

How far (in terms of angle) should the AC voltage lead the current for an inductor? As it was for the capacitor, the answer is remarkable: 90°. Why would this angle be such a convenient value? The answer comes from power and energy for the inductor, which is covered in Chapter 10. An ideal inductor does not dissipate energy, it stores energy. For this to happen, the mathematics of AC voltage, current, and power require the voltage and current to be 90° out of phase.

In summary, the inductor opposes an instantaneous change of current. When a DC signal is applied to a circuit with an inductor, the inductor current increases (or decreases)

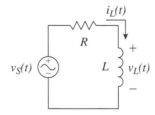

Figure 6.22 An AC Circuit with an Inductor

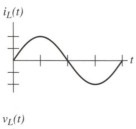

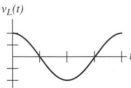

Figure 6.23 AC Voltage and Current Waveforms for an Inductor

---

[6] The discussion in this section is repeated in Chapter 8 for those who will study AC later.

over time until it reaches the DC steady-state value. In an inductive circuit with an AC signal, the inductor voltage leads the inductor current by 90°.

There is a mnemonic to help remember the phase shift relationships between the voltage and the current for inductors and capacitors. The mnemonic is *ELI the ICE man,* where *E* stands for voltage here (again, *E* has been used in the past to stand for electromotive force, which is voltage). In ELI, the voltage (*E*) comes first (leads) in an inductor (*L*) relative to the current (*I*)—voltage leads current in an inductor. Conversely, in ICE, current (*I*) comes first (leads) in a capacitor (*C*) relative to the voltage (*E*)—current leads voltage in a capacitor.

Caution: This mnemonic may help you learn, but it should not substitute for understanding the reasons behind the phase relationships. Understanding removes the need for mnemonics and memorizing: understanding is the path to permanent learning, whereas memorizing often results in short-term, temporary learning.

CHAPTER REVIEW

6.1 What Is Magnetism? What Is Electromagnetism?
- Moving charges create a magnetic field.
- Magnetic flux lines are used to visualize magnetic fields.
- Magnetic flux lines are solenoidal.
- Like poles repel and unlike poles attract.
- The right-hand rule (RHR) for wires and coils relates the current direction and magnetic flux direction.
- The property of magnetism of magnetic materials arises from electron charge motion at the atomic level.

6.2 What Is Inductance?
- Inductance is the property of storing electrical energy in the magnetic field.
- Inductance is defined as the magnetic flux created per unit of current through the inductor ($L = \Phi/I$).
- The energy stored in the magnetic field of an inductor is $E = \frac{1}{2}LI^2$ and is proportional to the total magnetic flux and current established in the inductor.

6.3 Inductance of a Coil
- The inductance of a coil is directly proportional to the permeability of the core, the number of wire turns squared, and the cross-sectional area of the core. It is inversely proportional to the length of the coil (core).
- The permeability expresses the magnetic effect of the core material in an inductor on the inductance.
- The relative permeability is the factor by which the permeability of the core material is greater than the permeability of air.

6.4 Combinations of Inductors
- The inductance of series inductors is the sum of the individual inductances because more

magnetic flux and magnetic field energy is stored for a given applied current.
- The inductance of parallel inductors is less than the inductor with the least inductance because the total current is divided among the inductors and there is less magnetic flux and magnetic field energy per inductor.

6.5 The Fundamental Action of Inductors in DC Circuits
- Faraday's law: Relative movement between a magnetic field and a conductor *induces* a voltage in the conductor if the conductor crosses magnetic flux lines.
- A time-changing magnetic field *induces* a voltage in the conductor.
- Lenz's law: The effect of the induced voltage always opposes the original cause of the induced voltage.
- Inductors oppose an instantaneous change of current.
- When DC is switched on to a circuit with inductance, the voltages and currents in the circuit go through a transition from the initial values to the DC steady-state values. This transition is called the (DC) transient.

6.6 The Fundamental Action of Inductors in AC Circuits
- Inductors oppose an instantaneous change of current.
- When AC is applied to a circuit with inductance, there is a 90° phase shift between the AC voltage and current: voltage leads current by 90°. The voltage must "come first" because of the rate of change of the magnetic field that induces the voltage across the inductor (Faraday's law) that opposes the current change (Lenz's law).

HOMEWORK PROBLEMS

Proper completion of the following questions will demonstrate your ability to describe a magnetic field and the relationship of the magnetic field to current and to define and describe inductance from a magnetic flux and from an energy viewpoint.

6.1   If a magnet is split into four pieces, describe the number and type of poles that will result.

6.2   What type of field is formed when current flows through a coil of wire?
  a. Sketch the resulting field, including the field direction.
  b. What is the source of the field?
  c. Describe what would happen if the energized coil was placed close to a permanent magnet.

6.3   Based on the material presented in this chapter, what would happen if you tried to suddenly switch the current in an inductor? Justify your response.

6.4   Both an inductor and a resistor are considered to be electrical components. What is the difference between the two?

6.5   The energy stored in an inductor is stored in what form? Where is the energy stored?

6.6   Could a straight piece of wire be considered an inductor? Explain.

Proper completion of the following questions will demonstrate your understanding of the physical factors that affect the amount of inductance in an inductor. It will also demonstrate your ability to calculate the inductance of ideal and practical inductors.

6.7   What physical parameters affect the value of an inductor?

6.8   What happens to the total flux in a toroid if the number of turns is doubled but the current in the turns is reduced by half? Explain.

6.9   If the inductance of an inductor is 40 mH and the current through the inductor is 10 mA, what is the value of the resulting flux? The value of the energy stored in the magnetic field?

6.10   A coil has 400 turns, diameter of 0.8 cm, length of 4 cm, and a ferrite core with a relative permeability of 60. Determine the inductance of this inductor.

6.11   For the same coil of problem 6.10 (400 turns, diameter of 0.8 cm, and length of 4 cm) determine the inductance if the ferrite core is removed. Which equation is most appropriate to use for the calculation?

6.12   A coil has a diameter of 0.5 cm, length of 2 cm, and a ferrite core with a relative permeability of 100. Determine the number of turns needed to achieve an inductance of 200 mH.

6.13   A toroid has a cross-sectional radius of 0.4 cm, a circumference of 10 cm, and a nickel core. Determine the number of turns needed to achieve an inductance of 100 mH.

6.14   The normal useful limit of Wheeler's equation is for a coil length of 0.8 of the radius. Calculate the inductance of an air-core coil at this limit for 500 turns and a length of 3 in.

6.15   A coil has a value of inductance of 10 mH and a current of 20 mA going through it. If the present number of turns is 400, what would be the new number of turns required to maintain the same amount of magnetic flux if the current is cut in half, all else being the same (hint: form a ratio)?

6.16   Calculate the cross-sectional area of a toroid with an inductance of 2 H, 1000 turns, a core with a relative permeability of 200, and a core radius of 10 cm. Is such an area realistically feasible? Convert the area to the corresponding radius. What is your opinion now on the feasibility of fabricating the toroid?

6.17   Calculate the length for a coil that would be required to achieve an inductance of 1 mH if the cross-sectional area is 0.4 cm^2, the core is nickel, and it has 500 turns.

6.18   For the same coil of problem 6.17 (desired inductance of 1 mH, cross-sectional area of 0.4 cm^2, and 500 turns), what would be the inductor length if the core is air?

Proper completion of the following questions will demonstrate your ability to determine the total inductance of series, parallel, and series–parallel combinations of inductors.

6.19 Determine the total equivalent inductance of the inductor combination shown in Figure P6.1.

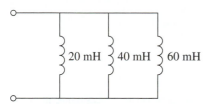

Figure P6.1 Inductor Combination to be Evaluated in Problem 6.19

6.20 Determine the total equivalent inductance of the inductor combination shown in Figure P6.2.

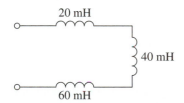

Figure P6.2 Inductor Combination to be Evaluated in Problem 6.20

6.21 Determine the total equivalent inductance of the inductor combination shown in Figure P6.3.

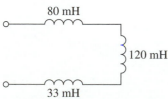

Figure P6.3 Inductor Combination to be Evaluated in Problem 6.21

6.22 Determine the total equivalent inductance of the inductor combination shown in Figure P6.4.

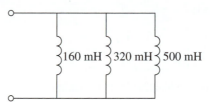

Figure P6.4 Inductor Combination to be Evaluated in Problem 6.22

6.23 Determine the total equivalent inductance of the inductor combination shown in Figure P6.5.

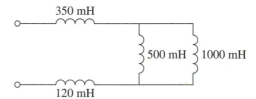

Figure P6.5 Inductor Combination to be Evaluated in Problem 6.23

6.24 Determine the total equivalent inductance of the inductor combination shown in Figure P6.6.

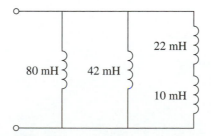

Figure P6.6 Inductor Combination to be Evaluated in Problem 6.24

6.25   Determine the total equivalent inductance of the inductor combination shown in Figure P6.7.

6.26   Determine the total equivalent inductance of the inductor combination shown in Figure P6.8.

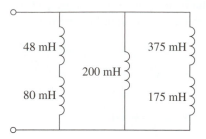

Figure P6.7 Inductor Combination to be Evaluated in Problem 6.25

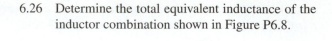
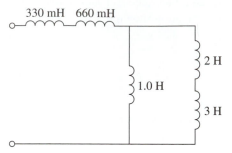

Figure P6.8 Inductor Combination to be Evaluated in Problem 6.26

Proper answers to the following questions will demonstrate your ability to describe the fundamental action of inductors in a DC and an AC circuit.

6.27   What does an inductor oppose an instantaneous change of? Why?

6.28   What is the difference between the transient response and the steady-state response in a circuit with an inductor?

6.29   What role does a resistor play in analyzing a circuit with an inductor?

6.30   Describe the change in current through an inductor, i.e., does it change instantaneously?

6.31   What is the phase relationship between the AC voltage and the current for an inductor?

6.32   Why is there a difference in the phase relationship between the AC voltage and the current for an inductor?

Proper answers to the following questions will demonstrate your understanding of the specifications of an inductor, including its basic construction.

6.33   Why is the permeability of the inductor's core so important?

6.34   When is Wheeler's equation used to calculate inductance?

6.35   What are the basic physical configurations for constructing an inductor?

6.36   What are the a) similarities and b) differences of an inductor and a permanent magnet?

RESPONSE OF RL AND RC CIRCUITS TO DC SIGNALS

As a result of successfully completing this chapter, you should be able to:

1. Describe the following signals: steady-state DC, steady-state AC, DC step, and exponential decay and buildup.
2. Describe how the exponential function produces exponential decay and buildup behavior in series RL and RC circuits.
3. State and explain the steady-state DC replacements for capacitors and inductors.
4. Explain the time constant relationship to the component values in series RC and RL circuits.
5. Determine and explain the transient expressions and plots for the current and the voltages for both the charging and the discharging of a series RC circuit.
6. Determine and explain the transient expressions and plots for the current and the voltages for both the buildup and the decay of a series RL circuit.

7.1 THE EXPONENTIAL SIGNAL

We have learned that an electric circuit consists of electrical components, conductors, and insulators that are connected in a complete path(s) in order to achieve a useful function. The components include sources, resistances, capacitors, inductors, and switches. A signal is a voltage or a current in the circuit. A signal can be the voltage (or current) that is applied to the circuit from a source, or it can be the voltage (or current) of a component or group of components. Because we are starting to accumulate a few signal types, we will pause here to list and reflect on them.

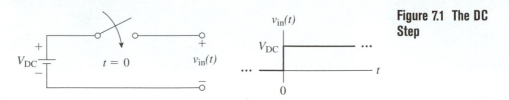

Figure 7.1 The DC Step

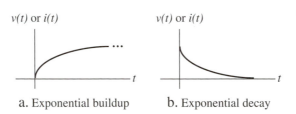

a. Exponential buildup   b. Exponential decay

Figure 7.2 Exponential Signals

Although there are an infinite number of signal types, only some are used extensively. The signals that are used most often are those that we have examined in the previous chapters: steady-state DC and sinusoidal steady-state AC. A new signal type was introduced at the ends of Chapter 5 and Chapter 6. When the DC source was switched on, the DC voltage jumped from zero volts up to the value of the DC source, as shown in Figure 7.1. This signal is called a DC step, or simply a *step*. It is often used in circuit analysis to study how a circuit responds when a DC voltage is suddenly applied to a circuit. A practical example is found in almost every battery-operated electronic device or appliance that must function properly when the switch is turned on to apply the battery voltage.

Two other signal types were examined at the ends of Chapter 5 and Chapter 6, as shown in Figure 7.2. In the first case, the voltage (or current) started at zero and rose in a "curve" toward a steady-state DC value as time elapsed, as shown in Figure 7.2a. It corresponded to the buildup of the voltage across the capacitor as the capacitor was charging during the capacitive transient. Another example was the buildup of the current through the inductor during the inductive transient.

In the second case, the voltage (or current) started at some value, and decreased in a "curve" toward zero, again as time elapsed, as shown in Figure 7.2b. It corresponded to the decay of the capacitor current as the capacitor was charging during the capacitive transient, for example. Another example was the decay of the inductor voltage during the inductive transient. Both of these signals, the buildup and the decay, are called *exponential signals* because they are described by the mathematical exponential function e^x. The exponential function will be examined in this section so that we are ready to use it in the capacitive or inductive circuit analysis when a DC step is applied to a resistor–capacitor (RC) or resistor–inductor (RL) circuit.

The exponential function is a function of the form of:

$$y = f(x) = b^x \tag{7.1}$$

where b is the base, x is the independent variable, and y is the dependent variable. One of the most useful bases in science and engineering is the irrational number $e = 2.7182818 \ldots$, which is the base of the natural logarithm (ln). This number is like π in the sense that it is a constant that frequently occurs in natural phenomena. If e is inserted as the base in Equation (7.1),

$$y = f(x) = e^x = (2.7182818 \ldots)^x \tag{7.2}$$

The exponential function is a standard key on scientific calculators and is a standard function in spreadsheet and mathematical calculation programs. In general, there may be a multiplicative constant (A), usually shown in front of the exponential function, and a multiplicative constant in the exponent (a):

$$y = Ae^{ax} \tag{7.3}$$

TABLE 7.1	x	e^{+x}
Exponential Function with a Positive Exponent	0	1.0
	1	2.7
	2	7.4
	3	20.1
	4	54.6
	5	148.4

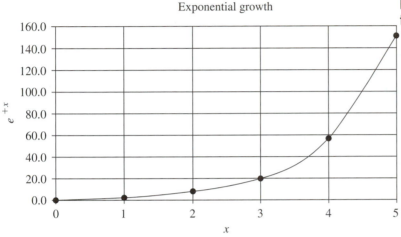

Exponential growth

Figure 7.3 Exponential Growth Due to a Positive Exponent

Three different expressions with the exponential function will be plotted in order to observe some of its general properties. First, let $A = 1$ and $a = 1$:

$$y = 1e^{+1x} = e^{+x} \tag{7.4}$$

The exponent is positive. The exponential function is calculated for a few values of x (see Table 7.1) that are plotted in Figure 7.3. Notice the behavior of the curve. It grows with an increasing slope as the x-value increases. This behavior is called an *exponential growth*. It is not common in electronics because it usually represents an unstable situation. The function grows without limit. It can represent the voltage and/or current over the time period that a device approaches failure. The exponential growth function with a positive exponent is common in many scientific phenomena, especially the growth of populations, such as bacteria.

The next exponential expression is an exponential function with a negative exponent. Let $A = 1$ and $a = -1$:

$$y = 1e^{-1x} = e^{-x} \tag{7.5}$$

The values of the exponential function are calculated for a few values of x (see Table 7.2) and are plotted in Figure 7.4. Notice the behavior of the curve. It decreases with a decreasing slope as the x-value increases. Hence, the function approaches a steady-state value of zero. This behavior is called an *exponential decay*. It is common in electronics because it usually represents a decaying signal, such as the current in a series RC circuit as the capacitor charges, or the voltage across an inductor in a series RL circuit as the inductor current builds up to the DC steady-state value—compare Figure 7.4 and Figure 7.2b. The exponential function with a negative exponent is one of the fundamental expressions for the voltages and the currents in inductive or capacitive transients, and it will be used extensively in this chapter.

A third expression that incorporates the exponential function involves an additive constant. Let $A = -1$ and $a = -1$, but add $+1$ to the exponential function:

$$y = 1 - 1e^{-1x} = 1 - e^{-x} \tag{7.6}$$

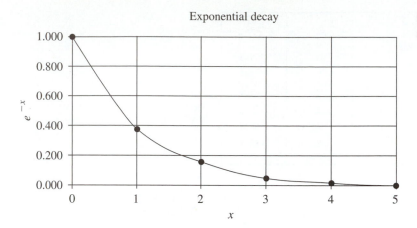

Figure 7.4 Exponential Decay Due to a Negative Exponent

TABLE 7.2		x	e^{-x}
Exponential Function with a Negative Exponent		0	1.000
		1	0.368
		2	0.135
		3	0.050
		4	0.018
		5	0.007

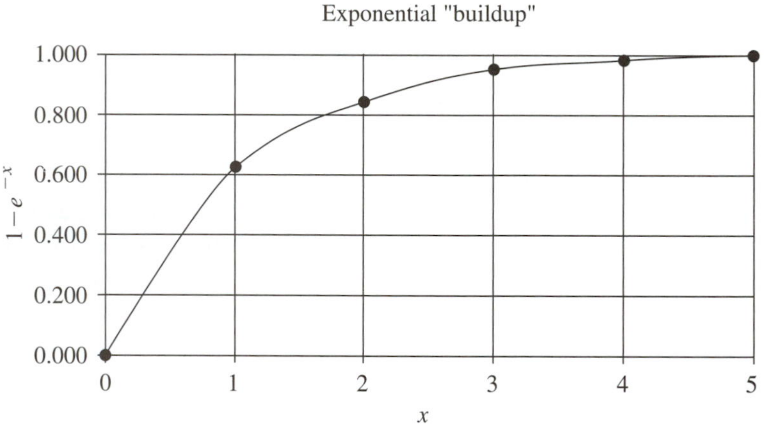

Figure 7.5 Exponential Buildup

Exponential "buildup"

TABLE 7.3		x	$1 - e^{-x}$
Exponential Function Expression for an Exponential Buildup		0	0.000
		1	0.632
		2	0.865
		3	0.950
		4	0.982
		5	0.993

This expression is calculated for a few values of *x* (see Table 7.3) that are plotted in Figure 7.5. Notice the behavior of the curve. It increases with a decreasing slope as the *x*-value increases. Hence, the function approaches some nonzero DC steady-state value. This

TABLE 7.4	transient type	fundamental expression	with constants
Common Exponential Function Expressions in DC Transients	exponential buildup	$y = 1 - e^{-x}$	$y = A(1 - e^{-ax})$
	exponential decay	$y = e^{-x}$	$y = Be^{-ax}$

behavior is called an *exponential buildup*. It is common in electronics because it usually represents a signal that builds up to a DC steady-state value, such as the voltage across the capacitor in a series RC circuit as the capacitor charges, or the current through a series RL circuit as the inductor current builds up to the DC steady-state value—compare Figures 7.5 and 7.2a. The exponential function in the expression of Equation (7.6) is one of the fundamental expressions for the voltages and the currents in inductive or capacitive transients, and it will be used extensively in this chapter.

The two expressions involving the exponential function that commonly occur in inductive or capacitive transients are summarized in Table 7.4. Generally, there are multiplicative constants in the exponential expressions, as shown in the third column. The response of inductors and capacitors to steady-state DC signals is examined in Section 7.2 because steady-state DC voltages and currents always result in RC and RL circuits after the transient is over. Then the exponential expressions will be utilized extensively in determining the response of voltages and currents in RC and RL circuits during the transient. These responses are covered in Section 7.3 and Section 7.4, respectively.

A major concept should be understood before leaving this section: *a given circuit may have several types of signals applied.* For example, an RC circuit may have a steady-state DC input, an AC input, a step input, or one of many other types of input signals. The signals within the circuit may, in general, be the same or different than the input signal, depending on the signal type applied. For example, a DC step applied to an RC circuit results in exponential voltage signals across the resistor and the capacitor. On the other hand, an AC signal input to an RC circuit results in AC voltage signals across the resistor and the capacitor—there is no change in signal type in this example. In general, one must distinguish signals and circuits: they are not the same thing, but both must be considered when analyzing a circuit.

7.2 RESPONSE OF RC AND RL CIRCUITS TO STEADY-STATE DC SIGNALS

The first topic to be addressed is what steady-state DC means. Every time a source is turned on, there is a transient during which the voltages and the currents must transition from zero to some steady-state DC value. Normally when "DC" is used without any modifier words, it is understood that the transient has occurred *and is over:* all voltages and currents have stabilized to their steady-state DC values. The phrase "steady-state DC" is used extensively in this chapter to distinguish it from transient DC signals.

What does a capacitor look like in a circuit when the source is a steady-state DC signal? The reasoning is based on the capacitor concepts that were established in Chapter 5. If the voltage across the capacitor has reached its steady-state DC value, then there is no change of charge on the capacitor plates. If there is no change of charge on the capacitor plates, then there is zero current because charge on the plates would be changing if current were present. What does a component with a finite voltage across it but zero current through it look like? It looks like an open. This fact is summarized by the following sentence:

A capacitor looks like an open to DC [steady-state].

The "steady-state" is usually not stated but is understood. However, the capacitor differs from an ideal open in a critical aspect: the capacitor stores energy that can be utilized later in the circuit. Although a practical open can be viewed as a capacitor with small plate areas (small, often negligible, capacitance[1]), an open is not usually intended to store charge and energy.

There is another aspect in which a capacitor may not be replaced by an open. If the leakage resistance of the capacitor is on the same order as or is smaller than the resistances in the circuit, then the capacitor should be replaced by a resistance equal to that leakage resistance instead of an open. Conveniently this is generally not the case, and the use of an open to represent a capacitor in a steady-state DC circuit is the most common case.

What does an inductor look like in a circuit when the source is a steady-state DC signal? The reasoning is based on the inductor concepts that were established in Chapter 6. If the current through the inductor has reached its steady-state DC value, then there is a constant magnetic field around the inductor. No change in magnetic field means that there is no voltage induced into the inductor (Faraday's law). What does a component with a finite current through it but zero voltage across it look like? A short. This fact is summarized by the following sentence:

An inductor looks like a short to DC [steady-state].

Again, the steady state is usually not stated but is understood. However, the inductor differs from an ideal short in two critical aspects. The inductor stores energy that can be utilized later in the circuit. Although a practical short can be viewed as an inductor with a short length of wire (small, often negligible, inductance[2]), a short is not usually intended to store energy.

The other aspect of why a practical inductor differs from an ideal short is the conductor resistance. Inductors often have a significant length of wire in them. This long wire has a finite, often non-negligible, resistance, as established in Chapter 2. Hence, it is common to use a resistance instead of a short to represent an inductor in a steady-state DC circuit. If the wire resistance is negligible compared to the other resistances in the circuit, then the short can be used.

Example 7.2.1

Simplify the circuit shown in Figure 7.6. The source has been on a long time (consider it a steady-state DC source).

Given: steady-state DC circuit shown in Figure 7.6

Desired: simplified circuit: replace *L* and *C* with their steady-state DC representations (models)

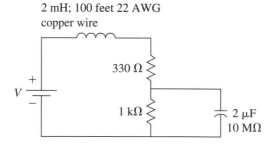

2 mH; 100 feet 22 AWG copper wire

330 Ω

1 kΩ   2 μF   10 MΩ

V

Figure 7.6 Circuit for Example 7.2.1

---

[1] As frequency increases for AC, the capacitance of an open is increasingly significant. It is one form of a *parasitic capacitance*—an interesting topic for further research.
[2] As frequency increases for AC, the inductance of a short length of conductor is increasingly significant. It is one form of a *parasitic inductance* (also called lead inductance)—another interesting topic for further research.

Strategy:

- Compare leakage R value of the capacitor with the circuit R values.
- Use an open if $R_{leak} \gg R_{ckt}$; otherwise, use R_{leak}.
- Determine the wire resistance.
- Compare the wire R value of the inductor with the circuit R values.
- Use a short if $R_{wire} \ll R_{ckt}$; otherwise, use R_{wire}.

Solution:

$$R_{leak} = 10 \text{ M}\Omega$$

$$R_{ckt} \sim 1 \text{ k}\Omega \text{ (the largest resistance in the circuit)}[3]$$

$$R_{leak} \gg R_{ckt} \Rightarrow \text{ignore } R_{leak}, \text{ and replace } C \text{ with an open}$$

22 AWG copper wire: $\rho_{Cu} = 10.4 \dfrac{\text{CM–}\Omega}{\text{ft}}, A_{22\text{ AWG}} = 640 \text{ CM}$

$$R = \frac{\rho\ell}{A} = \frac{\left(10.4 \dfrac{\text{CM–}\Omega}{\text{ft}}\right)(100 \text{ ft})}{(640 \text{ CM})} = 1.625 \ \Omega = R_{wire}$$

$$R_{ckt} \sim 330 \ \Omega \text{ (the lowest resistance in the circuit)}[4]$$

$$R_{wire} \ll R_{ckt} \Rightarrow \text{ignore } R_{wire}, \text{ and replace } L \text{ with a short}$$

The simplified circuit is shown in Figure 7.7, and steady-state DC circuit levels can be determined from it as needed.

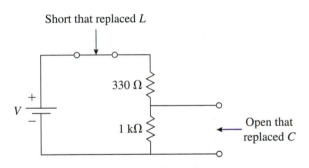

Short that replaced L

330 Ω

1 kΩ

V

Open that replaced C

Figure 7.7 Simplified Steady-state DC Circuit for Example 7.2.1

Thus, a capacitor usually can be replaced with an open in a steady-state DC circuit, or it is replaced by a resistance equal to the leakage resistance if the leakage resistance is not much greater than the circuit resistance levels. An inductor is replaced by a short in a steady-state DC circuit if the circuit resistance is much greater than the wire resistance of the inductor, or it is replaced by a resistance equal to the wire resistance if the circuit resistance is not much greater than the wire resistance of the inductor. Again, the steady-state DC responses of capacitors and inductors are the DC voltages and currents when the transient is over in an RC or RL circuit, i.e., when the circuit has reached DC steady state. Next, the time between when the circuit is turned on and when the circuit response reaches steady-state DC will be examined. Recall that this interval of time is called the *transient*.

---

[3] Technically, one should determine the Thévenin equivalent resistance as "seen" by the capacitor (Chapter 12); however, if the capacitor leakage resistance is more than 100 times the largest resistance in the circuit, R_{leak} can usually be ignored (1% or less error in a parallel circuit).

[4] Technically, one should determine the Thévenin equivalent resistance as "seen" by the inductor (Chapter 12); however, if the inductor wire resistance is less than 100 times the lowest resistance in the circuit, R_{wire} can usually be ignored (1% or less error in a series circuit).

7.3 RESPONSE OF RC SERIES CIRCUITS TO DC STEPS

The concepts behind the response of an RC circuit to the sudden turning on of a DC voltage (a step voltage, as described in Section 7.1) were established in Section 5.4. The main concept that was established is the fundamental action of capacitors in circuits:

The voltage across a capacitor cannot change instantaneously.

or the alternative equivalent statement:

Capacitors oppose an instantaneous change in voltage.

where the "voltage" is assumed to be the voltage across the capacitor. You should be able to explain why capacitors oppose an instantaneous change of voltage—please reread Section 5.4 if necessary. It was also stated in Section 5.4 that the resistance plays a role in this charging of the capacitor because the resistance will limit the current. This concept will become apparent in the discussion that follows. The approach to this discussion will be to examine an RC circuit with specific component values in order to establish the quantitative understanding of the transient voltages and current in a series RC circuit. The example contains a significant amount of reasoning and explanation—do not skip it. Then the general transient voltage and current expressions for a charging capacitor in an RC circuit will be extracted from the results of the example. The discharging of the capacitor is considered and illustrated through a continuation of the example. A summary table for charging and discharging RC circuits completes the section.

Consider the RC circuit shown in Figure 7.8. How long should the transient last once the switch is closed (at some arbitrary time that is labeled $t = 0$)? As was seen in Tables 7.2 and 7.3, and Figures 7.4 and 7.5, the exponential decay or buildup is within 1% of the steady-state value after the variable in the exponent reaches 5. This "less than 1%" is defined to be steady-state. Thus, the question, How long does the transient last? is really the question, How long does it take the exponent of the exponential function to reach -5? Hence, a thorough examination of the exponent of the exponential function for transients is necessary.

The general form of the exponent in the exponential function of transients is:

$$x = \frac{t}{\tau} \tag{7.7}$$

where t is the independent variable time and τ (the lowercase Greek letter tau) is referred to as the *time constant*. Thus, the exponent of the exponential function is a function of time. In the equations for the exponential decay and buildup in Table 7.4, $a = 1/\tau$, as shown in Table 7.5 (often, $-\frac{t}{\tau}$ is written as $-t/\tau$). How does the time constant influence the total time of the transient? If τ has a larger value, then t must become larger before the exponent becomes -5. Hence, the larger the time constant, the longer the transient lasts. How long does the transient last? If the exponent is -5 when steady-state is reached, then

$$-\frac{t}{\tau} = -5 \Rightarrow t = 5\tau \tag{7.8}$$

which means that the transient lasts for five time constants, or, equivalently, steady-state is reached after five time constants of time have elapsed. This "5τ" time interval from the start of the transient to steady-state is an engineering standard. It will be seen in RL circuits, too, in the next section.

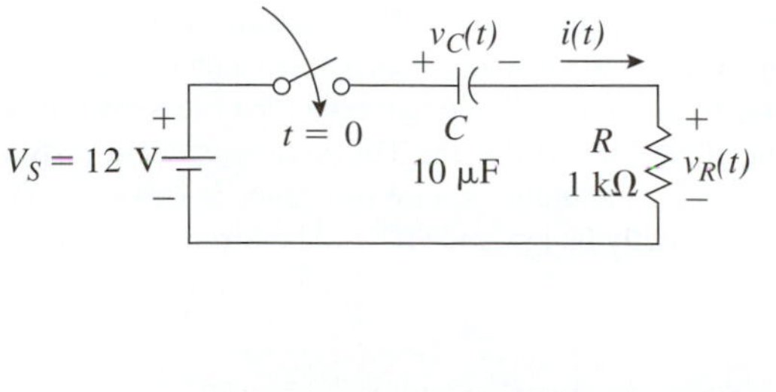

Figure 7.8 RC Circuit for Discussion of Transients

TABLE 7.5	transient type	general expression
Exponential Function Expressions in DC Transients	exponential buildup	$y = A(1 - e^{-t/\tau})$
	exponential decay	$y = Be^{-t/\tau}$

Thus, the transient lasts for 5τ. The value of the time constant needs to be determined next if any practical calculations are to be performed. The value of the time constant is mathematically derived using calculus and differential equations in transient circuit analysis. Here, qualitative arguments will be used to justify the expression for the time constant in RC circuits.

Refer again to Figure 7.8. First, upon what does the time constant depend? The time constant is a property of the circuit, not the source(s). Hence, the time constant should depend only on the resistance and capacitor component values. If the capacitance is increased, what should happen to the time constant? A larger capacitance means that more charge must flow onto the plates to reach a given voltage. Given the same limit of current because of the fixed value of series resistance, current must flow for a longer length of time. Hence, the time constant is directly proportional to capacitance.

If resistance is increased, what should happen to the time constant? A larger resistance means that the current is limited to a lower value. A smaller current means that current must flow for a longer length of time to build up the charge on the capacitor plates. Hence, the time constant is directly proportional to resistance. If the two proportionalities of this and the previous paragraph are put together into the same equation, then the time constant expression for an RC circuit results:

$$\boxed{\tau = RC} \tag{7.9}$$

Equation (7.9) has been boxed to emphasize its significance. The "RC time constant" is a well-known and often-used quantity in electronic circuits because instantaneous steps in voltages are avoided. It is even used in AC circuits where the time constant is related to how a circuit responds as frequency is changed (a topic that is treated in detail in Chapter 15). The value of τ can now be determined in a series RC circuit, and the exponential decay and buildup of signals can be calculated. The circuit in Figure 7.8 will be used to illustrate these calculations and to explain the concepts and reasoning behind the calculations. Notice how the multiplicative constants are determined in the next example.

___ **Example 7.3.1** _____

Determine, plot, and label the signals for the current, the capacitor voltage, and the voltage across the resistance in the circuit shown in Figure 7.8.

Given:

$$R = 1 \text{ k}\Omega$$

$$C = 10 \text{ }\mu\text{F}$$

$$V_S = 12 \text{ V}$$

Desired:

$$i(t) \text{ versus } t$$

$$v_R(t) \text{ versus } t$$

$$v_C(t) \text{ versus } t$$

Strategy:

■ Determine the time constant: $\tau = RC$.

■ Determine whether the quantity (voltage or current) is decaying or building up.

■ Select the appropriate exponential expression and insert the value for τ.

■ Determine the multiplicative constant from the circuit.

Solution: _____

$$\tau = RC = (1 \text{ k}\Omega)(10 \text{ }\mu\text{F}) = 10 \text{ ms}$$

$i(t)$:   The current decreases to zero as the capacitor charges $\rightarrow i(t)$ is an exponential decay:

$$i(t) = Ae^{-t/\tau}$$

The current is effectively zero after 5τ. Hence, the value of $i(t)$ at $t = 0$, i.e., $i(0)$, needs to be determined. It is emphasized that $i(0)$ is a constant: it is the value of the current at $t = 0$ only. At $t = 0$, there is zero voltage across the capacitor. Hence, by KVL, all the source voltage is across the resistance (true only at $t = 0$). The current at $t = 0$ is determined by Ohm's law:

$$i(0) = \frac{V_S}{R} = \frac{12 \text{ V}}{1 \text{ k}\Omega} = 12 \text{ mA}$$

Caution: Do not confuse the constant A with the "A" used for the unit of current.

But at $t = 0$, the current expression $i(t)$ has $t = 0$ inserted into the exponent:

$$i(0) = Ae^{-0/\tau} = Ae^0 = A(1) = A$$

Notice how the multiplicative constant was determined. When $t = 0$ is inserted into the exponent, the entire exponent is zero. Any nonzero constant to a power of 0 is equal to 1. This leaves the multiplicative constant equal to the value of the quantity (current in this case) at $t = 0$.

Hence, $i(0) = A = 12 \text{ mA}$, and the current expression is:

$$i(t) = 12e^{-t/10\text{ms}} \text{ mA}$$

The transient current signal is plotted and labeled in Figure 7.9a. The current decays from an initial value of 12 mA to 0 in an interval of 5 time constants. The initial current is equal to the source voltage divided by the series resistance because at $t = 0$, the capacitor has no charge on its plates and the capacitor voltage is 0. Thus, at $t = 0$, all voltage is across the resistance, and the initial current is V_S/R. The steady-state DC current is zero, which is the expected current in this circuit when the capacitor is replaced with an open in a steady-state DC circuit.

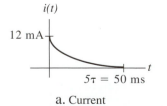

a. Current

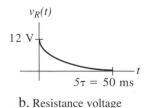

b. Resistance voltage

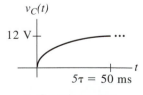

c. Capacitor voltage

Figure 7.9 Transient Signals for Example 7.3.1

$v_R(t)$:   The voltage across the resistance must track the current through the resistance (Ohm's law) → $v_R(t)$ is an exponential decay:

$$v_R(t) = Ae^{-t/\tau}$$

The voltage across the resistance is effectively zero after 5τ. Hence, the value of $v_R(t)$ at $t = 0$, i.e., $v_R(0)$, needs to be determined (again, $v_R(0)$ is a constant). At $t = 0$, there is zero voltage across the capacitor. Hence, by KVL, all the source voltage is across the resistance (true only at $t = 0$):

$$v_R(0) = V_S = 12 \text{ V}$$

Hence, $v_R(0) = A = 12$ V, and the resistance voltage expression is:

$$v_R(t) = 12e^{-t/10\text{ms}} \text{ V}$$

The transient voltage signal across the resistance is plotted and labeled in Figure 7.9b. Again, it must track the current due to Ohm's law. Hence, the voltage across the resistance decays from an initial value of 12 V to zero over an interval of five time constants. The initial value is 12 V because all the source voltage is across the resistance at $t = 0$ (there is no charge on the capacitor plates at $t = 0$ and, hence, zero voltage across the capacitor at $t = 0$). The steady-state DC voltage across the resistance is zero because the steady-state DC current is zero, as established previously.

$v_C(t)$:   The voltage across the capacitor builds up as time goes on because charge builds up on the capacitor plates due to the flow of current → $v_C(t)$ is an exponential buildup:

$$v_C(t) = A(1 - e^{-t/\tau})$$

At $t = 0$, there is zero voltage across the capacitor. The capacitor is effectively charged after 5τ. Hence, the value of $v_C(t)$ at $t = 5\tau$, i.e., $v_C(5\tau)$, needs to be determined (again, $v_C(5\tau)$ is a constant). By KVL, all the source voltage is across the capacitor at $t \geq 5\tau$:

$$v_C(5\tau) = 12 \text{ V}$$

$$v_C(5\tau) = A(1 - e^{-5\tau/\tau}) = A(1 - e^{-5})$$

But $e^{-5} = 0.007 \approx 0$; thus:

$$v_C(5\tau) = A(1 - 0) = A = 12 \text{ V}$$

The capacitor voltage expression is:

$$v_C(t) = 12(1 - e^{-t/10 \text{ ms}}) \text{ V}$$

The transient capacitor voltage signal is plotted and labeled in Figure 7.9c. The capacitor voltage builds up from zero to the source voltage over an interval of five time constants. The steady-state DC voltage across the capacitor equals the source voltage, which is the expected capacitor voltage in this circuit when the capacitor is replaced with an open in a steady-state DC circuit.

The expressions in Example 7.3.1 can now be generalized for an RC circuit when a DC voltage source is switched on at some defined time $t = 0$. The time constant is $\tau = RC$. The initial value of the current is $i(0)$, which is often symbolized by I_o. The reason for this symbol assignment is that an uppercase letter is usually a constant in an expression whereas lowercase letters are used to indicate variables. The subscript o indicates that the constant is an initial value. Hence, the general expression for the decaying current is:

$$i(t) = I_o e^{-t/\tau} \qquad (7.10)$$

The voltage across the resistance tracks the current. Hence, it will have the same mathematical form as the current, but the multiplicative constant is the initial voltage across the

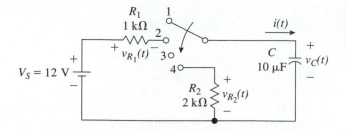

Figure 7.10 RC Circuit with Charging and Discharging Switch Positions

resistance, $v_R(0)$, which is symbolized by V_o. Hence, the general expression for the decaying voltage across the resistance in an RC circuit as the capacitor charges is:

$$v_R(t) = V_o e^{-t/\tau} \tag{7.11}$$

The voltage across the capacitor builds up as the capacitor charges. The capacitor voltage builds up from zero to some final steady-state DC value, $v_C(5\tau)$, the capacitor voltage at $t = 5\tau$. The transient ends and steady state begins at $t = 5\tau$. Again, a symbol is often assigned to $v_C(5\tau)$: V_f, the final value of the voltage across the capacitor, i.e., the steady-state DC capacitor voltage. Hence, the general expression for the voltage across the capacitor in an RC circuit as the capacitor charges is:

$$v_C(t) = V_f(1 - e^{-t/\tau}) \tag{7.12}$$

Thus, in Example 7.3.1, the constants had the following values:

$$\tau = RC = 10 \text{ ms}$$

$$I_o = 12 \text{ mA (initial circuit current)}$$

$$V_o = 12 \text{ V (initial voltage across the resistance)}$$

$$V_f = 12 \text{ V (final voltage across the capacitance)}$$

The charging of the capacitor in an *RC* circuit has now been characterized; however, the circuit is eventually turned off. Thus, the discharging of the capacitor is also practically important—every circuit is eventually turned off. How does the capacitor discharge? What are the signal waveforms during a capacitive discharge? Examine the circuit in Figure 7.10. It is the circuit from Figure 7.8 with a different switch arrangement. When the switch is in position 1, the circuit is off. When the circuit is moved to position 2, the capacitor charges and the circuit goes through a transient and reaches steady-state DC, as illustrated in Example 7.3.1. When the switch is moved to position 3, the current is zero—it was already zero in steady-state DC, but the capacitor voltage is still V_f. The charge is still on the capacitor plates, so the capacitor retains that voltage across it.

Finally, when the switch is moved to position 4, the capacitor is connected directly across another resistance (which may or may not be equal to the same value of the resistance during charging). Now the energy in the capacitor will be transferred to and converted in the resistance. Note that the current direction will reverse in discharge relative to the current direction during charging. The capacitor is returning the electrical energy to the circuit, so the capacitor behaves like a source of energy during the discharge transient. There is only finite energy stored in the electric field of the capacitor; hence, all quantities, voltages and currents, must decay during a discharge. Again, this interval of time for the decay transient is 5τ, where $\tau = R_2 C$ for the discharge. The general voltage and current expressions for the discharge are:[5]

---

[5] The subscripts on the initial voltage (of the discharge) for the capacitor and the resistance may need to be distinguished in general—the initial voltages may not be the same in all RC circuits; however, V_o is the same for $v_R(t)$ and $v_C(t)$ in the circuit of Figure 7.10.

$$i(t) = -I_o e^{-t/\tau} \tag{7.13}$$

$$v_R(t) = V_o e^{-t/\tau} \tag{7.14}$$

$$v_C(t) = V_o e^{-t/\tau} \tag{7.15}$$

where: $\tau = RC$ is the time constant (during discharge),

$I_o = i(0)$ is the initial value of the current,

$V_o = v_C(0)$ is the initial voltage across the capacitor, and

$V_o = v_R(0)$ is the initial voltage across the resistance.

The current is negative during the discharge because of the reverse current direction relative to the current direction during charging. The circuit in Figure 7.10 will be used in Example 7.3.2 to illustrate the discharge.

Example 7.3.2

Determine, plot, and label the signals for the current, the capacitor voltage, and the voltage across the resistance in the circuit shown in Fig. 7.10. The switch is moved from position 1 to 2 at $t = 0$. The switch remains in positions 2 and 3 for 100 ms each, and then remains in position 4 thereafter.

Given:

$$R_1 = 1\text{ k}\Omega$$

$$R_2 = 2\text{ k}\Omega$$

$$C = 10\text{ }\mu\text{F}$$

$$V_S = 12\text{ V}$$

Desired:

$$i(t) \text{ versus } t$$

$$v_R(t) \text{ versus } t$$

$$v_C(t) \text{ versus } t$$

Strategy: Determine the time constants: $\tau_c = R_1 C$ (charging time constant)

$\tau_d = R_2 C$ (discharging time constant)

For switch positions 2 and 4:

■   Determine whether the quantity (voltage or current) is decaying or building up.

■   Select the appropriate exponential expression and insert the value for τ.

■   Determine the multiplicative constant from the circuit.

Solution:

$$\tau_c = R_1 C = (1\text{ k}\Omega)(10\text{ }\mu\text{F}) = 10\text{ ms} \qquad \text{(charging time constant)}$$

$$\tau_d = R_2 C = (2\text{ k}\Omega)(10\text{ }\mu\text{F}) = 20\text{ ms} \qquad \text{(discharging time constant)}$$

switch position 1: all variables are zero

switch position 2: charging RC circuit: $v_C(t)$ builds up, $i(t)$ and $v_{R_1}(t)$ decay

$$i(t) = I_o e^{-t/\tau_c} = \frac{V_S}{R_1} e^{-t/\tau_c} = \frac{12\text{ V}}{1\text{ k}\Omega} e^{-t/\tau_c} = 12e^{-t/10\text{ ms}}\text{ mA}$$

$$v_{R_1}(t) = V_o e^{-t/\tau_c} = V_S e^{-t/\tau_c} = 12e^{-t/10\text{ ms}}\text{ V}$$

$$v_C(t) = V_f(1 - e^{-t/\tau_c}) = V_S(1 - e^{-t/\tau_c}) = 12(1 - e^{-t/10\text{ ms}})\text{ V}$$

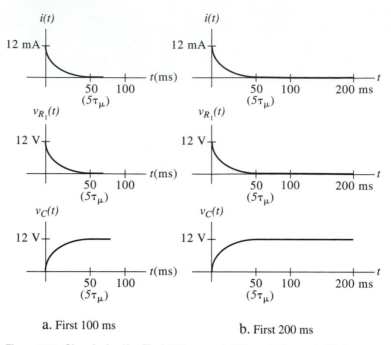

a. First 100 ms

b. First 200 ms

Figure 7.11 Signals for the First 100 ms and 200 ms in Example 7.3.2

The signals for the first 100 ms are plotted in Figure 7.11a. Note that steady-state DC is reached after $5\tau = 50$ ms.

switch position 3: All variables are zero except the capacitor voltage, which previously reached the steady-state DC value of 12 V. None of the voltages or currents change values while the switch is in position 3. The signals for the first 200 ms are plotted in Figure 7.11b.

switch position 4: discharging RC circuit: $v_C(t)$, $i(t)$ and $v_{R2}(t)$ decay

$$i(t) = -I_o e^{-t/\tau_d} = -\frac{V_S}{R_2}e^{-t/\tau_d} = -\frac{12\text{ V}}{2\text{ k}\Omega}e^{-t/\tau_d} = -6e^{-t/20\text{ ms}}\text{ mA}$$

$$v_{R_2}(t) = V_o e^{-t/\tau_d} = 12e^{-t/20\text{ ms}}\text{ V}$$

$$v_C(t) = V_o\, e^{-t/\tau_d} = 12e^{-t/20\text{ ms}}\text{ V}$$

where $t = 0$ for the time variable in these equations is really at $t = 200$ ms in this problem. The signals are plotted in Fig. 7.12. Note that steady-state DC is reached after $5\tau_c = 50$ ms for the charging interval, and is reached again after $5\tau_d = 100$ ms for the discharging interval. The current is negative during discharge because it is going in the opposite direction of that defined in Fig. 7.10, and it is 6 mA because there is a larger resistance in the discharging circuit relative to the charging circuit. Another plot had to be added for the voltage across R_2 because R_1 is not in the discharging circuit.

The key expressions and plots for charging and discharging RC circuits are summarized in Table 7.6 and Figure 7.13.

The capacitor can maintain a charge for a long time if its leakage resistance is low. Hence, many circuits actually switch the capacitor(s) to a resistor with a large resistance when they are switched off to allow the capacitors to slowly discharge. A resistor used in this manner is called a *bleeder resistor* because it "bleeds" off the charge on the capacitor(s).

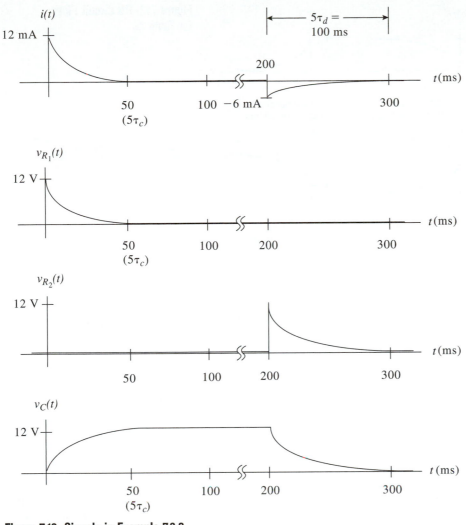

Figure 7.12 Signals in Example 7.3.2

TABLE 7.6	variable	charging RC circuit		discharging RC circuit	
		expression	plot	expression	plot
RC Circuit Transient Summary ($\tau = RC$)	capacitor voltage $v_C(t)$	$v_C(t) = V_f(1 - e^{-t/\tau})$	Fig. 7.13a	$v_C(t) = V_o e^{-t/\tau}$	Fig. 7.13d
	current $i(t)$	$i(t) = I_o e^{-t/\tau}$	Fig. 7.13b	$i(t) = -I_o e^{-t/\tau}$	Fig. 7.13e
	voltage across the resistance $v_R(t)$	$v_R(t) = V_o e^{-t/\tau}$	Fig. 7.13c	$v_R(t) = V_o e^{-t/\tau}$	Fig. 7.13f

7.4 RESPONSE OF RL SERIES CIRCUITS TO DC STEPS

The concepts behind the response of an RL circuit to the sudden turning on of a DC voltage (a step voltage, as described in Section 7.1) were established in Section 6.5. The main concept that was established is the fundamental action of inductors in circuits:

The current cannot change instantaneously through an inductor.

or the alternative equivalent statement:

Inductors oppose an instantaneous change of current.

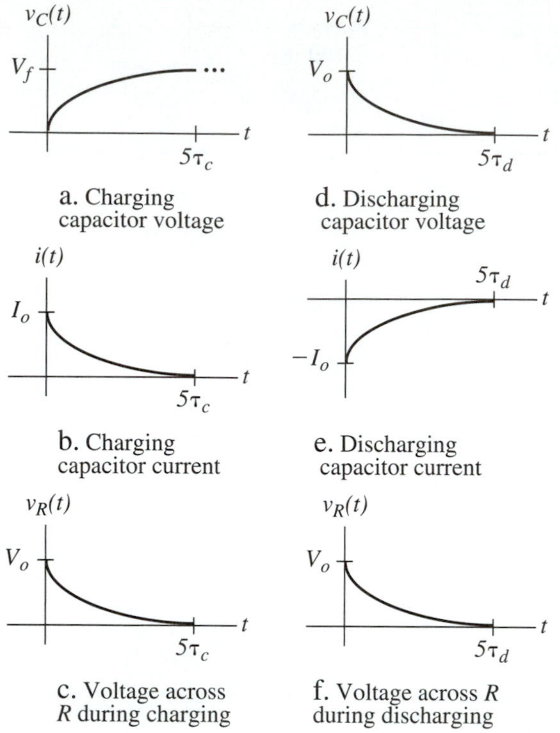

$v_C(t)$
V_f
$5\tau_c$
a. Charging capacitor voltage
$i(t)$
I_o
$5\tau_c$
b. Charging capacitor current
$v_R(t)$
V_o
$5\tau_c$
c. Voltage across R during charging
$v_C(t)$
V_o
$5\tau_d$
d. Discharging capacitor voltage
$i(t)$
$5\tau_d$
$-I_o$
e. Discharging capacitor current
$v_R(t)$
V_o
$5\tau_d$
f. Voltage across R during discharging

Figure 7.13 RC Circuit Plots for Table 7.6

where the current is understood to be through the inductor. You should be able to explain why inductors oppose an instantaneous change of current—please reread Section 6.5 if necessary. Resistance also plays a role in the RL circuit because the resistance will limit the current. This concept will become apparent in the discussion that follows. The approach to this discussion will be to examine an RL circuit with specific component values in order to establish the quantitative understanding of the transient voltages and currents in an RL circuit. The example contains a significant amount of reasoning and explanation—do not skip it. Then the general transient voltage and current expressions for the buildup of an inductor in an RL circuit will be extracted from the results of the example. The decay of the inductor is considered and illustrated through a continuation of the example. A summary table for the transient responses of RL circuits completes the section. It is assumed that you have studied Section 7.3 already.

Consider the RL circuit shown in Figure 7.14. How long should the transient last once the switch is closed (at some arbitrary time that is labeled $t = 0$)? As seen in Section 7.3, the transient lasts for five time constants (5τ), or, equivalently, steady state is reached after five time constants of time have elapsed. Thus, the question at this point is how to determine the value of the time constant in an RL circuit. Again, the value of the time constant is mathematically derived using calculus and differential equations in transient circuit analysis. Here, qualitative arguments will be used to justify the expression for the time constant in RL circuits.

Refer again to Figure 7.14. As with the RC circuit, the time constant is a property of the circuit, not the source(s). Hence, the time constant should depend only on the resistance and inductor component values. If the inductance is increased, what should happen to the time constant? A larger inductance means that more energy is stored in the magnetic field for a given current. Given the same limit of current because of the fixed value of

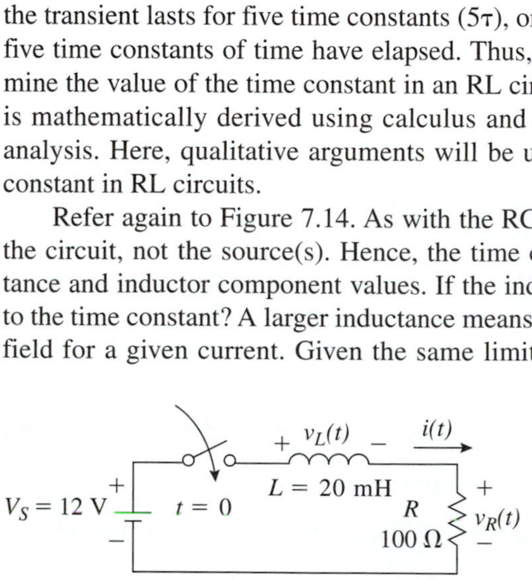

$v_L(t)$
$i(t)$
$L = 20$ mH
$V_S = 12$ V
$t = 0$
R
100 Ω
$v_R(t)$

Figure 7.14 RL Circuit for Discussion of Transients.

series resistance, current must build up over a longer length of time to reach the steady-state DC energy level in the magnetic field of the inductor. Hence, the time constant is directly proportional to inductance.

If resistance is increased, what should happen to the time constant? A larger resistance means that the current is lower. A smaller current means that there is less energy stored in the magnetic field. Consequently, it should take less time to build up or decay the energy in the magnetic field. Hence, the time constant is inversely proportional to resistance. If the two proportionalities of this and the previous paragraph are put together into the same equation, then the time constant expression for an RL circuit results:

$$\tau = \frac{L}{R}$$
(7.16)

Equation (7.16) has been boxed to emphasize its significance. The "RL time constant" is a well-known and often-used quantity in electronic circuits because instantaneous steps in currents are avoided. It is even used in AC circuits where the time constant is related to how a circuit responds as frequency is changed (a topic that is treated in detail in Chapter 15). The value of τ can now be determined in a series RL circuit, and the exponential decay and buildup of signals can be calculated. The circuit in Figure 7.14 will be used to illustrate these calculations and to explain the concepts and reasoning behind the calculations. Notice how the multiplicative constants are determined in this example.

Example 7.4.1

Determine, plot, and label the signals for the current, the inductor voltage, and the voltage across the resistance in the circuit shown in Figure 7.14.

Given:

$$R = 100 \ \Omega$$

$$L = 20 \ \text{mH}$$

$$V_S = 12 \ \text{V}$$

Desired:

$i(t)$ versus t

$v_R(t)$ versus t

$v_L(t)$ versus t

Strategy:

■   Determine the time constant: $\tau = L/R$.

■   Determine whether the quantity (voltage or current) is decaying or building up.

■   Select the appropriate exponential expression and insert the value for τ.

■   Determine the multiplicative constant from the circuit.

Solution:

$$\tau = \frac{L}{R} = \frac{20 \ \text{mH}}{100 \ \Omega} = 0.2 \ \text{ms} = 200 \ \mu\text{s}$$

$i(t)$:   The current increases from zero to the steady-state DC current value as the inductor magnetic field builds up → $i(t)$ is an exponential buildup:

$$i(t) = A(1 - e^{-t/\tau}) \ \text{amperes}$$

(Again, do not confuse the constant A with the unit of current.) The current effectively reaches steady-state DC after 5τ. Hence, the value of $i(t)$ at $t = 5\tau$, i.e., $i(5\tau)$, needs to be determined. It is emphasized that $i(5\tau)$ is a constant: it is the value of the current at $t = 5\tau$ only. At $t = 5\tau$, there is zero voltage across the ideal inductor. Hence, by KVL, all the source voltage is across the resistance (at $t = 5\tau$ and later in time). The current at $t = 5\tau$ is determined by Ohm's law:

$$i(5\tau) = \frac{V_S}{R} = \frac{12 \text{ V}}{100 \text{ } \Omega} = 120 \text{ mA} = I_f$$

where I_f is used to indicate the "final" value of the current, i.e., the steady-state DC value. But at $t = 5\tau$, the current expression $i(t)$ has $t = 5\tau$ inserted into the exponent:

$$i(5\tau) = A(1 - e^{-5\tau/\tau}) = A(1 - e^{-5}) \approx A(1) = A$$

Hence, $i(5\tau) = A = 120 \text{ mA} = I_f$, and the current expression is:

$$i(t) = I_f(1 - e^{-t/\tau}) = 120 \, (1 - e^{-t/200 \text{ } \mu s}) \text{ mA}$$

Notice how the multiplicative constant was determined. When $t = 5\tau$ is inserted into the exponent, the exponential function is effectively zero. This leaves the multiplicative constant equal to the value of the quantity (current in this case) at $t = 5\tau$.

The transient current signal is plotted and labeled in Figure 7.15a. The current builds up from zero to a final value of 120 mA in an interval of five time constants. The final current is equal to the source voltage divided by the series resistance because at $t \geq 5\tau$, the inductor has zero voltage across it. Thus, at $t = 5\tau$, all voltage is across the resistance, and the final current is V_S/R. This current is also the steady-state DC current, which is the expected current in this circuit when the inductor is replaced with a short in a steady-state DC circuit.

$v_R(t)$:   The voltage across the resistance must track the current through the resistance (Ohm's law) → $v_R(t)$ is an exponential buildup:

$$v_R(t) = A \, (1 - e^{-t/\tau})$$

The voltage across the resistance starts at 0 V and builds up to the source voltage after 5τ. Hence, the value of $v_R(t)$ at $t = 5\tau$, i.e., $v_R(5\tau)$, is V_S (again, $v_R(5\tau)$ is a constant). At $t = 5\tau$, there is zero voltage across the inductor. Hence, by KVL, all the source voltage is across the resistance (at $t = 5\tau$ and beyond):

$$v_R(5\tau) = V_S = 12 \text{ V}$$

Figure 7.15 Transient Signals for Example 7.4.1

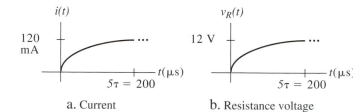

a. Current   b. Resistance voltage

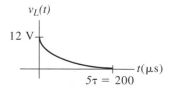

c. Inductor voltage

Hence, $v_R(5\tau) = A = 12\,\text{V} = V_f$, where V_f is the final voltage across the resistance at $t \geq 5\tau$, and the resistance voltage expression is:

$$v_R(t) = V_f(1 - e^{-t/\tau}) = 12(1 - e^{-t/200\,\mu s})\,\text{V}$$

The transient voltage signal across the resistance is plotted and labeled in Figure 7.15b. Again, it must track the current because of Ohm's law. Hence, the voltage across the resistance starts at zero volts and builds up to a value of 12 V over an interval of five time constants. The initial value is 0 V because all the source voltage is across the inductor at $t = 0$ (the inductor opposes an instantaneous change of current, so at $t = 0$, there is zero current and, hence, zero voltage across the resistance at $t = 0$). The steady-state DC voltage across the resistance is equal to the source voltage because the steady-state DC voltage across the inductor is zero, as established previously.

$v_L(t)$: The voltage across the inductor decays as time goes on because there is a slowing rate of change of current and the magnetic field $\rightarrow v_L(t)$ is an exponential decay:

$$v_L(t) = Ae^{-t/\tau}$$

At $t = 0$, there is a maximum voltage across the inductor because the inductor is opposing the maximum rate of change of current. In other words, the magnetic field has its maximum rate of change at $t = 0$ because the current is trying to increase from zero to some steady-state DC value. By Faraday's law, the highest rate of change of magnetic field induces the largest voltage. Thus, at $t = 0$, the voltage across the inductor must equal the source voltage and oppose the source voltage, polarity-wise, by Lenz's law. As the current increases it comes closer to its steady-state DC value; hence, there is less change in the magnetic field as the current and the magnetic field approach the steady-state DC values because less change is required to reach the steady-state DC values. The inductor has effectively reached steady-state DC after 5τ. Hence, the value of $v_L(t)$ at $t = 5\tau$, i.e., $v_L(5\tau)$, is zero. Thus:

$$v_L(0) = 12\,\text{V} = V_o$$

where V_o is assigned to the initial value of the voltage across the inductor. The expression for $v_L(t)$ at $t = 0$ becomes:

$$v_L(0) = Ae^{-0/\tau} = A(1) = A = V_o$$

The inductor voltage expression is:

$$v_L(t) = V_o e^{-t/\tau} = 12e^{-t/200\,\mu s}\,\text{V}$$

The transient inductor voltage signal is plotted and labeled in Figure 7.15c. The inductor voltage decays from the source voltage to zero over an interval of five time constants. The steady-state DC voltage across the inductor is zero, which is the expected inductor voltage in this circuit when the inductor is replaced with a short in a steady-state DC circuit.

In summary, the general expressions for the voltages and the currents in an RL circuit when a DC voltage source is switched on at some defined time $t = 0$ are:

$$i(t) = I_f(1 - e^{-t/\tau}) \tag{7.17}$$

$$v_R(t) = V_f(1 - e^{-t/\tau}) \tag{7.18}$$

$$v_L(t) = V_o e^{-t/\tau} \tag{7.19}$$

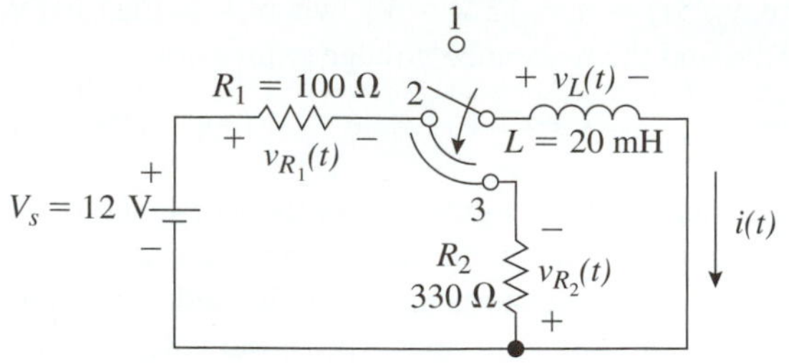

Figure 7.16 RL Circuit with Buildup and Decaying Switch Positions

where: $\tau = L/R$ is the time constant,

$I_f = i(5\tau)$ is the final value of the current,

$V_f = v_R(5\tau)$ is the final voltage across the resistance, and

$V_o = v_L(0)$ is the final voltage across the inductor.

Thus, in Example 7.4.1, the constants have the following values:

$$\tau = L/R = 200 \ \mu s$$

$$I_f = 120 \text{ mA (final circuit current)}$$

$$V_o = 12 \text{ V (initial voltage across the inductor)}$$

$$V_f = 12 \text{ V (final voltage across the resistance)}$$

The buildup of the inductor in an *RL* circuit has now been characterized; however, the circuit is eventually turned off. Thus, the decay of the inductor is practically important—every circuit is eventually turned off. How does the inductor decay? What are the signal waveforms during an inductive decay? Examine the circuit in Figure 7.16. It is the circuit from Figure 7.14 with a different switch arrangement. When the switch is in position 1, the circuit is off. When the circuit is moved to position 2, the magnetic field of the inductor builds up and the circuit goes through a transient and reaches steady-state DC as illustrated in Example 7.4.1. The current and voltages reach steady-state DC.

Finally, when the switch is moved toward position 3, notice that the switch makes contact with position 3 before it breaks contact with position 2. This is a special switch called a *make-before-break* switch. In other words, the switch makes contact with the next switch position before breaking contact with the previous switch position. Why would such a switch be needed? Consider the case of not using a make-before-break switch. When the switch is just moving off of position 2, there is a small gap through the air. However, the fundamental property of inductance is opposition to an instantaneous change of current. The magnetic field of the inductor would collapse in an attempt to keep the current going because the energy in the magnetic field has to go somewhere. Hence, the current actually arcs across the terminals of the switch. The energy to provide the high voltage required for the current to arc through the air comes from the magnetic field. The arcing time interval is short: $\tau = L/R$, and R through air is very high, so the time constant is extremely short. The arc is also detrimental to the switch because the contact materials are eroded with each arc. In some environments, arcs can ignite flammable gases, so this type of switch cannot be used. Hence, make-before-break switches are often used with inductors to prevent the arcing problem.

Thus, when the switch moves to position 3 in Figure 7.16, the inductor is connected directly across another resistance (which may or may not be equal to the value of the resistance during charging). Now the energy in the inductor will be transferred to and converted in this resistance. Note that the induced voltage across the inductor will have a reversed polarity during decay relative to the voltage polarity during buildup—by Lenz's law, the inductor opposes a change of current, so the voltage across the inductor will be such as to try to prevent the current from decreasing. This action corresponds to the collapsing of the magnetic field of the inductor. In fact, another way to look at this situation is from the magnetic flux viewpoint. The magnetic flux is moving in the opposite direction past the wire in the inductor during decay relative to the buildup case. Hence, the polarity of the induced voltage must be opposite to that of the buildup case.

During the decay transient, the inductor is returning the electrical energy stored in the magnetic field to the circuit, so the inductor behaves like a source of energy during the transient. There is only finite energy stored in the magnetic field of the inductor; hence, all quantities, voltages and currents, must decay during a decay. Again, this interval of time for the decay transient is 5τ, where $\tau = L/R_2$. The general voltage and current expressions for the discharge are:[6]

$$i(t) = I_o e^{-t/\tau} \tag{7.20}$$

$$v_R(t) = V_o e^{-t/\tau} \tag{7.21}$$

$$v_L(t) = -V_o e^{-t/\tau} \tag{7.22}$$

where the negative sign in Equation (7.22) accounts for the opposite polarity of the induced voltage during decay relative to buildup. The circuit in Figure 7.16 will be used in Example 7.4.2 to illustrate the decay.

Example 7.4.2

Determine, plot, and label the signals for the current, the inductor voltage, and the voltage across the resistance in the circuit shown in Figure 7.16. The switch is moved from position 1 to 2 at $t = 0$. The switch remains in positions 2 for 2 ms, and then is moved to position 3 and remains in that position thereafter.

Given:

$$R_1 = 100\ \Omega$$

$$R_2 = 330\ \Omega$$

$$L = 20\ \text{mH}$$

$$V_S = 12\ \text{V}$$

Desired:

$$i(t) \text{ versus } t$$

$$v_R(t) \text{ versus } t$$

$$v_L(t) \text{ versus } t$$

Strategy: Determine the time constants: $\tau_c = L/R_1$, $\tau_d = L/R_2$.

For switch positions 2 and 3:

■   Determine whether the quantity (voltage or current) is decaying or building up.
■   Select the appropriate exponential expression and insert the value for τ.
■   Determine the multiplicative constant from the circuit.

Solution:

$$\tau_c = \frac{L}{R_1} = \frac{0.02\ \text{H}}{100\ \Omega} = 200\ \mu\text{s} \qquad \text{(buildup time constant)}$$

$$\tau_d = \frac{L}{R_2} = \frac{0.02\ \text{H}}{330\ \Omega} = 60.606\ \mu\text{s} \quad \text{(discharging time constant)}$$

---

[6] The subscripts on the initial voltage (of the decay) for the inductor and the resistance may need to be distinguished in general—the initial voltages may not be the same in all RL circuits. However, V_o is the same for $v_R(t)$ and $v_L(t)$ in the circuit of Fig. 7.16.

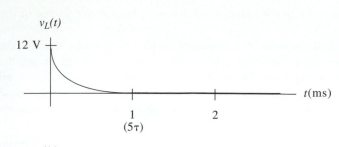

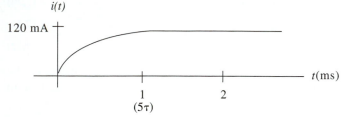

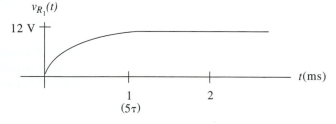

Figure 7.17 Signals for the First 200 ms in Example 7.4.2

switch position 1: all variables are zero

switch position 2: building RL circuit: $i(t)$ and $v_{R1}(t)$ build up, $v_L(t)$ decays

$$\tau = \tau_c = 200 \ \mu s$$

$$i(t) = I_f(1 - e^{-t/\tau_c}) = \frac{V_S}{R_1}(1 - e^{-1/\tau_c}) = \frac{12 \ V}{100 \ \Omega}(1 - e^{-t/\tau_c}) = 120(1 - e^{-t/200\mu s}) \ mA$$

$$v_{R_1}(t) = V_f(1 - e^{-t/\tau_c}) = V_S(1 - e^{-t/\tau_c}) = 12(1 - e^{-t/200\mu s}) \ V$$

$$v_L(t) = V_o e^{-t/\tau_c} = V_S e^{-t/\tau_c} = 12 e^{-t/200\mu s} \ V$$

The signals for the first 2 ms are plotted in Figure 7.17. Note that steady-state DC is reached after $5\tau = 1$ ms. All variables remain at the steady-state DC values until the switch is moved to position 3 at $t = 2$ ms.

switch position 3: decaying RL circuit: $v_L(t)$, $i(t)$ and $v_{R_2}(t)$ decay

$$\tau = \tau_d = 60.6 \ \mu s$$

$$i(t) = I_o e^{-t/\tau_d} = 120 e^{-t/60.6 \ \mu s} \ mA$$

$$v_{R_2}(t) = V_o e^{-t/\tau_d} = I_o R e^{-t/\tau_d} = (120 \ mA)(330 \ \Omega)e^{-t/\tau_d} = 39.6 e^{-t/60.6 \ \mu s} \ V$$

$$v_L(t) = -V_o e^{-t/\tau_d} = -39.6 e^{-t/60.6 \ \mu s} \ V$$

where $t = 0$ for the time variable in these equations is really at $t = 2$ ms in this problem. The signals are plotted in Figure 7.18. Note that steady-state DC is reached after $5\tau_c = 1$ ms for the charging interval, and is reached again after $5\tau_d = 0.303$ ms for the discharging interval. The inductor voltage is negative during decay because it has the opposite polarity of that defined in Figure 7.16. The value of the voltage is 39.6 V at $t = 2$ ms because the current must be the same immediately after the switch changes position and the current cannot change instantaneously through an inductor. With the same current and a larger resistance in the decaying circuit relative to the buildup circuit, the voltage at $t = 2$ ms will be larger than it was during buildup. Another plot had to be added for the voltage across R_2 because R_1 is not in the decaying circuit.

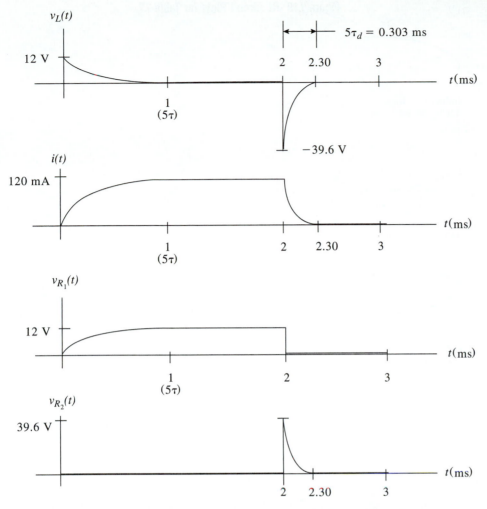

Figure 7.18 Signals in Example 7.4.2

The key expressions and plots for charging and discharging RL circuits are summarized in Table 7.7 and Figure 7.19.

TABLE 7.7	variable	building RL circuit		decaying RL circuit	
RL Circuit Transient Summary ($\tau = L/R$)		expression	plot	expression	plot
	inductor voltage $v_L(t)$	$v_L(t) = V_o e^{-t/\tau}$	Fig. 7.19a	$v_L(t) = -V_o e^{-t/\tau}$	Fig. 7.19d
	current $i(t)$	$i(t) = I_f(1 - e^{-t/\tau})$	Fig. 7.19b	$i(t) = I_o e^{-t/\tau}$	Fig. 7.19e
	voltage across the resistance $v_R(t)$	$v_R(t) = V_f(1 - e^{-t/\tau})$	Fig. 7.19c	$v_R(t) = V_o e^{-t/\tau}$	Fig. 7.19f

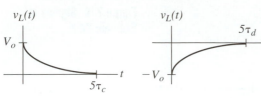

Figure 7.19 RL Circuit Plots for Table 7.7

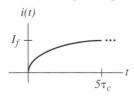

a. Inductor voltage
   during buildup

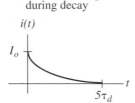

d. Inductor voltage
   during decay

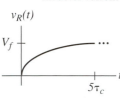

b. Building
   inductor current

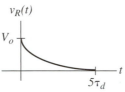

e. Decaying
   inductor current

c. Voltage across
   R during buildup

f. Voltage across
   R during decay

CHAPTER REVIEW

7.1   The Exponential Signal
- There are several signal types: steady-state DC, AC, step, exponential buildup, exponential decay, etc.
- The exponential buildup has the mathematical form $y = A(1 - e^{-ax})$.
- The exponential decay has the mathematical form $y = Ae^{-ax}$.
- Exponential functions are used to predict voltages and currents in RC and RL circuit transients.

7.2   Response of RC and RL Circuits to Steady-State DC Signals
- The ideal capacitor looks like an open to steady-state DC. A capacitor is replaced with an open in steady-state DC circuit analysis.
- If a capacitor has a leakage resistance that is not much larger than circuit resistance levels, then it should be replaced with a resistance equal to the leakage resistance in steady-state DC circuit analysis.
- The ideal inductor looks like a short to steady-state DC. An inductor is replaced by a short in steady-state DC circuit analysis.
- If an inductor has a wire resistance that is not much smaller than the circuit resistance levels, then it should be replaced with a resistance

equal to the wire resistance in steady-state DC circuit analysis.

7.3   Response of RC Series Circuits to DC Steps
- The time constant of a series RC circuit is $\tau = RC$ (charging or discharging).
- The transient lasts for 5τ. Voltages and currents in the RC circuit are within 1% of the steady-state DC values after 5τ.
- The voltage across the capacitor builds during charging: $v_C(t) = V_f(1 - e^{-t/\tau})$
- The current through and the voltage across the resistance decay during charging:

$$i(t) = I_o e^{-t/\tau} \qquad\qquad v_R(t) = V_o e^{-t/\tau}$$

- All voltages and currents decay during the discharge:

$$v_C(t) = V_o e^{-t/\tau} \qquad i(t) = -I_o e^{-t/\tau}$$
$$v_R(t) = V_o e^{-t/\tau}$$

- The current direction in discharge is the opposite of the current direction during charging.

7.4   Response of RL Series Circuits to DC Steps
- The time constant of a series RL circuit is $\tau = L/R$ (building or decaying).
- The transient lasts for 5τ. Voltages and currents in the RL circuit are within 1% of the steady-state DC values after 5τ.

- The current through and the voltage across the resistance build during charging:

$$i(t) = I_f(1 - e^{-t/\tau}) \qquad v_R(t) = V_f(1 - e^{-t/\tau})$$

- The voltage across the inductor decays during the buildup: $\quad v_L(t) = V_o e^{-t/\tau}$
- All voltages and currents exponentially decrease during the decay:

$$v_L(t) = -V_o e^{-t/\tau} \quad i(t) = I_o e^{-t/\tau}$$
$$v_R(t) = V_o e^{-t/\tau}$$

- The voltage polarity across the inductor during the decay is the opposite of the voltage polarity during the buildup.

HOMEWORK PROBLEMS

The Exponential Signal: Completion of the following problems will demonstrate your understanding of the exponential function and why it produces exponential decay and buildup behavior.

7.1 Calculate and plot the values of $f(x) = 25e^x$ for $x = 0, 0.5, 1.0, 1.5, 2.0,$ and 2.5.
    a. Describe what is happening to $f(x)$.
    b. Explain whether or not this is a stable function for positive values of x.

7.2 Calculate and plot the values of $f(x) = 6e^{-x}$ for $x = 0, 1.0, 2.0, 3.0, 4.0,$ and 5.0.
    a. Describe what is happening to $f(x)$.
    b. Explain whether or not this is a stable function for positive values of x.

7.3 Calculate and plot the values of $f(x) = 6(1 - e^{-x})$ for $x = 0, 1.0, 2.0, 3.0, 4.0,$ and 5.0.
    a. Describe what is happening to $f(x)$.
    b. Explain whether or not this is a stable function for positive values of x.

7.4 Calculate and plot the values of $f(t) = e^{-t/10}$ for $t = 0, 10.0, 20.0, 30.0, 40.0,$ and 50.0.
    a. Describe what is happening to $f(t)$.

    b. Explain whether or not this is a stable function for positive values of t.

7.5 Calculate and plot the values of $f(t) = 33e^{-t/3.3}$ for $t = 0, 1.0, 4.0, 7.0, 14.0,$ and 17.0.
    a. Describe what is happening to $f(t)$.
    b. Explain whether or not this is a stable function for positive values of t.

7.6 Calculate and plot the values of $f(t) = 12(1 - e^{-t/20})$ for $t = 0, 15.0, 25.0, 38.0, 66.0, 77.0, 91.0,$ and 108.0.
    a. Describe what is happening to $f(t)$.
    b. Explain whether or not this is a stable function for positive values of t.

7.7 Calculate and plot the values of $f(t) = 15(1 - e^{-t/144})$ for $t = 0, 120.0, 260.0, 430.0, 570.0,$ and 725.0.
    a. Describe what is happening to $f(t)$.
    b. Explain whether or not this is a stable function for positive values of t.

Steady-State Response of RC and RL Circuits to DC Signals: Completion of the following problems will demonstrate your ability to explain the DC replacements for capacitors and inductors.

7.8 Provide short answers to the following questions:
    a. When a circuit is in DC steady state, what does a capacitor look like? Explain.
    b. When a circuit is in DC steady state, what does an inductor look like? Explain.
    c. Does wire size make a difference in question (a)? Explain.
    d. Does wire size make a difference in question (b)? Explain.

7.9 Draw the equivalent steady-state DC circuit for the circuit shown in Figure P7.1. Assume there is no leakage resistance.

7.10 Draw the equivalent steady-state DC circuit for the circuit shown in Figure P7.2. Assume there is no wire resistance.

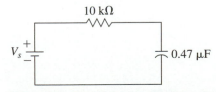

Figure P7.1 Circuit to be Evaluated in Problem 7.9

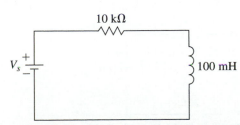

Figure P7.2 Circuit to be Evaluated in Problem 7.10

7.11 Draw the equivalent steady-state DC circuit for the circuit shown in Figure P7.3. Assume there is no leakage or wire resistance.

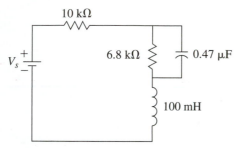

Figure P7.3 Circuit to be Evaluated in Problem 7.11

7.12 Draw the equivalent steady-state DC circuit for the circuit shown in Figure P7.4. Assume there is no leakage or wire resistance.

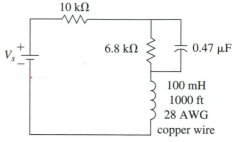

Figure P7.4 Circuit to be Evaluated in Problems 7.12 and 7.13

7.13 Draw the equivalent steady-state DC circuit for the circuit shown in Figure P7.4. Assume a 100 kΩ leakage resistance and assume that the wire used for the inductor is 28 AWG copper wire.

Response of RC and RL Circuits to DC Steps: Completion of the following problems will demonstrate your ability to determine the transient expressions and plots for the current and the voltages for both the buildup and the decay of a series RC circuit and of a series RL circuit.

7.14 Determine the expressions for $v_c(t)$, $i(t)$, and $v_R(t)$ for the circuit shown in Figure P7.5, and plot the results from time $t = 0$ until steady-state is reached.

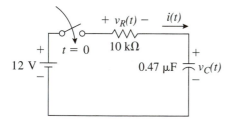

Figure P7.5 Circuit to be Evaluated in Problem 7.14

7.15 Determine the expressions for $v_c(t)$, $i(t)$, and $v_R(t)$ for the circuit shown in Figure P7.6, and plot the results from time $t = 0$ until steady-state is reached.

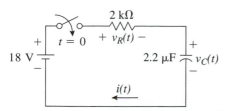

Figure P7.6 Circuit to be Evaluated in Problem 7.15

7.16 Determine the expressions for $v_c(t)$, $i(t)$, and $v_R(t)$ for the circuit shown in Figure P7.7, and plot the results from time $t = 0$ until steady-state is reached. Assume that $v_c(0) = 20.0$ V.

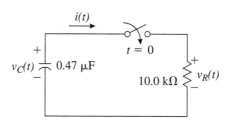

Figure P7.7 Circuit to be Evaluated in Problem 7.16

7.17 Determine the following for the circuit shown in Figure P7.8. Assume that $v_c(0) = 12.5$ V.
  a. The expressions for $v_c(t)$, $i(t)$, and $v_R(t)$.
  b. How much time will it take to complete a 75% change in $v_c(t)$, $i(t)$, and $v_R(t)$?

c. Plot $v_c(t)$, $i(t)$, and $v_R(t)$ from time $t = 0$ until steady-state is reached.

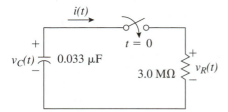

Figure P7.8 Circuit to be Evaluated in Problem 7.17

7.18   Determine the expressions for $v_L(t)$, $i(t)$, and $v_R(t)$ for the circuit shown in Figure P7.9, and plot the results from time $t = 0$ until steady-state is reached.

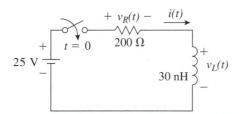

Figure P7.9 Circuit to be Evaluated in Problem 7.18

7.19   Determine the following for the circuit shown in Figure P7.10.
   a. The expressions for $v_L(t)$, $i(t)$, and $v_R(t)$.
   b. How much time will it take to complete a 75% change in $v_L(t)$, $i(t)$, and $v_R(t)$?
   c. Plot $v_L(t)$, $i(t)$, and $v_R(t)$ from time $t = 0$ until steady-state is reached.

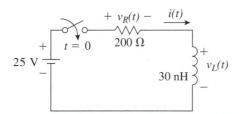

Figure P7.10 Circuit to be Evaluated in Problem 7.19

7.20   Determine the expressions for $v_L(t)$, $i(t)$, and $v_R(t)$ for the circuit shown in Figure P7.11 after the switch is moved to position 2, and plot the results from time $t = 0$ until the parameters reach steady state. Assume that the circuit is in steady

state when the switch is moved from position 1 to position 2 at $t = 0$, and also assume that the switch is a make-before-break switch.

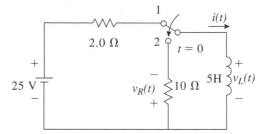

Figure P7.11 Circuit to be Evaluated in Problem 7.20

7.21   Determine the following for the circuit shown in Figure P7.12. Assume that the circuit is in steady state when the switch is moved from position 1 to position 2 at $t = 0$, and also assume that the switch is a make-before-break switch.
   a. The expressions for $v_L(t)$, $i(t)$, and $v_R(t)$.
   b. How much time will it take to complete a 55% change in $v_L(t)$, $i(t)$, and $v_R(t)$?
   c. Plot $v_L(t)$, $i(t)$, and $v_R(t)$ from time $t = 0$ until the parameters reach steady state.

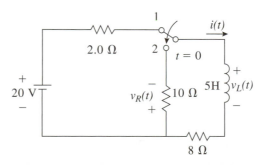

Figure P7.12 Circuit to be Evaluated in Problem 7.21

7.22   Determine the following for the circuit shown in Figure P7.13. Assume that the circuit is in steady state when the switch is moved from position 1 to position 2 at $t = 0$, and also assume that the switch is a make-before-break switch.
   a. The expressions for $v_L(t)$, $i(t)$, and $v_R(t)$.
   b. How much time will it take to complete a 55% change in $v_L(t)$, $i(t)$, and $v_R(t)$?

c. Plot $v_L(t)$, $i(t)$, and $v_R(t)$ from time $t = 0$ until steady-state is reached.

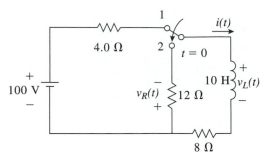

Figure P7.13 Circuit to be Evaluated in Problem 7.22

AC Sinusoidal Steady-State Signals, Phasors, and Impedance

As a result of successfully completing this chapter, you should be able to:

1. Explain why complex numbers are used in AC circuit analysis.
2. Explain what a complex number is, what j is, and what the polar and rectangular forms of complex numbers are.
3. Perform complex number arithmetic calculations, including polar–rectangular form conversions.
4. Describe what a phasor is.
5. Convert back and forth between time domain and phasor form for AC voltages and currents.
6. Describe and determine the reactances of inductors and capacitors.
7. Describe and determine the impedances of resistances, inductors, and capacitors.
8. Use phasors to determine AC voltages, currents, and impedances of components and the entire circuit for RC and RL series circuits.
9. Describe magnitude and phase relationships for voltages, currents, and impedances in RC and RL series circuits.

8.1 THE NEED FOR COMPLEX NUMBERS

The first purpose of this section is to establish the motivation for the use of complex numbers in AC electric circuits. The starting point is the AC (alternating current sinusoidal steady-state) signal, such as the example that is shown in the graph in Figure 8.1, which can be expressed generally in the *time domain* (as a function of time) as:

$$v(t) = V_P \sin(\omega t + \theta) \tag{8.1}$$

where: $v(t)$ is the voltage as a function of time t (v is the dependent variable and t is the independent variable),

V_P is the peak voltage in volts (V),

ω is the radian frequency in radians per second (rad/s), and

θ is the phase shift (relative to another sine wave) in radians (rad) or degrees (°).

For the sine wave shown in Figure 8.1, determine the peak voltage, radian frequency, and phase shift values to form the particular expression. You should arrive at approximately:

$$v(t) = 155 \sin(377t + 35°)$$

Now, consider the general AC circuit illustrated in Figure 8.2. As an example, let $v_1(t)$ and $v_2(t)$ be sinusoidal signals and let the voltage expressions for each sine wave be:

$$v_1(t) = 180 \sin(62{,}832t + 2.356)$$

$$v_2(t) = 180 \sin(62{,}832t + 1.571)$$

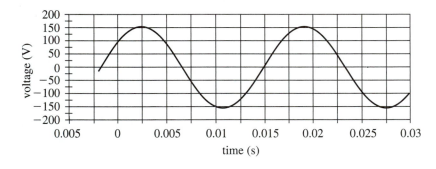

Figure 8.1 Sine Wave Example

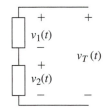

Figure 8.2 A General AC Circuit

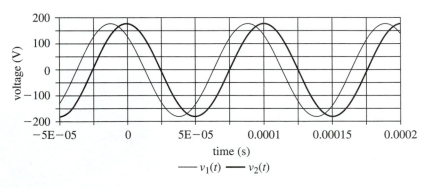

Figure 8.3 The Two Sinusoidal Voltages in Figure 8.2.

where the phase shifts have been expressed in radians. The sine waves are plotted in Figure 8.3. By KVL, the total voltage is the sum of the two voltages:

$$v_T(t) = v_1(t) + v_2(t)$$

$$v_T(t) = 180 \sin(62,832t + 2.356) + 180 \sin(62,832t + 1.571)$$

$$= 180[\sin(62,832t + 2.356) + \sin(62,832t + 1.571)]$$

To find the expression for the total voltage, the following trigonometric identity:

$$\sin(A) + \sin(B) = 2 \sin\left[\frac{1}{2}(A + B)\right]\cos\left[\frac{1}{2}(A - B)\right]$$

is applied to $v_T(t)$, resulting in:

$$v_T(t) = (180)(2) \sin\left\{\frac{1}{2}[(62,832t + 2.356) + (62,832t + 1.571)]\right\}$$

$$\times \cos\left\{\frac{1}{2}[(62,832t + 2.356) - (62,832t + 1.571)]\right\}$$

$$v_T(t) = 360 \sin(62,832t + 1.964) \cos(0.3925)$$

The cosine argument is just a constant; hence, the cosine expression can be evaluated and multiplied with the constant in front (360) to give a peak voltage of 333. The initial phase shift, $\theta = 1.964$ radians, can be converted into 112.5 degrees. The final expression for the total voltage is

$$v_T(t) = 333 \sin(62,832t + 112.5°).$$

The two original voltages and the total voltage are graphed in Figure 8.4.

This was a significant amount of work for the sum of two sinusoidal expressions. The problem would have been even harder if the peak amplitudes of the two original voltages were different, as is the usual case. Furthermore, there may be more than two voltages to add. In short, although the trigonometric approach works, it involves a large amount of labor. It is a valid but inefficient approach.

Is there a more efficient means to perform the mathematics of AC circuit analysis? The answer is yes, and it involves complex numbers. Recall from algebra that a complex number c is

$$c = a + bi \qquad\qquad (8.2)$$

Figure 8.4 The Original and Total Voltage Waveforms

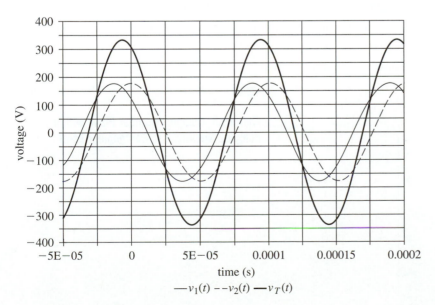

voltage (V) vs time (s)

$-v_1(t) \ --v_2(t) \ -v_T(t)$

where a is the real part and b is the imaginary part. The i is the square root of -1 (why? stay tuned) and indicates which part of the number is the imaginary part.

In electricity and electronics, i is the symbol for current. Hence, a different symbol is used to indicate the imaginary part: j. It is usually placed in front of the imaginary part:

$$c = a + jb \tag{8.3}$$

Complex numbers will be covered next with an emphasis on the significance and meaning (for example, the imaginary part is really *not* imaginary). Then complex number arithmetic and algebra will be covered. Finally, the use of complex numbers in AC circuit calculations will be illustrated.

8.1.1 Complex Numbers

What is a complex number? First, here is what a complex number is not: a *complex number is* not *a vector*, even though the math may be similar. A complex number has two parts: a real part and an imaginary part.

$$\text{complex number} = \text{real part} + j(\text{imaginary part})$$

Hence, complex numbers are "two-dimensional" numbers: $a + jb$ or, as used in electronics, $R + jX$. They contain two pieces of information. Why are two "dimensions" needed in AC circuits? The answer starts with the plot on the *complex number plane,* as shown in Figure 8.5.

What is the imaginary part of a complex number? The imaginary part is the part of the complex number that is plotted along the imaginary axis. Equivalently, the imaginary part is plotted ±90° with respect to the real part in the complex number plane. What is j? Mathematically, it is the square root of -1, but that has no meaning (yet). In a complex number, j indicates the imaginary part and means that it is ±90° with respect to the real part. Does the ±90° sound familiar? Recall the voltage-current phase relationships in inductors and capacitors with AC signals. As will be seen shortly, j means a lot more than the square root of -1.

Thus, a complex number $\tilde{Z}$, where the tilde over the symbol indicates that the number is complex, is written in *rectangular form* as:

$$\tilde{Z} = R + jK \tag{8.4}$$

where: $\tilde{Z}$ is a complex number, and the tilde indicates that the number is complex,

R is the real part of the complex number,

X is the imaginary part of the complex number, and

j indicates that the imaginary part has a phase shift of ±90° with respect to the real part.

What is the polar form of a complex number?

$$\tilde{Z} = Z \angle \theta \tag{8.5}$$

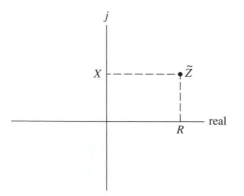

Figure 8.5 Complex Number Plane with a Complex Number $\tilde{Z}$ in Rectangular Form

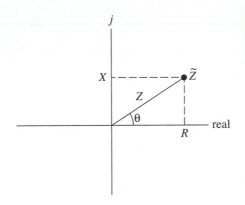

where: $\tilde{Z}$ is a complex number, and the tilde indicates that the number is complex,

Z is the magnitude of the complex number, and

θ is the phase shift (or phase, or angle, or phase angle) of the complex number.

The polar form of the complex number is plotted in Figure 8.6 along with the rectangular form. Thus, the answer to the question, Why are two *dimensions* needed in AC circuits? is that there are two pieces of information in an AC variable: the *magnitude,* peak or effective, and the *phase.* Thus, an AC signal can be represented by one complex number. However, both pieces of information must be retained separately within the complex number, either as magnitude and phase or real part and imaginary part.

The plot in Figure 8.6 leads one to ask how to convert between the rectangular and polar forms of complex numbers. The notation for complex numbers is:

$$\tilde{Z} = R + jX = Z\angle\theta \tag{8.6}$$

because this notation is commonly used in electric circuits, as expressed later in this chapter. The real part is assigned the symbol R and the imaginary part is assigned the symbol X. The magnitude is Z and the angle is θ. Given the rectangular form $R + jX$, how is the polar form $\tilde{Z} = Z\angle\theta$ determined? Apply the Pythagorean theorem to Figure 8.6. The magnitude of the complex number is:

$$Z = \sqrt{R^2 + X^2} \tag{8.7}$$

Apply trigonometry to Figure 8.6. The phase of the complex number is:

$$\theta = \tan^{-1}\left(\frac{X}{R}\right) = \arctan\left(\frac{X}{R}\right) \tag{8.8}$$

and the complex number in polar form is then $\tilde{Z} = Z\angle\theta$.

Conversely, given the polar form of the complex number, how is the rectangular form determined? Again, apply trigonometry to Figure 8.6:

$$R = Z\cos\theta \tag{8.9}$$

$$X = Z\sin\theta \tag{8.10}$$

and the complex number in rectangular form is $R + jX$.

Before modern calculators or computers, one had to perform these conversions manually. With modern calculators, such as the TI-86™ or HP-48G™, there is no need to perform the polar–rectangular conversions manually—the calculators perform the calculations in Equations (8.7), (8.8), (8.9), and (8.10) internally. However, if complex expressions with variables are being used (as in Chapters 15 and 16), then one must manually perform the complex number operations. Hence, it is important to understand the manual methods of performing complex number operations.

__ Example 8.1.1 _____

Convert $9 - j6$ into polar form manually. Check the result with your scientific calculator.

Given: $\tilde{Z} = R + jX = 9 - j6$

Desired: $\tilde{Z} = Z\angle\theta$

Strategy: $Z = \sqrt{R^2 + X^2}, \quad \theta = \tan^{-1}\left(\dfrac{X}{R}\right)$

Solution: _____

$$Z = \sqrt{R^2 + X^2} = \sqrt{9^2 + 6^2} = 10.817$$

$$\theta = \tan^{-1}\left(\frac{X}{R}\right) = \tan^{-1}\left(\frac{-6}{+9}\right) = -33.69°$$

Notice that the signs of the numerator and the denominator are important so that the proper quadrant of the angle can be determined.

$$\tilde{Z} = Z\angle\theta = 10.8\angle-33.7°$$

Check:

■ Enter the complex number $9 - j6$ into your calculator in the format required by your calculator.

■ Enter the proper keystroke(s) to perform the rectangular–polar conversion.

$$\text{result: } \tilde{Z} = Z\angle\theta = 10.817\angle-33.690° = 10.8\angle-33.7°$$

__ Example 8.1.2 _____

Convert $10.817\angle-33.690°$ into rectangular form manually. Check the result with your scientific calculator.

Given: $\tilde{Z} = Z\angle\theta = 10.817\angle-33.690°$

Desired: $\tilde{Z} = R + jX$

Strategy: $R = Z\cos\theta, X = Z\sin\theta$

Solution: _____

$$R = Z\cos\theta = 10.817\cos(-33.690°) = +9.0003$$

$$X = Z\sin\theta = 10.817\sin(-33.690°) = -6.0002$$

$$\tilde{Z} = R + jX = +9.0003 - j6.0002 = 9.00 - j6.00$$

Check:

■ Enter the complex number $10.817\angle-33.690°$ into your calculator in the format required by your calculator.

■ Enter the proper keystroke(s) to perform the polar-to-rectangular conversion.

$$\text{result: } \tilde{Z} = R + jX = +9.0003 - j6.0002 = 9.00 - j6.00$$

8.1.2 Complex-Number Operations

How does one perform addition, subtraction, multiplication, and division with complex numbers? The complex number operations shown below illustrate how complex numbers are evaluated in addition, subtraction, multiplication, and division if the calculations are performed manually. However, a scientific calculator will give the result no matter if the numbers are entered in rectangular or polar form.

■ Operation: addition and subtraction

■ Process: Use rectangular form; add or subtract the corresponding real and imaginary parts.

$$\tilde{Z}_1 \pm \tilde{Z}_2 = (R_1 + jX_1) \pm (R_2 + jX_2) = (R_1 \pm R_2) + j(X_1 \pm X_2) \tag{8.11}$$

Example 8.1.3

$$\text{If } \tilde{Z}_1 = 5 + j7, \tilde{Z}_2 = 2 + j9; \text{ find } \tilde{Z}_1 - \tilde{Z}_2.$$

$$\tilde{Z}_1 - \tilde{Z}_2 = (R_1 - R_2) + j(X_1 - X_2) = (5 - 2) + j(7 - 9) = +3 - j2$$

■ Operation: multiplication

■ Process: Use polar form, multiply the magnitudes and algebraically add the angles.

$$\tilde{Z}_1\tilde{Z}_2 = (Z_1 \angle \theta_1)(Z_2 \angle \theta_2) = Z_1 Z_2 \angle (\theta_1 + \theta_2) \tag{8.12}$$

Example 8.1.4

$$\text{If } \tilde{Z}_1 = 5 + j7, \tilde{Z}_2 = 2 + j9; \text{ find } \tilde{Z}_1\tilde{Z}_2.$$

$$\tilde{Z}_1\tilde{Z}_2 = (5 + j7)(2 + j9) = (8.6023\angle + 54.462°)(9.2195\angle + 77.471°)$$

$$= [(8.6023)(9.2195)]\angle(+ 54.462° + 77.471°) = 79.309\angle131.933°$$

$$\tilde{Z}_1\tilde{Z}_2 = 79.3\angle131.9°$$

■ Operation: division

■ Process: Use polar form; divide the magnitude of the numerator by the magnitude of the denominator; algebraically subtract the angle of the denominator from the angle of the numerator.

$$\frac{\tilde{Z}_1}{\tilde{Z}_2} = \frac{Z_1\angle\theta_1}{Z_2\angle\theta_2} = \frac{Z_1}{Z_2}\angle(\theta_1 - \theta_2) \tag{8.13}$$

Example 8.1.5

$$\text{If } \tilde{Z}_1 = 5 + j7, \tilde{Z}_2 = 2 + j9; \text{ find } \frac{\tilde{Z}_1}{\tilde{Z}_2}$$

$$\frac{\tilde{Z}_1}{\tilde{Z}_2} = \frac{(5 + j7)}{(2 + j9)} = \frac{(8.6023\angle+54.462°)}{(9.2195\angle+77.471°)} = \left[\frac{(8.6023)}{(9.2195)}\right]\angle(+54.462° - 77.471°)$$

$$= 0.933055\angle-23.009° = 0.933\angle-23.0°$$

8.1.3 The Meaning of j

The real meaning of j can be identified with a knowledge of the complex number operations covered in the previous section. Consider the complex number $0 + j1$, which is in rectangular form. Convert it into polar form: $1\angle +90°$. One times any quantity equals that quantity: $1j = j$. Hence:

$$\boxed{j = 1\angle +90°}\qquad\qquad (8.14)$$

In fact, complex numbers can be written as

$$\tilde{Z} = R\angle 0° + X\angle 90° \qquad\qquad (8.15)$$

but this is cumbersome and takes a lot of additional writing or typing. Hence, $\tilde{Z} = R + jX$ is normally written. However, knowing what j is allows us to evaluate expressions such as those illustrated by the following examples.

___ **Example 8.1.6** _____

Evaluate $\dfrac{1}{j}$.

$$\frac{1}{j} = \frac{1\angle 0°}{1\angle +90°} = 1\angle(0° - 90°) = 1\angle -90° = -j$$

___ **Example 8.1.7** _____

Evaluate $\dfrac{-1}{j10}$.

$$\frac{-1}{j10} = \frac{1\angle 180°}{10\angle +90°} = 0.1\angle(180° - 90°) = 0.1\angle +90° = +j0.1$$

Now, the question, Why does $j = \sqrt{-1}$? can be answered. Multiply j by j in polar form:

$$j^2 = (1\angle +90°)(1\angle +90°) = 1\angle +180° = -1 \qquad\qquad (8.16)$$

Take the square root of both sides:

$$\sqrt{j^2} = j = \sqrt{-1} \qquad\qquad (8.17)$$

Although j (or i in mathematics notation) is often *defined* to be the square root of -1, it really does equal the square root of -1, as has been demonstrated with the polar form of complex numbers.

8.1.4 Phasors

It has been previously shown that (1) working with AC voltages and currents in the time domain is cumbersome, (2) complex numbers contain two pieces of information, and (3) that complex number math is algebra-like but with rules for handling the real and imaginary parts in addition and subtraction and rules for handling the magnitudes and the angles in multiplication and division. It is possible to make use of complex numbers in AC circuit analysis. To do so, it is necessary to establish how AC signals are represented with complex numbers. Start with a general AC signal:

$$v_P(t) = V_P \sin(\omega t + \theta_V) \qquad\qquad (8.18)$$

$$i(t) = I_P \sin(\omega t + \theta_I) \qquad\qquad (8.19)$$

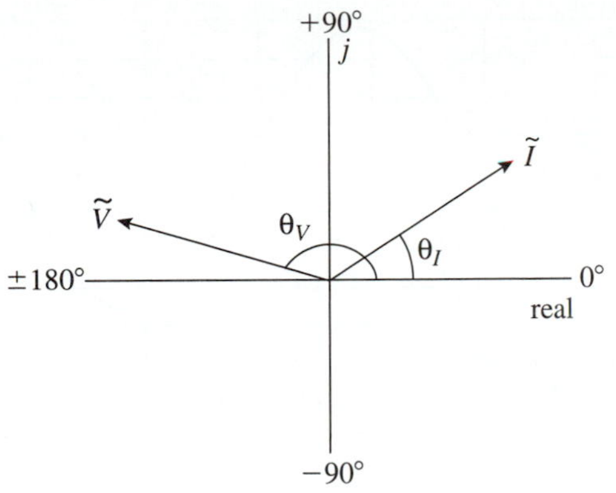

Figure 8.7 Phasor Voltage and Current Potted on the Complex Number Plane

where θ_V is the phase shift of the voltage with respect to a reference sine wave, and θ_I is the phase shift of the current with respect to the same reference sine wave. Freeze time at $t = 0$. Plot the magnitude and angle of $v(t)$ and $i(t)$ as a complex number in polar form on the complex number plane, as shown in Figure 8.7. Arrows are customarily shown, but again, they are not vectors, they are lines locating the complex numbers on the complex number plane.

A *phasor* is defined as the complex number that represents an AC sinusoidal steady-state voltage or current. The magnitude of the complex number is the peak or RMS (effective) value. The angle of the complex number is the phase shift. Hence, the phasor, presented as a complex number, contains two key pieces of information about the sinusoidal signal. In symbol form, the phasor notation is:

$$\tilde{V} = V_P \angle \theta_V, \tilde{I} = I_P \angle \theta_I$$

or

$$\tilde{V} = V_{RMS} \angle \theta_V, \tilde{I} = I_{RMS} \angle \theta_I \tag{8.20}$$

You must know or determine whether peak or RMS phasors are being used in any given problem. Because phasors are complex numbers, they may be converted into rectangular form if needed for numerical calculations.

Notice that the variable of time does *not* occur in a phasor. It is assumed that when phasors are being used, the time dependence is sinusoidal. Also, when using phasors in a given problem, *all signals* must *be at the same frequency*. Otherwise, the phase relationship between two AC signals does not remain constant as time goes on (recall the discussion in Section 4.4). A slogan is useful to learn what a phasor is:

A phasor is a complex number representation of an AC sinusoidal steady-state signal with the sinusoidal time dependence removed but assumed.

Notice how the phasor diagram shows the *relative* phases of the signals.

Example 8.1.8

a. Determine and plot the voltage and current phasors for the signals shown in Figure 8.8. Note that the x-axis is labeled in degrees, not time.
b. Determine which signal is leading and by how many degrees.
c. Determine the expressions for $v(t)$ and $i(t)$.

Given: voltage and current plots per Figure 8.8

Desired: a. $\tilde{V} = V_P \angle \theta_V, \tilde{I} = I_P \angle \theta_I$, plot in complex number plane

   b. which signal is leading, and the phase difference θ between v and i

   c. $v(t)$, $i(t)$

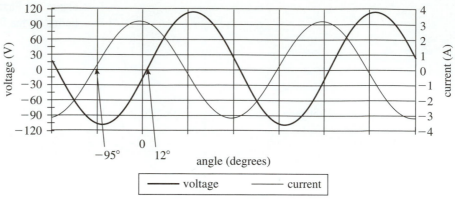

Figure 8.8 Voltage and Current Plots for Example 8.1.8

Strategy: a. V_P, I_P, and θ's for v and i from graph
   b. Let $\theta = \theta_V - \theta_I$; if v leads i, θ is positive; if i leads v, θ is negative.
   c. Insert results from (a) into time-domain expressions.

Solution:

a. $\tilde{V} = 110\angle -12°$ V. For $\tilde{I}$, note the current scale on the right side of the graph in Figure 8.8. $\tilde{I} = 3.2\angle +95°$ A; see Figure 8.9 for a plot of these phasors.

b. $\theta = \theta_V - \theta_I = -12° - (+95°) = -107°$ Thus, $i(t)$ is leading $v(t)$.

c. There is not enough information given to determine frequency, so leave ω as a variable in the time-domain expressions:

$$v(t) = V_P \sin(\omega t + \theta) = 110 \sin(\omega t - 12°)$$

$$i(t) = I_P \sin(\omega t + \theta) = 3.2 \sin(\omega t + 95°)$$

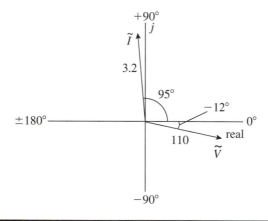

Figure 8.9 Voltage and Current Phasors in the Complex Plane for Example 8.1.8

To illustrate how useful complex numbers are in AC circuit calculations, determine $v_T(t)$ for the circuit in Figure 8.2 again.

Example 8.1.9

Find the expression for the total voltage in the circuit shown in Figure 8.2.

Given: $v_1(t) = 180 \sin(62{,}832t + 2.356)$, $v_2(t) = 180 \sin(62{,}832t + 1.571)$

Desired: $v_T(t)$

Strategy:

- ■ Determine the phasors from $v_1(t)$ and $v_2(t)$
- ■ Add the phasors
- ■ Return to the time domain.

Solution:

(Note the use of both radians and degrees in the following equations.)

$$v_1(t) = 180 \sin(62{,}832t + 2.356), \quad \tilde{V}_1 = 180\angle +135°$$

$$v_2(t) = 180 \sin(62{,}832t + 1.571), \quad \tilde{V}_1 = 180\angle +90°$$

$$\tilde{V}_T = \tilde{V}_1 + \tilde{V}_2 = 180\angle 135° + 180\angle 90°$$

$$= (-127.3 + j127.3) + (0 + j180) = -127.3 + j307.3$$

$$\tilde{V}_T = 332.6\angle + 112.5°$$

$$v_T(t) = 333 \sin(62{,}832t + 112.5°)$$

(Note: The third-to-last line is not necessary with modern calculators—try it!)

This result matches the previous result obtained earlier in this section. The complex number approach is significantly easier than the trigonometric approach. Hence, the use of complex numbers in AC circuit calculations has become a standard approach:

Phasors are used to add and subtract AC voltages or currents.

A primary example of this approach is in KVL and KCL calculations for AC circuits. One uses phasors for the voltages and currents, respectively. The use of phasors can be extended to AC circuit analysis in general.

General principle for AC circuit analysis

In general, AC circuit analysis is performed as it was in DC except one must use phasors for the voltages and the currents and complex numbers for the V/I ratio (*impedance*) for R, L, and C. (Impedance is covered in the next two sections of this chapter.)

Mathematical note: The mathematical justification of the multiplication and division operations for complex numbers is based on Euler's identity.

The important question in preparation for AC circuit analysis is, What are the voltage–current relations of resistances, capacitances, and inductances for AC signals? Capacitance will be examined first in the next section. Then the results for capacitance will be compared to the results for resistance to highlight the important concepts. Then inductance will be examined in Section 8.3.

8.2 Capacitive Reactance and Impedance

Consider the case of a capacitor and an AC signal. The voltage and current are always changing. Hence, the capacitor will always be reacting. This case is very special. Recall that phase shift was stated and determined in Chapter 4, but the reason that a sine wave might have a phase shift was not explained. We are now ready to understand why phase shift happens. Start with the AC circuit shown in Figure 8.10. The AC source delivers a sinusoidal voltage and current. The voltage across and the current through the resistance are always in phase, as established in Chapter 4.

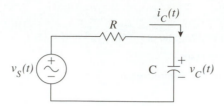

Figure 8.10 An AC Circuit with a Capacitor

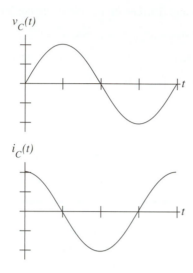

Figure 8.11 AC Voltage and Current Waveforms for a Capacitor

For the capacitor in Figure 8.10, however, the voltage can only change if the charge on the capacitor plates changes. Current is the rate of charge flow, not just the number of charges. Hence, current must be "on" first before any charges can build up (or leave) the capacitor plates. Thus, one would expect the current to *lead* the voltage, i.e., the current is advanced in phase relative to the voltage waveform, as shown in Figure 8.11. The current leads in order to deposit charge on (or remove charge from) the capacitor plates, and the voltage changes in response to the change of charge on the plates. Notice that when the current is zero, the capacitor voltage has leveled out at a maximum point (in either polarity)—zero current means no charge movement on to or off of the plates, and this means the voltage should not be changing (at that instant). Conversely, when the current is maximum, the voltage is changing the fastest—note the slope of the voltage waveform at the times when current is a maximum.

How far (in terms of angle) should the AC current lead the voltage for a capacitor? The answer is remarkable: 90°. Why would this angle be such a convenient value? The answer comes from power and energy for the capacitor, which is covered in Chapter 10. An ideal capacitor does not dissipate energy; it stores energy. For this to happen, the mathematics of AC voltage, current, and power will require the voltage and current to be 90° out of phase.

Thus, a capacitor always opposes an instantaneous change of voltage and this effect results in current leading the voltage by 90° for AC signals. What is the *i–v* relation for capacitors? Here is what it is *not:*

$$v(t) \neq C\, i(t) \tag{8.21}$$

Hence, the voltage-current (*i–v*) relationship[1] for capacitors is not an Ohm's law–like relationship for time-domain voltages and currents as it is for resistances.

Take a step back and ask, What are the two aspects of how a capacitor responds to an AC signal? One aspect has already been identified: *i leads v* by 90° for a capacitor. What are the two quantities that are important in the use of complex numbers with AC signals?

---

[1] People often say "voltage–current relationship" but abbreviate it in reverse as *i–v.*

Magnitude and phase. Hence, magnitude is the other aspect that must be addressed in the
i–v relationship for capacitors. What is the magnitude relationship between the voltage
across and the current through a capacitor? The ratio of voltage magnitude to current magnitude for capacitors is *capacitive reactance:*

$$X_C = \frac{V_P}{I_p} = \frac{V_{RMS}}{I_{RMS}}$$

(8.22)

where X_C = capacitive reactance (Ω).

Notice that the *magnitudes* of the voltage and current form an Ohm's law–like relationship. The ratio is not resistance, however, because electrical energy is not being converted into another form of energy. Instead the energy is being stored in the electric field of
the capacitor. The current is leading the voltage or, equivalently, the voltage is lagging the
current. In other words the capacitor is *reacting* to the AC signal by causing a 90° phase
shift between the voltage and the current and by setting a ratio between the voltage and
current magnitudes. Thus, the term *reactance* is used to describe the voltage-to-current
ratio in a component that stores energy, such as a capacitor (and an inductor, too, as covered in the next section).

There is another important and interesting aspect of capacitive reactance. What if the
frequency of the AC signal changes in a circuit with a capacitor? A higher frequency means
that the voltage and the current are alternating through cycles more quickly. For a given
current, which is the *rate* of charge flow, there is less time per half-cycle for charge to build
up on the capacitor plates. Less separated charge means a lower voltage is across the
capacitor. The voltage-to-current ratio is lower, i.e., the reactance is smaller. Hence, what
happens to capacitive reactance as frequency increases? Capacitive reactance decreases. In
fact this trend actually relates capacitive reactance to frequency:

$$X_C \propto \frac{1}{f}$$

(8.23)

Is capacitive reactance directly or inversely proportional to capacitance? If the capacitance increases, then more charge will be stored on the capacitor plates for the same voltage. Hence, a larger current had to flow to provide more charge relative to the case of a
smaller capacitance. Thus, a smaller voltage-to-current ratio results when the capacitance
increases, and capacitive reactance is inversely proportional to frequency. Thus, capacitive
reactance in terms of frequency and capacitance is (the factor of 2π comes out of the
derivation using calculus):

$$\boxed{X_C = \frac{1}{2\pi f C} = \frac{1}{\omega C}}$$

(8.24)

How can both the magnitude and phase relationships for capacitance be incorporated
into one equation? *Impedance* is defined as the *complex number ratio of the phasor V to
the phasor I* in AC circuits:

$$\boxed{\tilde{Z} = \frac{\tilde{V}}{\tilde{I}}}$$

(8.25)

The phasor voltage and the current can be expanded to show the magnitudes and phase
angles (use either peak or RMS values):

$$\tilde{Z} = \frac{\tilde{V}}{\tilde{I}} = \frac{V \angle \theta_V}{I \angle \theta_I} = \frac{V}{I} \angle(\theta_V - \theta_I) = Z\angle\theta$$

(8.26)

Thus, the magnitude and phase of the impedance are:

$$Z = \frac{V}{I}, \quad \angle\theta = \frac{\angle\theta_V}{\angle\theta_I} = \angle(\theta_V - \theta_I)$$

(8.27)

where $\tilde{Z}$ is the impedance (Ω),

      Z is the *magnitude* of the impedance (Ω), and

      θ is the *angle* (also called *phase*, *phase angle*, and *phase shift*) of the impedance, which is the angle between the voltage and the current.

Impedance is *not* resistance. Equation (8.25) looks like Ohm's law, but Ohm's law is strictly applicable to resistances only because Ohm's law represents the voltage-to-current ratio of the conversion of electrical energy to another form of energy; however, people often will casually refer to Equation (8.25) as Ohm's law. One might think of it as the AC form of Ohm's law. One can think of impedance as being an extension of resistance. In fact the units of impedance are ohms. Nonetheless, impedance is not to be confused as being solely resistance.

Henceforth, the symbol θ will normally be reserved for the angle (phase shift) between the voltage and the current, i.e., $\theta = \theta_V - \theta_I$. Incorporate the capacitive phase relationship $\theta_I = \theta_V + 90°$ into the impedance of a capacitor:

$$\tilde{Z}_C = \frac{V \angle \theta_V}{I \angle (\theta_V + 90°)} = \frac{V}{I} \angle -90° = X_C \angle -90° = 0 - jX_C = \frac{1}{j2\pi fC} = -j\frac{1}{2\pi fC} \quad (8.28)$$

Thus, the important result for the impedance of a capacitance is:

$$\boxed{\tilde{Z}_C = \frac{\tilde{V}_C}{\tilde{I}_C} = X_C \angle -90° = -jX_C = \frac{1}{j2\pi fC} = -j\frac{1}{2\pi fC} = -j\frac{1}{\omega C}} \quad (8.29)$$

The magnitude of the impedance of a capacitance is X_C, i.e., $Z_C = X_C$ (notice no tilde is over Z_C). What is the real part of the impedance of a capacitance? Zero. What is the imaginary part of the impedance of a capacitance? X_C. Thus, the impedance of a capacitance is purely imaginary. The magnitude of the impedance of the capacitance and the imaginary part of the impedance of a capacitance are the same: X_C. What does an imaginary impedance represent? Energy storage. In the case of the capacitor, the energy is stored in the electric field.

Why is the negative sign required for the angle? This is where a comparison to the *i–v* relation for resistance is insightful. By Ohm's law, *v* and *i* are *in phase* for a resistance:

$$R = \frac{V_{\text{RMS}}}{I_{\text{RMS}}} = \frac{V_P}{I_P} \quad (8.30)$$

Apply the definition of impedance:

$$\tilde{Z}_R = \frac{\tilde{V}_R}{\tilde{I}_R} = \frac{V \angle \theta_V}{I \angle \theta_I} = \frac{V}{I} \angle (\theta_V - \theta_I) = Z \angle \theta \quad (8.31)$$

But $\theta_V = \theta_I$, so the angles cancel, leaving

$$\boxed{\tilde{Z}_R = \frac{\tilde{V}_R}{\tilde{I}_R} = \frac{V}{I} \angle 0° = R = R + j0 = R \angle 0°} \quad (8.32)$$

The impedance of a resistance is purely real and represents the conversion of electrical energy into another form of energy. R is the magnitude of the impedance of a resistance. The imaginary part of the impedance of a resistance is ideally zero. The real part of the impedance equals R. The angle is zero because the phase shift between the voltage and current is zero degrees for a resistance.

In a capacitor, the phase shift is $-90°$ between the voltage and the current, i.e., $\theta_V - \theta_I = -90°$. Hence the negative sign in the phase means that the voltage is lagging the current. This nonzero angle is a fundamental difference between a resistance and a reactance. In resistances, voltage and current are in phase and the phase shift is zero. The corresponding electrical phenomenon is energy conversion. In reactances the voltage and the current are 90° out of phase and the phase shift is 90° (−90° for the capacitor). The corresponding electrical phenomenon is energy storage.

____ **Example 8.2.1** _____

Determine the (a) reactance and (b) impedance of a capacitor if the effective current is 25 mA when the peak-to-peak voltage is 18 V.

Given: $I_{RMS} = 25$ mA

$V_{P\text{-}P} = 18$ V

component: C

Desired: X_C

$\tilde{Z}_C$

Strategy: Convert V and I to peak (or RMS—pick one and be consistent) values.

$$X_C = \frac{V_P}{I_P}, \quad \tilde{Z}_C = X_C\angle{-90°}$$

Solution: _____

$$V_P = \frac{V_{P\text{-}P}}{2} = \frac{18}{2} = 9 \text{ V}$$

$$I_P = I_{RMS}\sqrt{2} = (25 \text{ mA})\sqrt{2} = 35.355 \text{ mA}$$

$$X_C = \frac{V_P}{I_P} = \frac{9 \text{ V}}{35.355 \text{ mA}} = 254.6 \text{ }\Omega = 255 \text{ }\Omega$$

$$\tilde{Z}_C = X_C\angle{-90°} = 255\angle{-90°} \text{ }\Omega = -j\,255 \text{ }\Omega$$

8.3 INDUCTIVE REACTANCE AND IMPEDANCE

Consider the case of the inductor and an AC signal. The voltage and current are always changing. Hence, the inductor will always be reacting. This case is also very special. Just as the case with the capacitor, the phase shift between the voltage across and the current through the inductor may now be understood. Start with the AC circuit shown in Figure 8.12. The AC source delivers a sinusoidal voltage and current. The voltage across and the current through the resistance are in phase, as established in Chapter 4.

When the current is at zero and increasing, its rate of change is the fastest (note the slope of the sine wave at $t = 0$). The magnetic field is consequently changing at its fastest rate. Thus, the induced voltage across the inductor should be at the maximum (peak) value—see $v_L(t)$ in Figure 8.13. As the current increases, its rate of change lessens. Hence, the induced voltage across the inductor should decrease. When the current reaches the maximum of the sine wave, the rate of change and the slope are zero at that time instant. Therefore, the induced voltage should be zero, as shown in Figure 8.13. The remainder of the cycle could be similarly argued. Hence, the inductor voltage is *leading* the inductor

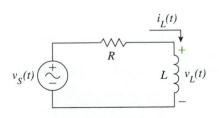

Figure 8.12 An AC Circuit with an Inductor

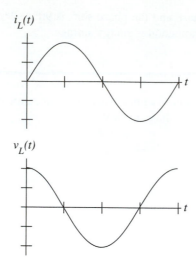

$i_L(t)$

$v_L(t)$

Figure 8.13 AC Voltage and Current Waveforms for an Inductor

current because the inductor voltage is proportional to the rate of change of the magnetic field or, equivalently, is directly proportional to the rate of change of the current (not the actual value of the current). The voltage is induced "ahead of" the current change and thus opposes the current change.

How far (in terms of angle) should the AC voltage lead the current for an inductor? As it was for the capacitor, the answer is remarkable: 90°. Why would this angle be such a convenient value? The answer comes from power and energy for the inductor, which is covered in Chapter 10. An ideal inductor does not dissipate energy, it stores energy. For this to happen, the mathematics of AC voltage, current, and power require the voltage and current to be 90° out of phase.

Thus an inductor always opposes an instantaneous change of current, and this effect results in voltage leading the current by 90° for AC signals. What is the i–v relation for inductors? As with capacitors, it is *not* a relationship between time-domain voltage and current:

$$v(t) \neq L\, i(t) \tag{8.33}$$

Analogous to the situation with capacitors, *inductive reactance* is the ratio of voltage magnitude to current magnitude for an inductor:

$$X_L = \frac{V_P}{I_P} = \frac{V_{\mathrm{RMS}}}{I_{\mathrm{RMS}}} \tag{8.34}$$

where X_L = inductive reactance.

As with capacitive reactance, there is no energy conversion in inductive reactance. Instead, there is energy storage in the magnetic field. What is the effect on inductive reactance as the frequency changes? Assume that frequency increases. The voltage and current are changing faster. The magnetic field is changing more quickly, and, by Faraday's law, this induces a larger voltage across the inductor. By Lenz's law, this larger voltage is opposing the current change, so the voltage-to-current ratio is larger. Hence, inductive reactance increases as frequency increases:

$$X_L \propto f \tag{8.35}$$

Is inductive reactance directly or inversely proportional to inductance? If the inductance increases, then there is more magnetic flux per unit current. This increased magnetic flux will induce more voltage per unit current. Hence, the voltage-to-current ratio and the inductive reactance are larger as inductance increases. Thus, the inductive reactance in terms of frequency and inductance is (a factor of 2π comes from the calculus derivation):

$$\boxed{X_L = 2\pi f L = \omega L} \tag{8.36}$$

Also, as with capacitance, impedance can be defined for inductance as the ratio of phasor voltage to phasor current by incorporating the inductive phase relationship $\theta_V = \theta_I + 90°$ into the impedance of an inductor:

$$\tilde{Z}_L = \frac{V\angle\,(\theta_I + 90°)}{I\angle\theta_I} = \frac{V}{I}\,\angle+90° = X_L\angle+90° = 0 + jX_L = j2\pi fL \qquad (8.37)$$

$$\boxed{\tilde{Z}_L = \frac{\tilde{V}_L}{\tilde{I}_L} = X_L \angle\,+90° = +jX_L = +j2\pi fL = +j\omega L} \qquad (8.38)$$

The magnitude of the impedance of an inductance is X_L, i.e., $Z_L = X_L$. What is the real part of the impedance of an inductance? Zero. What is the imaginary part of the impedance of a inductance? X_L. Thus, the impedance of an inductance is purely imaginary. The magnitude of the impedance of an inductance equals the imaginary part of the impedance of an inductance. What does an imaginary impedance represent? Energy storage. In the case of the inductor, the energy is stored in the magnetic field.

Why is the positive sign required for the angle? In an inductor, the phase shift is $+90°$ between the voltage and the current, i.e., $\theta_V - \theta_I = +90°$. Hence, the positive sign in the phase means that the voltage is leading the current. Again, this nonzero angle is a fundamental difference between a resistance and a reactance. In resistances, voltage and current are in phase, and the phase shift is zero. The corresponding electrical phenomenon is energy conversion. In reactances, the voltage and current are 90° out of phase, and the phase shift is ±90°. The corresponding electrical phenomenon is energy storage.

Example 8.2.1 will be reworked with the component being an inductor instead of a capacitor. Compare and explain the similarities and the differences between Examples 8.2.1 and 8.3.1.

___ **Example 8.3.1** _____

Determine the (*a*) reactance and (*b*) impedance of an inductor if the effective current is 25 mA when the peak-to-peak voltage is 18 V.

Given:

$$I_{RMS} = 25 \text{ mA}$$

$$V_{P-P} = 18 \text{ V}$$

component: L

Desired: X_L

$\tilde{Z}_L$

Strategy: convert *V* and *I* to peak (or RMS) values.

$$X_L = \frac{V_P}{I_P}, \quad \tilde{Z}_L = X_L\angle+90°$$

Solution: _____

$$V_P = \frac{V_{P-P}}{2} = \frac{18}{2} = 9 \text{ V}$$

$$I_P = I_{RMS}\sqrt{2} = (25 \text{ mA})\sqrt{2} = 35.355 \text{ mA}$$

$$X_L = \frac{V_P}{I_P} = \frac{9 \text{ V}}{35.355 \text{ mA}} = 254.6 \ \Omega = 255 \ \Omega$$

$$\tilde{Z}_L = X_L \angle +90° = 255 \angle +90° \ \Omega = +j255 \ \Omega$$

The similarity is the reactance value, i.e., the magnitude of the impedance value. The difference is the phase shift or, equivalently, the sign on the imaginary part of the impedance.

The general graphs of inductive reactance and capacitive reactance versus frequency are shown in Figure 8.14. Note that the reactance of a capacitor approaches infinity as frequency approaches zero (a DC open circuit), and approaches zero as frequency increases (an AC "short" circuit). The reactance of an inductor is zero at DC (a DC short) and approaches infinity as frequency increases (an AC open). In both cases, practical inductors and capacitors have nonzero resistance, which must be considered if it has a significant magnitude relative to the reactance. Often the resistance of a capacitor can be ignored, but less often can the resistance of an inductor be ignored. The wire resistance is usually significant.

A generic plot of the impedances of a resistance, an inductor, and a capacitor in the complex number plane is shown in Figure 8.15. The plot clearly shows the voltage-to-current phase relationships between the three components.

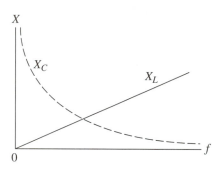

Figure 8.14 Inductive and Capacitive Reactance Versus Frequency

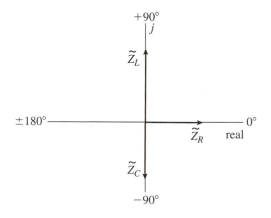

Figure 8.15 Resistive, Inductive, and Capacitive Impedances in the Complex Plane

8.4 Response of RL and RC Series Circuits to AC

The impedances for resistors (resistances in general), inductors, and capacitors have been established as the AC phasor voltage-to-current ratio. Each impedance value contains two critical pieces of information: the magnitude of the voltage-to-current ratio and the phase difference between the voltage and the current. Although resistors, inductors, and capacitors are interesting in themselves, they become even more interesting when used together. One may begin the study of AC circuits by first examining series circuits with a resistor

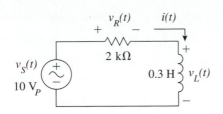

Figure 8.16 RL Circuit for Example 8.4.1

and either an inductor or capacitor. The objectives are to see how AC phasor voltages, phasor currents, and impedances are determined and to see the magnitude and phase relationships between them. The basis for the general series–parallel circuit analysis techniques is then established for use in Chapter 9.

Recall in DC circuits that resistances in series add. Similarly, *impedances in series add.*[2] For example, draw an AC voltage source, resistor and inductor in series, as shown in Figure 8.16. What is the total impedance of the circuit?

$$\tilde{Z}_T = \tilde{Z}_R + \tilde{Z}_L = (R + j0) + (0 + jX_L) = R + jX_L \tag{8.39}$$

Notice that the impedances must be added as *complex numbers.*

Example 8.4.1

Determine (*a*) the total impedance, (*b*) all phasor voltages, (*c*) the phasor current, and (*d*) the time-domain expressions for the voltages and currents of a 2 kΩ resistor in series with a 0.3 H inductor (as per Figure 8.16) at a frequency of 40 krads/s.

Given:

$$R = 2 \text{ k}\Omega$$

$$L = 0.3 \text{ H}$$

$$\omega = 40 \text{ krads/s}$$

$$V_S = 10 \text{ V (peak)}$$

Desired:

a. $\tilde{Z}_T$
b. $\tilde{V}_R, \tilde{V}_L, \tilde{V}_S$
c. $\tilde{I}$
d. $v_S(t), v_L(t), v_R(t),$ and $i(t)$

Strategy:

a. $X_L = \omega L, \tilde{Z}_R = R, \tilde{Z}_L = X_L\angle + 90°, \tilde{Z}_T = \tilde{Z}_R + \tilde{Z}_L$

b. $\tilde{I} = \dfrac{\tilde{V}_S}{\tilde{Z}_T}$ (finding current first is easiest here)

c. $\tilde{V}_S$ given, $\tilde{V}_R = \tilde{I}\tilde{Z}_R, \tilde{V}_L = \tilde{I}\tilde{Z}_L$

d. $v_S(t) = V_S \sin(\omega t + \theta_S),$ etc. (fill in the specific constants in the general time-domain expressions)

Solution:

a. $X_L = \omega L = (40 \text{ krads/s})(0.3 \text{ H}) = 12 \text{ k}\Omega$

---

[2] In Chapter 9, DC circuit analysis approaches will be extended to AC circuits.

$$\tilde{Z}_L = X_L \angle + 90° = 12\ \text{k}\Omega \angle + 90°\Omega = +j12\ \text{k}\Omega$$

$$\tilde{Z}_T = \tilde{Z}_R + \tilde{Z}_L = R + jX_L = (2 + j12)\ \text{k}\Omega$$

b. c. Note that the phase shift of the source was not given. The source is often the reference sine wave for phase shifts. That assumption will be made in this problem. Hence,

$$\tilde{V}_S = 10\angle 0°\ \text{V (peak)}$$

$$\tilde{I} = \frac{\tilde{V}_S}{\tilde{Z}_T} = \frac{10\angle 0°\ \text{V}}{(2 + j12)\ \text{k}\Omega} = 0.821995\ \text{mA}\angle -80.538° = 0.822\angle -80.5°\ \text{mA}$$

$$V_R = \tilde{I}\,\tilde{Z}_R = (0.821995\ \text{mA}\angle -80.538°)(2000\ \Omega)$$
$$= 1.6440\angle -80.538° = 1.64\angle -80.5°\ \text{V}$$

$$V_L = \tilde{I}\,\tilde{Z}_L = (0.821995\ \text{mA}\angle -80.538°)\ (12\ \text{k}\Omega\angle +90°)$$
$$= 9.8639\angle +9.462° = 9.86\angle +9.5°\ \text{V}$$

These phasors are all peak phasors because the source voltage magnitude is a peak voltage.

d. $v_S(t) = V_S \sin(\omega t + \theta_S) = 10.0 \sin(40{,}000t)\ \text{V}$
$v_R(t) = V_R \sin(\omega t + \theta_R) = 1.64 \sin(40{,}000t - 80.5°)\ \text{V}$
$v_L(t) = V_L \sin(\omega t + \theta_L) = 9.86 \sin(40{,}000t + 9.5°)\ \text{V}$
$i(t) = I_P \sin(\omega t + \theta_I) = 0.822 \sin(40{,}000t - 80.5°)\ \text{mA}$

Analysis:

1. $v_L(t)$ leads $i(t)$ for the inductor by 90°.

2. $v_R(t)$ and $i(t)$ are in phase for the resistor.

3. *The sum of the peak voltages for the inductor and the resistor,* 11.5 V, *does not equal the source voltage* (10 V): phasors, not just magnitudes, must be used in AC circuit calculations.

Extra: KVL

$$+\tilde{V}_S - \tilde{V}_R - \tilde{V}_L = 0;\ +\tilde{V}_S = +\tilde{V}_R + \tilde{V}_L$$

$$\tilde{V}_S = 1.6440\angle -80.538° + 9.8639\angle +9.462° = 10.0\angle 0.00°\ \text{V}$$

KVL and the other series–parallel circuit laws and analysis techniques will be covered in Chapter 9. What is the expression for the impedance of a resistance and capacitance in series?

$$\tilde{Z}_T = \tilde{Z}_R + \tilde{Z}_C = (R + j0) + (0 - jX_C) = R - jX_C \qquad \textbf{(8.40)}$$

▬ Example 8.4.2

Determine the total impedance of a 20 Ω resistor in series with a 3 μF capacitor at a frequency of 4000 Hz.

Given:

$$R = 20\ \Omega$$

$$C = 3\ \mu\text{F}$$

$$f = 4000\ \text{Hz}$$

Desired: $\tilde{Z}_T$

Strategy:

$$\omega = 2\pi f, X_C = \frac{1}{\omega C}, \quad \tilde{Z}_R = R, \quad \tilde{Z}_C = X_C\angle -90°, \quad \tilde{Z}_T = \tilde{Z}_R + \tilde{Z}_C$$

Solution: _____

$$\omega = 2\pi f = 2\pi (4000) = 25{,}133 \text{ rads/s}$$

$$X_C = \frac{1}{\omega C} = \frac{1}{(25{,}133)(3 \times 10^{-6})} = 13.263 \ \Omega$$

$$\tilde{Z}_C = X_C\angle -90° = 13.263\angle -90° \ \Omega = -j13.263 \ \Omega$$

$$\tilde{Z}_T = \tilde{Z}_R + \tilde{Z}_C = R - jX_C = 20 - j13.263 = 20.0 - j13.3 \ \Omega$$

CHAPTER REVIEW

8.1 The Need for Complex Numbers
- Mathematical operations with AC voltages and currents are cumbersome in the time domain. They are simplified when complex numbers are used.
- Complex numbers contain two pieces of numerical information: magnitude and phase (polar form) or, equivalently, the real and imaginary parts (rectangular form).
- The conversion from the polar form of a complex number to the rectangular form, or vice versa, is based on trigonometry. Modern scientific calculators perform these conversions automatically.
- The processes to manually perform mathematical operations with complex numbers are:
  - addition and subtraction: use rectangular form; add or subtract the corresponding real and imaginary parts
  - multiplication: use polar form, multiply the magnitudes and algebraically add the angles
  - division: use polar form; divide the magnitude of the numerator by the magnitude of the denominator; algebraically subtract the angle of the denominator from the angle of the numerator
- $j = 1\angle + 90°$.
- A phasor is a complex-number representation of an AC sinusoidal steady-state signal with the sinusoidal time dependence removed but assumed.

- AC circuit analysis is performed as it was in DC except that phasors must be used for the voltages and the currents, and complex numbers for the V/I ratio (impedance) for resistances, inductors, and capacitors.

8.2 Capacitive Reactance and Impedance
- Capacitive reactance is a real, positive number that is the magnitude ratio of the voltage to current for a capacitor:

$$X_C = \frac{V_P}{I_P} = \frac{V_{\text{RMS}}}{I_{\text{RMS}}}.$$

- Capacitive reactance is inversely proportional to frequency and capacitance:

$$X_C = \frac{1}{2\pi f C} = \frac{1}{\omega C}$$

- The AC current leads the voltage by 90° for a capacitor.
- The impedance of a capacitor incorporates both the reactance and the phase relationship of the phasor voltage-to-current ratio for the capacitor:

$$\tilde{Z}_C = \frac{\tilde{V}_C}{\tilde{I}_C} = X_C\angle -90° = -jX_C = -j\frac{1}{\omega C}$$

- The impedance of a resistance incorporates both the resistance and the phase relationship of the phasor voltage-to-current ratio for the resistance:

$$\tilde{Z}_R = \frac{\tilde{V}_R}{\tilde{I}_R} = R = R \angle 0°$$

8.3   Inductive Reactance and Impedance
- Inductive reactance is a real, positive number that is the magnitude ratio of voltage to current for an inductor:

$$X_L = \frac{V_P}{I_P} = \frac{V_{RMS}}{I_{RMS}}$$

- Inductive reactance is directly proportional to frequency and inductance:

$$X_L = 2\pi f L = \omega L$$

- The AC voltage leads the current by 90° for an inductor.

- The impedance of an inductor incorporates both the reactance and the phase relationship of the phasor voltage-to-current ratio for the inductor:

$$\tilde{Z}_L = \frac{\tilde{V}_L}{\tilde{I}_L} = X_L\angle + 90° = + jX_L = j\omega L$$

8.4   Response of RL and RC Series Circuits to AC
- Impedances are used for resistances, inductors, and capacitors. Phasors are used for voltages and currents in AC circuit analysis.

HOMEWORK PROBLEMS

Review of Sinusoidal Signals: Completion of the following problems will demonstrate that you are able to express a sinusoidal signal as a function of time. You will demonstrate that you can determine magnitude, frequency, and phase of a time-domain sinusoidal function.

8.1   For the AC signal shown in Figure P8.1:

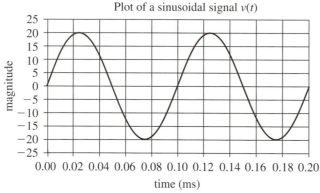

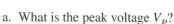

Figure P8.1 Graphic for Problem 8.1

a. What is the peak voltage V_P?
b. What is the frequency ω, in radians per second?
c. What is the phase shift in degrees?
d. Write the time-domain expression for the voltage $v(t)$.

8.2   For the AC signal shown in Figure P8.2:
a. What is the peak voltage V_P?

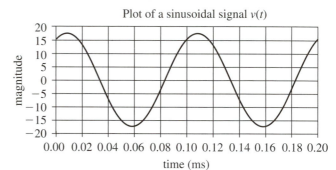

Figure P8.2 Graphic for Problem 8.2

b. What is the frequency ω, in radians per second?
c. What is the phase shift in degrees?
d. Write the time-domain expression for the voltage $v(t)$.

8.3   Graphically add the signals shown in Figures P8.1 and P8.2 and write the resulting time-domain expression. (Can you use a spreadsheet?)

8.4   Add the trigonometric functions that you determined in problems 8.1 and 8.2 using the appropriate trigonometric identities, and express your results as a function of time. Compare the result ot that of problem 8.3.

Completion of the following problems will demonstrate that you can explain why complex numbers are used in AC circuit analysis, what a complex number is, what j is, what the polar and rectangular forms of complex numbers are, and perform complex number arithmetic calculations, including polar-rectangular form conversions:

8.5   a. What does a complex number represent?

b. What does the *j* in a complex number represent?

c. Why are complex numbers used in AC circuit analysis?

8.6 For the following complex-number expression, determine:

$$(2.35 \angle 65.88°) + \frac{(3 - j6)}{(-9 - j8)}$$

a. the value in polar form
b. the value in rectangular form
c. Plot it on the complex-number plane (approximate sketch ok, but label values).

8.7 Given the following complex numbers: $A = 100 + j100$ and $B = 50 \angle -53.13°$, solve for

$$\frac{A \cdot B}{A + B}$$

8.8 Determine $\tilde{Z}$ as one complex number in both polar form and rectangular form if

a. $\tilde{Z} = \dfrac{(65.0 - j34.0)(62.5\angle -17.00°)}{40.8 \angle + 49.18°}$

b. $\tilde{Z} = \dfrac{(25.0 - j34.0)(69.5 \angle 17.00°)}{45.8 \angle - 49.18°}$

c. $\tilde{Z} = \dfrac{(120.0 - j60)(90\angle 60°)}{(100.0 - j100.0)}$

d. $\tilde{Z} = \dfrac{(120.0\angle 45°)(180.0 - j90.0)}{(100.0 - j100.0)}$

Time-Domain and Phasor Form for AC Voltages and Currents: Completion of the following problems will demonstrate that you can describe what a phasor is and can convert between time-domain and phasor form for AC voltages and currents.

8.9 Write the sinusoidal function for each of the waveforms in Figure P8.3. Convert the time-domain function into phasor form, perform the addition, and verify the resulting sum in the time domain.

8.10 Determine the RMS phasor current expression for: $i(t) = 5.28 \sin(377t - 37.35°)$ (Use correct notation too)

8.11 Determine the time-domain expression for $v(t)$ if $\tilde{V} = 27\angle -130°$.

8.12 a. What is a phasor?
b. What is the relationship between phasors and complex numbers?
c. What is the relationship between phasors and time domain expressions?

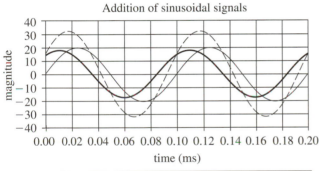

Figure P8.3 Graphic for Problem 8.9

Review of the Properties of Capacitors and Inductors: Completion of the problems in this section will demonstrate that you have a basic understanding of inductors and capacitors, can describe and determine reactance, and can describe and determine the impedance of resistances, inductors, and capacitors.

8.13 What is the peak voltage across a 0.33 µF capacitor at 20 kHz if the peak current is 0.15 mA?

8.14 Given the following time-domain voltage:

$$v(t) = 183 \sin(2562\, t + 29.82°)$$

determine:

a. $v(20\ ms)$
b. the RMS phasor expression
c. the peak phasor expression
d. the power dissipated in a 220 Ω resistor
For (e), (f), and (g), $v(t)$ is applied across a 2 µF capacitor; determine:
e. the reactance of the 2 µF capacitor

f. the impedance of the 2 µF capacitor
g. the expression for $i(t)$ through the 2 µF capacitor

8.15 a. What does a capacitor oppose a change of?
b. Why?
c. How does this property affect a circuit with a capacitor when there is a DC source?
d. How does this property affect a circuit with a capacitor when DC is turned on?
e. How does this property affect a circuit with a capacitor when there is an AC source?

8.16 a. What does an inductor oppose a change of?
b. Why?

c. How does this property affect a circuit with an inductor when there is a DC source?

d. How does this property affect a circuit with an inductor when DC is turned on?

e. How does this property affect a circuit with an inductor when there is an AC source?

8.17 How much power does an ideal inductor dissipate? Why?

8.18 How much power does an ideal capacitor dissipate? Why?

8.19 How much energy does a resistor store? Why?

8.20 What is the difference between reactance, resistance, and impedance? Be clear.

8.21 Does capacitive reactance increase or decrease as frequency decreases? Why?

8.22 Does inductive reactance increase or decrease as frequency decreases? Why?

Completion of the problems in this section will demonstrate that you have a basic understanding of how to use phasors to determine AC voltages, currents, and impedances of components and the entire circuit for RC *and* RL *series circuits. Proper completion of the problems will also demonstrate that you can determine the magnitude and phase relationships for voltages, currents, and impedances in* RC *and* RL *series circuits.*

8.23 A load has voltage $v(t) = 14.142 \sin(754t)$ V across it and current
$i(t) = 0.250 \sin(754t - 45°)$ mA through it.
a. What is the frequency (in hertz) of the voltage and the current?
b. Is the voltage leading or lagging the current?
c. What is the impedance expressed in both polar and rectangular forms?
d. Is this an inductive or a capacitive load?

8.24 Determine $v_1(t)$ in the circuit shown in Figure P8.4 using phasors. Use proper notation.

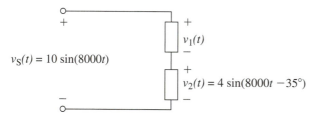

Figure P8.4 Graphic for Problem 8.24

8.25 What is the peak current in an 20 mH inductor at 5 kHz if the peak voltage is 16 V? What type of signal is being assumed with this question?

8.26 Determine the parameters listed for the circuit shown in Figure P8.5.
a. the impedance of the inductor
b. the impedance of the resistor
c. the total impedance
d. the current
e. the voltage drop across the inductor
f. the voltage drop across the resistor

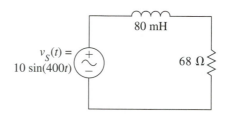

Figure P8.5 Circuit for Problem 8.26

8.27 Determine the quantities listed for the circuit shown in Figure P8.6.
a. the impedance of the capacitor
b. the impedance of the resistor
c. the total impedance
d. the current
e. the voltage drop across the capacitor
f. the voltage drop across the resistor

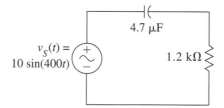

Figure P8.6 Circuit for Problem 8.27

8.28 Determine the quantities listed for the circuit shown in Figure P8.7.
a. the impedance of the capacitor
b. the impedance of the resistor
c. the total impedance
d. the current
e. the source voltage
f. the voltage drop across the resistor

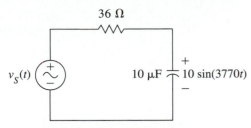

Figure P8.7 Circuit for Problem 8.28

8.29 Determine the quantities listed for the circuit shown in Figure P8.8.
   a. the impedance of the inductor
   b. the impedance of the resistor
   c. the total impedance
   d. the voltage of the source
   e. the voltage drop across the resistor
   f. the voltage drop across the inductor

8.30 Determine the quantities listed for the circuit shown in Figure 8.9.
   a. the impedance of the capacitor
   b. the impedance of the resistor
   c. the total impedance
   d. the voltage of the source
   e. the voltage drop across the resistor
   f. the voltage drop across the capacitor

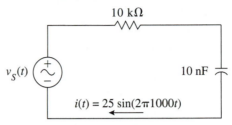

Figure P8.9 Circuit to be Analyzed in Problem 8.30

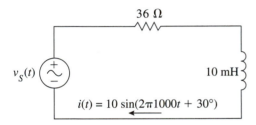

Figure P8.8 Circuit to be Analyzed in Problem 8.29

SERIES–PARALLEL ANALYSIS OF AC CIRCUITS

As a result of successfully completing this chapter, you should be able to:

1. Utilize phasors in AC series–parallel circuit voltage and current calculations.
2. Describe the fundamental properties of series and parallel circuits and subcircuits.
3. Calculate all voltages and currents in single source AC series, parallel, and series–parallel circuits.
4. Perform series–parallel circuit conversions.
5. Describe why and how superposition is used in multiple source AC series, parallel, and series–parallel circuits.
6. Calculate all voltages and currents in multiple-source AC series, parallel, and series–parallel circuits.
7. Explain what an AC current source is and how to analyze AC series–parallel circuits that contain a current source.

9.1 AC SERIES CIRCUITS

We have already learned how to analyze almost any DC series–parallel resistive circuit, even those with multiple sources. The series–parallel circuit analysis techniques were covered in Chapter 3. What if the signal sources are AC instead of DC and the circuit contains inductors and/or capacitors, too? There is good news: the circuit analysis *techniques* are the same. What is the catch? One must use complex numbers to represent phasors and impedances. This burden is not so bad. In fact it is not much of a burden at all. Modern scientific calculators and software usually have the capability to perform calculations with complex numbers. Moreover there is an interesting feature in AC circuit analysis. Recall that inductors and capacitors have a 90° phase shift between the voltage and the current, and that RC and RL circuits phase shift the AC signal between

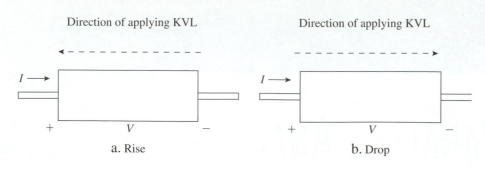

Direction of applying KVL   Direction of applying KVL   **Figure 9.1 Voltage Rises and Voltage Drops in General**

a. Rise   b. Drop

$0°$ and $\pm90°$. What would the phase shift be for AC signals in series, parallel, and series–parallel circuits with resistances, inductors, *and* capacitors? The answer is not obvious (it should not be because a circuit has not even been drawn yet). However, through the use of phasors and complex numbers, one can determine these phase shifts as well as the magnitudes of voltages, currents, and impedances using the same approach (techniques) as in DC circuit analysis.

How might one approach series–parallel circuit analysis with AC sources? One could review the circuit laws, concepts, and results for DC circuit analysis and then see if they can be extrapolated to the AC case. If there are any modifications because of AC signals, those changes could be incorporated as we proceed. Try this approach for series circuits.

The circuit laws, concepts, and results for DC series circuits are:

1. The current is the same through any element, source or load, in a series circuit.

$$I_1 = I_2 = I_3 = \ldots = I_N \qquad (9.1)$$

2. Kirchhoff's voltage law (KVL): the sum of the voltages around any closed path, including through the air, must equal zero.

$$\sum_{n=1}^{N} V_n = +V_{\text{rise1}} + V_{\text{rise2}} + V_{\text{rise3}} + \ldots - V_{\text{drop1}} - V_{\text{drop2}} - V_{\text{drop3}} - \ldots = 0 \quad (9.2)$$

where voltage rises are assigned a positive sign, and voltage drops are assigned a negative sign in this textbook. A summary of voltage rises and drops is given in Figure 9.1. Solid arrows indicate conventional current directions and dashed arrows indicate the direction that KVL is applied. The type of component does not matter in the determination of voltage drops or voltage rises. The key is the voltage polarity rise (negative to positive) or the voltage drop (positive to negative) in the direction that KVL is applied.

3. The total resistance of a series circuit is the sum of the individual resistances.

$$R_T = R_1 + R_2 + R_3 + \ldots + R_N = \sum_{n=1}^{N} R_n \qquad (9.3)$$

4. Voltage division occurs in series circuits. The voltage-divider rule (VDR) is a direct expression of this voltage division.

$$V_x = V_T \frac{R_x}{R_T} \qquad (9.4)$$

Which of these laws and concepts are applicable to AC series circuits? Constant current throughout a series circuit and KVL are fundamentals that do not change for any signal. Both must be true at any and all times, whether a DC signal, an AC signal, or any other signal. The current must be the same anywhere in the series circuit because there is no other place for the current to flow. KVL must hold because it is based on conservation of energy. These facts lead one to realize that the total impedance and VDR must also hold if impedances are used because both are based on items 1 and 2 and Ohm's law for AC circuits. This applicability of DC series circuit concepts to AC series circuits when impedances are used is valid: one can use the same circuit analysis techniques but must use impedances instead of just resistances.

The circuit laws, concepts, and results for AC series circuits are:

1. The current is the same through any element, source or load, in an AC series circuit.

$$\tilde{I}_1 = \tilde{I}_2 = \tilde{I}_3 = \ldots = \tilde{I}_N \qquad (9.5)$$

2. Kirchhoff's voltage law (KVL): the sum of the voltages around any closed path, including through the air, must equal zero.

$$\sum_{n=1}^{N} \tilde{V}_n = +\tilde{V}_{\text{rise1}} + \tilde{V}_{\text{rise2}} + \tilde{V}_{\text{rise3}} + \ldots - \tilde{V}_{\text{drop1}} - \tilde{V}_{\text{drop2}} - \tilde{V}_{\text{drop3}} - \ldots = 0 \qquad (9.6)$$

where voltage rises are assigned a positive sign, and voltage drops are assigned a negative sign in this textbook. The diagrams in Figure 9.1 are still valid, except each voltage and current symbol will have a tilde above it to indicate that it is a complex number. An interesting question should be forming in your mind. How does one assign a voltage polarity with AC signals when the polarity is alternating? The answer is *relative* polarities. If time is "frozen," and one voltage polarity is taken as the reference (usually, but not always, the voltage polarity of one source), the other voltage polarities and all current directions are *labeled* relative to that reference at that instant in time. However, one must remember that the voltage polarities are alternating in real time, just as the current direction is alternating.

3. The total impedance of a series circuit is the sum of the individual impedances.

$$\tilde{Z}_T = \tilde{Z}_1 + \tilde{Z}_2 + \tilde{Z}_3 + \ldots + \tilde{Z}_N = \sum_{n=1}^{N} \tilde{Z}_n \qquad (9.7)$$

4. Voltage division occurs in series circuits. The voltage-divider rule (VDR) is a direct expression of this voltage division.

$$\tilde{V}_x = \tilde{V}_T \frac{\tilde{Z}_x}{\tilde{Z}_T} \qquad (9.8)$$

Example 9.1.1

For the circuit shown in Figure 9.2, determine (*a*), $\tilde{Z}_T$, (*b*), $\tilde{I}$, (*c*) $\tilde{V}_R$ and $\tilde{V}_L$ by the VDR, and (*d*) $\tilde{V}_C$ by Ohm's law. (*e*) Verify KVL.

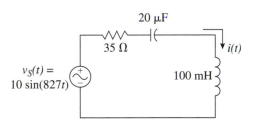

20 µF

35 Ω

$v_S(t) =$
10 sin(827t)

$i(t)$

100 mH

Figure 9.2 Series Circuit for Example 9.1.1

Given: $v_S(t) = 10 \sin(827t)$ V

   $R = 35 \ \Omega$

   $C = 20 \ \mu\text{F}$

   $L = 100 \ \text{mH}$

Desired: a. $\tilde{Z}_T$

   b. $\tilde{I}$

   c. $\tilde{V}_R$ and $\tilde{V}_L$ by the VDR

   d. $\tilde{V}_C$ by Ohm's law

   e. Verify KVL

Strategy: Determine $\tilde{Z}_R$, $\tilde{Z}_L$, and $\tilde{Z}_C$.

$$\tilde{Z}_T = \tilde{Z}_R + \tilde{Z}_C + \tilde{Z}_L$$

Label circuit diagram with voltage polarities and current directions.

$$\tilde{I} = \frac{\tilde{V}_S}{\tilde{Z}_T}$$

$$\tilde{V}_R = \tilde{V}_S \frac{\tilde{Z}_R}{\tilde{Z}_T}, \quad \tilde{V}_L = \tilde{V}_S \frac{\tilde{Z}_L}{\tilde{Z}_T}$$

$$\tilde{V}_C = \tilde{I} \tilde{Z}_C$$

KVL around series circuit

Solution: _____

$$\tilde{Z}_R = R = 35 \ \Omega$$

$$\tilde{Z}_C = -j\frac{1}{\omega C} = -j\frac{1}{(827)(20\mu)} = -j60.460 \ \Omega$$

$$\tilde{Z}_L = +j\omega L = +j(827)(0.1) = +j82.7 \ \Omega$$

a. $\tilde{Z}_T = \tilde{Z}_R + \tilde{Z}_C + \tilde{Z}_L = 35 - j60.46 + j82.7$

$\qquad = 35.000 + j22.240 \ \Omega = 35.0 + j22.2 \ \Omega$

Note: All voltages and currents are expressed as peak values in this example.

b. $\tilde{I} = \dfrac{\tilde{V}_S}{\tilde{Z}_T} = \dfrac{10\angle 0°}{35 + j22.24} = 0.24115\angle -32.434° = 0.241\angle -32.4°\text{A}$

c. $\tilde{V}_R = \tilde{V}_S \dfrac{\tilde{Z}_R}{\tilde{Z}_T} = 10\angle 0° \dfrac{35}{35 + j22.24} = 8.4402\angle -32.433° = 8.44\angle -32.4° \text{ V}$

$\qquad \tilde{V}_L = \tilde{V}_S \dfrac{\tilde{Z}_L}{\tilde{Z}_T} = 10\angle 0° \dfrac{+j82.7}{35 + j22.24} = 19.943\angle +57.567° = 19.9\angle +57.6° \text{ V}$

d. $\tilde{V}_C = \tilde{I}\tilde{Z}_C = (0.24115\angle -32.434°)(-j60.46)$

$\qquad = 14.5797\angle -122.434° = 14.6\angle -122.4° \text{ V}$

e. (Refer to Figure 9.3) $+\tilde{V}_S - \tilde{V}_R - \tilde{V}_C - \tilde{V}_L = 0$

$\qquad +\tilde{V}_S = +\tilde{V}_R + \tilde{V}_C + \tilde{V}_L$

$\qquad\qquad = +8.4402\angle -32.433° + 19.943\angle +57.567° + 14.5797\angle -122.434°$

$\qquad\qquad = 9.9999\angle -0.001° = 10.0\angle 0.0 \text{ V}$

KVL is satisfied.

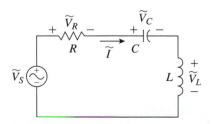

Figure 9.3 Phasor Voltages and Current for Example 9.1.1

9.2 AC PARALLEL CIRCUITS

As with AC series circuits, we start off with a review of the important laws, concepts, and results that were found for DC parallel circuits.

The circuit laws, concepts, and results for DC parallel circuits are:

1. The voltage across each component in a parallel circuit is the same.

$$V_S = V_1 = V_2 = V_3 = \ldots V_N \tag{9.9}$$

2. Kirchhoff's current law (KCL): the sum of the currents entering and leaving a node must equal zero.

$$\sum_{n=1}^{N} I_n = +I_1 + I_2 + I_3 + \ldots + I_N = 0 \tag{9.10}$$

where the current is assigned to be positive if it is entering the node and is assigned to be negative if it is leaving the node.

3. The total resistance of a DC parallel circuit is:

$$R_T = \frac{1}{\dfrac{1}{R_1} + \dfrac{1}{R_2} + \ldots + \dfrac{1}{R_N}} \qquad \text{in general} \tag{9.11}$$

$$R_T = R_1 \parallel R_2 = \frac{R_1 R_2}{R_1 + R_2} \qquad \text{for two parallel resistors.} \tag{9.12}$$

4. Current division occurs in parallel circuits. The current-divider rule (CDR) is a direct expression of this current division.

$$I_x = \frac{I_T R_T}{R_x} \qquad \text{in general} \tag{9.13}$$

$$I_1 = \frac{I_T R_2}{R_1 + R_2} \qquad I_2 = \frac{I_T R_1}{R_1 + R_2} \qquad \text{for two parallel resistors.} \tag{9.14}$$

5. Conductance is 1 over resistance. Conductance is useful in parallel circuits.

$$G = \frac{I}{V} = \frac{1}{R} \tag{9.15}$$

6. The total conductance of a parallel circuit is the sum of the individual conductances.

$$G_T = G_1 + G_2 + \ldots + G_N \tag{9.16}$$

Which of these laws and concepts are applicable to AC parallel circuits? Constant voltage across a parallel circuit and KCL are fundamentals that do not change for any other signal. Both must be true at any and all times, whether a DC signal, an AC signal, or any signal. The electrical potential energy difference per unit charge, i.e., voltage, is the same across all elements in the parallel circuit, assuming negligible conductor resistance, because the conductor on the "top" of the parallel circuit has no voltage across itself nor does the conductor on the "bottom." KCL is based on net zero charge entering and leaving the node at all times; hence, the signal type does not matter. These facts lead one to realize that the total impedance and CDR must also hold if impedances are used because both are based on items 1 and 2 and Ohm's law. This applicability of DC parallel circuit concepts to AC parallel circuits when impedances are used is good: one can use the same circuit analysis techniques but must use impedances instead of just resistances.

The circuit laws, concepts, and results for AC parallel circuits are:

1. The voltage across each component in a parallel circuit is the same.

$$\tilde{V}_S = \tilde{V}_1 = \tilde{V}_2 = \tilde{V}_3 = \ldots \tilde{V}_N \tag{9.17}$$

2. Kirchhoff's current law (KCL): the sum of the currents entering and leaving a node must equal zero.

$$\sum_{n=1}^{N} \tilde{I}_n = +\tilde{I}_1 + \tilde{I}_2 + \tilde{I}_3 + \ldots + \tilde{I}_N = 0 \qquad (9.18)$$

with the same sign convention as used in DC circuit analysis with KCL.

3. The total impedance of an AC parallel circuit is:

$$\tilde{Z}_T = \cfrac{1}{\cfrac{1}{\tilde{Z}_1} + \cfrac{1}{\tilde{Z}_2} + \ldots + \cfrac{1}{\tilde{Z}_N}} \qquad \text{in general} \qquad (9.19)$$

$$\tilde{Z}_T = \tilde{Z}_1 \parallel \tilde{Z}_2 = \frac{\tilde{Z}_1 \tilde{Z}_2}{\tilde{Z}_1 + \tilde{Z}_2} \qquad \text{for two parallel impedances} \qquad (9.20)$$

4. Current division occurs in parallel circuits. The current-divider rule (CDR) is a direct expression of this current division.

$$\tilde{I}_X = \frac{\tilde{I}_T \tilde{Z}_T}{\tilde{Z}_X} \qquad \text{in general} \qquad (9.21)$$

$$\tilde{I}_1 = \frac{\tilde{I}_T \tilde{Z}_2}{\tilde{Z}_1 + \tilde{Z}_2} \qquad \tilde{I}_2 = \frac{\tilde{I}_T \tilde{Z}_1}{\tilde{Z}_1 + \tilde{Z}_2} \qquad \text{for two parallel impedances.} \qquad (9.22)$$

Example 9.2.1

For the circuit shown in Figure 9.4, determine (a) $\tilde{Z}_T$, (b) $\tilde{V}$, (c) $\tilde{I}_R$ by the CDR, and (d) $\tilde{I}$ by Ohm's law. (e) Verify KCL.

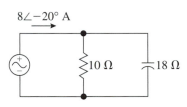

Figure 9.4 Parallel Circuit for Example 9.2.1

Given: $R = 10\ \Omega$
$X_C = 18\ \Omega$
$\tilde{I}_T = 8\angle{-20°}\ \text{A}$

Desired: a. $\tilde{Z}_T$
b. $\tilde{V}$
c. $\tilde{I}_R$ by the CDR
d. $\tilde{I}_C$ by Ohm's law
e. Verify KCL

Strategy: Determine $\tilde{Z}_R$ and $\tilde{Z}_C$.

$$\tilde{Z}_T = \frac{\tilde{Z}_R \tilde{Z}_C}{\tilde{Z}_R + \tilde{Z}_C}$$

Label circuit diagram with voltage polarities and current directions.

$$\tilde{V} = \tilde{I}_T \tilde{Z}_T$$

$$\tilde{I}_R = \frac{\tilde{I}_T \tilde{Z}_T}{\tilde{Z}_R} \quad \text{or} \quad \tilde{I}_1 = \frac{\tilde{I}_T \tilde{Z}_2}{\tilde{Z}_R + \tilde{Z}_C} \qquad \text{(the first equation is easier—we have the numerator from the previous step)}$$

$$\tilde{I}_C = \frac{\tilde{V}}{\tilde{Z}_C}$$

KCL at "top" node

Solution:

$$\tilde{Z}_R = R = 10 \ \Omega$$

$$\tilde{Z}_C = -jX_C = -j18 \ \Omega$$

a. $\tilde{Z}_T = \dfrac{\tilde{Z}_R\tilde{Z}_C}{\tilde{Z}_R + \tilde{Z}_C} = \dfrac{(10)(-j18)}{10 - j18} = 8.7416\angle -29.055° = 8.74\angle -29.1° \ \Omega$

b. (See Figure 9.5 for voltage polarity and current directions.)

$$\tilde{V} = \tilde{I}_T\tilde{Z}_T = (8\angle -20°)(8.7416\angle -29.055°) = 69.933\angle -49.055°$$

$$= 69.9\angle -49.1° \ V$$

c. $\tilde{I}_R = \dfrac{\tilde{I}_T\tilde{Z}_T}{\tilde{Z}_R} = \dfrac{(8\angle -20°)(8.7416\angle -29.055°)}{10}$

$$= 6.9933\angle -49.055° = 6.99\angle -49.1° \ A$$

d. $\tilde{I}_C = \dfrac{\tilde{V}}{\tilde{Z}_C} = \dfrac{69.933\angle -49.055°}{-j18} = 3.8851\angle +40.945° = 3.89\angle +40.9° \ A$

e. KCL at top node: $\quad +\tilde{I}_T - \tilde{I}_R - \tilde{I}_C = 0$

$$+\tilde{I}_T = +\tilde{I}_R + \tilde{I}_C = +6.9933\angle -49.055° + 3.8851\angle +40.945° = 8.00\angle -20.0° \ A$$

KCL is satisfied.

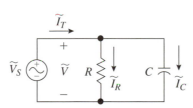

Figure 9.5 Voltage Polarity and Current Directions for the Solution in Example 9.2.1

Items 5 and 6 in "the circuit laws, concepts, and results for AC parallel circuits" are interesting. Although conductance being the inverse of resistance ($G = 1/R$) has been established, what is the inverse of reactance? What is the inverse of impedance? AC parallel circuit analysis needs to be expanded beyond using conductances. The inverse of reactance is called *susceptance* and is given the symbol *B:*

$$B = \frac{1}{X} \tag{9.23}$$

(Do not confuse susceptance *B* with magnetic flux density *B*—you should know from the context in which it is used.) Recall that as resistance increases, conductance decreases, and vice versa. Similarly, as reactance increases, susceptance decreases, and vice versa. Subscripts are often used to differentiate between inductive and capacitive susceptances:

$$B_L = \frac{1}{X_L} \qquad B_C = \frac{1}{X_C} \tag{9.24}$$

If susceptance is the inverse of reactance, what is the inverse of impedance? The inverse of impedance is called *admittance* and is given the symbol $\tilde{Y}$. Admittance must be a complex number because impedance is a complex number:

$$\tilde{Y} = \frac{1}{\tilde{Z}} \tag{9.25}$$

How does admittance ($\tilde{Y}$) relate to conductance (G) and susceptance (B)? Recall that resistance is the real part of impedance and reactance is the imaginary part of impedance:

$$\tilde{Z} = R \pm jX \tag{9.26}$$

where the plus sign is used with inductive reactance because voltage leads current, and the minus sign is used with capacitive reactance because current leads voltage. Similarly, conductance is the real part of admittance, and susceptance is the imaginary part of admittance:

$$\tilde{Y} = G \pm jB \tag{9.27}$$

where the plus sign is used with capacitive susceptance, and the minus sign is used with inductive susceptance. Can you show why these signs are true? Start with the definition of admittance and express it in terms of the phasor voltages and currents:

$$\tilde{Y} = \frac{1}{\tilde{Z}} = \frac{\tilde{I}}{\tilde{V}} = \frac{I}{V} \angle(\theta_I - \theta_V) \tag{9.28}$$

For a capacitor, current leads voltage by 90°, so $\theta_I - \theta_V$ is +90°, which corresponds to $+j$ for admittance in rectangular form. Similarly, for an inductor, voltage leads current by 90°, so $\theta_I - \theta_V$ is −90°, which corresponds to $-j$ for admittance in rectangular form.

One other concern should be the units of admittance. Just as ohms are the units for resistance, reactance, and impedance, the unit of conductance, siemens (S), is the unit for conductance, susceptance, and admittance. Thus, items 5 and 6 can now be stated for AC circuits:

5.   Admittance is 1 over impedance. Admittance is useful in parallel circuits.

$$\tilde{Y} = \frac{\tilde{I}}{\tilde{V}} = \frac{1}{\tilde{Z}} \quad \text{(S)} \tag{9.29}$$

6.   The total admittance of a parallel circuit is the sum of the individual admittances.

$$\tilde{Y}_T = \tilde{Y}_1 + \tilde{Y}_2 + \ldots + \tilde{Y}_N = \sum_{n=1}^{N} \tilde{Y}_n \tag{9.30}$$

Information Research Exercise 9.2.1   (web, library, and/or vendor catalogs) _____

a.   What contributions did Werner von Siemens make to electricity?
b.   What products does the Siemens company manufacture?

_____

___ Example 9.2.2 _____

For the circuit shown previously in Figure 9.4, determine (a) $\tilde{Y}_T$ from G and B, and (b) $\tilde{Y}_T$ from $\tilde{Z}_T$

Given: $R = 10 \ \Omega$

$X_C = 18 \ \Omega$

$\tilde{Z}_T = 8.7416 \angle -29.055° \ \Omega = 7.6415 - j4.24535 \ \Omega$ (from Example 9.2.1)

Desired: a. $\tilde{Y}_T$ from G and B
b. $\tilde{Y}_T$ from $\tilde{Z}_T$

Strategy: a. $G = 1/R$, $B = 1/X$, $\tilde{Y}_T = G + jB$

b. $\tilde{Y}_T = 1/\tilde{Z}_T$

Solution:

a. $G = \dfrac{1}{R} = \dfrac{1}{10} = 0.1$ S

$B_C = \dfrac{1}{X_C} = \dfrac{1}{18} = 0.055556$ S

$\tilde{Y} = G + jB = 0.1 + j0.05556 = 0.100 + j0.0556$ S

b. $\tilde{Y}_T = \dfrac{1}{\tilde{Z}} = \dfrac{1}{8.7416\angle -29.055°} = 0.114396\angle +29.055°$

$= 0.0999993 + j0.055556 = 0.100 + j0.0556$ S

The results from parts (a) and (b) match, as expected.

There is an interesting observation to make about the impedance–admittance relationship. Start with the total impedance of the circuit in Figure 9.4:

$$\tilde{Z}_T = 7.6415 - j4.24535 \ \Omega = R - jX$$

A common error made by new electronics students is to determine the conductance (G) and the susceptance (B) from the resistance (R) and the reactance (X) of the total impedance:

$$G = \frac{1}{R} = \frac{1}{7.6415} = 0.13086 = 0.131 \text{ S} \tag{9.31}$$

$$B_C = \frac{1}{X_C} = \frac{1}{4.24535} = 0.23555 = 0.236 \text{ S} \tag{9.32}$$

Compare these results with those of Example 9.2.2:

$$G = \frac{1}{R} = \frac{1}{10} = 0.1 = 0.100 \text{ S} \tag{9.33}$$

$$B_C = \frac{1}{X_C} = \frac{1}{18} = 0.055556 = 0.0556 \text{ S} \tag{9.34}$$

These answers do not match! Why not? The reason is subtle yet significant. The conductance and susceptance in Equation (9.33) are the inverses of the *parallel* resistance and reactance of the circuit, respectively. However, the incorrect conductance and susceptance in Equations (9.31) and (9.32) are the inverses of the resistance and reactance that are in the rectangular form of impedance. But the rectangular form of impedance is the sum of individual component impedances that are in *series*. Hence, the conductance and susceptance obtained in Equation (9.31) are in error. You must be able to distinguish between proper and invalid determinations of conductance and susceptance from resistance and reactance (or vice versa).

The previous discussion suggests an even more interesting possibility: one can determine the series equivalent circuit of a given parallel circuit or the parallel equivalent circuit of a given series circuit. The following examples illustrate the proper methods to make theses series–parallel conversions. Then a derivation of the general series–parallel conversions is given. The results are general handbook equations for series–parallel conversions.

Example 9.2.3

Determine the (*a*) admittance and (*b*) parallel equivalent circuit for the series circuit shown in Figure 9.6a.

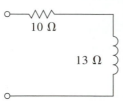
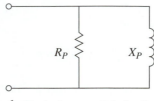

a. Series circuit   b. Equivalent parallel circuit

Figure 9.6 Circuits for Example 9.2.3

Given: $R_S = 10 \ \Omega$

$X_S = 13 \ \Omega$ (inductive)

Desired: a. $\tilde{Y}_T$

b. R_P and X_P

Strategy: a. $\tilde{Y}_T = \dfrac{1}{\tilde{Z}_T}$

b. $R_P = 1/G_P$, and $X_P = 1/B_P$

Solution:

a. $\tilde{Y}_T = \dfrac{1}{\tilde{Z}_T} = \dfrac{1}{10 + j13} = 0.037175 - j0.048327 \ S = 0.0372 - j0.0483 \ S$

b. $R_P = \dfrac{1}{G_P} = \dfrac{1}{0.037175} = 26.9 \ \Omega$

$X_P = \dfrac{1}{B_P} = \dfrac{1}{0.048327} = 20.7 \ \Omega$

Example 9.2.4

Determine the component values of the series equivalent circuit for the parallel circuit shown in Figure 9.7a.

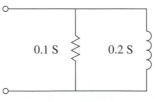
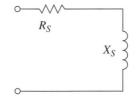

a. Parallel circuit   b. Equivalent series circuit

Figure 9.7 Circuits for Example 9.2.4

Given: $G_P = 0.1 \ S$

$B_P = 0.2 \ S$ (inductive)

Desired: R_S and X_S

Strategy: $\tilde{Y}_T = G_P - jB_P, \ \tilde{Z}_T = \dfrac{1}{\tilde{Y}_T} = R_S + jX_S$

Solution: _____

$$\tilde{Y}_T = 0.1 - j0.2 \text{ S}$$

$$\tilde{Z}_T = \frac{1}{\tilde{Y}_T} = \frac{1}{0.1 - j0.2} = 4.4721\angle +63.435° = 2.000 + j4.000 \text{ }\Omega$$

$$R_S = 2.000 \text{ }\Omega; X_S = 4.000 \text{ }\Omega$$

Example 9.2.5 _____

Determine the component values for the parallel equivalent circuit of the series circuit shown in Figure 9.8a if $\omega = 20$ krads/s.

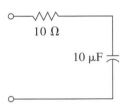
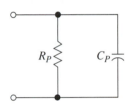

Figure 9.8 Circuits for Example 9.2.5

10 Ω

10 μF

R_P C_P

a. Series circuit   b. Equivalent parallel circuit

Given: $R_S = 10 \text{ }\Omega$
          $C_S = 10 \text{ }\mu\text{F}$

Desired: R_P and C_P

Strategy: $\tilde{Z}_C = -j\dfrac{1}{\omega C}; \tilde{Z}_T = \tilde{Z}_R + \tilde{Z}_C; \tilde{Y}_T = \dfrac{1}{\tilde{Z}_T} = G_P - jB_P$

$$R_P = 1/G_P, \text{ and } C_P = B_P/\omega$$

Solution: _____

$$\tilde{Z}_C = -j\frac{1}{\omega C} = -j\frac{1}{(20 \text{ krad/s})(10 \text{ }\mu\text{F})} = -j5 \text{ }\Omega$$

$$\tilde{Y}_T = \frac{1}{\tilde{Z}_T} = \frac{1}{10 - j5} = 0.08000 - j0.04000 \text{ S}$$

$$R_P = \frac{1}{G_P} = \frac{1}{0.08} = 12.5 \text{ }\Omega$$

$$C_P = \frac{B_P}{\omega} = \frac{0.04}{20 \text{ k}} = 2.00 \text{ }\mu\text{F}$$

Example 9.2.6 _____

For the circuit in Figure 9.9, determine (a) the total admittance, and (b) the values for the components of the series equivalent circuit. The frequency is 1 kHz.

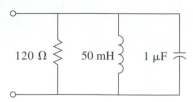

Figure 9.9 Parallel Circuit for Example 9.2.6

120 Ω   50 mH   1 μF

Given: $R_P = 120 \ \Omega$
          $C_P = 1 \ \mu\text{F}$
          $L_P = 50 \ \text{mH}$
          $f = 1 \ \text{kHz}$

Desired: a. $\tilde{Y}_T$
           b. R_S and X_S

Strategy: a. $\tilde{Y}_T = \tilde{Y}_R + \tilde{Y}_C + \tilde{Y}_L = G_P + jB_C - jB_L = \dfrac{1}{R_P} + j\omega C - j\dfrac{1}{\omega L}$

           b. $\tilde{Z}_T = \dfrac{1}{\tilde{Y}_T} = R_S + jX_S, \ X_S \rightarrow C$

Solution:

a. $\tilde{Y}_T = \dfrac{1}{R_P} + j\omega C - j\dfrac{1}{\omega L} = \dfrac{1}{120} + j\,(2\pi 1000)(10^{-6}) - j\,\dfrac{1}{(2\pi 1000)(0.050 \ \text{m/H})}$

$$\tilde{Y}_T = (8.3333 + j3.1001) \ \text{mS} = (8.33 + j3.10) \ \text{mS}$$

b. $\tilde{Z}_T = \dfrac{1}{\tilde{Y}_T} = \dfrac{1}{8.3333 \ \text{mS} + j3.1001 \ \text{mS}} = 105.41 - j39.215 \ \Omega$

The imaginary part of the impedance is negative; hence, the equivalent series impedance is capacitive.

$$R_S = 105 \ \Omega; \quad X_S = 39.215 \ \Omega \ \text{(capacitive)}$$

$$C_S = \dfrac{1}{2\pi f X_S} = \dfrac{1}{2\pi(1000)(39.215)} = 4.06 \ \mu\text{F}$$

The series equivalent circuit consists of a resistance of 105 Ω in series with a capacitance of 4.06 μF.

Note that the series equivalent circuit in Example 9.2.6 contains only two components, not three (as in the original circuit). The two reactive components are accounted for by the single series reactive component in the series equivalent circuit.

The series-to-parallel and parallel-to-series conversions can be developed in general for use in computer software, such as a spreadsheet. These developments are presented next.

Generalization of Series–Parallel Conversions to "Handbook" Equations

Case 1: Series-to-Parallel Conversion—see Figure 9.10.

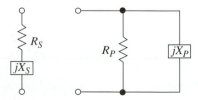

Figure 9.10 Series-to-Parallel Circuit Conversion

R_S

jX_S

R_P jX_P

$$\tilde{Z}_T = R_S \pm jX_S, \ \tilde{Y}_T = G_P \mp jB_P \qquad\qquad (9.35)$$

Invert the total series impedance to obtain the total admittance:

$$\tilde{Y}_T = \frac{1}{(R_S \pm jX_S)} \tag{9.36}$$

Multiply the numerator and the denominator by the complex conjugate of the denominator to make the resultant denominator real:

$$\tilde{Y}_T = \frac{1}{(R_S \pm jX_S)} \cdot \frac{(R_S \mp jX_S)}{(R_S \mp jX_S)} = \frac{R_S \mp jX_S}{R_S^2 + X_S^2} \tag{9.37}$$

Break the result into real and imaginary parts:

$$\tilde{Y}_T = \frac{R_S}{R_S^2 + X_S^2} \mp j\frac{X_S}{R_S^2 + X_S^2} \tag{9.38}$$

The real part of admittance is conductance; the inverse is the parallel resistance:

$$G_P = \frac{R_S}{R_S^2 + X_S^2}, \qquad \boxed{R_P = \frac{R_S^2 + X_S^2}{R_S}} \tag{9.39}$$

The imaginary part of admittance is susceptance; the inverse is the parallel reactance:

$$\mp jB_P = \mp \frac{jX_S}{R_S^2 + X_S^2}, \qquad \boxed{X_P = \frac{R_S^2 + X_S^2}{X_S}} \tag{9.40}$$

Case 2: Parallel-to-Series Conversion–see Figure 9.11.

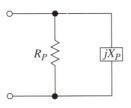

Figure 9.11 Parallel-to-Series Circuit Conversion

Try to label the steps (as was done for the series-to-parallel conversion):

$$\tilde{Z}_T = R_P \| (\pm jX_P), \tilde{Z}_T = R_S \pm jX_S \tag{9.41}$$

$$\tilde{Z}_T = \frac{R_P(\pm jX_P)}{(R_P \pm jX_P)} \tag{9.42}$$

$$\tilde{Z}_T = \frac{R_P(\pm jX_P)}{(R_P \pm jX_P)} \cdot \frac{(R_P \mp jX_P)}{(R_P \mp jX_P)} = \frac{\pm jX_P R_P^2 \mp jX_P(\pm jX_P) R_P}{R_P^2 + X_P^2} \tag{9.43}$$

$$\tilde{Z}_T = \frac{R_P X_P^2 \pm jR_P^2 X_P}{R_P^2 + X_P^2} = \frac{R_P X_P^2}{R_P^2 + X_P^2} \pm j\frac{R_P^2 X_P}{R_P^2 + X_P^2} \tag{9.44}$$

$$\boxed{R_S = \frac{R_P X_P^2}{R_P^2 + X_P^2}} \tag{9.45}$$

$$\pm jX_S = \pm j\frac{R_P^2 X_P}{R_P^2 + X_P^2} \qquad \boxed{X_S = \frac{R_P^2 X_P}{R_P^2 + X_P^2}} \tag{9.46}$$

Check the results of Examples 9.2.3 through 9.2.6 using these series–parallel conversion equations.

9.3 AC SERIES–PARALLEL CIRCUITS

For AC series circuits, KVL, total impedance, the VDR, and the fact that the current is the same throughout a series circuit have been established. For AC parallel circuits, KCL, total impedance and admittance, the CDR, and the fact that the voltage is the same throughout a parallel circuit have been established. AC circuit analysis of *series–parallel circuits* naturally comes next. Some general guidelines for AC series–parallel circuit analysis are extended from the DC case:

1. The *general approach* to analyzing a given AC series–parallel circuit is to identify the groups of impedances that are in series and those that are in parallel and to apply the appropriate series and parallel circuit laws to the impedances in that group. You must be able to identify groups of impedances as being in series or in parallel with other groups of impedances and apply the appropriate series and parallel circuit laws to those groups.
2. There are usually multiple approaches to analyzing an AC series–parallel circuit, but if you take some time to *think* about your strategy, you will usually come up with an *efficient* approach.
3. A single-source series–parallel circuit is fundamentally a series circuit or fundamentally a parallel circuit. The source is the key. If a single impedance is in series with the source on either side (or both sides) of the source, then the circuit is fundamentally a series circuit. If the current path from the source divides on both sides of the source, then the circuit is fundamentally a parallel circuit.
4. A group of series impedances, usually a "leg" in a parallel circuit, is called a *branch*.
5. In the VDR and the CDR, the total voltage or current becomes the total voltage or the current for that group of impedances, not necessarily the entire circuit. Also, KVL and KCL are always valid and can be used where appropriate.

Several examples are now presented to illustrate series–parallel analysis of AC circuits.

___ **Example 9.3.1** _____

Determine the total admittance of the circuit in Figure 9.12.

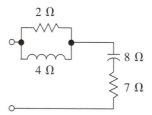

Figure 9.12 Circuit for Example 9.3.1

Given: circuit in Figure 9.12
Desired: $\tilde{Y}_T$

Strategy: $\tilde{Y}_1 = \dfrac{1}{2} + \dfrac{1}{j4}$, $\tilde{Z}_T = \dfrac{1}{\tilde{Y}_1} - j8 + 7$, $\tilde{Y}_T = \dfrac{1}{\tilde{Z}_T}$

Solution: _____

$$\tilde{Y}_1 = \frac{1}{2} + \frac{1}{+j4} = 0.5 - j0.25 \text{ S}$$

$$\tilde{Z}_T = \frac{1}{0.5 - j0.25} - j8 + 7 = 8.6000 - j7.2000 \ \Omega$$

$$\tilde{Y}_T = \frac{1}{\tilde{Z}_T} = \frac{1}{8.6 - j7.2} = 0.08916\angle +39.936°$$

$$= 0.0892\angle +39.9° \text{ S} = 0.0684 - j0.0572 \text{ S}$$

Example 9.3.2

Determine (a) $\tilde{I}_x$ and (b) $\tilde{V}_o$ for the circuit shown in Figure 9.13.

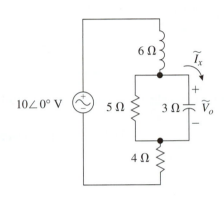

Figure 9.13 Circuit for Example 9.3.2

Given: circuit in Figure 9.13

Desired: a. $\tilde{I}_X$

b. $\tilde{V}_o$

Strategy: $\tilde{Z}_x = [5 \parallel (-j3)]$, $\tilde{Z}_T = +j6 + \tilde{Z}_x + 4$

$$\tilde{I}_T = \frac{\tilde{V}_S}{\tilde{Z}_T}$$

$$\tilde{V}_o = \tilde{I}_T \tilde{Z}_x$$

$$\tilde{I}_x = \frac{\tilde{V}_o}{\tilde{Z}_C}$$

Solution:

$$\tilde{Z}_x = \frac{(5)(-j3)}{5 - j3} = 2.5725\angle -59.036° \ \Omega$$

$$\tilde{Z}_T = +j6 + 2.5725\angle -59.036° + 4 = 6.5372\angle +35.478° \ \Omega$$

$$\tilde{I}_T = \frac{\tilde{V}_S}{\tilde{Z}_T} = \frac{10\angle 0°}{6.5372\angle +35.478°} = 1.5297\angle -35.478° \text{ A}$$

$$\tilde{V}_o = \tilde{I}_T \tilde{Z}_X = (1.5297\angle -35.478°)(2.5725\angle -59.036°)$$

$$= 3.9352\angle -94.514° = 3.94\angle -94.5° \text{ V}$$

$$\tilde{I}_x = \frac{\tilde{V}_o}{\tilde{Z}_C} = \frac{3.9352\angle -94.514°}{3\angle -90°} = 1.3117\angle -4.514° \text{ A} = 1.31\angle -4.5° \text{ A}$$

___ **Example 9.3.3** _____

Redraw the circuit in Figure 9.14, clearly showing series and parallel groups of components.

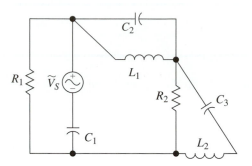

Figure 9.14 Circuit for Example 9.3.3

Solution: _____

The redrawn circuit for Example 9.3.3 is shown in Figure 9.15.

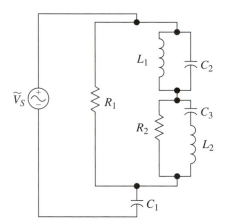

Figure 9.15 Redrawn Circuit for Example 9.3.3

___ **Example 9.3.4** _____

Determine (a) $\tilde{I}_x$ and (b) $\tilde{V}_{ab}$ for the circuit shown in Figure 9.16.

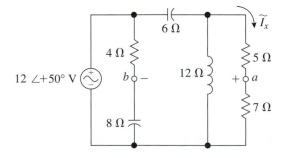

Figure 9.16 Circuit for Example 9.3.4

Given: circuit in Figure 9.16
Desired: a. $\tilde{I}_x$
          b. $\tilde{V}_{ab}$

Strategy: Redraw and label the circuit (Figure 9.17).

$$\tilde{Z}_P = \left[+j12 \parallel (5 + 7) \right], \ \tilde{Z}_2 = -j6 + \tilde{Z}_P$$

a. $\tilde{I}_2 = \dfrac{\tilde{V}_S}{\tilde{Z}_2}, CDR \rightarrow \tilde{I}_x$

b. $VDR \rightarrow \tilde{V}_y, \ \tilde{V}_x = \tilde{I}_x(7), \ KVL \rightarrow \tilde{V}_{ab}$

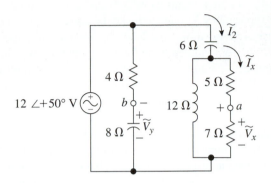

Figure 9.17 Redrawn Circuit for Example 9.3.4

Solution: _____

$$\tilde{Z}_P = \frac{(+j12)(5 + 7)}{+j12 + 5 + 7} = 8.4853\angle +45.000° \ \Omega$$

$$\tilde{Z}_2 = -j6 + 8.4853\angle +45.000° = 6.0000\angle +0° \ \Omega$$

Comment: This convenient number for the impedance of branch 2 is purely a coincidence, but coincidences do happen.

a. $\tilde{I}_2 = \dfrac{\tilde{V}_S}{\tilde{Z}_2} = \dfrac{12\angle +50°}{6\angle 0°} = 2\angle +50° \ A$

$$\tilde{I}_x = \frac{(2\angle +50°)(+j12)}{(5 + 7) + (+j12)} = 1.4142\angle +95.000° = 1.41\angle +95.0° \ A$$

b. $\tilde{V}_x = \tilde{I}_x(7) = (1.4142\angle +95.000°)(7) = 9.8995\angle +95.000° \ V$

$$\tilde{V}_y = \frac{\tilde{V}_S(-j8)}{4 - j8} = \frac{(12\angle +50°)(-j8)}{4 - j8} = 10.733\angle +23.435° \ V$$

KVL: (Refer to Figure 9.17) $+\tilde{V}_{ab} - \tilde{V}_x + \tilde{V}_y = 0$

$$+\tilde{V}_{ab} = +\tilde{V}_x - \tilde{V}_y = +9.8995\angle +95.000° - 10.733\angle +23.435°$$

$$= 12.083\angle +152.425° = 12.1\angle +152.4° \ V$$

9.4 ANALYSIS OF MULTIPLE-SOURCE AC CIRCUITS USING SUPERPOSITION

So far we have learned how to analyze complex AC series–parallel circuits that contain several impedances. As in the DC case, the next step is to analyze the circuit that contains *several sources* as well as several impedances. But there is one condition that must hold true if superposition with phasors is to be used: *the sources must be of the same frequency.* If they are not of the same frequency, then the superposition of voltage or current phasors is not valid and a different approach must be used (covered in Chapter 15). Why cannot phasors be added if the frequencies are different? Think about reactances in the circuit. The reactance of an inductor or a capacitor depends on frequency. Thus the impedances would not be the same for different frequencies, and voltages and currents could not be calculated under the same circuit impedance conditions. Addition of phasors breaks down if the frequencies are different.

Superposition is a circuit analysis technique to handle multiple-source AC circuits. As in the DC case, the circuit is analyzed one source at a time. The other sources are "deactivated." How does one deactivate an AC voltage source? Ask this question as in the DC case: How can a voltage source be made to zero volts? *Replace the voltage source with a short.* A short guarantees that the voltage will be zero across that short. Similarly an AC current source may be in a circuit. How is an AC current source deactivated? The current source must be replaced by an *open* to make the current equal zero.

Thus, to use superposition, first label the desired quantities to be found. Assign an arbitrary polarity to each voltage to be found and an arbitrary direction to each current to be found. Deactivate all sources except one. Analyze the circuit for the desired quantities using series–parallel circuit analysis. Be sure to note voltage polarities and current directions carefully. Then, deactivate that first source and reactivate another source. Analyze the circuit for the desired quantities using series–parallel circuit analysis. Again, be sure to note voltage polarities and current directions carefully. Continue with this process until all sources have been accounted for. Then, add the corresponding resultant phasor voltages and currents from the analysis of each source, but be sure to assign the proper sign to each quantity: positive if it agrees with the voltage polarity or current direction that was initially assigned, and negative if opposite. This last step is actually the superposition step: the phasor voltages and currents are *superposed.*

Example 9.4.1

Determine the current $\tilde{I}_a$ and $i_a(t)$ in Figure 9.18.

Given: circuit in Figure 9.18

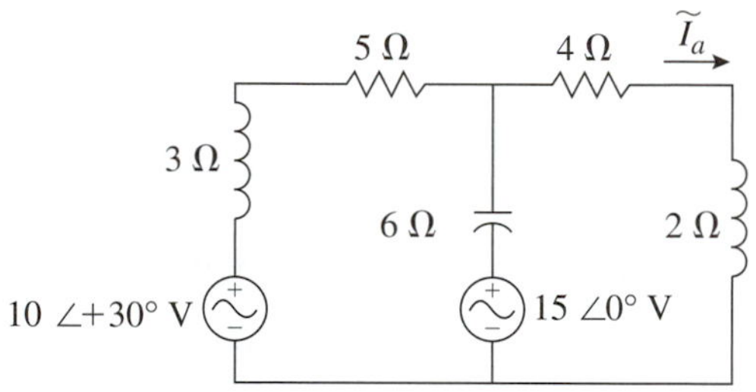

Figure 9.18 Circuit for Example 9.4.1

Desired: $\tilde{I}_a$ and $i_a(t)$
           Note: $\tilde{I}_a$ uses the peak value of $i_a(t)$.

Strategy: superposition
           Peak values for sources are shown.

Solution:

Deactivate the 15 V source; redraw the circuit (see Figure 9.19).

$$\text{substrategy: } \tilde{Z}'_T = j3 + 5 + \left[-j6 \,\|\, (4 + j2)\right]$$

$$\tilde{I}'_T = \frac{10\angle +30°}{\tilde{Z}'_T}$$

$$\text{CDR} \rightarrow \tilde{I}'_a$$

$$\tilde{Z}'_T = 5 + j3 + \cfrac{1}{\cfrac{1}{-j6} + \cfrac{1}{4 + j2}} = 5 + j3 + 4.7434\angle -18.435° = 9.6177\angle +8.973° \ \Omega$$

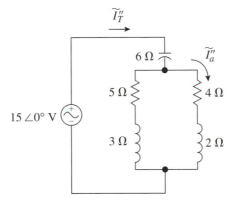

Figure 9.19 Redrawn Circuit with One Source for Example 9.4.1

$$\tilde{I}'_T = \frac{10\angle+30°}{9.6177\angle+8.973°} = 1.0398\angle+21.027° \text{ A}$$

$$\tilde{I}'_a = \frac{(1.0398\angle+21.027°)(6\angle-90°)}{(-j6) + (4 + j2)} = 1.1028\angle-23.973° \text{ A}$$

Deactivate the 10 V source; redraw the circuit (see Figure 9.20).

Figure 9.20 Redrawn Circuit with the Other Source for Example 9.4.1

substrategy: $\tilde{Z}''_T = -j6 + \left[(5 + j3) \parallel (4 + j2)\right]$

$$\tilde{I}''_T = \frac{15\angle0°}{\tilde{Z}''_T}$$

$$CDR \rightarrow \tilde{I}''_a$$

$$\tilde{Z}''_T = -j6 + \frac{1}{\dfrac{1}{5 + j3} + \dfrac{1}{4 + j2}} = -j6 + 2.5328\angle+28.474° = 5.2844\angle-65.082° \ \Omega$$

$$\tilde{I}''_T = \frac{15\angle0°}{5.2844\angle-65.082°} = 2.8386\angle+65.082° \text{ A}$$

$$\tilde{I}''_a = \frac{(2.8386\angle+65.082°)(5 + j3)}{(5 + j3) + (4 + j2)} = 1.6076\angle+66.991° \text{ A}$$

superposition: $\tilde{I}_a = \tilde{I}'_a + \tilde{I}''_a = +1.1028\angle-23.973° + 1.6076\angle+66.991°$

$$\tilde{I}_a = 1.9342\angle + 32.234° = 1.93\angle + 32.2° \text{ A}$$

$$i_a(t) = 1.93 \sin(\omega t + 32.2°) \text{ A} \text{ (no specific frequency information given)}$$

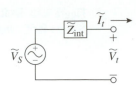

Figure 9.21 Model of a Practical AC Voltage Source

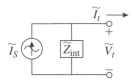

Figure 9.22 Schematic Symbols of AC Current Sources

a. Ideal current source   b. Model of a practical current source

Before leaving this chapter, we will examine some practical AC sources. The properties of ideal AC sources are:

An ideal AC voltage source has a constant voltage and can deliver any current value.
An ideal AC current source delivers a constant current at any voltage value.

The "constant voltage" and "constant current" referred to in the preceding statements are the peak (or effective) values.

As in the case of practical DC voltage sources, practical AC voltage sources have an internal series impedance. In other words, a practical voltage source does have some internal loss mechanisms that depend on the type of source that it is as well as a net internal reactance, either capacitive or inductive. Hence, a practical AC voltage source has a *model* of an ideal voltage source in series with an internal impedance as shown in Figure 9.21. The application of KVL to Figure 9.21 results in an expression for the terminal voltage that is similar to the DC case (apply KVL and develop this result).

$$\tilde{V}_t = \tilde{V}_S - \tilde{V}_{Z_{int}} = \tilde{V}_S - \tilde{I} \pm \tilde{Z}_{int} \qquad (9.47)$$

Similarly, ideal and practical AC current sources are shown in Figure 9.22. The terminal current is developed from KCL, as it was in the DC case (apply KCL and develop this result).

$$\tilde{I}_t = \tilde{I}_S - \tilde{I}_{Z_{int}} = \tilde{I}_S - \frac{\tilde{V}_t}{\tilde{Z}_{int}} \qquad (9.48)$$

The analysis of a circuit that contains an AC current source is performed using the same general approach as with DC circuits. Again, we have leveraged our knowledge and understanding of DC circuit analysis to AC circuit analysis. Did you notice which quantity was not calculated in this chapter that was determined in previous chapters (conspicuous by its absence)? Power. In the next chapter, we shall leverage our understanding of power, stored energy, and AC circuits to understand and determine the power in AC circuits that contain resistances, inductors, and capacitors—stay tuned.

CHAPTER REVIEW

9.1   AC Series Circuits
- The current is the same throughout a series circuit.
- Kirchhoff's voltage law (KVL): the sum of the phasor voltages around any closed path, including through the air, must equal zero.
- The total impedance in a series circuit is the sum of the individual impedances.

- Voltage division occurs in series circuits. The voltage-divider rule (VDR) is a direct expression of this voltage division.

9.2   AC Parallel Circuits
- The voltage is the same throughout a parallel circuit.
- Kirchhoff's current law (KCL): the sum of the phasor currents entering and leaving a node must equal zero.

- Admittance is 1 over impedance. Admittance is useful in parallel circuits.
- The total admittance of a parallel circuit is the sum of the individual admittances.
- Current division occurs in parallel circuits. The current-divider rule (CDR) is a direct expression of this current division.

9.3 AC Series–Parallel Circuits
- Series–parallel circuits are analyzed by applying the analysis techniques for series and parallel circuits to series groups (branches) and parallel groups of impedances within the series–parallel circuit.

9.4 Analysis of Multiple-Source AC Circuits Using Superposition
- Superposition can be used to determine voltages and currents in the circuit if the frequency of all sources is the same.

- Superposition is the addition (with proper sign) of the corresponding phasor voltages and currents due to each source in the circuit. Series–parallel circuit analysis is used to analyze the circuit for each source.
- In superposition a voltage source is deactivated by replacing it with a short to guarantee zero voltage.
- In superposition a current source is deactivated by replacing it with an open to guarantee zero current.
- The model of a practical AC voltage source is an ideal voltage source in series with an internal impedance.
- The model of a practical AC current source is an ideal current source in parallel with an internal impedance.

HOMEWORK PROBLEMS

Your answers to the following questions will demonstrate that you can describe the fundamental properties of series and parallel circuits and subcircuits.

9.1 What is the total impedance of a series circuit?

9.2 What can be said about the total voltage in a series circuit?

9.3 What can be said about the total current in a series circuit?

9.4 List at least three circuit laws, concepts, and results for AC series circuits.

9.5 What is the total admittance of a parallel circuit?

9.6 What can be said about the total voltage in a parallel circuit?

9.7 What can be said about the total current in a parallel circuit?

9.8 List at least three circuit laws, concepts, and results for AC parallel circuits.

Your solutions to the following problems will demonstrate that you can calculate all voltages and currents in single-source AC series and parallel circuits and can perform series–parallel circuit conversions.

9.9 Calculate the total impedance for the series circuit shown in Figure P9.1 and determine the equivalent parallel circuit.

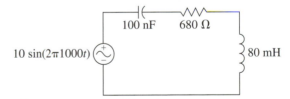

Figure P9.2 Circuit for Problem 9.10

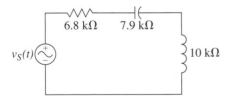

Figure P9.1 Circuit for Problem 9.9

9.10 Calculate the total impedance for the series circuit shown in Figure P9.2 and determine the equivalent parallel circuit.

9.11 Determine the quantities listed for the circuit shown in Figure P9.3.
 a. the impedance of the capacitor
 b. the impedance of the resistor
 c. the total impedance
 d. the voltage of the source in phasor form and as a function of time
 e. the phasor current through the resistor
 f. the phasor current through the capacitor

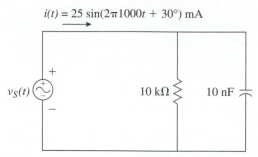

$i(t) = 25 \sin(2\pi 1000t + 30°)$ mA

Figure P9.3 Circuit to be Analyzed in Problem 9.11

9.12 Determine the quantities listed for the circuit shown in Figure P9.4.
   a. the impedance of the capacitor
   b. the impedance of the resistor
   c. the total impedance
   d. the total source current in phasor form and as a function of time $i(t)$
   e. the current through the resistor in phasor form and as a function of time
   f. the current through the capacitor in phasor form and as a function of time

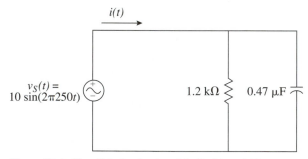

$i(t)$

$v_S(t) = 10 \sin(2\pi 250t)$

Figure P9.4 Circuit to be Analyzed in Problem 9.12

9.13 Determine the quantities listed for the circuit shown in Figure P9.5.
   a. the impedance of the inductor
   b. the impedance of the resistor
   c. the total impedance
   d. the total source current in phasor form and as a function of time $i(t)$
   e. the current through the resistor in phasor form and as a function of time
   f. the current through the inductor in phasor form and as a function of time

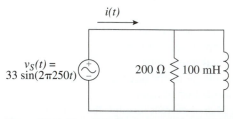

$i(t)$

$v_S(t) = 33 \sin(2\pi 250t)$

Figure P9.5 Circuit to be Analyzed in Problem 9.13

9.14 Determine the quantities listed for the circuit shown in Figure P9.6.
   a. the impedance of the inductor
   b. the impedance of the resistor
   c. the total impedance
   d. the source voltage in phasor form and as a function of time
   e. the current through the resistor in phasor form and as a function of time
   f. the current through the inductor in phasor form and as a function of time

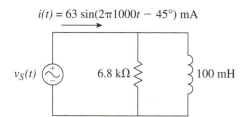

$i(t) = 63 \sin(2\pi 1000t - 45°)$ mA

Figure P9.6 Circuit to be Analyzed in Problem 9.14

9.15 Calculate the total admittance for the parallel circuit shown in Figure P9.7 and determine the equivalent series circuit.

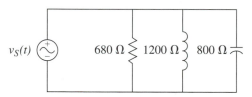

Figure P9.7 Circuit for Problem 9.15

9.16 Calculate the total admittance for the parallel circuit shown in Figure P9.8 and determine the equivalent series circuit.

$10 \sin(2\pi 400t)$

Figure P9.8 Circuit for Problem 9.16

Your solutions to the following problems will demonstrate that you can utilize phasors in voltage and current calculations in AC series and parallel circuits.

9.17   Calculate the current, $\tilde{I}_T$, and the voltages $\tilde{V}_C$, $\tilde{V}_R$, and $\tilde{V}_L$ for the series circuit shown in Figure P9.9, and verify KVL.

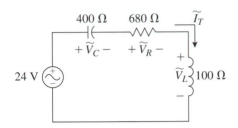

Figure P9.9 Circuit for Problem 9.17

9.18   Calculate current $\tilde{I}_T$ and the voltages $\tilde{V}_C$, $\tilde{V}_R$, and $\tilde{V}_L$ for the series circuit shown in Figure P9.10, and verify KVL.

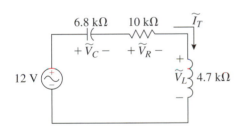

Figure P9.10 Circuit for Problem 9.18

9.19   Calculate the total current, $\tilde{I}_T$, and the currents $\tilde{I}_C$, $\tilde{I}_R$, and $\tilde{I}_L$ for the parallel circuit shown in Figure P9.11, and verify KCL.

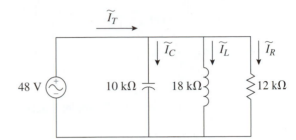

Figure P9.11 Circuit for Problem 9.19

9.20   Calculate the total current, $\tilde{I}_T$, and the currents $\tilde{I}_C$, $\tilde{I}_R$, and $\tilde{I}_L$ for the parallel circuit shown in Figure P9.12, and verify KCL.

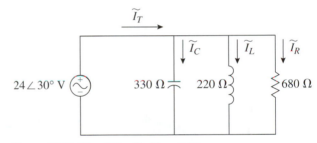

Figure P9.12 Circuit for Problem 9.20

Your solutions to the following problems will demonstrate that you can calculate all voltages and currents in single-source AC series–parallel circuits.

9.21   Calculate voltage $\tilde{V}_{ab}$ and the currents $\tilde{I}_T$, $\tilde{I}_1$, and $\tilde{I}_2$ for the series–parallel circuit shown in Figure P9.13.

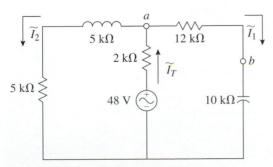

Figure P9.13 Circuit for Problem 9.21

9.22   Calculate the voltage $\tilde{V}_{ab}$ and the currents $\tilde{I}_T$, $\tilde{I}_1$, and $\tilde{I}_2$ for the series–parallel circuit shown in Figure P9.14.

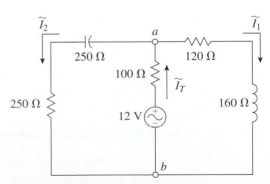

Figure P9.14 Circuit for Problem 9.22

9.23 Calculate voltage $\tilde{V}_{ab}$ and the currents $\tilde{I}_T$, $\tilde{I}_1$, and $\tilde{I}_2$ for the series–parallel circuit shown in Figure P9.15.

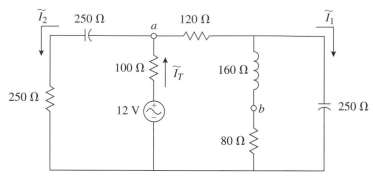

Figure P9.15 Circuit for Problem 9.23

9.24 Calculate voltage $\tilde{V}_{ab}$ and the currents $\tilde{I}_T$, $\tilde{I}_1$, and $\tilde{I}_2$ for the series–parallel circuit shown in Figure P9.16.

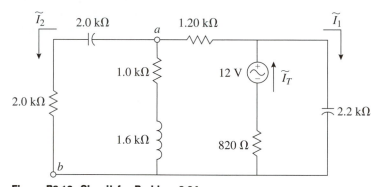

Figure P9.16 Circuit for Problem 9.24

Your solutions to the following problems will demonstrate your ability to apply superposition to calculate voltages and currents in multiple-source AC and series–parallel circuits.

9.25 Determine the voltage $\tilde{V}_{ab}$ and the currents $\tilde{I}_1$ and $\tilde{I}_2$ for the multiple-source series–parallel circuit shown in Figure P9.17.

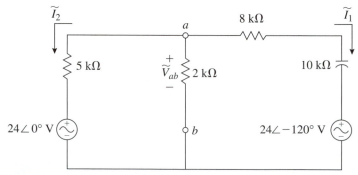

Figure P9.17 Circuit for Problem 9.25

9.26 Determine the voltage $\tilde{V}_{ab}$ and the currents $\tilde{I}_1$ and $\tilde{I}_2$ for the multiple-source series–parallel circuit shown in Figure P9.18.

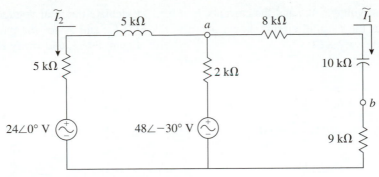

Figure P9.18 Circuit for Problem 9.26

9.27 Determine the voltage $\tilde{V}_{ab}$ and the current $\tilde{I}_1$ for the multiple-source series–parallel circuit shown in Figure P9.19.

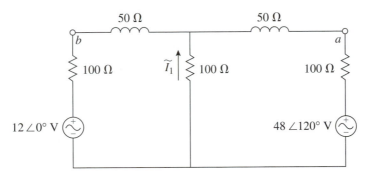

Figure P9.19 Circuit for Problem 9.27

9.28 Determine the voltage $\tilde{V}_{ab}$ and the current $\tilde{I}_1$ for the series–parallel circuit shown in Figure P9.20.

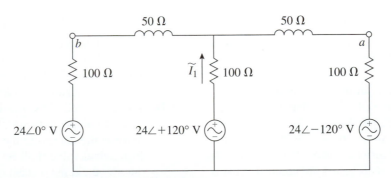

Figure P9.20 Circuit for Problem 9.28

Your solutions to the following problems will demonstrate your ability to explain what an AC current source is and your ability to analyze AC series–parallel circuits that contain a current source.

9.29 How does an AC current source differ from a DC current source?

9.30 What is the difference between an ideal current source and a practical current source? Draw a schematic for a practical AC current source.

9.31   Calculate the total voltage, $\tilde{V}_T$, and the currents $\tilde{I}_C$, $\tilde{I}_R$, and $\tilde{I}_L$ for the parallel circuit shown in Figure P9.21, and verify KCL.

9.32   Calculate the total voltage, $\tilde{V}_T$, and the currents $\tilde{I}_C$, $\tilde{I}_R$, and $\tilde{I}_L$ for the parallel circuit shown in Figure P9.22, and verify KCL.

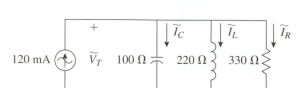

Figure P9.21   Circuit for Problem 9.31

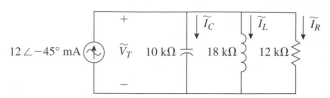

Figure P9.22   Circuit for Problem 9.32

9.33   Calculate the total voltage, $\tilde{V}_T$, and the voltages $\tilde{V}_C$, $\tilde{V}_R$, and $\tilde{V}_L$ for the series circuit shown in Figure P9.23, and verify KVL.

9.34   Calculate the total voltage, $\tilde{V}_T$, and the voltages $\tilde{V}_C$, $\tilde{V}_R$, and $\tilde{V}_L$ for the series circuit shown in Figure P9.24, and verify KVL.

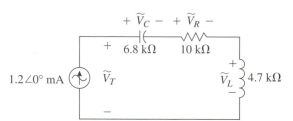

Figure P9.23   Circuit for Problem 9.33

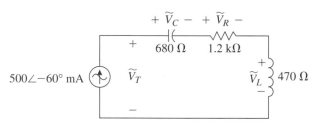

Figure P9.24   Circuit for Problem 9.34

The following problems ask that you use a computer software circuit simulation package to verify the results of prior calculations. The problems will also ask you to demonstrate that you understand the significance of the simulation results.

9.35   Verify the results in the odd-numbered problems 9.17 through 9.33 using Electronic Workbench or PSpice®.

9.36   Verify the results in the even-numbered problems 9.18 through 9.34 using Electronic Workbench or PSpice®.

9.37   For the circuit shown in Figure P9.15, assume that the 250 Ω resistor has a 5.0% tolerance. If all other components values are nominal, what is the greatest change in the magnitude of $\tilde{I}_T$ from its nominal value? Which variation in the resistance, positive or negative, caused this greatest change?

9.38   For the circuit shown in Figure P9.16, assume that the 1 kΩ resistor has a 5.0% tolerance. If all other components values are nominal, what is the greatest change in the magnitude of $\tilde{I}_T$ from its nominal value? Which variation in the resistance, positive or negative, caused this greatest change?

9.39   For the circuit shown in Figure P9.19, assume that all components are at their nominal value. What is the change in the magnitude of $\tilde{I}_1$ from its nominal value if all resistors are out of tolerance by the same percentage of:
   a.   +10%
   b.   −10%
   c.   +20%
   d.   −20%

9.40 For the circuit shown in Figure P9.20, assume that all components are at their nominal value. What is the change in the magnitude of $\tilde{V}_{ab}$ from its nominal value if all resistors are out of tolerance by the same percentage of:

a. $+10\%$
b. -10%
c. $+20\%$
d. -20%

POWER IN AC CIRCUITS

As a result of successfully completing this chapter, you should be able to:

1. Describe why complex power is needed to express power in AC circuits.
2. Describe complex power, apparent power, real power, reactive power, power factor angle, and power factor and the differences between them.
3. Calculate complex power, apparent power, real power, reactive power, power factor angle, and power factor for components, groups of components, and entire circuits using two approaches:
   a. complex power equation in terms of phasor voltage and phasor current, and
   b. summing real or reactive powers of individual components.
4. Describe what power factor correction is and why it is important.
5. Determine the parallel reactance and component value required for power factor correction.

10.1 COMPLEX POWER IN CIRCUITS WITH AC SIGNALS

Recall that in Section 4.3, the power of an AC signal applied to a resistive load was determined. The AC voltage $v(t)$ was graphically multiplied with the AC current $i(t)$ to produce an instantaneous power $p(t) = v(t) \, i(t)$. (See Figure 4.11) Although $p(t)$ "pulsated," there was a net *average* power:

$$P_{ave} = \frac{V_p I_p}{2} = V_{RMS} I_{RMS} = V_{eff} I_{eff} \qquad (10.1)$$

where the RMS and eff subscripts have identical meaning (RMS will be used in this chapter, as it was in Chapter 4). Again, this development of average power (also called true power and real power) was performed for a resistance. The voltage and current were in phase. We have since learned that the voltage and current for inductors and capacitors are $\pm 90°$ out of phase. Thus, you should again perform the graphical multiplication of $v(t)$ and $i(t)$ when they are $\pm 90°$ out of phase, as shown in Figure 10.1. Which case is being shown: inductor or capacitor (is voltage or current leading by 90°)?[1]

---

[1] Current is leading the voltage; hence, the capacitor case is being shown.

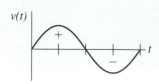

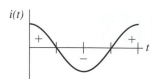

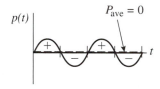

Figure 10.1 Multiplication of *v*(*t*) and *i*(*t*) with 90° Relative Phase Shift for a Capacitor

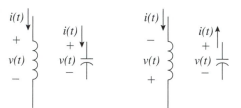

Figure 10.2 Energy Storage and Return in Inductors and Capacitors

a. Energy storage   b. Energy return

Unlike the $p(t)$ result for the resistive load, the $p(t)$ for the voltage and current ±90° out of phase *has zero average power*. In fact, the power is positive as much as it is negative. What is the significance of positive and negative power? Consider the passive sign convention. When the passive sign convention is obeyed in a resistance, the electrical energy is converted into another form of energy and the power is defined to be positive. However, in inductors and capacitors, energy is not converted but is stored. Electrical energy is stored in the magnetic field of inductors and is stored in the electric field of capacitors. Electrical power is defined to be positive when power is supplied to the inductor or capacitor from the circuit. Hence, the voltage polarity sign and the current direction still obey the passive sign convention, as shown in Figure 10.2a.

What is negative power for inductors and capacitors? It must be the "reverse" of positive power, i.e., the stored electrical energy in the magnetic field of inductors or the electric field of capacitors is returned to the circuit. This would correspond to the magnetic field collapsing in an inductor ("defluxing") or a capacitor discharging, with the appropriate reversal of either the voltage polarity or current direction, as shown in Figure 10.2b.

The net result of this discussion is that ideal inductors and capacitors have zero average power, i.e., no electrical energy is being converted into another form. While there is electrical energy stored in inductors and capacitors in half of a cycle of the AC signal, all of the energy is returned to the circuit in the other half of the cycle. Thus, the real question is how to specify a powerlike quantity, that corresponds to energy storage, for inductors and capacitors. There is a need for a method to express both electrical energy conversion and electrical energy storage with AC signals, and the solution to that need is called *complex power*.

In order to develop the expressions that are used to calculate complex power, start with the equations for DC energy storage in inductors and capacitors:

$$E_C = \tfrac{1}{2}\,CV^2 \qquad\qquad E_L = \tfrac{1}{2}\,LI^2 \qquad\qquad \textbf{(10.2)}$$

One needs calculus to derive the stored energy expressions for AC signals.[2] An alternative explanation, not precisely justified, is presented here. Recall that in AC average power calculations, the effective (RMS) values for voltage and current were used. Thus, in AC energy calculations, effective values of voltage and current are normally used:

$$E_{C_{(ave)}} = {}^1\!/_2 \, CV_{RMS}^2 \qquad\qquad E_{L_{(ave)}} = {}^1\!/_2 \, LI_{RMS}^2 \qquad\qquad (10.3)$$

Henceforth the capital letter for E *shall be used to indicate DC energy or average energy in the AC case.* Let us experiment with these equations by converting the voltage or current squared into voltage times current:

$$E_C = {}^1\!/_2 \, CV_{RMS}^2 = {}^1\!/_2 \, CV_{RMS}I_{RMS}X_C = {}^1\!/_2 \, CV_{RMS}I_{RMS}\frac{1}{\omega C} = \frac{V_{RMS}I_{RMS}}{2\omega} \qquad (10.4)$$

$$E_L = {}^1\!/_2 \, LI_{RMS}^2 = {}^1\!/_2 \, L\frac{V_{RMS}}{X_L}I_{RMS} = {}^1\!/_2 \, L\frac{V_{RMS}}{\omega L}I_{RMS} = \frac{V_{RMS}I_{RMS}}{2\omega} \qquad (10.5)$$

The general result is the same. The average energy stored in an inductor or a capacitor is directly proportional to the "power" $V_{RMS}I_{RMS}$ for that component (and inversely proportional to radian frequency). Power is put in quotation marks because it does not represent energy conversion, but instead represents energy storage in inductors and capacitors. Hence, $(V_{RMS}I_{RMS})/(2\omega)$ is an energy expression that contains a powerlike quantity that represents energy storage in inductors and capacitors.

Real power is $V_{RMS}I_{RMS}$ when the AC voltage and current are in phase. Therefore, an additional interpretation of "power" is needed for $V_{RMS}I_{RMS}$ that represents stored energy when voltage and current are $\pm 90°$ out of phase. Recall that impedance had real and imaginary parts. The real part represents energy conversion in a resistance and the voltage across and the current through the resistance are in phase. The imaginary part of impedance relates to energy storage in inductors and capacitors, and the voltage across and current through the reactive components are $\pm 90°$ out of phase. Similarly, power will be complex and is defined as:

$$\tilde{S} = P + jQ \qquad\qquad (10.6)$$

where P = real (true, average) power in watts (W),

   Q = reactive power in volt-amperes reactive (VARs), and

   $\tilde{S}$ = complex power in volt-amperes (VA).

This quantity is a complex number, just as impedance was a complex number. It is called *complex power* and is given the symbol $\tilde{S}$. The real part of Equation (10.6) is called *real* (or *true* or *average*) power and corresponds to energy conversion in a resistance. Real power is that part of $V_{RMS}I_{RMS}$ for which the voltage and current are in phase. Real power is still given the symbol P. The imaginary part of Equation (10.6) is called the *reactive power* and corresponds to the energy storage in the magnetic field of an inductor or the electric field of a capacitor, both of which have the property of reactance. Reactive power is that part of $V_{RMS}I_{RMS}$ for which the voltage and the current are $\pm 90°$ out of phase. This $\pm 90°$ phase shift is the reason that j is present in this term. Reactive power is traditionally given the symbol Q (do not confuse reactive power Q with electric charge Q—you are supposed to know from the context of the situation which one is appropriate). In a word-equation form:

$$\tilde{S} = (\text{in-phase part of } V_{RMS}I_{RMS})\angle 0° + (90° \text{ out-of-phase part of } V_{RMS}I_{RMS})\angle \pm 90°$$
$$= (\text{in-phase part of } V_{RMS}I_{RMS}) \pm j(90° \text{ out-of-phase part of } V_{RMS}I_{RMS}) \qquad (10.7)$$

The real power is that part of $V_{RMS}I_{RMS}$ for which the voltage and the current are in phase and reactive power is that part of $V_{RMS}I_{RMS}$ for which the voltage and the current are

---

[2] See Hayt, Kemmerly, and Durbin, *Engineering Circuit Analysis,* 6th ed., pp. 177 through 187, for example.

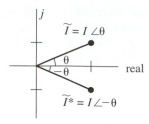 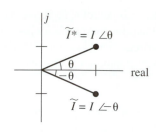

Figure 10.3 The Complex-Conjugate Operation Shown in the Complex-Number Plane

a. With positive original angle   b. With negative original angle

$\pm 90°$ out of phase. How does one separate the $V_{\text{RMS}}I_{\text{RMS}}$ product up into real and imaginary parts? Start with the RMS voltage and the current phasors and convert into rectangular form:

$$\tilde{S} = \tilde{V}_{\text{RMS}}\tilde{I}_{\text{RMS}} = (V_{\text{RMS}}\angle\theta_V)(I_{\text{RMS}}\angle\theta_I) = V_{\text{RMS}}I_{\text{RMS}}\angle(\theta_V + \theta_I) \text{ ???}$$

Stop. Something is wrong with this equation—try to identify it. The magnitude is the same as before, so something must be wrong with the phase angle. A qualitative argument follows. The total phase angle in this equation, $(\theta_V + \theta_I)$, is the sum of the voltage and current phases, which has not appeared up to now. The phase shift *between* the voltage and the current is physically significant for impedance and admittance. Hence, the phase angle should be the *difference*, not the sum, between the voltage and current phase angles:

$$\tilde{S} = V_{\text{RMS}}I_{\text{RMS}}\angle(\theta_V - \theta_I) \tag{10.8}$$

How is the negative (opposite sign) of an angle obtained with complex numbers? The negative of the angle in a complex number is found by taking the *complex conjugate*.[3] The complex conjugate of a complex number is indicated by a superscript asterisk:

$$\tilde{I}^*_{\text{RMS}} = (I_{\text{RMS}}\angle\theta_I)^* = I_{\text{RMS}}\angle-\theta_I \tag{10.9}$$

The complex-conjugate operation is shown in Figure 10.3. The phasor current and the complex conjugate of the phasor current are plotted in the complex-number plane. Note how the angle has the opposite sign of that in the original complex number, and, equivalently, the imaginary part has the opposite sign of that in the original complex number. If the original phase angle is positive, the complex conjugate produces a negative phase angle, as shown in Figure 10.3a, as well as a negative imaginary part. If the original phase angle is negative, the complex conjugate produces a positive phase angle, as shown in Figure 10.3b (the negative of a negative angle is positive), as well as a positive imaginary part.

Now we can proceed with converting the complex power expression into rectangular form. First, the voltage and the current in the complex power expression are expressed in polar form:

$$\tilde{S} = \tilde{V}_{\text{RMS}}\tilde{I}^*_{\text{RMS}} = (V_{\text{RMS}}\angle\theta_V)(I_{\text{RMS}}\angle\theta_I)^* = V_{\text{RMS}}\angle\theta_V(I_{\text{RMS}}\angle-\theta_I) \tag{10.10}$$

Then, the magnitudes and the angles are combined according to the multiplication rule for complex numbers:

$$\tilde{S} = V_{\text{RMS}}I_{\text{RMS}}\angle(\theta_V - \theta_I) = V_{\text{RMS}}I_{\text{RMS}}\angle\theta \tag{10.11}$$

where θ is defined as the angle between the voltage and the current $(\theta_V - \theta_I)$, just as it is in impedance (more will be said about this angle and its relationship to impedance later in this chapter). Then the polar form is converted to rectangular form using trigonometry. Refer to Figure 10.4. How is the real part found? We use the cosine to find the horizontal leg of the right triangle. Similarly, we use the sine to determine the imaginary part:

$$\tilde{S} = V_{\text{RMS}}I_{\text{RMS}} \cos\theta + jV_{\text{RMS}}I_{\text{RMS}} \sin\theta = P + jQ \tag{10.12}$$

$$P = \text{Re}\,(\tilde{S}) = V_{\text{RMS}}I_{\text{RMS}} \cos\theta \tag{10.13}$$

---

[3] Most college algebra textbooks cover this topic.

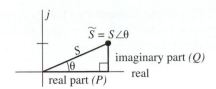

Figure 10.4 Real and Imaginary Parts of Complex Power

$$Q = \text{Im}\,(\tilde{S}) = V_{\text{RMS}}I_{\text{RMS}}\sin\theta \tag{10.14}$$

where *Re* means "take the real part of" the complex number and *Im* means "take the imaginary part of" the complex number.

The math in Equation (10.12) illustrates how the phase angle is the key to determining the part of $V_{\text{RMS}}I_{\text{RMS}}$ for which the voltage and the current are in phase and the part of $V_{\text{RMS}}I_{\text{RMS}}$ for which the voltage and the current are out of phase, as clarified by showing the polar form of each term in Equation (10.12):

$$\tilde{S} = V_{\text{RMS}}I_{\text{RMS}}\cos\theta + jV_{\text{RMS}}I_{\text{RMS}}\sin\theta$$

$$= (V_{\text{RMS}}I_{\text{RMS}}\cos\theta)\angle 0° + (V_{\text{RMS}}I_{\text{RMS}}\sin\theta)\angle 90° \tag{10.15}$$

$$= P + jQ$$

The conversion to rectangular form, as shown in Equation (10.15), is easily performed with the polar-to-rectangular conversion function on scientific calculators. Hence,

$$\tilde{S} = P + jQ = S\angle\theta \tag{10.16}$$

where: $\tilde{S}$ = complex power in volt-amperes (VA),

  P = real (true, average) power in watts (W)

    = real part of complex power in rectangular form,

  Q = reactive power in volt-amperes reactive (VARs)

    = imaginary part of complex power in rectangular form,

  S = apparent power in volt-amperes (VA)

    = magnitude of complex power in polar form, and

  θ = power factor angle in degrees or radians

    = phase of complex power in polar form.

The unit watt is reserved for power that represents energy conversion. Since reactive, apparent, and complex power do not represent solely energy conversion, their units are given different names.

Thus, the expression for complex power is the mathematical combination of real power (energy conversion) and reactive power (stored energy) for AC sinusoidal steady-state signals of a single frequency. The key complex power expression is:

$$\boxed{\tilde{S} = \tilde{V}\tilde{I}^* = VI\angle(\theta_V - \theta_I) = S\angle\theta = P + jQ} \tag{10.17}$$

where the phasor voltage and current use effective (RMS) values. The apparent power S is the effective voltage magnitude multiplied by the effective current magnitude. In DC circuits, (VI) is normally considered to be real power, but the phase shift between the voltage and the current has an impact, as previously demonstrated. Thus, for AC signals, VI is called *apparent power*. Also note that θ, the power factor angle, is the angle between the voltage and the current:

$$\theta = \theta_V - \theta_I \tag{10.18}$$

The reactive power, Q, is the *imaginary* part of the complex power, and is the difference between the inductive and capacitive reactive powers:

$$Q = Q_L - Q_C \tag{10.19}$$

where the positive sign for Q_L and the negative sign for Q_C arise from the phase shifts of the inductor and the capacitor, respectively (developed in detail in the next section). Equation (10.17) is a fundamental result that is the key to all complex-number calculations, as illustrated in the next section.

10.2 How to Calculate Complex Power

There are two ways to calculate the total complex power provided by a source to a circuit:

a. Determine the total phasor voltage and current supplied by the source to the circuit (single source circuits only), and then use Equation (10.17).

or

b. Determine the real or reactive power of each component in the circuit, add all the real powers to obtain P_{TOTAL}, add up all the reactive powers to obtain Q_{TOTAL} (positive Q for inductors, negative Q for capacitors), and then form the total complex power per Equation (10.17). (Note: For multiple source circuits, perform superposition first.)

Examples will be used to illustrate both techniques. Example 10.2.1 illustrates the first method.

Example 10.2.1

Determine the total complex power provided by the source to the circuit shown in Figure 10.5.

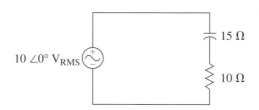

Figure 10.5 Circuit for Examples 10.2.1 and 10.22

Given: $\tilde{V}_S = 10\angle0°$ V$_{RMS}$
$\qquad$ $R = 10\ \Omega$
$\qquad$ $X_C = 15\ \Omega$

Desired: $\tilde{S}$

Strategy: $\tilde{Z}_T, \tilde{I}, \tilde{S} = \tilde{V}\tilde{I}^*$

Solution:

$$\tilde{Z}_T = R - jX_C = 10 - j15\ \Omega$$

$$\tilde{I} = \frac{\tilde{V}_S}{\tilde{Z}_T} = \frac{10\angle0°}{10 - j15} = 0.55470\angle+56.310°\ A$$

$$\tilde{S} = \tilde{V}\tilde{I}^* = (10\angle0°)(0.55470\angle+56.310°)^* = (10\angle0°)(0.55470\angle-56.310°)$$

$$\tilde{S} = 5.5470\angle-56.310 = 3.0769 - j4.6154$$

$$\tilde{S} = 5.55\angle-56.3\ VA = (3.08 - j4.62)\ VA = 3.08\ W - j4.62\ VAR$$

Both real power and reactive power are present, as expected, because both resistance and reactance are present in the circuit.

In the second method, the real or reactive power for each individual component must be determined first. These expressions are obtained from the key complex power result as expressed in Equation (10.17). All phasors are assumed to have RMS magnitudes throughout this discussion. For a resistance,

$$\tilde{S}_R = \tilde{V}_R\tilde{I}_R^* = V_RI_R \angle \theta_V - \theta_I = V_RI_R \angle 0° = S_R \angle 0° = P_R + j0 = P_R \qquad \textbf{(10.20)}$$

because the voltage and the current are in phase. The R subscript is used to emphasize that it is the phasor voltage across and the phasor current through the resistance that apply to this equation. Hence, real power corresponds to energy conversion. The real power in any resistance may be calculated from the *magnitude only* of the voltage across and/or the current through the resistance, and the reactive power is zero:

$$\boxed{P_R = V_RI_R = \frac{V_R^2}{R} = I_R^2R \qquad Q_R = 0} \qquad \text{(resistances only)} \qquad \textbf{(10.21)}$$

It is emphasized that the voltage and the current in Equation (10.21) are magnitudes only (note no tildes are present). For a capacitor,

$$\tilde{S}_C = \tilde{V}_C\tilde{I}_C^* = V_CI_C \angle(\theta_V - \theta_I) = V_CI_C \angle -90° = S_C \angle -90° = 0 - jQ_C = -jQ_C \quad \textbf{(10.22)}$$

because the voltage and the current are 90° out of phase [$i(t)$ leads $v(t)$]. The C subscript is used to emphasize that it is the phasor voltage across and the phasor current through the capacitor that apply to this equation. Note that Equation (10.22) illustrates why Q has a negative j in the complex power expression for capacitors. Thus the reactive power of any capacitance (Q_c) may be calculated from the *magnitude only* of the voltage across and/or the current through the capacitance, and the real power is zero:

$$\boxed{P_C = 0 \qquad Q_C = V_CI_C = \frac{V_C^2}{X_C} = I_C^2X_C} \qquad \text{(capacitors only)} \qquad \textbf{(10.23)}$$

Similarly for an inductance,

$$\tilde{S}_L = \tilde{V}_L\tilde{I}_L^* = V_LI_L \angle(\theta_V - \theta_I) = V_LI_L \angle +90° = S_L \angle +90° = 0 + jQ_L = +jQ_L \quad \textbf{(10.24)}$$

because the voltage and the current are 90° out of phase [$v(t)$ leads $i(t)$]. The L subscript is used to emphasize that it is the phasor voltage across and the phasor current through the inductor that apply to this equation. Note that Equation (10.24) illustrates why Q has a positive j in the complex power expression for inductors. Thus the reactive power of any inductance (Q_L) may be calculated from the *magnitude only* of the voltage across and/or the current through the inductance, and the real power is zero:

$$\boxed{P_L = 0 \qquad Q_L = V_LI_L = \frac{V_L^2}{X_L} = I_L^2X_L} \qquad \text{(inductors only)} \qquad \textbf{(10.25)}$$

The total complex power for any circuit may then be determined from the sum of the individual real and reactive powers:

$$\boxed{\tilde{S} = \tilde{V}\tilde{I}^* = \Sigma P + j[\Sigma Q_L - \Sigma Q_C]} \qquad \textbf{(10.26)}$$

where the summation sign is designated by the uppercase Greek letter sigma (Σ). Thus, ΣP is the sum of all of the real powers, ΣQ_L is the sum of all of the inductive reactive powers, and ΣQ_C is the sum of all of the capacitive reactive powers.

Example 10.2.2

Determine the complex power in the circuit shown in Figure 10.5.

Given: $\tilde{V}_S = 10\angle 0°\ V_{RMS}$

$R = 10\ \Omega$

$X_C = 15\ \Omega$

Desired: $\tilde{S}$

Strategy: $\tilde{Z}_T, \tilde{I}, P = I^2R, Q_C = I^2X_C, \tilde{S} = P - jQ_C$

Solution:

$$\tilde{Z}_T = R - jX_C = 10 - j15\ \Omega$$

$$\tilde{I} = \frac{\tilde{V}_S}{\tilde{Z}_T} = \frac{10\angle 0°}{10 - j15} = 0.55470\angle +56.310°\ A$$

$$P = I^2R = (0.55470)^2\ (10\ \Omega) = 3.0769\ W$$

$$Q_C = I^2X_C = (0.55470)^2\ (15\ \Omega) = 4.6154\ VAR$$

$$\tilde{S} = P - jQ_C = 3.08\ W - j4.62\ VAR = 3.08 - j4.62\ VA = 5.55\angle -56.3°\ VA$$

The complex power results of Examples 10.2.1 and 10.2.2 match, as expected. The real power, which represents energy conversion in the resistance, and the reactive power, which represents energy storage in the reactance (capacitance in this case), can be visualized using the *power triangle*. The power triangle is obtained from a plot in the complex number plane and is defined by three quantities: the origin, the real power on the real axis, and the reactive power on the imaginery axis (see Figure 10.6 in Example 10.2.3). The apparent power S and the power factor angle θ are also conveniently labeled on the power triangle, as illustrated in Example 10.2.3.

Example 10.2.3

Sketch and label the power triangle for the circuit in Figure 10.5.

Given: from Examples 10.2.1 and 10.2.2:

$$\tilde{S} = 3.08\ W - j4.62\ VAR = 5.55\angle -56.3°\ VA$$
$$P = 3.08\ W$$
$$Q_C = 4.62\ VAR$$
$$S = 5.55\ VA$$
$$\theta = -53.3°$$

Desired: power triangle

Strategy: Plot P and Q_C in approximate proportion.

Sketch the triangle.

Label P, Q, S, θ.

Solution:

See Figure 10.6.

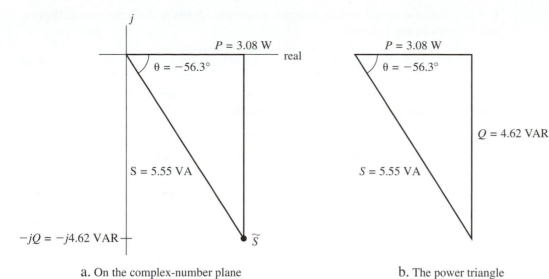

a. On the complex-number plane   b. The power triangle

Figure 10.6 Power Triangle for Example 10.2.3

If the circuit is inductive rather than capacitive, the triangle "flips up" because the j for inductive reactive power is positive. Both the inductive and capacitive cases for power triangles are shown in Figure 10.7. If a circuit has both inductance and capacitance, the nature of the total impedance, inductive or capacitive, determines the power triangle.

There is another aspect of complex power of which to be aware. See if you can determine what that aspect is from the following two equations:

$$\tilde{S} = \tilde{V}\tilde{I}^* = VI\angle(\theta_V - \theta_I) = S\angle\theta \tag{10.27}$$

$$\tilde{Z} = \frac{\tilde{V}}{\tilde{I}} = \frac{V\angle\theta_V}{I\angle\theta_I} = \frac{V}{I}\angle(\theta_V - \theta_I) = Z\angle\theta \tag{10.28}$$

What is the point of this exercise? *The angle of complex power and the angle of impedance are identical!* Check out the results in Examples 10.2.1 and 10.2.2:

$$\tilde{S}_T = 3.08 \text{ W} - j4.62 \text{ VAR} = S\angle\theta = 5.55\angle-56.3° \text{ VA}$$

$$\tilde{Z}_T = R - jX_C = 10 - j15 \text{ }\Omega = Z\angle\theta = 18.0\angle-56.3° \text{ }\Omega$$

The angle of the total circuit impedance equals the angle of the total complex power of the circuit. This fact is a useful check in complex power calculations.

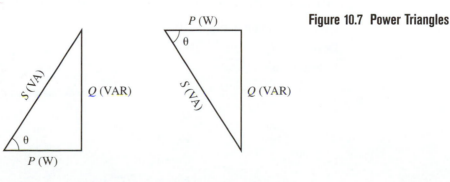

a. Inductive circuit   b. Capacitive circuit

Figure 10.7 Power Triangles

As a summary of this section, complex power can *always* be determined from the general complex power equation:

$$\tilde{S} = \tilde{V}\tilde{I}^* = VI\angle(\theta_V - \theta_I) = S\angle\theta = P + jQ \qquad \textbf{(10.29)}$$

where the voltage and current phasors are RMS values. Complex power can also be determined by summing individual component powers:

$$P_R = V_R I_R = \frac{V_R^2}{R} = I_R^2 R \qquad\qquad Q_R = 0 \qquad \textbf{(10.30)}$$

$$P_C = 0 \qquad\qquad Q_C = V_C I_C = \frac{V_C^2}{X_C} = I_C^2 X_C \qquad \textbf{(10.31)}$$

$$P_L = 0 \qquad\qquad Q_L = V_L I_L = \frac{V_L^2}{X_L} = I_L^2 X_L \qquad \textbf{(10.32)}$$

$$\tilde{S} = \tilde{V}\tilde{I}^* = \Sigma P + j\left[\Sigma Q_L - \Sigma Q_C\right] \qquad \textbf{(10.33)}$$

again, with RMS voltage and current phasors. The use of Equation (10.29) has the advantage of a direct approach to the total complex power of the circuit. The use of Equations (10.30) through (10.33) has the advantage of knowing the powers of individual components as well as the total complex power of the circuit. These fundamental relationships for complex power are also applicable to series–parallel AC circuits in general, as illustrated in the next section.

10.3 COMPLEX POWER CALCULATIONS IN SERIES–PARALLEL CIRCUITS

The fundamental relationships for complex power that were developed and applied to a series AC circuit in the previous section also apply to series, parallel, and series–parallel AC circuits with a single source or multiple sources, as illustrated by the next two examples.

Example 10.3.1

Determine the complex power provided by the source to the circuit shown in Figure 10.8 by (*a*) determining and summing the individual powers of the components, and (*b*) $\tilde{S}_T = \tilde{V}\tilde{I}^*$.

Given: circuit in Figure 10.8

Desired: $\tilde{S}_T$

Strategy: a.

$\tilde{Z}_T, \tilde{I}, \tilde{V}_x$ *by series–parallel circuit analysis*

$Q_C = I^2 X_C$

$Q_L = V_x^2 / X_L$

$Q_L = V_x^2 / R$

$\tilde{S}_T = P + j(Q_L - Q_C)$

b. $\tilde{S}_T = \tilde{V}_S \tilde{I}^*$

Figure 10.8 Circuit for Example 10.3.1

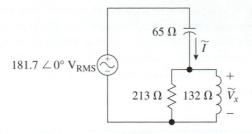

Solution:

a.

$$\tilde{Z}_x = \frac{(\tilde{Z}_R)(\tilde{Z}_L)}{\tilde{Z}_R + \tilde{Z}_L} = \frac{(213)(+j132)}{213 + j132} = +112.20\angle 58.213°$$

$$\tilde{Z}_T = \tilde{Z}_C + \tilde{Z}_x = -j65 + 112.20\angle 58.213° = 66.451\angle +27.198° \ \Omega$$

$$\tilde{I} = \frac{\tilde{V}_S}{\tilde{Z}_T} = \frac{181.7\angle 0°}{66.451\angle 27.198°} = 2.7343\angle -27.198° \ \mathrm{A_{RMS}}$$

$$\tilde{V}_X = \tilde{I}\tilde{Z}_X = (2.7343\angle -27.198°)(112.20\angle 58.213°) = 306.79\angle +31.015° \ \mathrm{V_{RMS}}$$

$$Q_C = I^2 X_C = (2.7343)^2(65) = 485.97 \ \mathrm{VAR}$$

$$Q_L = \frac{V_x^2}{X_L} = \frac{306.79^2}{132} = 713.03 \ \mathrm{VAR}$$

$$P = \frac{V_x^2}{R} = \frac{306.79^2}{213} = 441.88 \ \mathrm{W}$$

$$\tilde{S}_T = P + j(Q_L - Q_C) = 441.88 + j(713.03 - 485.97)$$

$$= 441.88 \ \mathrm{W} + j227.06 \ \mathrm{VAR} = 442 \ \mathrm{W} + j227 \ \mathrm{VAR}$$

b.

$$\tilde{S}_T = \tilde{V}\tilde{I}^* = (181.7\angle 0°)(2.7343\angle -27.198°)^* = (181.7\angle 0°)(2.7343\angle +27.198°)$$

$$= 496.83\angle +27.198° \ \mathrm{VA} \ 441.90 + j227.08 = 442 \ \mathrm{W} + j227 \ \mathrm{VAR}$$

The total complex power of the circuit is the same from either approach. The circuit is inductive, as shown by the inductive total circuit impedance or the inductive total complex power.

The techniques of Example 10.3.1 can be extended to any circuit. If there are three inductances, then three Q_L terms are summed. Resistances and capacitances are handled in a corresponding manner.

However, what if there is more than one source? Can the powers in each component due to each source be summed, or are the voltages (or currents) summed? If all sources are at the same frequency,[4] then it is required first to determine the total phasor voltage (or current) of each component due to all sources (recall the reasoning in the superposition discussion of Section 9.4). Then the complex power *for each component* is determined from that total phasor voltage (or current) for that component. Finally, the total complex power of the circuit is determined by summing the complex powers of the individual components.

---

[4] The case of different frequencies is addressed in Chapter 15.

Note that $\tilde{S}_T = \tilde{V}_T \tilde{I}_T^*$ can *not* be used because there is no total voltage or total current for the circuit in general, when there are multiple sources. Also, one can *not* use $\tilde{S}_T = \tilde{V}_T \tilde{I}_T^*$ for each source and sum those complex powers. When there are multiple sources of the same frequency, the only method to determine the total complex power of the circuit is to sum the complex powers of all components after the total phasor voltage (or current) has been found for each component.

___ Example 10.3.2 ___

Determine the total complex power provided by the source to the circuit in Example 9.4.1. The circuit schematic is repeated in Figure 10.9.

Given: The circuit from Figure 9.18 is repeated in Figure 10.9 with the current in each branch labeled.

Desired: $\tilde{S}_T$

Strategy: There is no "total" voltage or current for this circuit because the two sources are not in series or parallel. Hence, the only approach is:

■ Determine total voltage or current for each component using superposition.

■ Calculate individual component powers.

■ Sum powers for total complex power of the circuit.

Solution: ___

Use the results from Example 9.4.1. For the $10\angle +30°$ source (Figure 10.10):

$$\tilde{I}_T' = 1.0398\angle +21.027° \text{ A} = \tilde{I}_b'$$

$$\tilde{I}_a' = 1.1028\angle -23.973° \text{ A}$$

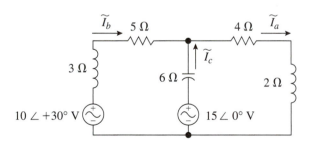

Figure 10.9 Circuit from Example 9.4.1 with Currents in Each Branch Labeled

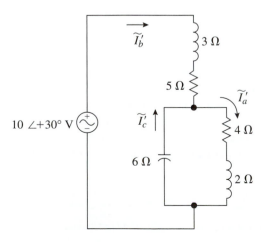

Figure 10.10 Redrawn Circuit with One Source for Example 10.3.2

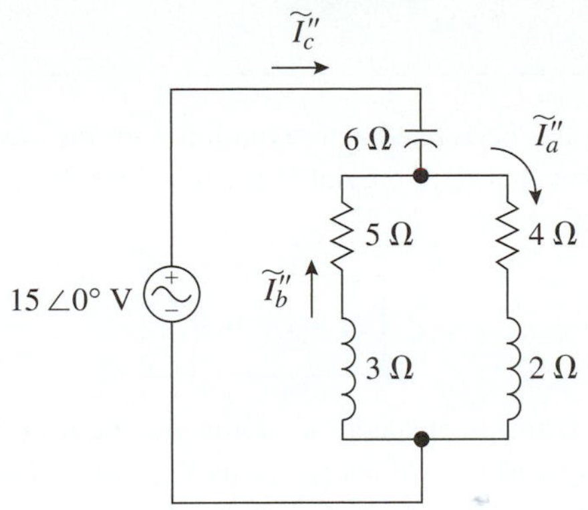

Figure 10.11 Redrawn Circuit with the Other Source for Example 10.3.2

By KCL,

$$\tilde{I}_c' = \tilde{I}_a' - \tilde{I}_b' = 0.8220\angle-87.413° \text{ A}$$

From Example 9.4.1: for the 15 $\angle0°$ source (Figure 10.11):

$$\tilde{I}_T'' = 2.8386\angle+65.082° \text{ A} = \tilde{I}_c''$$

$$\tilde{I}_a'' = 1.6076\angle+66.991° \text{ A}$$

By KCL,

$$\tilde{I}_b'' = \tilde{I}_a'' - \tilde{I}_c'' = 1.2331\angle-117.407° \text{ A}$$

superposition:

$$\tilde{I}_a = \tilde{I}_a' + \tilde{I}_a'' = +1.1028\angle-23.973° + 1.6076\angle+66.991°$$

$$\tilde{I}_a = 1.9342\angle+32.233° \text{ A}$$

similarly,

$$\tilde{I}_b = 0.8265\angle-60.818° \text{ A}$$

$$\tilde{I}_c = 2.1434\angle+54.880° \text{ A}$$

intermediate check—KCL:

$$\tilde{I}_a = +\tilde{I}_b + \tilde{I}_c = 1.9341\angle+32.233° \text{ A} \qquad \text{ok}$$

For convenience, assign subscripts for the power calculations:

6 Ω capacitive reactance: subscript 1
2 Ω inductive reactance: subscript 2
3 Ω inductive reactance: subscript 3
4 Ω resistance: subscript 4
5 Ω resistance: subscript 5

Determine the power of each component from the total current through that component:

$$Q_1 = I_c^2 X_1 = (2.1434)^2 \, (6) = 27.5650 \text{ VAR (capacitive)}$$

$$Q_2 = I_a^2 X_2 = (1.9342)^2 \, (2) = 7.4823 \text{ VAR (inductive)}$$

$$Q_3 = I_b^2 X_3 = (0.8265)^2 \, (3) = 2.0493 \text{ VAR (inductive)}$$

$$P_4 = I_a^2 R_4 = (1.9342)^2 \, (4) = 14.965 \text{ W}$$

$$P_5 = I_b^2 R_5 = (0.8265)^2 \, (5) = 3.4155 \text{ W}$$

Finally, sum the individual component complex powers to obtain the total circuit complex power:

$$\tilde{S} = \Sigma P + j\,[\Sigma Q_L - \Sigma Q_c] = (P_4 + P_5) + j(Q_2 + Q_3 - Q_1)$$

$$= (3.4155 + 14.9645) + j(7.4823 + 2.0493 - 27.5650)$$

$$= 18.3800 - j18.0334 = 18.4 \text{ W} - j18.0 \text{ VAR}$$

10.4 POWER FACTOR AND pf CORRECTION

The term *power factor* has been used in association with the power factor angle, θ. What is the power factor? Power factor, abbreviated pf, is a definition with physical significance. The power factor is defined as:

$$\text{pf} \equiv \cos\theta = \frac{P}{S}$$

(10.34)

The *P/S* ratio comes from an inspection of the power triangle—see Figure 10.7. The physical significance lies in the values for pf as the power factor angle is varied. A few values are tabulated in Table 10.1. Notice that when the power factor angle is 0°, the power factor is 1. When the power factor angle is ±90°, the power factor is 0. Hence, the power factor is an indication of the amount of real power relative to the total apparent power. When the power factor is near 1, almost all the complex power is real. When the power factor is near 0, almost all the complex power is reactive. When the power factor is in the "middle region" between 1 and 0, there is a mixture of real and reactive powers.

Notice from Table 10.1 that given the power factor, one cannot tell whether the power factor angle is positive or negative. In practice there is a word amended to the power factor to indicate the type of load. The word is either *leading* or *lagging,* and it is applied to the current relative to the voltage. If the power factor is leading, then the current *leads* the voltage; hence the circuit is capacitive. If the power factor is lagging, then the current *lags* the voltage; hence the circuit is inductive.

How is power factor used? The following discussion will illustrate the use of power factor through the effect of power factor correction on a circuit. Start with the circuit of Examples 10.2.1 and 10.2.2. Treat the circuit as a "block" with complex power, as shown in Figure 10.12. The *Q* is nonzero because there is reactive power present, capacitive in this case. What is the power factor of the circuit in Figure 10.12?

TABLE 10.1	power factor angle	power factor
Power Factor for a Few Power Factor Angle Values	+90°	0
	+60°	0.5000
	+45°	0.7071
	+30°	0.8660
	+0°	1.0000
	−30°	0.8660
	−45°	0.7071
	−60°	0.5000
	−90°	0

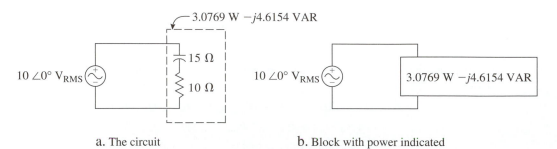

a. The circuit   b. Block with power indicated

Figure 10.12 Circuit in Figure 10.5 with the Load Treated as a Single Unit

$$\text{pf} = \frac{P}{S} = \frac{3.0769}{|3.0769 - j4.6154|} = \frac{3.0769}{5.5470} = 0.55470 = 0.555 \text{ leading}$$

where the magnitude of the complex power (the denominator) is the apparent power. The 0.555 indicates that both real and reactive power are significantly present in the circuit.

What if another component was connected to the circuit that added a Q to the circuit such that the total reactive power of the circuit would be zero? For example, if one added $+ j4.6154$ VA (4.6154 VARs inductive) to the circuit in Figure 10.12, then:

$$\tilde{S} = \Sigma P + j\Sigma Q = 3.0769 + j(+ 4.6154 - 4.6154) = 3.0769 + j0 \text{ VA} = 3.08 \text{ W}$$

The total complex power equals the real power, and the reactive power is zero. What is the new power factor?

$$\text{pf} = \frac{P}{S} = \frac{3.0769}{|3.0769 - j0|} = \frac{3.0769}{3.0769} = 1.00$$

This change in the total complex power and power factor has a profound effect on the circuit. This effect will be illustrated by calculating the total circuit current:

$$\tilde{S} = \tilde{V}\tilde{I}^* \rightarrow \tilde{I} = \left(\frac{\tilde{S}}{\tilde{V}}\right)^* = \left(\frac{3.0769}{10\angle 0°}\right)^* = 0.30769\angle 0° \text{ A} = 0.308\angle 0° \text{ A}$$

Compare this current in the circuit with power factor correction to the original current in the circuit without pf correction (from Example 10.2.1 or 10.2.1):

$$\tilde{I} = 0.555\angle + 56.3° \text{ A}$$

What are the primary effects of power factor correction? (1) The power factor equals 1, i.e., *the total circuit voltage and current are in phase.* (2) *The circuit current is reduced.* The reduced current has very significant technical and economic impacts: smaller wire sizes, less power dissipation in wires, and so on. In short, power factor correction saves money by not wasting resources required to handle larger current levels.

How should the component be connected into the circuit? The component is usually added in parallel with the original circuit, as sketched in Figure 10.13, because the load is treated as a single unit or is remote from the source. Often, power lines feed a load, so breaking the circuit to add the component in series is impractical. In fact, if one were to attempt to add a component in series, the entire circuit current and power values would change from the initial values, and the correction would have to be adjusted in an iterative manner.

What type of component is added in parallel with the load in this example? It must be an inductor because the original circuit was capacitive. If the load were inductive, as is often the case with electric motors that are used to drive industrial machinery and production lines, then a capacitor would be added in parallel.

What is the value of the component to add in parallel in Figure 10.13? Another piece of information is required to answer this question. Attempt to identify it by examining the following equations, determining what is known, and hence what is not known:

$$Q_L = \frac{V_S^2}{X_L} \tag{10.35}$$

$$X_L = \omega L \tag{10.36}$$

Figure 10.13 Parallel Connection of a Component to Add Opposite Q

The known quantities are the source voltage and the reactive power required for power factor correction. Frequency must be known to determine the component value, L in this case. Assume that the frequency of the source is 60 Hz for the circuit in Figure 10.13. Then,

$$X_L = \frac{V_S^2}{Q_L} = \frac{10^2}{4.6154} = 21.667 \ \Omega$$

$$L = \frac{X_L}{\omega} = \frac{21.667}{2\pi(60)} = 57.5 \ \text{mH}$$

Thus, a 57.5 mH (or the closest standard value available—search vendor catalogs on the web) must be placed in parallel with the load for power factor correction. The other inductor ratings (frequency, voltage, current, temperature, and so on) must also be met. Two more examples follow to illustrate the technique of power factor correction.

Example 10.4.1

Determine (a) the reactance for power factor correction in the circuit shown in Figure 10.8. (b) Determine the component value if the frequency is 60 Hz. (c) Determine the pf both before and after pf correction.

Given: $\tilde{S}_T = 441.90 \ \text{W} + j227.08 \ \text{VAR}$ for the circuit in Figure 10.8 (see Example 10.3.1)
$\tilde{V}_S = 181.7\angle 0° \ \text{V}_{\text{RMS}}$ (from Figure 10.8)
$f = 60 \ \text{Hz}$

Desired:

a. X for pf correction
b.   value of L or C for pf correction
c.   pf before pf correction
      pf after pf correction

Strategy: $Q_{\text{pf correction}} = \text{Im} (\tilde{S}_T)$ (inductive or capacitive, as appropriate)

$$X_C = \frac{V_s^2}{Q_C}$$

$$C = \frac{1}{\omega X_C}$$

$$\text{pf} = \frac{P}{S}$$

Solution:

$$Q_{\text{pf correction}} = \text{Im} (\tilde{S}_T) = 227.08 \ \text{VAR (capacitive)}$$

$$X_C = \frac{V_S^2}{Q_C} = \frac{181.7^2}{227.15} = 145.39 = 145 \ \Omega$$

$$C = \frac{1}{\omega X_C} = \frac{1}{2\pi 60(145.34)} = 1.8245 \times 10^{-5} = 18.2 \ \mu\text{F}$$

$$\text{pf}_{\text{before}} = \frac{P}{S} = \frac{441.90}{|441.90 \ \text{W} + j227.08|} = \frac{441.90}{496.83} = 0.88944 = 0.889$$

$$\text{pf}_{\text{after}} = 1.000$$

Thus, a capacitor of approximately 18 μF would be connected in parallel with the circuit, as shown in Figure 10.14, and is required for pf correction. You should check this result by

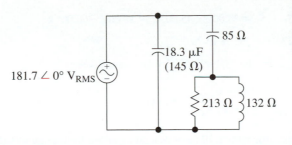

calculating the total power of the circuit with this capacitor in place. The complex power should be real only.

Power factor correction can be extended to systems where the individual component values may not be known, but the complex power is known, as illustrated by the next example.

Example 10.4.2

Determine the reactance for power factor correction in the circuit shown in Figure 10.15.

Given: load 1:   pf = 0.7071 lagging

I = 200.0 A (magnitude only, phase unknown)

load 2:   pf = 0.5000 leading

I = 100.0 A (magnitude only, phase unknown)

$\tilde{V}_S$ = 120.0∠0° V_{RMS}

Desired: X for pf correction

Strategy: Determine θ for each load from pf: $\theta = \cos^{-1}$ (pf).

Determine $\tilde{S}$ for each load from V, I, θ, and leading/lagging status:

$$\tilde{S} = VI \angle \theta$$

$$\tilde{S}_T = \tilde{S}_1 + \tilde{S}_2$$

$$Q_{\text{pf correction}} = \text{Im}\left(\tilde{S}_T\right) \text{ (inductive or capacitive, as appropriate)}$$

$$X = \frac{V_S^2}{Q}$$

Solution:

$\theta_1 = \cos^{-1}(\text{pf}_1) = \cos^{-1}(0.7071) = 45.00°$ (positive due to lagging pf)

$\theta_2 = \cos^{-1}(\text{pf}_2) = \cos^{-1}(0.5000) = 60.00°$ (negative due to leading pf)

$\tilde{S}_1 = V_S I_1 \angle \theta_1 = (120.0)(200.0)\angle +45.00° = 24.00\angle +45.00°$ kVA

$\tilde{S}_2 = V_S I_2 \angle \theta_2 = (120.0)(100.0)\angle -60.00° = 12.00\angle -60.00°$ kVA

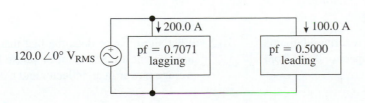

Figure 10.15 Circuit for Example 10.4.2 (Only Current Magnitudes are Given.)

$$\tilde{S}_T = \tilde{S}_1 + \tilde{S}_2 = 24.00\angle +45.00° \text{ k} + 12.00\angle -60.00° \text{ k}$$

$$= 22.97 + j6.578 \text{ kVA}$$

$$Q_{\text{pf correction}} = \text{Im}\left(\tilde{S}_T\right) = 6.600 \text{ kVAR (capacitive)}$$

$$X_C = \frac{V_S^2}{Q_C} = \frac{120.0^2 \text{ V}_{\text{RMS}}}{6.600 \text{ kVAR}} = 2.19 \text{ } \Omega$$

Thus, a capacitor with 2.19 Ω of reactance at the frequency of the source must be added in parallel with the circuit for pf correction.

CHAPTER REVIEW

10.1 Complex Power in Circuits with AC Signals
- The 90° phase relationship between voltage and current for ideal inductors and capacitors results in zero real power, i.e., no energy conversion.
- $V_{\text{RMS}}I_{\text{RMS}}$ for inductors and capacitors is directly proportional to the average energy stored in these reactive components.
- Reactive power in inductors corresponds to energy storage in the magnetic field.
- Reactive power in capacitors corresponds to energy storage in the electric field.
- Complex power is the mathematical combination of real and reactive power.
- $\tilde{S} = \tilde{V}\tilde{I}^* = \tilde{V}\tilde{I} \angle (\theta_V - \theta_I) = S \angle \theta = P + jQ$

10.2 How to Calculate Complex Power
- There are two ways to calculate complex power in a circuit:
  a. From total RMS phasor voltage and current: $\tilde{S} = \tilde{V}\tilde{I}*$
  b. From summing individual component powers: $\tilde{S} = \Sigma P + j\left[\Sigma Q_L - \Sigma Q_C\right]$
- The power triangle is a visualization of the complex power relationships.

10.3 Complex Power Calculations in Series–Parallel Circuits (see concepts in 10.2)
- Some multiple source circuits do not have a total voltage and current; hence it is necessary to sum individual component powers after the total voltage and/or current for each component has been determined.

10.4 Power Factor and pf Correction
- Power factor, pf, expresses the amount of real power relative to the apparent power.
- $\text{pf} \equiv \cos\theta = \dfrac{P}{S}$
- Power factor correction is the addition of a reactive component in parallel to the circuit. The reactance adds reactive power equal to the reactive power of the circuit but of the opposite type. Hence the total reactive power is 0, and the power factor is 1 after pf correction.
- Total circuit current is reduced after power factor correction.

HOMEWORK PROBLEMS

Your answers to the following questions will demonstrate that you can describe why complex power is needed to express power in AC circuits. You will also demonstrate that you can describe apparent power, real power, reactive power, power factor angle, and power factor and the differences between them.

10.1 How does the power in the reactive elements such as the inductor and the capacitor differ from the power in a resistive load?

10.2 What significance does the fact that the voltage and current are ±90° out of phase have on the average power in an inductor and a capacitor?

10.3 What is the significance of positive and negative instantaneous power in an inductor and a capacitor?

10.4 What does $V_{RMS}\,I_{RMS}$ represent when the AC voltage and current are in phase?

10.5 What does $V_{RMS}\,I_{RMS}$ represent when voltage and current are ±90° out of phase?

10.6 Describe the components of complex power.

10.7 What are the units for each of the components of complex power?

10.8 Why is VI in AC signals called apparent power?

Your answers to the following questions will demonstrate that you can calculate complex power for components, groups of components, and entire circuits using the equations that define complex power in terms of phasor voltage and phasor current in series circuits. You will also demonstrate your ability to calculate apparent power, real power, reactive power, power factor angle, and power factor and demonstrate that you can calculate complex power by summing the real and reactive powers of individual components in series circuits.

10.9 Determine (*a*) total complex power, (*b*) *P* in the resistor, (*c*) *P* in the capacitor, (*d*) *Q* in the resistor, and (*e*) *Q* in the capacitor for the circuit shown in Figure P10.1. Sketch the power triangle for the total complex power.

10.11 Determine (*a*) total complex power, (*b*) *P* in the resistor, (*c*) *P* in the inductor, (*d*) *P* in the capaci-

tor, (*e*) *Q* in the resistor, (*f*) *Q* in the inductor, and (*g*) *Q* in the capacitor for the circuit shown in Figure P10.3. Sketch the power triangle for the total complex power.

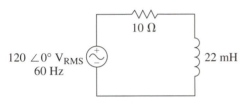

Figure P10.1 Circuit for Problem 10.9

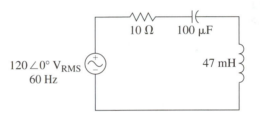

Figure P10.3 Circuit for Problem 10.11

10.10 Determine (*a*) total complex power, (*b*) *P* in the resistor, (*c*) *P* in the inductor, (*d*) *Q* in the resistor, and (*e*) *Q* in the inductor for the circuit shown in Figure P10.2. Sketch the power triangle for the total complex power.

10.12 Determine (*a*) total complex power, (*b*) *P* in the resistor, (*c*) *P* in the inductor, (*d*) *P* in the capacitor, (*e*) *Q* in the resistor, (*f*) *Q* in the inductor, and (*g*) *Q* in the capacitor for the circuit shown in Figure P10.4. Sketch the power triangle for the total complex power.

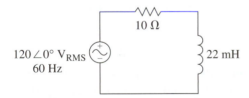

Figure P10.2 Circuit for Problem 10.10

10.11 Determine (*a*) total complex power, (*b*) *P* in the resistor, (*c*) *P* in the inductor, (*d*) *P* in the capaci-

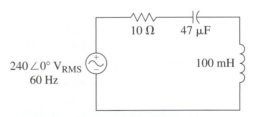

Figure P10.4 Circuit for Problem 10.12

Your answers to the following questions will demonstrate that you can calculate complex power for components, groups of components, and entire circuits using the equations that define complex power in terms of phasor voltage and phasor current in series–parallel circuits. You will also demonstrate your ability to calculate apparent power, real power, reactive power, power factor angle, and power factor and demonstrate that you can calculate complex power by summing the real and reactive powers of individual components in series–parallel circuits.

10.13 Determine (*a*) total complex power, (*b*) *P* in the resistor, (*c*) *P* in the inductor, (*d*) *P* in the capacitor, (*e*) *Q* in the resistor, (*f*) *Q* in the inductor, and (*g*) *Q* in the capacitor for the circuit shown in Figure P10.5. Sketch the power triangle for the total complex power and determine the power factor.

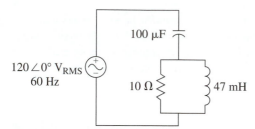

Figure P10.5 Circuit for Problem 10.13

10.14 Determine (*a*) total complex power, (*b*) *P* in each resistor, (*c*) *P* in the inductor, (*d*) *P* in the capacitor, (*e*) *Q* in each resistor, (*f*) *Q* in the inductor, and (*g*) *Q* in the capacitor for the circuit shown in Figure P10.6. Sketch the power triangle for the total complex power and determine the power factor.

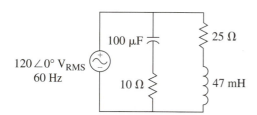

Figure P10.6 Circuit for Problem 10.14

10.15 Determine the total complex power provided by the source in the circuit shown in Figure P10.7. Determine the power factor.

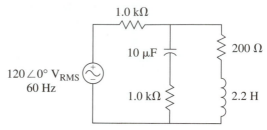

Figure P10.7 Circuit for Problem 10.15

10.16 Determine the total complex power provided by the source in the circuit shown in Figure P10.8. Determine the power factor.

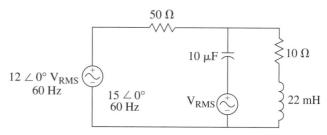

Figure P10.8 Circuit for Problem 10.16

Your answers to the following questions will demonstrate that you can determine the parallel reactance and the component value required for power factor correction and can determine the power factor before and after correction.

10.17 Determine the reactance for power factor correction for the circuit shown in Figure P10.9. Determine the power factor before and after correction.

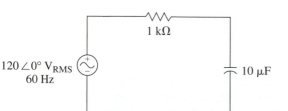

Figure P10.9 Circuit for Problem 10.17

10.18 Determine the reactance for power factor correction for the circuit shown in Figure P10.10. Determine the power factor before and after correction.

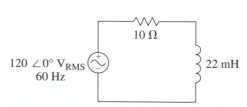

Figure P10.10 Circuit for Problem 10.18

10.19 Determine the reactance for power factor correction for the circuit shown in Figure P10.11. Determine the power factor before and after correction.

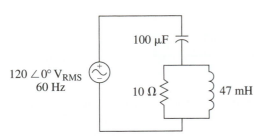

Figure P10.11 Circuit for Problem 10.19

10.20 Determine the reactance for power factor correction for the circuit shown in Figure P10.12. Determine the power factor before and after correction.

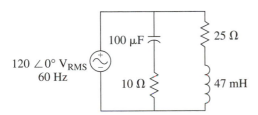

Figure P10.12 Circuit for Problem 10.20

10.21 A 440 V 60 Hz parallel circuit is shown in Figure P10.13 and consists of a motor that draws 10 kVA with a power factor of 0.7071 lagging (load 1) and a load that dissipates 10 kW (load 2).
   a. What is the overall power factor for the circuit?

   b. What is the value for reactance, and how must it be connected to the circuit to correct the power factor to 1.00?

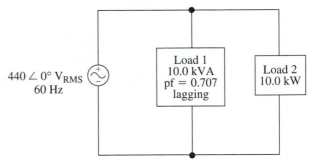

Figure P10.13 Circuit for Problem 10.21

10.22 A 440 V 60 Hz parallel circuit is shown in Figure P10.14 and consists of a motor that draws 25 kVA with a power factor of 0.50 lagging (load 1) and a load that dissipates 10 kW and has a power factor pf = 0.90 leading (load 2).
   a. What is the overall power factor for the circuit?
   b. What is the value for reactance, and how must it be connected to the circuit to correct the power factor to 1.00?

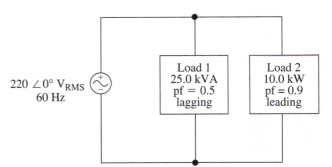

Figure P10.14 Circuit for Problem 10.22

10.23 For the circuit shown in Figure P10.15, calculate the
   a. apparent power,
   b. real power and reactive power,
   c. power factor angle, and power factor.
   d. Is the circuit net capacitive or net inductive?
   e. What is the value for reactance, and how must it be connected to the circuit to correct the power factor to 1.00?

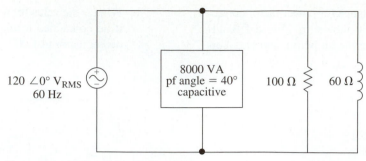

Figure P10.15 Circuit for Problem 10.23

10.24 For the circuit shown in Figure P10.16, calculate the
  a. apparent power,
  b. real power and reactive power,
  c. power factor angle, and power factor.

  d. Is the circuit net capacitive or net inductive?
  e. What is the value for reactance, and how must it be connected to the circuit to correct the power factor to 1.00?

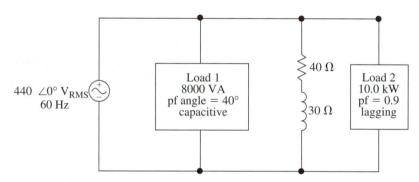

Figure P10.16 Circuit for Problem 10.24

10.25 What are the primary effects of power factor correction?

10.26 How should the reactive component for power factor correction be connected into the circuit, and why is the way that it is connected important?

ADVANCED METHODS OF DC AND AC CIRCUIT ANALYSIS

As a result of successfully completing this chapter, you should be able to:

1. Explain why more sophisticated methods of circuit analysis are required.
2. Solve for voltages and currents in circuits using the mesh analysis circuit technique.
3. Solve for voltages and currents in circuits using the nodal (node voltage) analysis circuit technique.
4. Perform delta–wye conversions.
5. Explain why bridge circuits are used and determine if a bridge circuit is balanced.

From previous math courses, you should be able to:

a. Set up simultaneous algebraic equations through application of the appropriate circuit laws.
b. Use a calculator to obtain the solution to a set of simultaneous equations. A knowledge of the mechanics of solving simultaneous equations is assumed from previous courses.
c. Insert the solutions back into the original or other appropriate equations to check the solution. This step will reinforce the concept of a simultaneous solution.

11.1 THE NEED FOR MORE SOPHISTICATED ANALYSIS TECHNIQUES

Recall from Chapter 3 (and Chapter 9 for AC) that superposition is used to analyze a circuit that has multiple sources. Superposition is the straightforward application of series–parallel circuit analysis techniques for each source, with the other sources properly deactivated. Then the results are superposed (added with proper signs). Although straightforward, superposition can be cumbersome. The purpose of the next example is to illustrate how cumbersome superposition can become in order to establish the need for the more sophisticated and efficient circuit analysis techniques that will be covered in this chapter. The advanced methods of circuit analysis will be illustrated with the same example so that a direct comparison of the results and techniques can be made.

___ **Example 11.1.1** _____

Determine the current through each resistance in Figure 11.1.

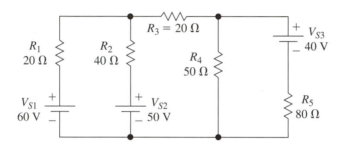

Figure 11.1 Multisource Circuit for Example 11.1.1

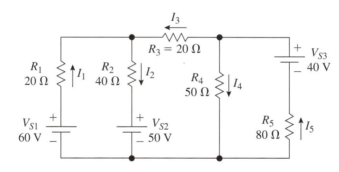

Figure 11.2 Current Direction Assignment in Example 11.1.1

Given: DC multisource circuit in Figure 11.1

Desired: I_1, I_2, I_3, I_4, and I_5, as shown in Figure 11.2, including current direction in each R

Strategy: superposition, series–parallel circuit analysis

Solution: _____

First, arbitrarily assign current directions for each resistance—see Figure 11.2.

Apply superposition. For source V_{S1}, refer to Figure 11.3. Notice how the current direction assignments must be maintained when the circuit is redrawn. Analyze the circuit for the current through each resistance due to V_{S1}.

$$R'_T = R_1 + R_2 \,\|\, [R_3 + (R_4 \,\|\, R_5)] = 20 + \cfrac{40\left[20 + \dfrac{(50)(80)}{(50+80)}\right]}{40 + \left[20 + \dfrac{(50)(80)}{50+80}\right]}$$

$$= 20 + \frac{40(20 + 30.769)}{40 + (20 + 30.769)} = 20 + \frac{40(50.769)}{40 + (50.769)} = 20 + 22.373 = 42.373\ \Omega$$

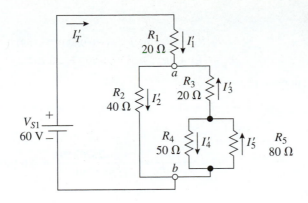

Figure 11.3 Circuit in Figure 1.2 Redrawn with Only V_{S1} Activated

$$I'_T = \frac{V_{S1}}{R'_T} = \frac{60}{42.373} = 1.4160 \text{ A}$$

$$I'_1 = +I'_T = +1.4160 \text{ A}$$

CDR:

$$I'_3 = \frac{-I'_T R_2}{R_2 + [R_3 + (R_4 \parallel R_5)]} = \frac{(-1.4610)(40)}{40 + 50.769} = -0.6240 \text{ A}$$

KCL at node a:

$$+I'_1 - I'_2 + I'_3 = 0$$

$$I'_2 = +I'_1 + I'_3 = +1.4160 - 0.6240 = +0.7920 \text{ A}$$

CDR:

$$I'_4 = \frac{-I'_3 R_5}{R_4 + R_5} = \frac{-(-0.6240)(80)}{80 + 50} = +0.3840 \text{ A}$$

KCL at node b:

$$I'_5 = +I'_3 + I'_4 = -0.6240 + 0.3840 = -0.2400 \text{ A}$$

The next step requires that you solve for the currents due to the second V source.

For source V_{S2}, refer to Figure 11.4. Analyze the circuit for the current through each resistance due to V_{S2}.

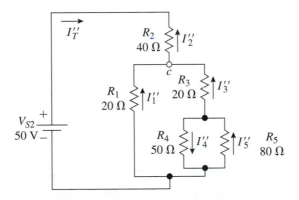

Figure 11.4 Circuit in Figure 11.2 Redrawn with Only V_{S2} Activated

$$R''_T = R_2 + R_1 \parallel [R_3 + (R_4 \parallel R_5)] = 40 + \frac{20[20 + 30.769]}{20 + [20 + 30.769]}$$

$$= 40 + \frac{20[50.769]}{20 + [50.769]} = 40 + 14.348 = 54.348 \ \Omega$$

$$I''_T = \frac{V_{S2}}{R''_T} = \frac{50}{54.348} = 0.9200 \text{ A}$$

$$I''_2 = -I''_T = -0.9200 \text{ A}$$

CDR:

$$I_1'' = \frac{-I_T''[R_3 + (R_4 \parallel R_5)]}{R_1 + [R_3 + (R_4 \parallel R_5)]} = \frac{(-0.9200)(50.769)}{20 + 50.769} = -0.6600 \text{ A}$$

KCL at node c:

$$I_3'' = -I_1'' + I_2'' = -(-0.9200) + (-0.6600) = -0.2600 \text{ A}$$

CDR:

$$I_4'' = \frac{-I_3'' R_5}{R_4 + R_5} = \frac{-(-0.2600)(80)}{50 + 80} = +0.1600 \text{ A}$$

CDR:

$$I_5'' = \frac{I_3'' R_5}{R_4 + R_5} = \frac{(-0.2600)(50)}{50 + 80} = -0.1000 \text{ A}$$

The next step requires that you solve for the currents due to the third source.

For source V_{S3}, refer to Figure 11.5. Analyze the circuit for the current through each resistance due to V_{S3}. It is left to you to perform the series–parallel circuit analysis with source 3. The results are:

$$I_1''' = -0.1600 \text{ A}$$

$$I_2''' = +0.0800 \text{ A}$$

$$I_3''' = +0.2400 \text{ A}$$

$$I_4''' = +0.1600 \text{ A}$$

$$I_5''' = +0.4000 \text{ A}$$

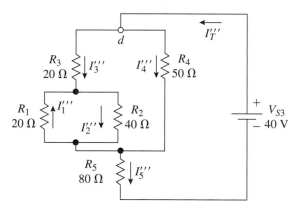

Figure 11.5 Circuit Figure 11.2 Redrawn with Only V_{S3} Activated

Finally, the superposition of the currents leads to total values in the circuit:

$$I_1 = +I_1' + I_1'' + I_1''' = +1.4160 - 0.6600 - 0.1600 = +0.596 \text{ A}$$

$$I_2 = +I_2' + I_2'' + I_2''' = +0.7920 - 0.9200 + 0.0800 = -0.048 \text{ A}$$

$$I_3 = +I_3' + I_3'' + I_3''' = -0.6240 - 0.2600 + 0.2400 = -0.644 \text{ A}$$

$$I_4 = +I_4' + I_4'' + I_4''' = +0.3840 + 0.1600 + 0.1600 = +0.704 \text{ A}$$

$$I_5 = +I_5' + I_5'' + I_5''' = -0.2400 - 0.1000 + 0.4000 = +0.060 \text{ A}$$

Check the results using KCL in Figure 11.2:

$$+I_1 - I_2 + I_3 = 0 \rightarrow I_2 = +I_1 + I_3$$

$$I_2 = +0.596 + (-0.644) = -0.048 \text{ A} \qquad \text{ok}$$

$$-I_3 - I_4 + I_5 = 0 \rightarrow I_5 = +I_3 + I_4$$

$$I_5 = +(-0.644) + 0.704 = +0.060 \text{ A} \qquad \text{ok}$$

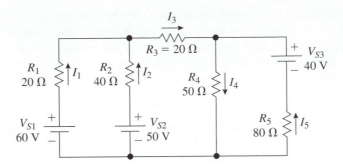

Figure 11.6 Actual Current Directions in Example 11.1.1 (all currents positive)

On the bottom node:

$$-I_1 + I_2 + I_4 - I_5 = 0$$

$$-(+0.596) + (-0.048) + 0.704 - (+0.060) = 0 \text{ A} \qquad \text{ok}$$

Some of the current values are negative. Those currents flow in a direction opposite the originally selected directions. The actual current directions are labeled in Figure 11.6. The current values are all positive with the current directions shown in Figure 11.6.

The preceding example illustrates how cumbersome superposition can be when there are several sources and several voltages and/or currents are desired. More powerful circuit analysis techniques exist for analyzing such circuits. The most common advanced circuit analysis techniques are called *mesh* analysis and *nodal* (also called node voltage) analysis. These analysis techniques are generally more straightforward than superposition. The circuit analysis strategy in superposition depends on the particular circuit, and each circuit is different. In mesh or nodal analysis, a standard strategy is utilized in each technique (the mesh analysis technique differs from the nodal analysis technique). The particular features of the given circuit are incorporated into the application of the standard procedures of each analysis technique. Another advantage of these powerful circuit analysis techniques is that the uniform procedures are more easily adopted in computer software. Thus, knowledge of these techniques gives one insight into the working of circuit simulation software.

In this chapter both sections on mesh and nodal circuit analysis are divided into two subsections. The first subsection introduces the technique and illustrates its use with DC circuits. The second section illustrates the use of the technique with AC circuits. The first subsection is required before the second subsection. Read only the first subsection if you are covering only DC circuits. Read both parts if you are covering both DC and AC circuits.

11.2 MESH ANALYSIS

11.2.1 DC Signals

Mesh analysis is really an organized method of applying KVL. The currents in each branch of a circuit are broken down into *mesh currents*. A mesh current is a complete current path in each "window" of a circuit (see Figure 11.7 in Example 11.2.1). A window is an open space in the schematic of the circuit that is completely surrounded by electrical components and wires. Each window, formally named a *mesh*, is labeled with a mesh current. The mesh current will be assumed to flow around that complete path of electrical components and wires that surround the window. Then KVL is applied to each mesh. Each voltage in the KVL equations is expressed in terms of the mesh currents. If a component has two mesh currents, both must be expressed in that voltage term with proper signs for voltage polarities. The resulting KVL equations are a set of simultaneous equations that are solved for the mesh currents. Then all other voltages and currents in the circuit can be determined from the mesh currents. Notice that mesh analysis is the organized application of KVL to solve for mesh *currents*.

The previous paragraph suggests the standard procedure (approach) that characterizes mesh analysis. Let us expand the description of the details in the procedure and then apply mesh analysis to the circuit in Example 11.1.1 to see if it is indeed less cumbersome than superposition. First, a review of appropriate KVL concepts is given:

■   KVL: the sum of the voltage rises and drops around a closed path is zero.

■   A voltage rise is considered to go from negative to positive across a component in the direction around the closed path that KVL is applied and from positive to negative for a voltage drop. The rises and drops are assigned independent of current directions. Voltage rises are assigned to be positive, and voltage drops are assigned to be negative.

The general mesh analysis approach is:

1.   Assign a mesh current (including direction) to each mesh.
2.   Assign one branch current to each branch, generally with the mesh current directions in mind. Assign only one voltage polarity across each passive component. It will correspond to the branch current direction.
3.   Write KVL for each mesh.
4.   Express the voltage across each passive component as the branch current times the resistance in the KVL equations.
5.   Express each branch current in terms of the appropriate mesh current(s).
6.   Insert the "branch currents in terms of mesh currents" expressions into the KVL equations and simplify.
7.   Solve the simultaneous KVL equations for the mesh currents.
8.   Determine the branch currents from the mesh currents. Determine component voltages from the branch currents.

____ Example 11.2.1 ____

Determine the current through each resistance in Figure 11.1.

Given: DC multisource circuit in Figure 11.1

Desired: I_1, I_2, I_3, I_4, and I_5, including current direction in each R

Strategy: mesh analysis

Solution: ____

1.   A mesh current direction is assigned to each mesh in the circuit shown in Figure 11.1—see Figure 11.7. The mesh current direction, CW or CCW, is arbitrary.
2.   A branch current, including direction, is assigned to each branch—also shown in Figure 11.7. The voltage polarities across the resistances, corresponding to the assigned branch current direction, are labeled.
3.   KVL is applied to each mesh:

$$\text{Mesh A: } +V_{S1} - V_1 - V_2 - V_{S2} = 0$$
$$\text{Mesh B: } +V_{S2} + V_2 - V_3 - V_4 = 0$$
$$\text{Mesh C: } -V_5 + V_{S3} - V_4 = 0$$

Figure 11.7 Mesh Current and Branch Current Assignments, and Voltage Labels

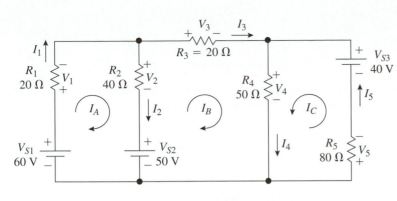

4. The voltage across each resistance is expressed in terms of branch currents (insert the values for the resistances and the sources, too):

Mesh A: $+60 - I_1 20 - I_2 40 - 50 = 0$

Mesh B: $+50 + I_2 40 - I_3 20 - I_4 50 = 0$

Mesh C: $-I_5 80 + 40 - I_4 50 = 0$

5. Express each branch current in terms of the mesh currents:

$$I_1 = +I_A$$

$$I_2 = +I_A - I_B$$

Note that branch current I_2 is comprised of two mesh currents (I_A and I_B). I_A is in the same direction as I_2; hence it is positive. I_B is in the opposite direction of I_2; hence it is negative. For the other branches:

$$I_3 = +I_B$$

$$I_4 = +I_B + I_C$$

Note that I_4 is determined using the same approach as was used to determine I_2.

$$I_5 = +I_C$$

6. Insert the "branch currents in terms of mesh currents" expressions into the KVL equations:

Mesh A:

$$+60 - I_A 20 - (+I_A - I_B)40 - 50 = 0$$

Mesh B:

$$+50 + (+I_A - I_B)40 - I_B 20 - (+I_B + I_C)50 = 0$$

Mesh C:

$$-I_C 80 + 40 - (+I_B + I_C)50 = 0$$

Part of this step is to arrange the KVL equations into standard form for solving simultaneously. All terms with variables (mesh currents) are aligned on the left-hand sides of the equations and the constants are placed on the right-hand sides of the equations. Keep the signs with the constants:

Mesh A:

$$I_A (-20 - 40) + I_B (+40) + I_C (0) = +50 - 60$$

$$I_A (-60) + I_B (+40) + I_C (0) = -10$$

Mesh B:

$$I_A (+40) + I_B (-40 - 20 - 50) + I_C (-50) = -50$$

$$I_A (+40) + I_B (-110) + I_C (-50) = -50$$

Mesh C:

$$I_A(0) + I_B(-50) + I_C(-80-50) = -40$$
$$I_A(0) + I_B(-50) + I_C(-130) = -40$$

7. Solve the simultaneous equations for the mesh currents. This capability is standard in most scientific calculators. Math software can also be used.

Mesh A:

$$I_A(-60) + I_B(+40) + I_C(0) = -10$$

Mesh B:

$$I_A(+40) + I_B(-110) + I_C(-50) = -50$$

Mesh C:

$$I_A(0) + I_B(-50) + I_C(-130) = -40$$

$$I_A = +0.59600\ A$$
$$I_B = +0.64400\ A$$
$$I_C = +0.06000\ A$$

8. Determine the branch currents from the mesh currents (using the expressions obtained during the mesh analysis setup):

$$I_1 = +I_A = +0.59600\ A = +0.596\ A$$
$$I_2 = +I_A - I_B = +0.59600 - 0.64400 = -0.048\ A$$
$$I_3 = +I_B = +0.64400 = +0.644\ A$$
$$I_4 = +I_B + I_C = +0.64400 + 0.06000 = +0.704\ A$$
$$I_5 = +I_C = +0.06000 = +0.060\ A$$

These branch current results match the currents obtained in the superposition approach in Example 11.1.1 (branch current I_3 has the opposite sign because this current was labeled in opposite directions in the superposition analysis of Example 11.1.1 and in the mesh analysis of Example 11.2.1).

Computer simulation results for Example 11.2.1 are illustrated in Figures 11.8 and 11.9. Compare the simulation output results with those of Example 11.2.1. In Figure 11.8, the DC currents are indicated. They are positioned on the end of the component (including

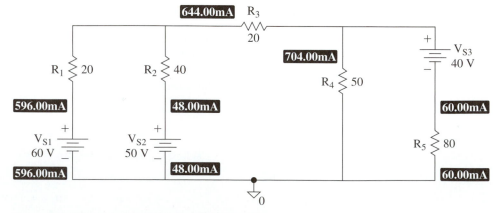

Figure 11.8 PSpice® Computer Simulation Results
PSpice simulation output, used with permission of Cadence Design Systems, Inc.

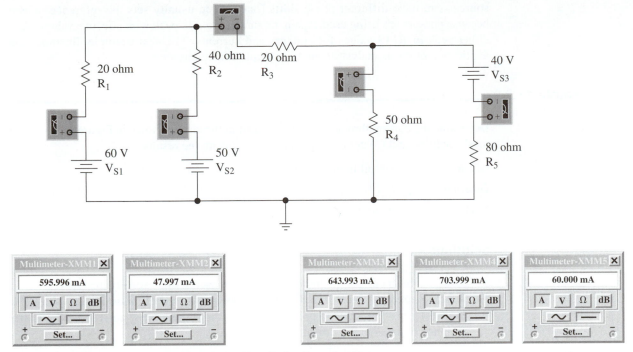

Figure 11.9 MultiSIM® Computer Simulation Results for Example 11.2.1
Printed Courtesy of Electronics Workbench

the sources) into which the current enters. Hence, branch current $I_1 = +0.596\ A$ agrees with that current through R_1 in Figure 11.8 because they were defined in the same direction. Branch current I_2, on the other hand, was defined to go down in the previous mesh analysis. The mesh analysis result was $I_2 = -0.048\ A$. Hence the current actually goes up through R_2. This conclusion is in agreement with the simulation results in Figure 11.8. The other currents similarly match between the mesh analysis results and the computer simulation results.

Now, when you examine computer simulation results, you will have some idea of what is behind the solution to complex circuits that are analyzed using software. The computer simulation results are not "magic" but are based on circuit laws and analysis techniques that you know and could perform using paper and calculator (if necessary). The use of a computer is just many times faster. More importantly, by knowing the systematic circuit analysis techniques, you are developing insight into circuit analysis and solutions that will help you detect errors (especially erroneous data input). It also becomes part of your conceptual foundation that is needed to understand future advances and new ideas in electronics and circuits.

Thus the approach in mesh analysis is systematic. The approach in mesh analysis is always the same and does not depend on the particular circuit. Another organized approach, nodal analysis, will be covered in Section 11.3. For those who are also studying mesh analysis with AC signals, continue in subsection 11.2.2. For those who are studying DC now and AC later, skip subsection 11.2.2 and continue in Section 11.3.1.

11.2.2 AC Signals

The procedure for mesh analysis with AC signals is the same as for DC signals; however, impedances and phasors must be used. Also, the same restriction as with superposition in multiple-source AC circuits applies: all of the sources must be of the same frequency. The sources can have different phase shifts (one source usually sets the reference phase) because phasors are being used. Again, be sure not to mix peak with effective values—use either peak *or* RMS values for voltages and currents. The next example illustrates the application of mesh analysis to an AC multiple-source circuit.

Example 11.2.2

Determine the current through each component in the circuit shown in Figure 11.10 by (*a*) mesh analysis, and (*b*) computer simulation. Compare the results.

Given: AC circuit in Figure 11.10
Determine: $\tilde{I}_{L1}, \tilde{I}_{L2}, \tilde{I}_{R1}, \tilde{I}_{R2}, \tilde{I}_{C1}$
Strategy: mesh analysis

Solution:

1. Assign a mesh current (including direction) to each mesh—see Figure 11.11.
2. A branch current, including direction, is assigned to each branch—also shown in Figure 11.11. The voltage polarities across the impedances, corresponding to the assigned branch current directions, are labeled.
3. KVL is applied to each mesh:

   Mesh A:

   $$+\tilde{V}_{S1} - \tilde{V}_{L1} - \tilde{V}_{R1} + \tilde{V}_{C1} - \tilde{V}_{S2} = 0$$

   Mesh B:

   $$+\tilde{V}_{S2} - \tilde{V}_{C1} - \tilde{V}_{R2} - \tilde{V}_{L2} = 0$$

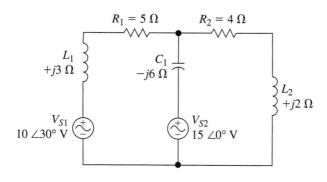

Figure 11.10 AC Circuit for Example 11.2.2

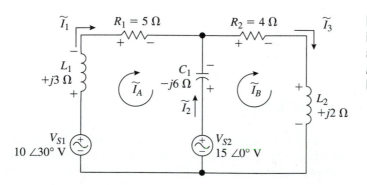

Figure 11.11 AC Circuit in Figure 11.10 with Mesh and Branch Currents Assigned and Voltage Polarities Labeled.

4. The voltage across each impedance is expressed in terms of branch currents (insert the values for the impedances and sources, too):

Mesh A:

$$+10\angle+30° - \tilde{I}_1\,(+j3) - \tilde{I}_1\,(5) + \tilde{I}_2\,(-j6) - 15\angle0° = 0$$

Mesh B:

$$+15\angle0° - \tilde{I}_2\,(-j6) - \tilde{I}_3\,(4) - \tilde{I}_3\,(+j2) = 0$$

5. Express each branch current in terms of the appropriate mesh currents:

$$\tilde{I}_1 = +\tilde{I}_A$$
$$\tilde{I}_2 = +\tilde{I}_B - \tilde{I}_A$$
$$\tilde{I}_3 = +\tilde{I}_B$$

6. Insert the "branch currents in terms of mesh currents" expressions into the KVL equations:

Mesh A:

$$+10\angle+30° - \tilde{I}_A\,(+j3) - \tilde{I}_A\,(5) + (+\tilde{I}_B - \tilde{I}_A)(-j6) - 15\angle0° = 0$$

Mesh B:

$$+15\angle0° - (+\tilde{I}_B - \tilde{I}_A)(-j6) - \tilde{I}_B\,(4) - \tilde{I}_B\,(+j2) = 0$$

Again, arrange the KVL equations into standard form for solving simultaneously (be sure to keep the signs with the constants):

Mesh A:

$$\tilde{I}_A\,(-j3 - 5 + j6) + \tilde{I}_B\,(-j6) = +15\angle0° - 10\angle+30°$$
$$\tilde{I}_A\,(-5 + j3) + \tilde{I}_B\,(-j6) = 8.0742\angle-38.262°$$

Mesh B:

$$\tilde{I}_A\,(-j6) + \tilde{I}_B\,(+j6 - 4 - j2) = -15\angle0°$$
$$\tilde{I}_A\,(-j6) + \tilde{I}_B\,(-4 + j4) = -15 + j0$$

7. Solve the simultaneous KVL equations for the mesh currents:

Mesh A:

$$\tilde{I}_A\,(-5 + j3) + \tilde{I}_B\,(-j6) = 8.0742\angle-38.262°$$

Mesh B:

$$\tilde{I}_A\,(-j6) + \tilde{I}_B\,(-4 + j4) = -15 + j0$$

$$\tilde{I}_A = 0.82641\angle-60.818° \text{ A}$$
$$\tilde{I}_B = 1.9342\angle+32.235° \text{ A}$$

8. Determine the branch currents from the mesh currents:

$$\tilde{I}_{L1} = \tilde{I}_{R1} = \tilde{I}_1 = +\tilde{I}_A = 0.82641\angle-60.818° = 0.826\angle-60.8° \text{ A}$$
$$\tilde{I}_{C1} = \tilde{I}_2 = +\tilde{I}_B - \tilde{I}_A = +1.9342\angle+32.235° - 0.82641\angle-60.818° = 2.14\angle+54.9° \text{ A}$$
$$\tilde{I}_{L2} = \tilde{I}_{R2} = \tilde{I}_3 = +\tilde{I}_B = 1.9342\angle+32.235° = 1.93\angle+32.2° \text{ A}$$

In order to use computer simulation software, one normally must enter the component values; however, only the reactance values were given for the inductors and the capacitors in this example. How is circuit simulation software used if reactances are given instead of component values? This is an example of a situation in which understanding the principles behind circuit analysis pays off. The reactances of inductors and capacitors are, respectively:

$$X_L = \omega L \tag{11.1}$$

$$X_C = \frac{1}{\omega C} \tag{11.2}$$

If the radian frequency is set to one radian per sec ($\omega = 1$ r/s, $f = 0.159155$ Hz), then the reactances become:

$$X_L = L \tag{11.3}$$

$$X_C = \frac{1}{C} \tag{11.4}$$

Thus, the component values to enter into the simulation software for this example are:

$$L_1 = X_{L1} = 3$$
$$L_2 = X_{L2} = 2$$
$$C_1 = \frac{1}{X_{C1}} = \frac{1}{6} = 0.16667$$

The resistance values are entered as is. The schematic in Figure 11.12 on page 265 has these component values for the inductors and the capacitor.

The computer simulation results shown in Figure 11.12 match those obtained above using mesh analysis. The "IPRINT" parts were placed into the circuit such that the directions for the currents matched the branch current directions that were defined in Figure 11.9. IM(V_PRINT1) is the magnitude of the current through IPRINT PRINT1. IP(V_PRINT1) is the phase of the current through IPRINT PRINT1. The other currents are similarly identified.

Again, note the systematic procedure behind mesh analysis. In the next section of this chapter, another systematic circuit analysis technique, nodal analysis, will be examined.

11.3 NODAL ANALYSIS

11.3.1 DC Signals

Nodal (also called node voltage) circuit analysis is just an organized method of applying KCL to solve for voltages at nodes. First, *nontrivial nodes* are identified. A nontrivial node has three or more components connected to the node. The junction between only two components, which would be in series, is a node but is a "trivial" node. Then one of the nontrivial nodes is selected as the *reference node*. The reference node selection may be arbitrary, but usually it is the ground connection or the node with the largest number of components connected to it. Recall that a voltage is measured across something, i.e., because voltage is electrical potential energy difference, there must be two points involved. The reference node is one of the two points for each voltage and is the reference point for all node voltages. Thus, a *node voltage* is the voltage between a nontrivial node and the reference node.

Once the nontrivial nodes are identified and the reference node is selected, each branch current is labeled. KCL is applied to each nontrivial node except the reference node. Then each branch current is expressed in terms of node voltages. Some of these expressions are straightforward and some require the application of KVL to help determine the branch current expression in terms of node voltages. These "branch current in

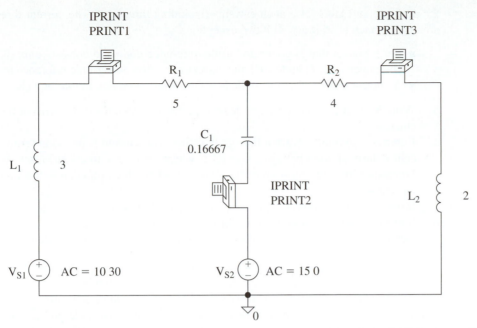

The following results were copied from the "Analysis/Examine Output" toolbar selection in PSpice (output file):

```
****11/22/101 15:43:19******** NT Evaluation PSpice (July 1997)************
*D:\PSPICE\ex11-2-2.sch

****  AC Analysis                TEMPERATURE = 27.000 DEG C
*****************************************************
  FREQ    IM(V_PRINT1)IP(V_PRINT1)
 1.592E-01  8.264E-01  -6.082E+01
  FREQ    IM(V_PRINT2)IP(V_PRINT2)
 1.592E-01  2.143E+00 5.488E+01
  FREQ    IM(V_PRINT3)IP(V_PRINT3)
 1.592E-01  1.934E+00  3.223E=+01
```

Figure 11.12 PSpice® Computer Simulation Schematic and Results for Example 11.2.2
PSpice simulation output, used with permission of Cadence Design Systems, Inc.

terms of the node voltage" expressions are substituted back into the KCL equations. The equations are manipulated into standard form and this set of simultaneous equations is solved for the node voltages. All other voltages and currents in the circuit can be determined from the node voltages.

The previous paragraph suggests the standard procedure (approach) that characterizes nodal analysis. Let us expand the description of the details in the procedure and then apply nodal analysis to the circuit in Example 11.1.1 to see if it is also less cumbersome than superposition. First, a review of appropriate KCL concepts is given:

■ KCL: the sum of the currents entering and leaving a node is zero.

■ A current entering a node is considered to be positive, and a current leaving a node is considered to be negative. The current signs are assigned independent of voltage polarities.

The general nodal analysis approach is:

1. Identify and label all nontrivial nodes, including a reference node. The reference node is usually ground of the circuit or the node with the greatest number of connections.

2. Indicate and label all branch currents (including direction). The current direction in each branch is arbitrary with two exceptions:

   ■ For a branch that is connected to the reference node and consists only of passive components (*R, L,* and/or *C*), the current must flow toward the reference node.
   ■ A current source in a branch dictates the current direction in that branch.

3. Write KCL at each nontrivial node (except the reference node) in terms of the branch currents.
4. Express each branch current (except branches with current sources) in terms of adjacent nontrivial node voltages. Use KVL where necessary to assist in this. For those branches with a current source, the value of the branch current equals the value of the current source.
5. Insert the branch current expressions into the KCL equations.
6. Solve the simultaneous equations for the node voltages.
7. Determine other voltages and the branch currents from these node voltages.

There is a reason that a branch current must flow toward the reference node when the branch is connected between a nontrivial node and the reference node and the branch contains only passive components (resistances, inductors, and/or capacitors). The polarity of the node voltage is assumed to be positive at the nontrivial node and negative at the reference node. By the passive sign convention, the current must flow toward the reference node. This scenario will be illustrated in the next example.

Example 11.3.1

Determine the node voltages in the circuit shown in Figure 11.1 using (*a*) nodal analysis, and (*b*) computer simulation. (*c*) Determine the branch currents through all the resistances from the node voltages.

Given: DC multisource circuit in Figure 11.1

Desired: node voltages

$I_1, I_2, I_3, I_4,$ and I_5

Strategy: nodal analysis → node voltages

node voltages → $I_1, I_2, I_3, I_4,$ and I_5

Solution:

1. Identify and label all nontrivial nodes, including a reference node. The reference node is arbitrarily selected. There are three nontrivial nodes (including the reference), so V_1 and V_2 are to be found using nodal analysis. They are labeled on Figure 11.13.

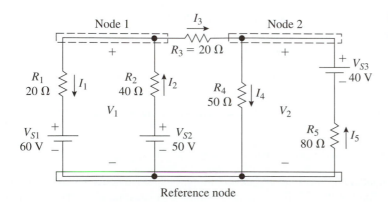

Figure 11.13 Figure 11.1 with Nodes and Branch Currents Labeled

2. Indicate and label all branch currents (including direction)—see Figure 11.13. The current direction in each branch is arbitrary except through R_4: it must flow toward the reference node because the branch consists only of a passive component (R) and it is connected directly between a nontrivial node and the reference node.

3. Write KCL at each nontrivial node (except the reference node) in terms of the branch currents.

    Node 1:

    $$-I_1 + I_2 - I_3 = 0$$

    Node 2:

    $$+I_3 - I_4 + I_5 = 0$$

4. Express each branch current in terms of adjacent nontrivial node voltages. Refer to Figure 11.14 for use of KVL to determine branch currents. Notice that the branches are shown in isolation with the node voltage(s) labeled either across that branch or on both sides of the branch (see I_3).

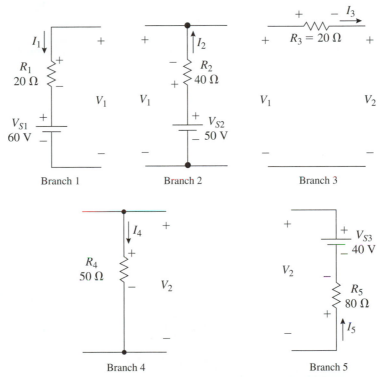

Figure 11.14 Branch Current Determination in Terms of Node Voltage Using KVL

Branch 1:

$$\text{KVL: } +V_1 - I_1R_1 - V_{S1} = 0$$

$$+V_1 - I_1 20 - 60 = 0$$

$$I_1 = \frac{+V_1 - 60}{20} = +\frac{V_1}{20} - 3$$

Branch 2:

$$\text{KVL: } +V_1 + I_2R_2 - V_{S2} = 0$$

$$+V_1 + I_2 40 - 50 = 0$$

$$I_2 = \frac{+50 - V_1}{40} = +1.25 - \frac{V_1}{40}$$

Branch 3:

$$\text{KVL:} +V_1 - I_3 R_3 - V_2 = 0$$

$$+V_1 - I_3 20 - V_2 = 0$$

$$I_3 = \frac{+V_1 - V_2}{20} = +\frac{V_1}{20} - \frac{V_2}{20}$$

Branch 4: V_2 is directly across R_4; hence, by the passive sign convention, I_4 must flow toward the reference node with the underlying assumption of the polarity of V_2 as labeled in Figure 11.13.

$$I_4 = +\frac{V_2}{R_4} = +\frac{V_2}{50}$$

Branch 5:

$$\text{KVL:} +V_2 - V_{S3} + I_5 R_5 = 0$$

$$+V_2 - 40 + I_5 80 = 0$$

$$I_5 = \frac{+40 - V_2}{80} = +0.5 - \frac{V_2}{80}$$

5. Insert the branch current expressions into the KCL equations.

Node 1:

$$-I_1 + I_2 - I_3 = 0$$

$$-\left(+\frac{V_1}{20} - 3\right) + \left(+1.25 - \frac{V_1}{40}\right) - \left(+\frac{V_1}{20} - \frac{V_2}{20}\right) = 0$$

$$V_1\left(\frac{-1}{20} - \frac{1}{40} - \frac{1}{20}\right) + V_2\left(\frac{+1}{20}\right) = -3 - 1.25$$

$$V_1(-0.125) + V_2(+0.05) = -4.25$$

Node 2:

$$+I_3 - I_4 + I_5 = 0$$

$$+\left(+\frac{V_1}{20} - \frac{V_2}{20}\right) - \left(\frac{+V_2}{50}\right) + \left(+0.5 - \frac{V_2}{80}\right) = 0$$

$$V_1\left(\frac{+1}{20}\right) + V_2\left(\frac{-1}{20} - \frac{1}{50} - \frac{1}{80}\right) = -0.5$$

$$V_1(+0.05) + V_2(-0.0825) = -0.5$$

6. Solve the simultaneous equations for the node voltages.

Node 1:

$$V_1(-0.125) + V_2(+0.05) = -4.25$$

Node 2:

$$V_1(+0.05) + V_2(-0.0825) = -0.5$$

$$V_1 = +48.080 \text{ V}$$

$$V_2 = +35.200 \text{ V}$$

The results from computer simulation are shown in Figures 11.15 and 11.16. The node voltages match those of the nodal analysis. Note that the voltages for other nodes are automatically generated in Figure 11.15. The reference node, which is indicated by the ground symbol, is at 0 V. The voltage of the reference node with respect to itself should be zero volts.

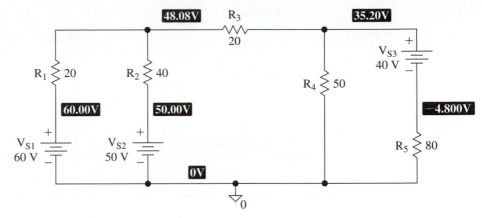

Figure 11.15 PSpice® Computer Simulation Results for Example 11.3.1
PSpice simulation output, used with permission of Cadence Design Systems, Inc.

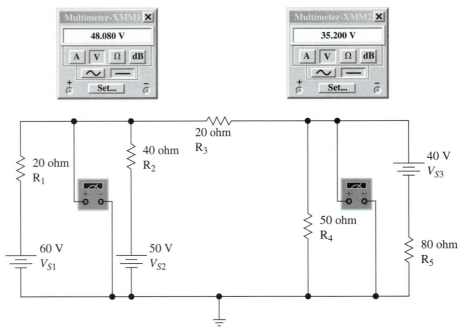

Figure 11.16 MultiSIM™ Computer Simulation Results for Example 11.3.1
Printed courtesy of Electronics Workbench

7. Determine other voltages and the branch currents from these node voltages.
    Branch 1:

$$I_1 = +\frac{V_1}{20} - 3 = +\frac{48.080}{20} - 3 = -0.596 \text{ A}$$

Branch 2:

$$I_2 = +1.25 - \frac{V_1}{40} = +1.25 - \frac{48.080}{40} = +0.048 \text{ A}$$

Branch 3:

$$I_3 = +\frac{V_1 - V_2}{20} = +\frac{48.080 - 35.200}{20} = +0.644 \text{ A}$$

Branch 4:

$$I_4 = +\frac{V_2}{50} = +\frac{35.200}{50} = +0.704 \text{ A}$$

Branch 5:

$$I_5 = \frac{+40 - V_2}{80} = \frac{+40 - 35.200}{80} = +0.060 \text{ A}$$

These results match those of the mesh analysis given in Example 11.2.1. Notice that the directions of branch currents I_1 and I_2 are labeled oppositely between Example 11.2.1 and Example 11.3.1; hence, the results are the negatives of each other.

11.3.2 AC Signals

The procedure for nodal analysis with AC signals is the same as for DC signals; however, impedances and phasors must be used. Also, the same restriction as with superposition in multiple-source AC circuits applies: all the sources must be of the same frequency. The sources can have different phase shifts (one source usually sets the reference phase) because phasors are being used. Again, be sure not to mix peak with effective values—use either peak *or* RMS values for voltages and currents. The next example illustrates the application of nodal analysis to an AC multiple-source circuit.

Example 11.3.2

Determine the voltages across the 8 Ω and 4 Ω resistances and the current through the capacitor in the circuit shown in Figure 11.17. Check the results using simulation.

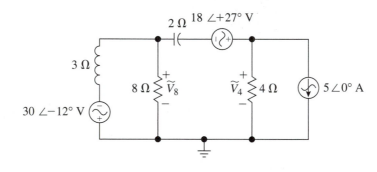

Figure 11.17 AC Circuit for Example 11.3.2

Given: circuit in Figure 11.17

Desired: $\tilde{V}_{8\Omega}, \tilde{V}_{4\Omega}, \tilde{I}_C$

Strategy: nodal analysis

node voltages $\rightarrow \tilde{V}_{8\Omega}, \tilde{V}_{4\Omega}, \tilde{I}_C$

Solution:

1. Identify and label all nontrivial nodes. The reference node is determined by the ground in this circuit. See Figure 11.18. Note that $\tilde{V}_{8\Omega} = \tilde{V}_1$ and $\tilde{V}_{4\Omega} = \tilde{V}_2$.

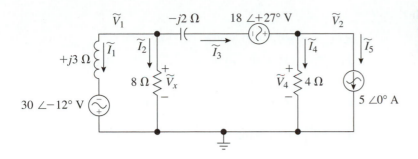

Figure 11.18 The Circuit in Figure 11.17 with the Node Voltages and Branch Currents Identified

2. Indicate all branch currents. The current direction in each branch is arbitrary *except* for the current in the 8 Ω and 4 Ω resistances—they must flow toward the reference node. The branch current direction in the current source branch must be the same as the current source current direction—see Figure 11.18.

3. Write KCL at each nontrivial node (except the reference node) in terms of the branch currents:

   Node 1:

   $$-\tilde{I}_1 - \tilde{I}_2 - \tilde{I}_3 = 0$$

   Node 2:

   $$+\tilde{I}_3 - \tilde{I}_4 - \tilde{I}_5 = 0$$

   All terms in the equation for node 1 can be multiplied by (-1) for convenience:

   Node 1:

   $$+\tilde{I}_1 + \tilde{I}_2 + \tilde{I}_3 = 0$$

   Node 2:

   $$+\tilde{I}_3 - \tilde{I}_4 - \tilde{I}_5 = 0$$

4. Express each branch current (except branches with current sources) in terms of adjacent nontrivial node voltages. Use KVL where necessary.

   Three branches are straightforward: the node voltages are directly across the 8 Ω and 4 Ω resistances, and the branch current in the rightmost branch equals the current source current:

   $$\tilde{I}_2 = +\frac{\tilde{V}_1}{8} \qquad \tilde{I}_4 = +\frac{\tilde{V}_2}{4} \qquad \tilde{I}_5 = 5\angle 0°$$

   Use KVL to help determine the branch currents in the other branches (refer to Figure 11.19):

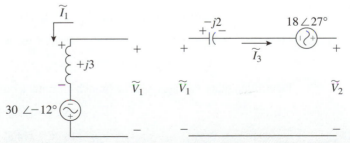

Figure 11.19 Branch Current Determination in Terms of Node Voltages Using KVL

Branch 1:

$$\text{KVL: } +\tilde{V}_1 - \tilde{I}_1 (+j3) + 30\angle -12\,° = 0$$

$$\tilde{I}(+j3) = +\tilde{V}_1 + 30\angle -12°$$

$$\tilde{I}_1 = \frac{+\tilde{V}_1 + 30\angle -12°}{+j3} = \frac{+\tilde{V}_1}{+j3} + \frac{+30\angle -12°}{+j3}$$

$$\tilde{I}_1 = \frac{+\tilde{V}_1}{+j3} + 10\angle -102°$$

Branch 3:

$$\text{KVL: } +\tilde{V}_1 - \tilde{I}_3 (-j2) + 18\angle +27° - \tilde{V}_2 = 0$$

$$\tilde{I}_3(-j2) = +\tilde{V}_1 + 18\angle +27° - \tilde{V}_2$$

$$\tilde{I}_3 = \frac{+\tilde{V}_1 + 18\angle +27° - \tilde{V}_2}{-j2}$$

$$\tilde{I}_3 = \frac{+\tilde{V}_1}{-j2} + 9\angle +117° + \frac{\tilde{V}_2}{+j2}$$

5. Insert the branch current expressions into the KCL equations.

   Node 1:

$$+\tilde{I}_1 + \tilde{I}_2 + \tilde{I}_3 = 0$$

$$+\left(\frac{+\tilde{V}_1}{+j3} + 10\angle -102°\right) + \left(+\frac{\tilde{V}_1}{8}\right) + \left(\frac{+\tilde{V}_1}{-j2} + 9\angle +117° + \frac{\tilde{V}_2}{+j2}\right) = 0$$

$$+\tilde{V}_1\left(\frac{1}{+j3} + \frac{1}{8} + \frac{1}{-j2}\right) + \tilde{V}_2\left(\frac{1}{+j2}\right) = -10\angle -102° - 9\angle +117°$$

$$+\tilde{V}_1 (0.20833\angle +53.130°) + \tilde{V}_2 (0.5\angle -90°) = 6.4120\angle +15.954°$$

   Node 2:

$$+\tilde{I}_3 - \tilde{I}_4 - \tilde{I}_5 = 0$$

$$\left(\frac{+\tilde{V}_1}{-j2} + 9\angle +117° + \frac{\tilde{V}_2}{+j2}\right) - \left(+\frac{\tilde{V}_2}{4}\right) - 5\angle 0° = 0$$

$$\tilde{V}_1\left(\frac{1}{-j2}\right) + \tilde{V}_2\left(\frac{1}{+j2} - \frac{1}{4}\right) = -9\angle +117° + 5\angle 0°$$

$$\tilde{V}_1 (0.5\angle +90°) + \tilde{V}_2 (0.55902\angle -116.565°) = 12.1185\angle -41.431°$$

6. Solve the simultaneous equations for the node voltages.

   Node 1:

$$+\tilde{V}_1 (0.20833\angle +53.130°) + \tilde{V}_2 (0.5\angle -90°) = 6.4120\angle +15.954°$$

   Node 2:

$$\tilde{V}_1 (0.5\angle +90°) + \tilde{V}_2 (0.55902\angle -116.565°) = 12.1185\angle -41.431°$$

$$\tilde{V}_1 = 15.655\angle +169.175° = 15.7\angle +169.2° \text{ V}$$

$$\tilde{V}_2 = 18.892\angle +114.770° = 18.9\angle +114.8° \text{ V}$$

7. Determine other voltages and the branch currents from these node voltages.

$$\tilde{V}_1 = \tilde{V}_{8\Omega} = 15.7\angle +169.2° \text{ V}$$

$$\tilde{V}_2 = \tilde{V}_{4\Omega} = 18.9\angle +114.8° \text{ V}$$

Branch current 3 is the current through the 2 Ω capacitive reactance:

$$\tilde{I}_3 = \frac{+\tilde{V}_1}{-j2} + 9\angle+117° + \frac{\tilde{V}_2}{+j2}$$

$$= \frac{15.655\angle+169.175°}{-j2} + 9\angle+117° + \frac{18.892\angle+114.770°}{+j2} =$$

$$\tilde{I}_3 = 5.2457\angle+54.838° = 5.25\angle+54.8° \text{ A}$$

For the computer simulation, the inductor and capacitor values were adjusted so that at $\omega = 1$ r/s ($f = 0.159155$ Hz), the proper reactance values were present. The simulation schematic and results are shown in Figure 11.20. The node voltages were labeled V_1 and V_2 on the schematic in Figure 11.20. The simulation results match those of the nodal analysis given. Branch currents were also determined (except in the branch with the current

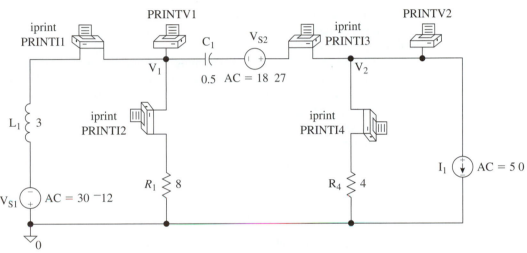

The following results were copied from the "Analysis/Examine Output" toolbar selection in PSpice (output file):

```
****11/22/101 16:50:01******** NT Evaluation PSpice (July 1997)************
*D:\PSPICE\ex11-3-2.sch
****  AC Analysis            TEMPERATURE = 27.000 DEG C
****************************************************************
  FREQ    VM(V1)     VP(V1)
  1.592E-01  1.565E+01  1.692E+02
  FREQ    VM(V2)     VP(V2)
  1.592E-01  1.889E+01  1.148E+02
  FREQ    IM(V_PRINT1)IP(V_PRINT1)
  1.592E-01  4.784E+00  -1.033E=+02
  FREQ    IM(V_PRINT2)IP(V_PRINT2)
  1.592E-01  1.957E+00  1.692E=+02
  FREQ    IM(V_PRINT3)IP(V_PRINT3)
  1.592E-01  5.246E+00  5.484E=+01
  FREQ    IM(V_PRINT4)IP(V_PRINT4)
  1.592E-01  4.723E+00  1.148E=+02
```

Figure 11.20 PSpice® Computer Simulation Schematic and Results for Example 11.3.2
PSpice simulation output, used with permission of Cadence Systems, Inc.

source). The current in branch 3 also matches the I_3 that was determined from the node voltages in the nodal analysis.

11.4 Bridge Circuits and Delta–Wye Circuits

11.4.1 DC Signals

There is a common configuration of resistances that defies straightforward series–parallel circuit analysis. The configuration is shown in Figure 11.21 and is called a *bridge circuit*. The configuration is often shown in two formats: the "H" form, shown in Figure 11.21a, and the traditional bridge form, shown in Figure 11.21b. They are electrically identical. The bridge circuit is often used in measurement applications (to be examined later in this section). The analysis is not straightforward because the circuit cannot be simplified using series–parallel circuit analysis techniques—the "middle" resistance in the "H", i.e., across the bridge, is not in series nor in parallel. Mesh or nodal analysis could be used to analyze the circuit. There is another analysis technique that makes use of a configuration conversion, which is the topic of this section. Another motivation underlying this topic is its use in *three-phase* circuits, which are used in many power delivery and control circuits. Three-phase circuits are introduced in Chapter 14.

Buried within the bridge circuit in Figure 11.21 are two other component configurations, as shown in Figure 11.22 and Figure 11.23:

■ A triangle called the *delta,* which is named after the uppercase Greek letter Δ, and looks like a triangle, as shown in Figure 11.22.

■ A Y configuration called the *wye,* which is the noun that describes any arrangement shaped like the uppercase letter Y, as shown in Figure 11.23.

The analysis technique makes use of converting a wye configuration into a delta, or vice versa. The reason for making such a conversion is illustrated in Figure 11.24. If the upper delta within the bridge circuit is converted to a wye, then the resultant circuit is a series–parallel circuit that is straightforward to analyze.

Figure 11.21 Bridge Circuit Configuration

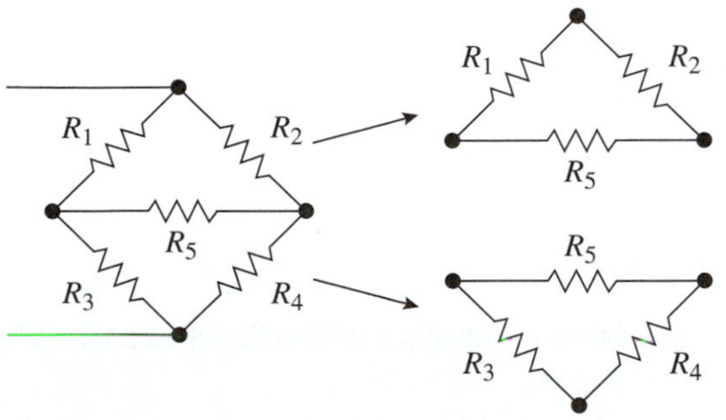

a. "H" form   b. Traditional bridge form

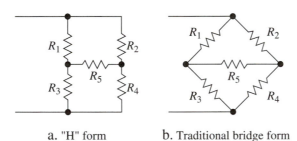

Figure 11.22 Delta Configurations Within the Bridge Circuit

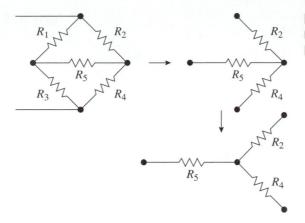

Figure 11.23 One of Two Wye Configurations Within the Bridge Circuit

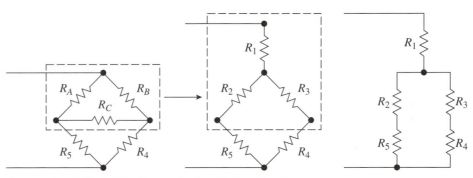

Figure 11.24 Delta-to-Wye Conversion in a Bridge Circuit

Note that resistance subscripts between wye and delta configurations must be distinguished. Henceforth, resistances in the delta will have letter subscripts and resistances in the wye will have number subscripts.

The conversion of delta to wye (or wye to delta) starts with identification of the three outer terminals, as shown in Figure 11.25. Lowercase letters will be used here, typically *a, b,* and *c.* The letter of each terminal is "opposite" the corresponding letter on the resistance in the delta configuration. The transformation may be visualized by superposing the wye and delta circuits, as shown in Figure 11.25. The locations of the resistances in relationship to one another are established in Figure 11.25 for use with the following transformation equations:[1]

■   For delta-to-wye conversions ($\Delta \rightarrow Y$), use:

$$R_Y = \frac{\text{product of the } R\text{'s in the } \Delta \text{ on each side of } R \text{ in the } Y}{\text{sum of } R\text{'s in the } \Delta} \qquad \textbf{(11.5)}$$

The resistances in the delta that are on each side of the resistance in the wye are in reference to the superposed delta and wye configurations in Figure 11.25. For example, resistance R_2 in the wye is calculated from the resistances in the delta using:

$$R_2 = \frac{R_A R_C}{R_A + R_B + R_C} \qquad \textbf{(11.6)}$$

■   For wye-to-delta conversions ($Y \rightarrow \Delta$), use:

---

[1] These transformation equations can be derived using algebra by equating the resistance between each set of corresponding terminals between the two configurations. See Irwin and Kearns, *Introduction to Electrical Engineering,* for example.

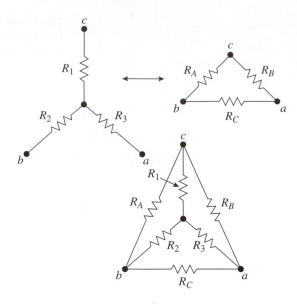

Figure 11.25 Delta–Wye Conversions

$$R_\Delta = \frac{\text{sum of the product of two } R\text{'s at a time in the Y}}{\text{the } R \text{ in the Y that is opposite the } R \text{ in the } \Delta} \qquad \textbf{(11.7)}$$

The resistance in the wye that is "opposite" the resistance in the delta is in reference to the superposed delta and wye configurations in Figure 11.25. For example, resistance R_C in the delta is calculated from the resistances in the wye using:

$$R_C = \frac{R_1R_2 + R_1R_3 + R_2R_3}{R_1} \qquad \textbf{(11.8)}$$

■ If the resistances are equal in one configuration, then they are equal in the other configuration. The transformation equations simplify to:[2]

$$R_\Delta = 3R_Y \qquad \textbf{(11.9)}$$

___ **Example 11.4.1** _____

Determine the delta configuration that is equivalent to the wye configuration in Figure 11.26.

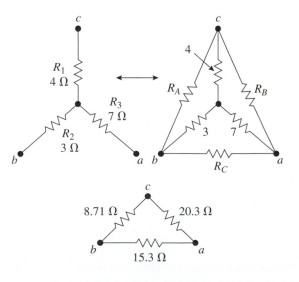

Figure 11.26 Delta–Wye Conversion for Example 11.4.1

_____
[2] Substitute $R_1 = R_2 = R_3$ into Equation (11.8) and simplify.

Given: wye configuration with:

$$R_1 = 4\,\Omega$$
$$R_2 = 3\,\Omega$$
$$R_3 = 7\,\Omega$$

Desired: R_A, R_B, R_C in the delta configuration

Strategy: $R_\Delta = \dfrac{\text{sum of the product of two }R\text{'s at a time in the Y}}{\text{the }R\text{ in the Y that is opposite the }R\text{ in the }\Delta}$

Solution:

$$R_A = \frac{R_1 R_2 + R_1 R_3 + R_2 R_3}{R_3} = \frac{(4)(3) + (4)(7) + (3)(7)}{7} = \frac{61}{7} = 8.7143 = 8.71\,\Omega$$

$$R_B = \frac{61}{3} = 20.333 = 20.3\,\Omega$$

$$R_C = \frac{61}{4} = 15.250 = 15.3\,\Omega$$

The resistances in the delta are labeled in Figure 11.26.
As a check, convert the delta configuration back into a wye (Figure 11.27):

Figure 11.27 Delta–Wye Conversion Check for Example 11.4.1

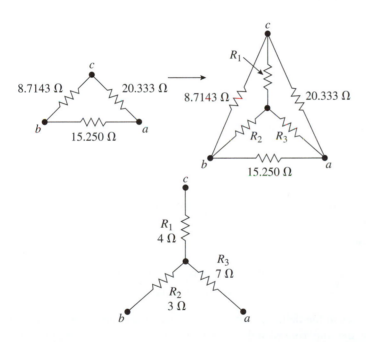

$$R_Y = \frac{\text{product of the }R\text{'s in the }\Delta\text{ on each side of }R\text{ in the Y}}{\text{sum of }R\text{'s in the }\Delta}$$

$$R_2 = \frac{R_A R_C}{R_A + R_B + R_C} = \frac{(8.7143)(15.250)}{8.7143 + 20.333 + 15.250} = 3.00003 = 3.00\,\Omega$$

You should check R_1 and R_3. The results are shown in Figure 11.27.

We return to the original motivation for delta and wye configurations, i.e., bridge circuits. A bridge circuit, as drawn in Figure 11.28, is useful for measurements. Suppose one of the branches, say branch 4, is a sensor (temperature, strain, and so on). The circle around R_4 is used to indicate that it is not a standard resistor, but that it is the resistance of

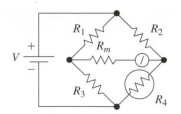

Figure 11.28 Bridge Circuit Use for Measurements

a sensor. The resistance across the bridge, R_m is the resistance of the meter circuit, which is indicated by the circle with the "indicator needle." The nominal value for R_4 of the sensor at some known state, such as room temperature for a thermistor (for measuring temperature) or no stress on a strain gauge (for measuring strain), is used to set the ratios between the resistances:

$$\frac{R_1}{R_3} = \frac{R_2}{R_4} \tag{11.10}$$

Under the condition of Equation (11.10), there is zero voltage across R_m and hence no current through the meter. The bridge is *balanced*. One can think of the voltage-divider rule across the R_1–R_3 branch and the R_2–R_4 branch. The voltages across R_3 and R_4 (or R_1 and R_2) are the same. The voltage across R_m is zero by KVL. Analytically, one could convert one of the deltas into a wye and show that the voltage across the terminals to which R_m was connected is zero.

When the sensor is placed into the environment for which measurements are desired, the value of R_4 changes. The bridge becomes *unbalanced* and a current will flow through R_m. This is where mesh analysis, nodal analysis, or delta–wye conversions could be used. The voltage across the terminals to which R_m was connected and the current through the meter could be determined. With proper calibration (such as temperature versus current through the meter), a useful instrument is available.

11.4.2 AC Signals

Delta–wye conversions are approached in the same manner as for DC circuits, except impedances are used instead of resistances:

■   For delta-to-wye conversions ($\Delta \rightarrow Y$), use:

$$\tilde{Z}_Y = \frac{\text{product of the } \tilde{Z}\text{'s in the } \Delta \text{ on each side of } \tilde{Z} \text{ in the Y}}{\text{sum of } \tilde{Z}\text{'s in the } \Delta} \tag{11.11}$$

The impedances in the delta that are on each side of the impedance in the wye is in reference to the superposed delta and wye configurations in Figure 11.29. For example, impedance $\tilde{Z}_2$ in the wye is calculated from the impedances in the delta using:

$$\tilde{Z}_2 = \frac{\tilde{Z}_A \tilde{Z}_C}{\tilde{Z}_A + \tilde{Z}_B + \tilde{Z}_C} \tag{11.12}$$

■   For wye-to-delta conversions ($Y \rightarrow \Delta$), use:

$$\tilde{Z}_\Delta = \frac{\text{product of two } \tilde{Z}\text{'s at a time in the Y}}{\text{the } \tilde{Z} \text{ in the Y that is opposite the } \tilde{Z} \text{ in the } \Delta} \tag{11.13}$$

The impedance in the wye that is "opposite" the impedance in the delta is in reference to the superposed delta and wye configurations in Figure 11.29. For example, impedance $\tilde{Z}_C$ in the delta is calculated from the impedances in the wye using:

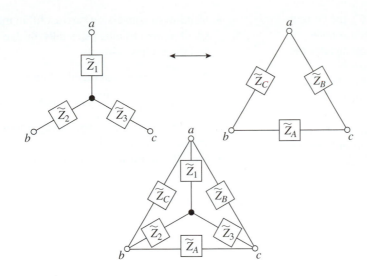

Figure 11.29 Delta–Wye Conversions with Impedances

$$\tilde{Z}_C = \frac{\tilde{Z}_1\tilde{Z}_2 + \tilde{Z}_1\tilde{Z}_3 + \tilde{Z}_2\tilde{Z}_3}{\tilde{Z}_3}$$ (11.14)

■ If the impedances are equal in one configuration, then they are equal in the other configuration. The transformation equations simplify to:

$$\tilde{Z}_\Delta = 3\tilde{Z}_Y$$ (11.15)

There are AC bridge circuits also, but this topic will be left for future studies. Delta–wye conversions in AC circuits are especially used in three-phase circuit analysis—see Chapter 14.

CHAPTER REVIEW

11.1   The Need for More Sophisticated Analysis Techniques
- Superposition becomes cumbersome when there are a large number of sources and/or number of voltages and/or currents to be determined.
- Mesh and nodal analysis techniques have standard procedures that are applicable to all circuits.
- The standard procedures of mesh and nodal analysis make these techniques usable for circuit simulation software.

11.2   Mesh Analysis
- The general mesh analysis approach is:
  1. Assign a mesh current (including direction) to each mesh.
  2. Assign one branch current to each branch, generally with the mesh current directions in mind. Assign only one voltage polarity across each passive component. It will correspond to the branch current direction.
  3. Write KVL for each mesh.
  4. Express the voltage across each passive component as the branch current times the

resistance (DC) or impedance (AC) in the KVL equations.
  5. Express each branch current in terms of the appropriate mesh current(s).
  6. Insert the "branch currents in terms of mesh currents" expressions into the KVL equations.
  7. Solve the simultaneous KVL equations for the mesh currents.
  8. Determine the branch currents from the mesh currents. Determine component voltages from the branch currents.

11.3   Nodal Analysis
- The general nodal analysis approach is:
  1. Identify and label all nontrivial nodes, including a reference node. The reference node is usually ground of the circuit or the node with the greatest number of connections.
  2. Indicate and label all branch currents (including direction). The current direction in each branch is arbitrary with two exceptions:
     - For a branch that is connected to the reference node and consists only of passive

components (R, L, and/or C), the current must flow toward the reference node.
- A current source in a branch dictates the current direction in that branch.

3. Write KCL at each nontrivial node (except the reference node) in terms of the branch currents.
4. Express each branch current (except branches with current sources) in terms of adjacent nontrivial node voltages. Use KVL where necessary to assist in this. For those branches with a current source, the value of the branch current equals the value of the current source.
5. Substitute the branch current expressions into the KCL equations.
6. Solve the simultaneous equations for the node voltages.
7. Determine other voltages and the branch currents from these node voltages.

11.4 Bridge Circuits and Δ–Y Circuits
- A bridge circuit has a resistance (or impedance) across two points in the circuit that makes straightforward series–parallel circuit analysis unusable.
- A delta (Δ) configuration of resistances (or impedances) is connected in a "triangle."
- A wye (Y) configuration of resistances (or impedances) is connected in a "Y."

- Delta–wye conversions are useful in the analysis of some circuits, especially bridge circuits and three-phase circuits.
- A DC bridge is balanced when the ratios of resistances $\dfrac{R_1}{R_3} = \dfrac{R_2}{R_4}$ are equal. Zero current flows through the bridge arm. Otherwise, the bridge is unbalanced.
- For DC delta-to-wye conversions (Δ → Y), use:

$$R_Y = \frac{\text{product of the } R\text{'s in the } \Delta \text{ on each side of } R \text{ in the Y}}{\text{sum of } R\text{'s in the } \Delta}$$

- For DC wye-to-delta conversions (Y → Δ), use:

$$R_\Delta = \frac{\text{sum of the product of two } R\text{'s at a time in the Y}}{\text{the } R \text{ in the Y that is opposite the } R \text{ in the } \Delta}$$

- If the resistances are equal in one configuration, then they are equal in the other configuration. Use: $R_\Delta = 3R_Y$
- The delta–wye conversion equations for AC circuits correspond to the DC delta–wye conversion equations except that impedances are used instead of resistances.

HOMEWORK PROBLEMS

Proper completion of the circuit analysis in the following section will demonstrate your ability to describe and perform mesh analysis in single and multiple-source DC series–parallel circuits.

11.1 Why is it important that mesh currents be defined? What is the difference between mesh currents and branch currents?
11.2 State the process that is followed when analyzing a circuit using mesh analysis.
11.3 How is KVL used in mesh analysis?
11.4 What is the difference between analyzing a circuit using mesh analysis and superposition?
11.5 Determine V_A in the circuit shown in Figure P11.1 using mesh analysis.
11.6 Determine voltage V_{ab} and current I_2 in the circuit shown in Figure P11.2 using mesh analysis.

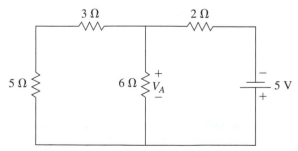

Figure P11.1 Schematic Drawing for Problems 11.5 and 11.19

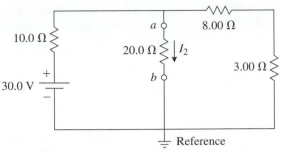

Figure P11.2 Schematic Drawing for Problems 11.6 and 11.20

11.7   Determine V_A in the circuit shown in Figure P11.3 using mesh analysis.

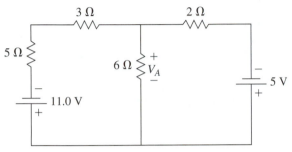

Figure P11.3 Schematic Drawing for Problems 11.7 and 11.21

11.8   Determine voltage V_{ab} and current I_2 in the circuit shown in Figure P11.4 using mesh analysis.

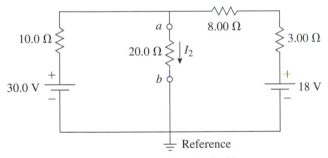

Figure P11.4 Schematic Drawing for Problems 11.8 and 11.22

11.9   Determine I_X in the circuit shown in Figure P11.5 using mesh analysis.

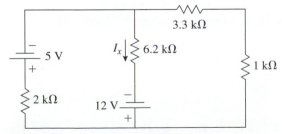

Figure P11.5 Schematic Drawing for Problems 11.9 and 11.23

11.10 Determine I_2 in the circuit shown in Figure P11.6 using mesh analysis.

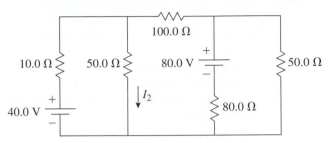

Figure P11.6 Schematic Drawing for Problems 11.10 and 11.24

11.11 For the circuit shown in Figure P11.7, determine (a) I_1, (b) I_2, (c) I_3, and (d) V_{ab} using mesh analysis.

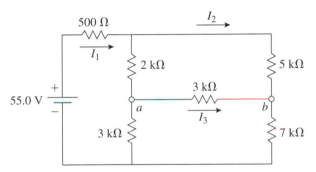

Figure P11.7 Schematic Drawing for Problems 11.11 and 11.25

11.12 For the circuit shown in Figure P11.8, determine (a) I_1, (b) I_2, (c) I_3, and (d) V_{ab} using mesh analysis.

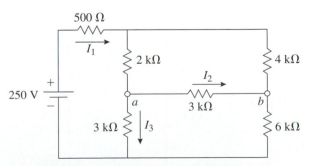

Figure P11.8 Schematic Drawing for Problems 11.12 and 11.26

Proper completion of the circuit analysis in the following section will demonstrate your ability to describe and perform nodal analysis in single and multiple-source DC series–parallel circuits and to state the advantages of using nodal analysis.

11.13 Why is it important that node voltages be defined?

11.14 State the process that is followed when analyzing a circuit using nodal analysis.

11.15 What is the difference between a trivial node and a nontrivial node?

11.16 What is the difference between analyzing a circuit using nodal analysis and mesh analysis?

11.17 How is KCL used in nodal analysis?

11.18 Are there times when it is better to use nodal analysis over mesh analysis?

11.19 a) Determine V_A in the circuit shown in Figure P11.1 using nodal analysis. b) Compare the results with those found in problem 11.5.

11.20 a) Determine voltage V_{ab} and current I_2 in the circuit shown in Figure P11.2 using nodal analysis. b) Compare the results with those found in problem 11.6.

11.21 a) Determine V_A in the circuit shown in Figure P11.3 using nodal analysis. b) Compare the results with those found in problem 11.7.

11.22 a) Determine voltage V_{ab} and current I_2 in the circuit shown in Figure P11.4 using nodal analysis. b) Compare the results with those found in problem 11.8.

11.23 a) Determine I_x in the circuit shown in Figure P11.5 using Nodal analysis. b) Compare the results with those found in problem 11.9.

11.24 a) Determine I_2 in the circuit shown in Figure P11.6 using nodal analysis. b) Compare the results with those found in problem 11.10.

11.25 For the circuit shown in Figure P11.7, determine (a) I_1, (b) I_2, (c) I_3, and (d) V_{ab} using nodal analysis. b) Compare the results with those found in problem 11.11.

11.26 a) For the circuit shown in Figure P11.8, determine (a) I_1, (b) I_2, (c) I_3, and (d) V_{ab} using nodal analysis. b) Compare the results with those found in problem 11.12.

Proper completion of the circuit analysis in the following section will demonstrate your ability to describe and perform mesh analysis in single and multiple-source AC series–parallel circuits.

11.27 Why is it important that mesh currents be defined? What is the difference between mesh currents and branch currents?

11.28 State the process that is followed when analyzing a circuit using mesh analysis.

11.29 How is KVL used in mesh analysis?

11.30 What is the difference between analyzing a circuit using mesh analysis and superposition?

11.31 Calculate voltage $\tilde{V}_{ab}$ and currents $\tilde{I}_T, \tilde{I}_1$, and $\tilde{I}_2$ for the series–parallel circuit shown in Figure P11.9 using mesh analysis.

11.32 Calculate voltage $\tilde{V}_{ab}$ and currents $\tilde{I}_T, \tilde{I}_1$, and $\tilde{I}_2$ for the series–parallel circuit shown in Figure P11.10 using mesh analysis.

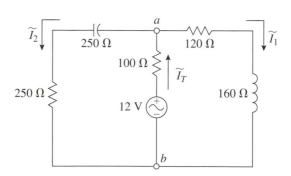

Figure P11.10 Circuit for Problems 11.32 and 11.40

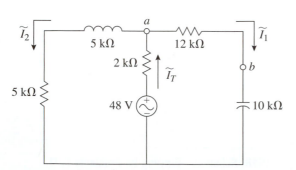

Figure P11.9 Circuit for Problems 11.31 and 11.39

11.33 Calculate currents $\tilde{I}_T$, $\tilde{I}_1$, and $\tilde{I}_2$ for the series–parallel circuit shown in Figure P11.11 using mesh analysis.

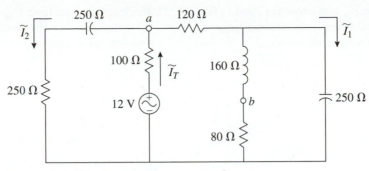

Figure P11.11 Circuit for Problems 11.33 and 11.41

11.34 Calculate voltage $\tilde{V}_{ab}$ and currents $\tilde{I}_T$, $\tilde{I}_1$, and $\tilde{I}_2$ for the series–parallel circuit shown in Figure P11.12 using mesh analysis.

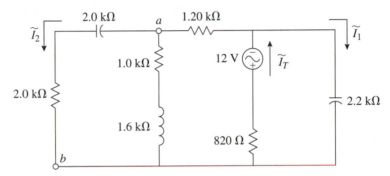

Figure P11.12 Circuit for Problems 11.34 and 11.42

11.35 Determine voltage $\tilde{V}_{ab}$ and currents $\tilde{I}_1$ and $\tilde{I}_2$ for the multiple-source series–parallel circuit shown in Figure P11.13 using mesh analysis.

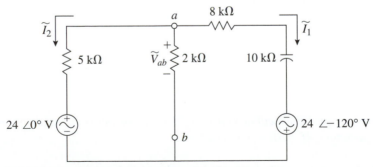

Figure P11.13 Circuit for Problems 11.35 and 11.43

11.36 Determine voltage $\tilde{V}_{ab}$ and currents $\tilde{I}_1$ and $\tilde{I}_2$ for the multiple-source series–parallel circuit shown in Figure P11.14 using mesh analysis.

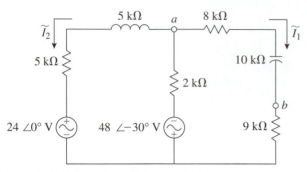

Figure P11.14 Circuit for Problems 11.36 and 11.44

11.37 Determine voltage $\tilde{V}_{Ac}$ and current $\tilde{I}_1$ for the multiple-source series–parallel circuit shown in Figure P11.15 using mesh analysis.

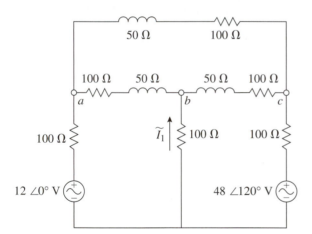

Figure P11.15 Circuit for Problems 11.37 and 11.45

11.38 Determine voltage $\tilde{V}_{Ac}$ and current $\tilde{I}_1$ for the series–parallel circuit shown in Figure P11.16 using mesh analysis.

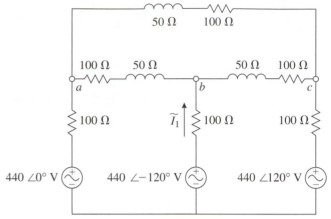

Figure P11.16 Circuit for problems 11.38 and 11.46

Your solutions to the following problems will demonstrate that you can perform nodal analysis to calculate all voltages and currents in single-source AC series–parallel circuits.

11.39 a) Calculate voltage $\tilde{V}_{ab}$ and currents $\tilde{I}_T$, $\tilde{I}_1$, and $\tilde{I}_2$ for the series–parallel circuit shown in Figure P11.9 using nodal analysis. b) Compare the results with those found in problem 11.31.

11.40 a) Calculate voltage $\tilde{V}_{ab}$ and currents $\tilde{I}_T$, $\tilde{I}_1$, and $\tilde{I}_2$ for the series–parallel circuit shown in Figure P11.10 using nodal analysis. b) Compare the results with those found in problem 11.32.

11.41 a) Calculate the currents $\tilde{I}_T$, $\tilde{I}_1$, and $\tilde{I}_2$ for the series–parallel circuit shown in Figure P11.11 using nodal analysis. b) Compare the results with those found in problem 11.33.

11.42 a) Calculate voltage $\tilde{V}_{ab}$ and currents $\tilde{I}_T$, $\tilde{I}_1$, and $\tilde{I}_2$ for the series–parallel circuit shown in Figure P11.12 using nodal analysis. b) Compare the results with those found in problem 11.34.

Your solutions to the following problems will demonstrate your ability to perform nodal analysis to calculate voltages and currents in multiple-source AC series–parallel circuits.

11.43 a) Determine voltage $\tilde{V}_{ab}$ and currents $\tilde{I}_1$ and $\tilde{I}_2$ for the multiple-source series–parallel circuit shown in Figure P11.13 using nodal analysis. b) Compare the results with those found in problem 11.35.

11.44 a) Determine voltage $\tilde{V}_{ab}$ and currents $\tilde{I}_1$ and $\tilde{I}_2$ for the multiple-source series–parallel circuit shown in Figure P11.14 using nodal analysis. b) Compare the results with those found in problem 11.36.

11.45 a) Determine voltage $\tilde{V}_{ac}$ and current $\tilde{I}_1$, for the multiple-source series–parallel circuit shown in Figure P11.15 using nodal analysis. b) Compare the results with those found in problem 11.37.

11.46 a) Determine voltage $\tilde{V}_{ac}$ and current $\tilde{I}_1$ for the series–parallel circuit shown in Figure P11.16 using nodal analysis. b) Compare the results with those found in problem 11.38.

Your solutions to the following problems will demonstrate your ability to perform and use Y–Δ and Δ–Y conversions in the analysis of circuits to calculate voltages and currents in DC series–parallel circuits.

11.47 Use a Y–Δ conversion to determine the Δ configuration of the circuit in Figure P11.17.

11.49 Use a Δ–Y or a Y–Δ conversion to solve for the current I_1 in Figure P11.19. (*Hint:* Be careful in selecting which Y or Δ you are converting.)

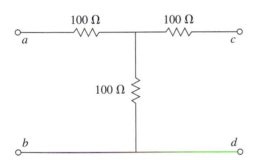

Figure P11.17 Schematic Drawing for Problem 11.47

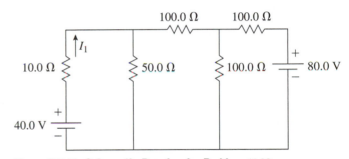

Figure P11.19 Schematic Drawing for Problem 11.49

11.48 Use a Δ–Y conversion to determine the Y configuration of the circuit in Figure P11.18.

11.50 Use a Y–Δ or a Δ–Y conversion to solve for currents I_1 and I_2 in Figure P11.20. Compare the results with those found in problem 11.11.

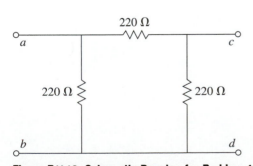

Figure P11.18 Schematic Drawing for Problem 11.48

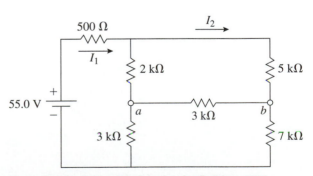

Figure P11.20 Schematic Drawing for Problem 11.50

Your solutions to the following problems will demonstrate your ability to perform AC Y–Δ and Δ–Y conversion.

11.51 Use a Y–Δ conversion to determine the Δ configuration of the circuit in Figure P11.21.

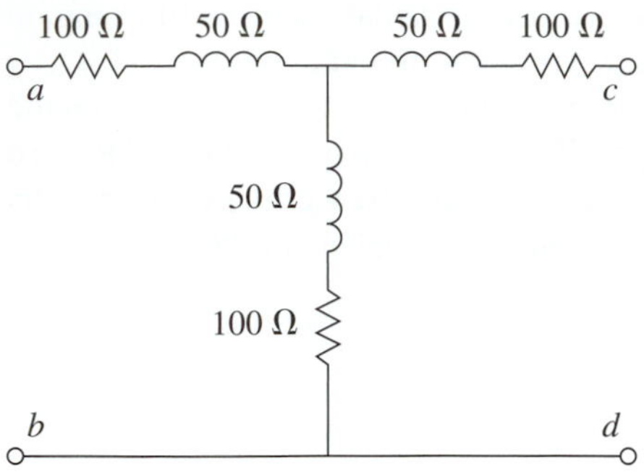

11.52 Use a Δ–Y conversion to determine the Y configuration of the circuit in Figure P11.22.

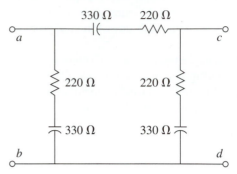

Figure P11.21 **Schematic Drawing for Problem 11.51**

Figure P11.22 **Schematic Drawing for Problem 11.52**

COMPUTER SIMULATION

Your instructor will select an appropriate set of problems for you to solve using computer simulation software.

11.53 How do you incorporate capacitive and inductive
reactances into a computer simulation when the
software inputs are capacitance and inductance?

11.54 At how many frequencies is a simulation solution valid? Explain.

THÉVENIN AND NORTON EQUIVALENT CIRCUITS

As a result of successfully completing this chapter, you should be able to:

1. Explain what an equivalent circuit is and why equivalent circuits are needed.
2. Determine the Thévenin equivalent circuit of a given circuit.
3. Determine the Norton equivalent circuit of a given circuit.
4. Convert between Thévenin and Norton equivalent circuits.
5. Solve for the voltages and currents for a given load using the equivalent circuit.
6. Determine the load resistance or impedance required for maximum power transfer.

12.1 WHAT IS AN EQUIVALENT CIRCUIT? WHY ARE EQUIVALENT CIRCUITS NEEDED?

The equivalent circuit concept is a powerful tool in the analysis and design of circuits. An example will be used to show the need for equivalent circuits and to explain what an equivalent circuit is. Start with the circuit in Figure 12.1. The load in this example is the resistance R_L. What does the arrow through R_L indicate? Resistance R_L is variable. What if we want to know the voltage across and the current through R_L as the resistance of R_L is varied? One could perform superposition, mesh, or nodal analysis for each value of R_L (or, equivalently, automate the analysis by simulating the circuit performance with circuit analysis software). Then, one could observe trends by studying a series of results as R_L is varied.

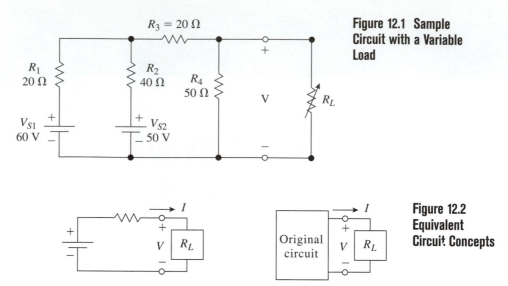

Figure 12.1 Sample Circuit with a Variable Load

Figure 12.2 Equivalent Circuit Concepts

a. Possible equivalent circuit with R_L connected   b. Original circuit with R_L

An alternative to this labor-intensive approach is to determine an *equivalent circuit.* An equivalent circuit results in the same voltage and current at the load as does the original circuit; however, the equivalent circuit is simpler than the original circuit, so the consequent trends as R_L is varied are much easier to obtain. The equivalent circuit can be as simple as a voltage source in series with a resistance, as shown in Figure 12.2a. This specific form of an equivalent circuit is called the *Thévenin equivalent circuit.*

What must be the key property of the equivalent circuit as compared with the original circuit? If the voltage across and the current through R_L are the same in both the original circuit and the equivalent circuit when connected to R_L, as shown in Figure 12.2a and b, then the equivalent circuit is a valid representation of the original circuit as far as R_L is concerned. As far as R_L is concerned, the original circuit and the equivalent circuit are indistinguishable. Thus,

> *An equivalent circuit is a simplified representation of the original circuit that results in the same voltage and current performance for a load as would have resulted in the original circuit.*

The Thévenin equivalent circuit is represented by a voltage source in series with a resistance.

In this chapter the sections on Thévenin and Norton (another representation) equivalent circuits are each divided into two subsections. The first one introduces the equivalent circuit and illustrates how it is determined and used with DC circuits. The second one illustrates the determination and use with AC circuits. The first subsection is required before the second subsection. Read only the first subsection if you are covering only DC equivalent circuits. Read both parts if you are studying both DC and AC equivalent circuits.

12.2 THÉVENIN EQUIVALENT CIRCUITS

12.2.1 DC Signals

The procedure to determine the Thévenin equivalent circuit is general—it does not change based on the circuit under consideration. First, the generic strategy will be discussed and summarized. A straightforward example will be used to illustrate the procedure. Then, the procedure will be applied to the circuit in Figure 12.1. Finally, the voltage across the load will be determined for the case of $R_L = 80\ \Omega$ in this example.

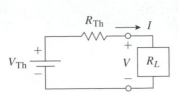

Figure 12.3 The Thévenin Equivalent Circuit with the Load Connected

There are two quantities to determine in a Thévenin equivalent circuit. Can you name them? Look at Figure 12.2a. What are the unknowns? The voltage source and the series resistance of the Thévenin equivalent circuit are the unknowns. Thus, the strategy to determine the Thévenin equivalent circuit consists of two substrategies: how is the voltage source of the Thévenin equivalent circuit, called the *Thévenin voltage,* determined? and how is the series resistance of the Thévenin equivalent circuit, called the *Thévenin resistance,* determined? These quantities are labeled in Figure 12.3.

The first step in determining the Thévenin equivalent circuit is to clearly determine the load in the original circuit. Label the terminals at each end of the load. The load may be one component or part of the circuit with several components—it depends on the particular application. Then, *remove* the load from the original circuit. Note that the load is *not* part of the equivalent circuit. Determine the voltage between the terminals using appropriate circuit analysis methods—this is the Thévenin voltage. Be careful to note the polarity. This voltage is often called the *open circuit voltage* between the terminals because there is an open left when the load is removed.

The next step is to determine the Thévenin resistance. Start by deactivating all sources in the original circuit (with R_L still removed). Do you recall how to deactivate sources? Sources had to be deactivated in the analysis of multiple-source circuits using superposition. Voltage sources were replaced with shorts to guarantee zero volts. Current sources were replaced with opens to guarantee zero amperes. Then, determine the total resistance between the marked terminals—this is the Thévenin resistance. Build the Thévenin equivalent circuit and reattach the load, as sketched in Figure 12.3. Then, determine the load voltage and the load current using straightforward circuit analysis. If the load is varied, the analysis on this simplified circuit is performed again.

In summary, the procedure to determine the Thévenin equivalent circuit is:

- Determine the load in the original circuit. Label the terminals at each end of the load.
- *Remove* the load from the original circuit.
- Label the polarity of the open circuit (Thévenin) voltage between the terminals.
- Determine the open circuit voltage between the terminals. This is V_{Th}.
- Deactivate all sources in the original circuit (with R_L still removed).
- Determine the total resistance between the terminals. This is R_{Th}.
- Build the Thévenin equivalent circuit and reattach the load per Figure 12.3.
- Analyze the simplified circuit for the load voltage and the load current.

___ **Example 12.2.1** _____

    a.   Determine the Thévenin equivalent circuit for the circuit shown in Figure 12.4 if the load is the 33 kΩ resistance.

    b.   Determine the voltage across the load.

    Given: Original circuit in Figure 12.4

                 The load is the 33 kΩ resistance

    Desired: Thévenin equivalent circuit

                 The voltage across the 33 kΩ load

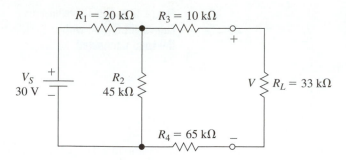

Figure 12.4 Example Circuit for Example 12.2.1

Strategy: Determine the Thévenin equivalent circuit.

Determine the load voltage using the Thévenin equivalent circuit.

Solution: _____

The load is defined. Remove the load, mark (label) the terminals, and label the voltage polarity for V_{Th}, as shown in Figure 12.5.

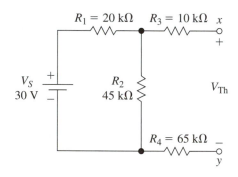

Figure 12.5 Original Circuit with the Load Removed, V_{Th} Labeled, and the Terminals Marked

The Thévenin voltage is recognized to be the voltage across the R_2 because the voltages across R_3 and R_4 are zero (the current through R_3 and R_4 is zero due to the open circuit between the x and y terminals). $V_{Th} = V_{R_2}$ is found by the voltage-divider rule:

$$V_{R_2} = \frac{V_S(R_2)}{R_1 + R_2} = \frac{30(45\ k\Omega)}{20\ k\Omega + 45\ k\Omega} = 20.769\ V = 20.8\ V = V_{Th}$$

$$V_{Th} = 20.8\ V$$

The voltage source is deactivated (replaced by a short) and the Thévenin resistance is found between the marked terminals (see Figure 12.6).

$$R_{Th} = R_3 + (R_1 \parallel R_2) + R_4 = 10\ k\Omega + \frac{(20\ k\Omega)(45\ k\Omega)}{20\ k\Omega + 45\ k\Omega} + 65\ k\Omega = 88.846\ k\Omega = 88.8\ k\Omega$$

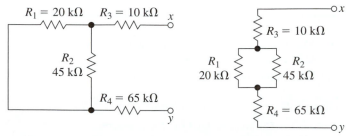

a. 30 V source replaced with a short   b. Redrawn circuit

Figure 12.6 Original Circuit with the Load Removed and All Sources Deactivated

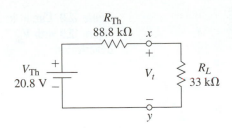

Figure 12.7 Thévenin Equivalent Circuit with the Load Reconnected

The Thévenin equivalent circuit is drawn, and the load is reattached (see Figure 12.7). The terminal voltage across the load, i.e., the voltage is found directly using the voltage-divider rule:

$$V_t = \frac{V_{Th}(R_L)}{R_{Th} + R_L} = \frac{20.769(33 \text{ k}\Omega)}{88.846 \text{ k}\Omega + 33 \text{ k}\Omega} = 5.625 \text{ V} = 5.63 \text{ V}$$

The circuit in Figure 12.1 will now be used to illustrate the procedure to determine and use the Thévenin equivalent circuit for a more complex original circuit.

Example 12.2.2

a. Determine the Thévenin equivalent circuit for the circuit shown in Figure 12.1 if the load is R_L.
b. Determine the voltage across the load if $R_L = 80 \ \Omega$.

Given: Original circuit is in Figure 12.1

The load is R_L

Desired: Thévenin equivalent circuit

The voltage across R_L if $R_L = 80 \ \Omega$

Strategy: Determine the Thévenin equivalent circuit—use superposition to determine the Thévenin voltage.

Determine the load voltage using the Thévenin equivalent circuit.

Solution:

a. The load is defined. Remove the load, mark (label) the terminals, and mark the polarity for V_{Th}, as shown in Figure 12.8.

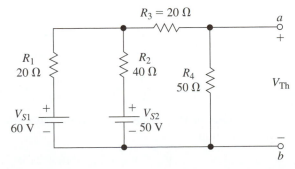

Figure 12.8 Original Circuit with the Load Removed, V_{Th} Labeled, and the Terminals Marked

Recognition: V_{Th} is the voltage across R_4 in this circuit.
Substrategy for the circuit in Figure 12.8: Use superposition to determine V_{Th}.
Redraw the circuit with V_{S1} deactivated (see Figure 12.9).
Substrategy for Figure 12.9: $R_x = R_1 \parallel (R_3 + R_4)$, voltage-divider rule → V_x → V'_{Th}

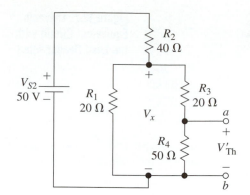

Figure 12.9 Circuit in Figure 12.8 with V_{S1} Deactivated

$$R_x = \frac{(R_1)(R_3 + R_4)}{R_1 + (R_3 + R_4)} = \frac{(20)(20 + 50)}{20 + (20 + 50)} = 15.556\ \Omega$$

$$V_x = \frac{V_{S2}\,(R_x)}{R_2 + R_x} = \frac{50(15.556)}{40 + 15.556} = 14.000\ \text{V}$$

$$V'_{Th} = \frac{V_x\,(R_4)}{R_3 + R_4} = \frac{14(50)}{20 + 50} = +10.000\ \text{V}$$

The polarity of V'_{Th} is the same as originally defined, so it is positive.
Redraw the circuit with V_{S1} deactivated (see Figure 12.10).
Substrategy for Figure 12.10: $R_y = R_2 \parallel (R_3 + R_4)$, voltage-divider rule $\rightarrow V_y \rightarrow V''_{Th}$

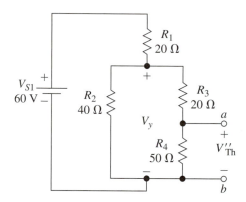

Figure 12.10 Circuit in Figure 12.8 with V_{S2} Deactivated

$$R_y = \frac{(R_2)(R_3 + R_4)}{R_2 + (R_3 + R_4)} = \frac{(40)(20 + 50)}{40 + (20 + 50)} = 25.4545\ \Omega$$

$$V_y = \frac{V_{S1}(R_y)}{R_1 + R_y} = \frac{60(25.4545)}{20 + 25.4545} = 33.600\ \text{V}$$

$$V''_{Th} = \frac{V_y\,(R_4)}{R_3 + R_4} = \frac{33.6(50)}{20 + 50} = +24.000\ \text{V}$$

The polarity of V''_{Th} is the same as originally defined, so it is positive.
Superposition:

$$V_{Th} = V'_{Th} + V''_{Th} = 10 + 24 = 34.000\ \text{V} = 34.0\ \text{V}$$

Deactivate all sources to determine R_{Th} (see Figure 12.11):

$$R_{Th} = 50 \parallel [20 + (20 \parallel 40)] = 20.000\ \Omega$$

The Thévenin equivalent circuit is shown in Figure 12.12 with the original load reattached.

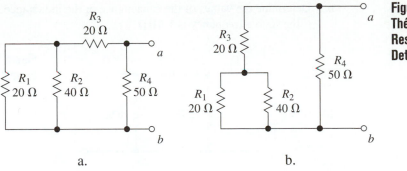

Figure 12.11
Thévenin
Resistance
Determination

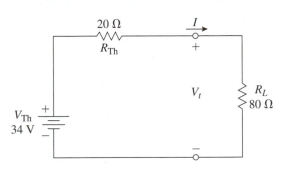

Figure 12.12 Thévenin
Equivalent Circuit with
the Load Reconnected

b. The terminal voltage V_t is determined from Figure 12.12.

Strategy: VDR

Solution: _____

$$V_t = \frac{V_{Th}\, R_L}{R_{Th} + R_L} = \frac{34(80)}{20 + 80} = 27.2 \text{ V}$$

Note that if the load is changed, it is significantly easier to reperform the V_t calculation using the Thévenin equivalent circuit than it is to reanalyze the original circuit. However, simulations could also be used to analyze the circuit as the load is changed.

12.2.2 AC Signals

The procedure for finding the Thévenin equivalent circuit with AC signals is the same as for DC signals; however, impedances and phasors must be used. The Thévenin resistance in DC becomes the Thévenin impedance in AC, and the voltage source becomes an AC source. The Thévenin equivalent circuit is valid only at the frequency analyzed because capacitors and inductors will have different impedances if the frequency is changed. Also, if more than one source is present, the same restriction as with superposition in multiple-source AC circuits applies: all of the sources must be of the same frequency. The sources can have different phase shifts (one source usually sets the reference phase) because phasors are being used. Again, be sure not to mix peak with effective values—use either peak *or* RMS values for voltages and currents. The next example illustrates the determination of the Thévenin equivalent circuit with an AC circuit.

___ **Example 12.2.3** _____

a. Determine the Thévenin equivalent circuit as "seen" by R_L for the circuit shown in Figure 12.13.

b. Determine the values of the capacitance or the inductance in the Thévenin impedance if the source frequency is 1 MHz.

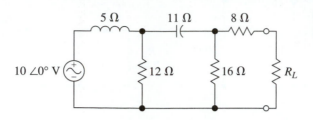

Figure 12.13 Circuit for Example 12.2.3

Given: Circuit in Figure 12.13 with the load identified

Desired: Thévenin equivalent circuit

            C or L value in the Thévenin impedance at 1 MHz

Strategy: The standard procedure for finding Thévenin equivalent circuit or.

            Use reactance equations to find L or C value in Thévenin impedance.

Solution:

The load is defined. Remove the load, mark the terminals, and mark the polarity for V_{Th}, as shown in Figure 12.14.

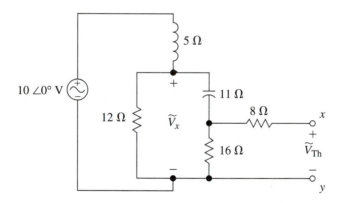

Figure 12.14 Determination of $\tilde{V}_{Th}$ in Example 12.2.3

Determine V_{Th} using series–parallel circuit analysis.
Substrategy: let $\tilde{Z}_x = 12 \,\|\, (-j11 + 16)$.

Voltage-divider rule $\rightarrow \tilde{V}_x$

Voltage-divider rule $\rightarrow \tilde{V}_{Th}$

$$\tilde{Z}_x = \frac{1}{\dfrac{1}{12} + \dfrac{1}{16 - j11}} = 7.7451\angle -13.061° \ \Omega$$

$$\tilde{V}_x = \frac{(10\angle 0°)\tilde{Z}_x}{+j5 + \tilde{Z}_x} = \frac{(10\angle 0°)(7.7451\angle -13.061°)}{+j5 + (7.7451\angle -13.061°)} = 9.4282\angle -36.364° \ \text{V}$$

How is the 8 Ω resistance taken into account? The current through the 8 Ω resistance is zero, and the voltage across it is zero. Hence, by KVL, the voltage across the 16 Ω resistance is the Thévenin voltage.

$$\tilde{V}_{Th} = \frac{(\tilde{V}_x)16}{16 - j11} = \frac{(9.4282\angle -36.364°)16}{16 - j11} = 7.7692\angle -1.855° \ \text{V}$$

$$\tilde{V}_{Th} = 7.77\angle -1.86° \ \text{V}$$

To determine the Thévenin impedance, replace the voltage source in Figure 12.13 with a short and determine the total impedance between the terminals (see Figure 12.15).

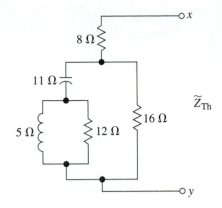

$$\tilde{Z}_{Th} = 8 + \{16 \parallel [-j11 + (12 \parallel +j5)]\}$$

$$= 8 + \cfrac{1}{\cfrac{1}{16} + \cfrac{1}{-j11 + \cfrac{1}{\left(\cfrac{1}{12} + \cfrac{1}{+j5}\right)}}} = 8 + \cfrac{1}{\cfrac{1}{16} + \cfrac{1}{6.9695\angle -75.244°}}$$

$$= 8 + 5.8660\angle -54.479° = 12.367\angle -22.710°$$

$$\tilde{Z}_{Th} = 11.408 - j4.77435\ \Omega = 11.4 - j4.77\ \Omega\ \text{at 1 MHz:}$$

The imaginary part of $\tilde{Z}_{Th}$ is negative; hence, it is a capacitance.

$$X_C = \frac{1}{2\pi fC} = 4.77435\ \Omega \rightarrow C = 3.3335 \times 10^{-8} = 33.3\ \text{nF}$$

The Thévenin equivalent circuit with component values is shown in Figure 12.16. It is valid only at 1 MHz.

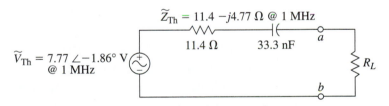

Figure 12.16 Thévenin Equivalent Circuit for Example 12.2.3

Of what use are the component values represented in a Thévenin equivalent circuit? One use is a side-by-side laboratory comparison of the equivalent circuit and the original circuit to verify that the equivalent circuit is indeed equivalent.

12.3 NORTON EQUIVALENT CIRCUITS

12.3.1 DC Signals

The *Norton equivalent circuit* is another equivalent circuit form. Instead of a series voltage source and resistance (the Thévenin equivalent circuit), the equivalent circuit consists of a parallel current source and resistance, as shown in Figure 12.17. This parallel combination will also provide the identical current through and voltage across the load as the original circuit. The use of a current source in an equivalent circuit is an important application of

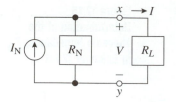

Figure 12.17 Norton Equivalent Circuit

current sources—it will be used in several subsequent electronics courses. For now the effort will be concentrated on how to determine the Norton equivalent circuit, how to use it to perform circuit calculations, and how it relates to the Thévenin equivalent circuit. First, the generic strategy will be discussed and summarized. Then, the procedure will be applied to the circuits in Figures 12.4 and 12.1.

The first step in determining the Norton equivalent circuit is to clearly determine the load in the original circuit. Label the terminals at each end of the load. The load may be one component or part of the circuit with several components—it depends on the particular application. Then, remove the load from the original circuit. So far, this procedure is the same as for the Thévenin equivalent circuit. The next step is where the difference occurs: place a short between the marked terminals. Arbitrarily label a direction for the current in this short. Determine the current through the short between the terminals—this is the Norton current. This current is often called the *short-circuit current* between the terminals because it is the current through the short that replaced the load. If the short-circuit current is going from x to y through the short, as shown in Figure 12.18a, then the Norton current source must go toward terminal x, as shown in Figure 12.18b. If a short were placed between the terminals in Figure 12.18b, the same short-circuit current direction would result. The converse also holds.

The next step is to determine the Norton resistance. This procedure is exactly the same as it was for the Thévenin equivalent circuit. Thus, $R_N = R_{Th}$. Start by deactivating all sources in the original circuit (with R_L still removed, but the short also removed). Then, determine the total resistance between the marked terminals—this is the Norton resistance. Build the Norton equivalent circuit and reattach the load, as sketched in Figure 12.17. Then, determine the load voltage and the load current using straightforward circuit analysis. If the load is varied, the analysis on this simplified circuit is performed again.

In summary, the procedure to determine the Norton equivalent circuit is:

■ Determine the load in the original circuit. Label the terminals at each end of the load.

■ Remove the load from the original circuit. Place a short between the terminals.

■ Label the direction of the short-circuit (Norton) current between the terminals.

■ Determine the short-circuit current (including direction) between the terminals. This is I_N.

■ Deactivate all sources in the original circuit (with R_L still removed and the short removed).

■ Determine the total resistance between the terminals. This is R_N.

■ Build the Norton equivalent circuit and reattach the load per Figure 12.17.

■ Analyze the simplified circuit for the load voltage and the current.

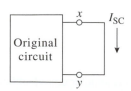

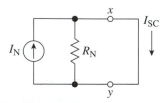

a. Original circuit with a short between terminals   b. Norton current source direction

Figure 12.18 Norton Current Source Direction Based on the I_{SC} Direction

Example 12.3.1

a.   Determine the Norton equivalent circuit for the circuit shown in Figure 12.4 if the load is the 33 kΩ resistance (note that this circuit is the same circuit that was analyzed in Example 12.2.1).

b.   Determine the voltage across the load.

Given: Original circuit in Figure 12.4

           The load is the 33 kΩ resistance.

Desired: Norton equivalent circuit

           The voltage across the 33 kΩ load

Strategy: Determine the Norton equivalent circuit.

           Determine the load voltage using the Norton equivalent circuit.

Solution:

The load is defined. Remove the load, mark (label) the terminals, place a short between the terminals, and mark the direction for I_{SC}, as shown in Figure 12.19.

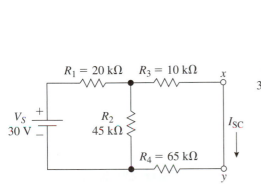

a. Short between the terminals

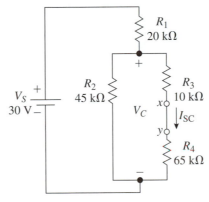

b. Redrawn

Figure 12.19 Original Circuit with the Load Removed, I_{SC} Labeled, and the Terminals Marked

Substrategy: $R_C = R_2 \parallel (R_3 + R_4)$, voltage-divider rule → V_C, Ohm's law → I_{SC}

$$R_C = \frac{(R_2)(R_3 + R_4)}{R_2 + (R_3 + R_4)} = \frac{(45\ k\Omega)(10\ k\Omega + 65\ k\Omega)}{45\ k\Omega + (10\ k\Omega + 65\ k\Omega)} = 28.125\ k\Omega$$

$$V_C = \frac{V_S(R_C)}{R_1 + R_C} = \frac{30\ (28.125\ k\Omega)}{20\ k\Omega + 28.125\ k\Omega} = 17.533\ V$$

$$I_{SC} = \frac{V_C}{R_3 + R_4} = \frac{17.533}{75\ k\Omega} = 0.23377\ mA$$

$$I_N = 0.234\ mA$$

R_N is found exactly as it was in Example 12.2.1;

$$R_N = 88.846\ k\Omega = 88.8\ k\Omega$$

Draw in the Norton equivalent circuit, reattach the load (see Figure 12.20), and determine V_{xy}. Does the result match that of Example 12.2.1?

$$V_{xy} = I_N(R_N \parallel R_L) = (0.23377\ mA)\frac{(88.846\ k\Omega)(33\ k\Omega)}{88.846\ k\Omega + 33\ k\Omega} = 5.6250\ V = 5.63\ V$$

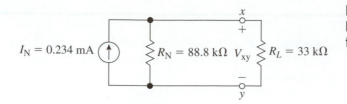

Figure 12.20 Norton Equivalent Circuit with the Load Reconnected

$I_N = 0.234$ mA $R_N = 88.8$ kΩ V_{xy} $R_L = 33$ kΩ

The circuit in Figure 12.1 will now be used to illustrate the procedure to determine and use the Norton equivalent circuit.

— Example 12.3.2

a. Determine the Norton equivalent circuit for the circuit shown in Figure 12.1 if the load is R_L (note that this is the same circuit that was analyzed in Example 12.2.2).
b. Determine the voltage across the load if $R_L = 80\ \Omega$.

Given: Original circuit in Figure 12.1

               The load is R_L.

Desired: Norton equivalent circuit

               The voltage across R_L if $R_L = 80\ \Omega$

Strategy: Determine the Norton equivalent circuit.

               Determine the load voltage using the Norton equivalent circuit.

Solution: _____

a. The load is defined. Remove the load, mark (label) the terminals, place a short between the terminals, and mark the direction for I_{SC}, as shown in Figure 12.21.

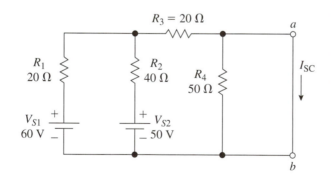

Figure 12.21 Original Circuit with the Load Removed, I_{SC} Labeled, and the Terminals Marked

$R_3 = 20\ \Omega$

R_1 $20\ \Omega$ R_2 $40\ \Omega$ R_4 $50\ \Omega$ I_{SC} a

V_{S1} 60 V V_{S2} 50 V b

Recognition: R_4 is shorted out in this circuit in the I_N determination.
Substrategy: Use superposition to determine I_{SC}.
Replace V_{S2} with a short. Redraw the circuit (Figure 12.22) and solve for I'_{SC}:

$$R_x = R_2 \parallel R_3 = \frac{(40)(20)}{40 + 20} = 13.333\ \Omega$$

$$V_x = \frac{V_{S1}R_x}{R_1 + R_x} = \frac{(60)(13.333)}{20 + 13.333} = 24.000\ \text{V}$$

$$I'_{SC} = \frac{V_x}{R_3} = \frac{24}{20} = +1.200\ \text{A}$$

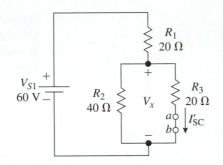

Figure 12.22 Circuit in Figure 12.21 with V_{S2} Deactivated

Replace V_{S1} with a short. Redraw the circuit (Figure 12.23) and solve for I''_{SC}:

$$R_y = R_1 \| R_3 = \frac{20}{2} = 10 \ \Omega$$

$$V_y = \frac{V_{S2}R_y}{R_2 + R_y} = \frac{(50)(10)}{40 + 10} = 10 \text{ V}$$

$$I''_{SC} = \frac{V_y}{R_3} = \frac{10}{20} = +0.500 \text{ A}$$

Superposition:

$$I_{SC} = +I'_{SC} + I''_{SC} = 1.2 + 0.5 = 1.700 \text{ A} = I_N$$

Norton resistance: found exactly the same way as in Example 12.2.2:

$$R_N = 20.0 \ \Omega$$

Draw the Norton equivalent circuit, reattach the load (see Figure 12.24).

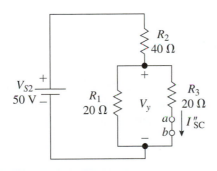

Figure 12.23 Circuit in Figure 12.15 with V_{S1} Deactivated

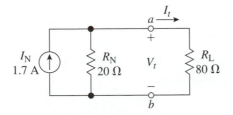

Figure 12.24 Norton Equivalent Circuit with the Load Reconnected

b. The voltage across the load, V_t, now needs to be determined.
    Strategy: $V_t = I_N (R_N \| R_L)$

Solution: _____

$$V_t = 1.7 \left(\frac{(20)(80)}{20 + 80} \right) = 27.200 = 27.2 \text{ V}$$

The result of 27.2 V agrees with the result of Example 12.2.2.

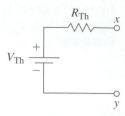
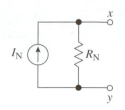

Figure 12.25 Side-by-side Comparison of Thévenin and Norton Equivalent Circuits

a. Thévenin equivalent circuit   b. Norton equivalent circuit

It would be convenient if one could convert back and forth between Thévenin and Norton equivalent circuits. Then one could find the equivalent circuit that is "easiest" to determine and then choose the equivalent circuit that is easiest to perform calculations with the load attached. Since the Thévenin and Norton equivalent circuits are both equivalent circuits of the same original circuit, they must be equivalent to each other. Hence, there must be conversion, back and forth between Thévenin and Norton equivalent circuits.

Examine the Thévenin and Norton equivalent circuits side by side, as shown in Figure 12.25. If a short is placed across the terminals of the Norton equivalent circuit, what is I_{SC}? It is equal to I_N because the Norton resistance is shorted out. If a short is placed across the terminals of the Thévenin equivalent circuit, what is I_{SC}? It is equal to V_{Th}/R_{Th} by Ohm's law. If the two equivalent circuits are equivalent to each other, then the short-circuit currents must be equal. Also, if one determines the Thévenin equivalent resistance of the Norton equivalent circuit (replace the current source with an open), the resistance between the terminals is R_N. Hence,

$$ I_N = \frac{V_{Th}}{R_{Th}} \qquad R_{Th} = R_N \qquad (12.1) $$

Similarly the open-circuit voltage of the Thévenin equivalent circuit is V_{Th} because there is zero current and zero voltage drop across R_{Th} if there is an open in the series path. Correspondingly the open-circuit voltage across the terminals of the Norton equivalent circuits is $I_N R_N$. These results are identical with Equation (12.1). Hence, the equations in Equation (12.1) are the fundamental equations for converting between a Thévenin equivalent circuit and a Norton equivalent circuit. In fact Equation 13.1 is *the general conversion between a voltage source with series resistance and a current source with parallel resistance,* and this conversion is called a *source conversion.* Source conversions are often useful in circuit analysis.

___ **Example 12.3.3** _____

Perform a source conversion of the results from Example 12.2.2 and check against the results of Example 12.3.2.

Given: Thévenin equivalent circuit in Figure 12.8

Desired: Norton equivalent circuit

Strategy:

$$ I_N = \frac{V_{Th}}{R_{Th}} \qquad R_{Th} = R_N $$

Solution: _____

$$ R_{Th} = R_N = 20\ \Omega $$

$$ I_N = \frac{V_{Th}}{R_{Th}} = \frac{34\ V}{20\ \Omega} = 1.70\ A $$

which matches the Norton equivalent circuit in Figure 12.24.

12.3.2 AC Signals

The procedure for finding the Norton equivalent circuit with AC signals is the same as for DC signals. However, impedances and phasors must be used. The Norton equivalent circuit is valid only at the frequency analyzed because capacitors and inductors will have different impedances if the frequency is changed. Also, if more than one source is present, the same restriction as with superposition in multiple-source AC circuits applies: all of the sources must be of the same frequency. The sources can have different phase shifts (one source usually sets the reference phase) because phasors are being used. Again, be sure not to mix peak with effective values—use either peak *or* RMS values for voltages and currents. The next example illustrates the determination of the Norton equivalent circuit with an AC circuit and the result is checked with a source conversion.

___ **Example 12.3.4** _____

    a.   Determine the Norton equivalent circuit as "seen" by R_L for the circuit shown in Figure 12.13.
    b.   Check the results against those of Example 12.2.3 using a source conversion.

    Given: circuit in Figure 12.13 with the load identified

    Desired: Norton equivalent circuit

               Check results using a source conversion.

    Strategy: Determine the Norton equivalent circuit.

               Convert the Norton equivalent circuit into a Thévenin equivalent circuit.

Solution: _____

The load is defined. Remove the load, mark the terminals, place a short between the terminals, and mark the direction for $\tilde{I}_{SC}$, as shown in Figure 12.26.

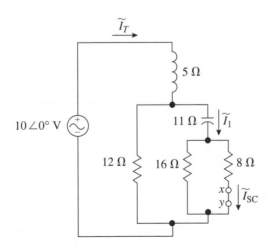

Figure 12.26
Determination of $\tilde{I}_N = \tilde{I}_{SC}$ in Example 12.3.4

    Substrategy: $\tilde{Z}_T, \tilde{I}_T$

                    CDR $\rightarrow \tilde{I}_1$

                    CDR $\rightarrow \tilde{I}_{SC}$

$$\tilde{Z}_T = +j5 + \{12 \parallel [-j11 + (16 \parallel 8)]\}$$

$$\text{Let } \tilde{Z}_D = [-j11 + (16 \parallel 8)] = -j11 + \cfrac{1}{\left(\cfrac{1}{16} + \cfrac{1}{8}\right)} = -j11 + 5.3333$$

$$\text{Let } \tilde{Z}_E = 12 \parallel \tilde{Z}_D = \frac{1}{\dfrac{1}{12} + \dfrac{1}{\tilde{Z}_D}} = 7.1458\angle -31.734°$$

$$\tilde{Z}_T = +j5 + \tilde{Z}_E = 6.2030\angle +11.545° \ \Omega$$

$$\tilde{I}_T = \frac{10\angle 0°}{\tilde{Z}_T} = 1.6121\angle -11.545° \ \text{A}$$

$$\tilde{I}_1 = \frac{\tilde{I}_T(12)}{12 + \tilde{Z}_D} = 0.94234\angle +20.855° \ \text{A}$$

$$\tilde{I}_{\text{SC}} = \frac{\tilde{I}_1(16)}{16 + 8} = 0.62823\angle +20.855° \ \text{A}$$

$$\tilde{I}_{\text{N}} = 0.62823\angle +20.855° \ \text{A} = 0.628\angle +20.9° \ \text{A}$$

The Norton equivalent circuit impedance, $\tilde{Z}_{\text{N}}$, is found exactly in the same manner as $\tilde{Z}_{\text{Th}}$ (see Example 12.2.3).

$$\tilde{Z}_{\text{N}} = \tilde{Z}_{\text{Th}} = 11.408 - j4.77435 = 11.4 - j4.77 \ \Omega$$

Draw the Norton equivalent circuit and reconnect the load—see Figure 12.27.

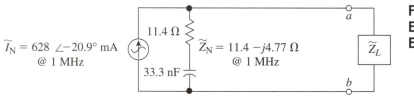

$\tilde{I}_{\text{N}} = 628 \ \angle -20.9° \ \text{mA}$ @ 1 MHz

11.4 Ω

33.3 nF

$\tilde{Z}_{\text{N}} = 11.4 - j4.77 \ \Omega$ @ 1 MHz

$\tilde{Z}_L$

a

b

Figure 12.27 Norton Equivalent Circuit for Example 12.3.4

As a check, convert the Norton equivalent circuit in Figure 12.27 into a Thévenin equivalent circuit:

$$\tilde{V}_{\text{Th}} = \tilde{I}_{\text{N}}\tilde{Z}_{\text{N}} = (0.62823\angle +20.855°)(11.408 - j4.77435)$$

$$\tilde{V}_{\text{Th}} = 7.77\angle -1.86° \ \text{V}$$

This Thévenin voltage result matches the result of Example 12.2.3 to within 0.01° (the effect of limited number resolution and rounding).

12.4 Models of Sources and Maximum Power Transfer

12.4.1 DC Signals

Thévenin and Norton equivalent circuits are used as models of actual sources. The models of DC sources were covered in Section 3.6. The Thévenin equivalent circuit matches the model of a practical voltage source—compare Figures 3.35 and 12.3. The Norton equivalent circuit matches the model of a practical current source—compare Figures 3.36 and 12.17.

A common situation in circuits is efficient power transfer from the source to the load. For example, one does not wish to dissipate an unnecessary percentage of the energy in a battery-operated electronic product in the resistance of the electronic circuitry of the source. This would just produce unused heat in the product. Hence, the appropriate question to ask is, What is the condition for maximum power transfer from the source to the load?

Consider the Thévenin equivalent circuit example shown in Figure 12.28. The load resistance is variable. If the power in the load is calculated and plotted as R_L is varied, the

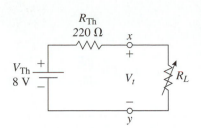

Figure 12.28 Circuit for Maximum Power Transfer Discussion

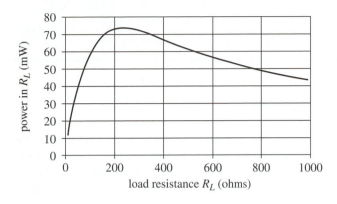

Figure 12.29 Power in Load Versus Load Resistance

graph in Figure 12.29 results. The value of the load resistance that resulted in the maximum amount of power in the load occurred when $R_L = 220\ \Omega$, i.e., when:

$$R_L = R_{\text{Th}} \tag{12.2}$$

Because half of the Thévenin voltage drops across R_L and the other half drops across R_{Th}, the amount of power that transfers from the source to the load is:

$$P_{\text{MAX}} = \frac{\left(\dfrac{V_{\text{Th}}}{2}\right)^2}{R_L} = \frac{V_{\text{Th}}^2}{4R_L} \tag{12.3}$$

An analysis of the Norton equivalent circuit gives the corresponding results:

$$R_L = R_{\text{N}} \tag{12.4}$$

$$P_{\text{MAX}} = \frac{I_{\text{N}}^2 R_L}{4} \tag{12.5}$$

Thus, the condition on the load resistance for maximum power transfer from the source to the load in a DC circuit is:

$$\boxed{R_L = R_{\text{Th}} = R_{\text{N}}} \tag{12.6}$$

___ Example 12.4.1 ___

For the circuit in Example 12.2.1, determine (*a*) the load resistance for maximum power transfer, and (*b*) the power dissipated in the load.

Given: From the Thévenin equivalent circuit that was determined in Example 12.2.1:

$$R_{\text{Th}} = 88.846\ \text{k}\Omega$$

$$V_{\text{Th}} = 20.769\ \text{V}$$

Desired:

a. R_L for maximum power transfer

b. $P_{L\text{-MAX}}$

Strategy:

$$R_L = R_{\text{Th}}$$

$$P_{\text{MAX}} = \frac{V_{\text{Th}}^2}{4R_L}$$

Solution: _____

$$R_L = R_{\text{Th}} = 88.846 \text{ k}\Omega = 88.8 \text{ k}\Omega$$

$$P_{\text{MAX}} = \frac{V_{\text{Th}}^2}{4R_L} = \frac{(20.769 \text{ V})^2}{4(88.846 \text{ k}\Omega)} = 1.2138 \text{ mW} = 1.21 \text{ mW}$$

12.4.2 AC Signals

The maximum power transfer condition on the load for AC circuits is similar to that of DC circuits, but the reactances in the circuit must be considered. Two types of loads will be examined. In the first type of load, the load impedance must be the complex conjugate of the Thévenin (or Norton) equivalent circuit impedance to achieve maximum power transfer:

$$\boxed{\tilde{Z}_L = \tilde{Z}_{\text{Th}}^* = \tilde{Z}_{\text{N}}^*} \qquad (12.7)$$

This reason that Equation (12.7) is the condition for maximum power transfer to the load in an AC circuit involves the complex conjugate.[1] The total impedance of the circuit in Figure 12.30a is:

$$\tilde{Z}_T = \tilde{Z}_L + \tilde{Z}_{\text{Th}} = \tilde{Z}_L + \tilde{Z}_L^* \qquad (12.8)$$

Because the load impedance is the complex conjugate of the Thévenin impedance, one can simplify:

$$\tilde{Z}_T = \tilde{Z}_L + \tilde{Z}_L^* = R_L + jX_L + R_L - jX_L = 2R_L \qquad (12.9)$$

which is the same condition for maximum power transfer as in a DC circuit. A similar analysis of the Norton equivalent circuit in Figure 12.30b would give the same result.

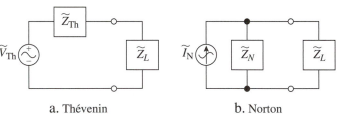

a. Thévenin   b. Norton

Figure 12.30 AC Equivalent Circuits with the Load Connected

_____ **Example 12.4.2** _____

For the circuit in Example 12.2.3, determine (*a*) the load impedance for maximum power transfer, and (*b*) the power dissipated in the load.

Given: From the Thévenin equivalent circuit that was determined in Example 12.2.3:

$$\tilde{Z}_{\text{Th}} = 11.408 - j4.77435 \ \Omega$$

_____

[1] The conditions for maximum power transfer in both DC and AC circuits can be formally derived using calculus.

$$V_{\text{Th}} = 7.7692\angle -1.855° \text{ V (assume it is RMS)}$$

Desired:

a. $\tilde{Z}_L$ for maximum power transfer

b. $P_{L\text{-MAX}}$

Strategy:

$$\tilde{Z}_L = \tilde{Z}^*_{\text{Th}}$$

$$P_{\text{MAX}} = \frac{V^2_{\text{Th}}}{4R_L}$$

Solution: _____

$$\tilde{Z}_L = \tilde{Z}^*_{\text{Th}} = (11.408 - j4.77435)^* = 11.408 + j4.77435 \ \Omega$$

$$P_{\text{MAX}} = \frac{V^2_{\text{Th}}}{4R_L} = \frac{7.7692^2}{4(11.408)} = 1.3228 \text{ W} = 1.32 \text{ W}$$

Note that only the real part of the load impedance is utilized in the real power calculation of P_{MAX}.

There is another type of load for maximum power transfer in AC circuits. What is the condition on the load if the load is resistive only (no reactance)? The condition on the load impedance for maximum power transfer from the source to the load when the load is a pure resistance is that the load resistance must equal the *magnitude* of the Thévenin (or Norton) equivalent circuit impedance:[2]

$$R_L = |\tilde{Z}_{\text{Th}}| = |\tilde{Z}_{\text{N}}| \qquad\qquad (12.10)$$

___ **Example 12.4.3** _____

For the circuit in Example 12.2.3, determine (*a*) the load impedance for maximum power transfer if the load is a pure resistance, and (*b*) the power dissipated in the load.

Given: From the Thévenin equivalent circuit that was determined in Example 12.2.3.

$$\tilde{Z}_{\text{Th}} = 11.408 - j4.77435 \ \Omega$$

$$\tilde{V}_{\text{Th}} = 7.7692\angle -1.855° \text{ V (assume it is RMS)}$$

Desired: a. $\tilde{R}_L$ for maximum power transfer for a purely resistive load

b. $P_{L\text{-MAX}}$

Strategy: $R_L = |\tilde{Z}_{\text{Th}}|$

$$\text{VDR} \rightarrow \tilde{V}_L, P_{\text{MAX}} = \frac{V^2_L}{R_L}$$

Solution: _____

$$R_L = |\tilde{Z}_{\text{Th}}| = |11.408 - j4.77435| = 12.367 = 12.4 \ \Omega$$

_____

[2] Calculus is required to derive this result.

$$\tilde{V}_L = \frac{\tilde{V}_{\mathrm{Th}}\, R_L}{\tilde{Z}_{\mathrm{Th}} + R_L} = \frac{(7.7692\angle -1.855°)(12.367)}{(11.408 - j4.77435) + 12.367} = 3.9622\angle +9.500° \text{ V}$$

$$P_{\mathrm{MAX}} = \frac{V_L^2}{R_L} = \frac{3.9622^2}{12.367} = 1.2694 \text{ W} = 1.27 \text{ W}$$

The maximum power in the load of Example 12.4.3 when the load impedance is a pure resistance is 1.27 W. When the load impedance was the complex conjugate of the Thévenin impedance, the maximum power in the load was 1.32 W. This is not an overwhelming difference, but it could have been. What would be the trend in this analysis as the reactive part of the Thévenin impedance increased relative to the real part of the Thévenin impedance?

It is useful to be aware that there are other Thévenin and Norton impedance determination techniques that are useful, especially in electronic circuits with semiconductor components such as transistors.

CHAPTER REVIEW

12.1 **What Is an Equivalent Circuit? Why Are Equivalent Circuits Needed?**
- An equivalent circuit is a simplified representation of the original circuit that results in the same voltage and current performance for a load as the original circuit.
- Equivalent circuits are useful when the voltage and/or current is desired for a load as the value of the load is being changed.

12.2 **Thévenin Equivalent Circuits**
12.2.1 *DC Signals*
- The Thévenin equivalent circuit consists of a voltage source in series with a resistance for DC circuits.
- The procedure to determine the DC Thévenin equivalent circuit is:
  - Determine the load in the original circuit. Label the terminals at each end of the load.
  - Remove the load from the original circuit.
  - Label the polarity of the open circuit (Thévenin) voltage between the terminals.
  - Determine the open-circuit voltage between the terminals. This is V_{Th}.
  - Deactivate all sources in the original circuit (with R_L still removed).
  - Determine the total resistance between the terminals. This is R_{Th}.
  - Build the Thévenin equivalent circuit and reattach the load.
  - Analyze the simplified circuit for the load voltage and current.

12.2.2 *AC Signals*
- The Thévenin equivalent circuit consists of a phasor voltage source in series with an impedance for AC circuits. The Thévenin equivalent circuit is valid only at the frequency of the source(s) in the original circuit.

- The same procedure as used for DC circuits is used to determine the AC Thévenin equivalent circuit except phasors and impedances are used.

12.3 **Norton Equivalent Circuits**
12.3.1 *DC Signals*
- The Norton equivalent circuit consists of a current source in parallel with a resistance for DC circuits. The procedure for determining the DC Norton equivalent circuit is:
  - Determine the load in the original circuit. Label the terminals at each end of the load.
  - Remove the load from the original circuit. Place a short between the terminals.
  - Label the direction of the short-circuit (Norton) current between the terminals.
  - Determine the short-circuit current (including direction) between the terminals. This is I_{N}.
  - Deactivate all sources in the original circuit (with R_L still removed and the short removed).
  - Determine the total resistance between the terminals. This is R_{N}.
  - Build the Norton equivalent circuit and reattach the load.
  - Analyze the simplified circuit for the load voltage and current.
  - Source conversions are used to convert between a voltage source with a series resistance and a current source with a parallel resistance.

12.3.2 *AC Signals*
- The Norton equivalent circuit consists of a phasor current source in parallel with an impedance for AC circuits. The Norton equivalent circuit is valid only at the frequency of the source(s) in the original circuit.

- The same procedure as used for DC circuits is used to determine the AC Norton equivalent circuit except phasors and impedances are used.

12.4 Models of Sources and Maximum Power Transfer
  - The model for a practical voltage source is the same as the Thévenin equivalent circuit.
  - The model for a practical current source is the same as the Norton equivalent circuit.

  12.4.1 *DC Signals*
  - The condition on the load resistance for maximum power transfer from the source to the load in a DC circuit is that the load resistance must equal the Thévenin (Norton) equivalent circuit resistance: $R_L = R_{Th} = R_N$.

12.4.2 *AC Signals*
  - The condition on the load impedance for maximum power transfer from the source to the load in an AC circuit is that the load impedance must equal the complex conjugate of the Thévenin (or Norton) equivalent circuit impedance: $\tilde{Z}_L = \tilde{Z}_{Th}^* = \tilde{Z}_N^*$
  - The condition on the load impedance for maximum power transfer from the source to the load in an AC circuit when the load is a pure resistance is that the load resistance must equal the magnitude of the Thévenin (or Norton) equivalent circuit impedance:

$$R_L = |\tilde{Z}_{Th}| = |\tilde{Z}_N|.$$

HOMEWORK PROBLEMS

The following questions will demonstrate that you are able to explain what an equivalent circuit is and why equivalent circuits are needed.

12.1 Why is the concept of the equivalent circuit a powerful tool in the analysis and design of circuits?

12.2 Explain how one could use an equivalent circuit while varying a load resistor R_L to observe output trends in a circuit.

12.3 When using an equivalent circuit to observe a varying load, what must be the key property of the equivalent circuit as compared to the original circuit?

12.4 Draw a simplified representation of the original circuit that results in the same voltage and current performance for a load as the original circuit.

The following questions will demonstrate that you are able to explain what a Thévenin equivalent circuit is and how to determine the Thévenin equivalent circuit for a given DC circuit.

12.5 Briefly state what a Thévenin equivalent DC circuit is.

12.6 Clearly describe the procedure that is used to determine a Thévenin equivalent DC circuit.

12.7 Determine the Thévenin equivalent DC circuit as seen by the 500 Ω resistor between terminals *a* and *b* in the circuit shown in Figure P12.1. What is the terminal voltage V_{ab} for the following load resistors: 300, 400, 500, 600, and 700 Ω?

12.8 Determine the Thévenin equivalent DC circuit as seen by the 500 Ω load between the terminals *a* and *b* in the circuit shown in Figure P12.2. What is the terminal voltage V_{ab} for the following load resistors: 300, 400, 500, 600, and 700 Ω?

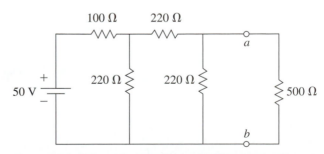

Figure P12.2 Schematic Drawing for Problems 12.8 and 12.16

12.9 Determine the Thévenin equivalent DC circuit as seen between the terminals *a* and *b* in the circuit shown in Figure P12.3. What is the terminal voltage V_{ab} and the load current I_L for the

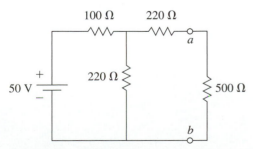

Figure P12.1 Schematic Drawing for Problems 12.7 and 12.15

following load resistors: 1.0 k, 1.5 k, 2.0 k, 2.5 k, and 3.0 kΩ?

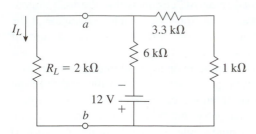

Figure P12.3 Schematic Drawing for Problems 12.9 and 12.17

12.10 Determine the Thévenin equivalent DC circuit as seen between the terminals a and b in the circuit shown in Figure P12.4. What is the terminal voltage V_{ab} and the load current I_L for the following load resistors: 2.0 k, 3.0 k, and 4.0 kΩ? (*Hint:* When determining the Thévenin resistance you may want to try a Δ–Y transformation.)

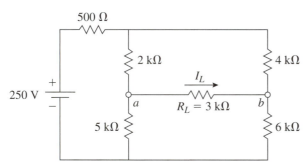

Figure P12.4 Schematic Drawing for Problems 12.10 and 12.18

12.11 Determine the Thévenin equivalent DC circuit as seen between the terminals a and b in the circuit shown in Figure P12.5. What is the terminal voltage V_{ab} and the load current I_L for the following load resistors: 1.0 k, 1.5 k, 2.0 k, 2.5 k, and 3.0 kΩ? Compare your results with those of problem 12.9. Are they the same? Can you explain why or why not?

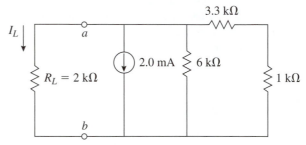

Figure P12.5 Schematic Drawing for Problems 12.11 and 12.19

12.12 Determine the Thévenin equivalent DC circuit as seen between the terminals a and b in the circuit shown in Figure P12.6. What is the terminal voltage V_{ab} and the load current I_L for the following load resistors: 2.0 k, 3.0 k, and 4.0 kΩ? (*Hint:* When determining the Thévenin resistance you may want to try a Δ–Y transformation.) Compare your results with those for problem 12.10. Are they the same? Can you explain why or why not?

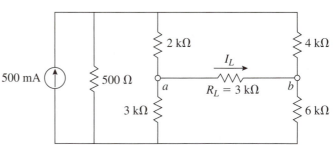

Figure P12.6 Schematic Drawing for Problems 12.12 and 12.20

The following questions will demonstrate that you are able to explain what a Norton equivalent DC circuit is and how to determine the Norton equivalent DC circuit for a given DC circuit.

12.13 Briefly state what a Norton equivalent DC circuit is.

12.14 Clearly describe the procedure that is used to determine a Norton equivalent DC circuit.

12.15 Determine the Norton equivalent DC circuit as seen by the 500 Ω resistor between terminals a and b in the circuit shown in Figure P12.1. What is the terminal voltage V_{ab} for the following load resistors: 300, 400, 500, 600, and 700 Ω? Compare your results with those of problem 12.7.

12.16 Determine the Norton equivalent DC circuit as seen between the terminals a and b in the circuit shown in Figure P12.2. What is the terminal voltage V_{ab} for the following load resistors: 300, 400, 500, 600, and 700 Ω? Compare your results with those of problem 12.8.

12.17 Determine the Norton equivalent DC circuit as seen between the terminals a and b in the circuit shown in Figure P12.3. What are the terminal voltage V_{ab} and the load current I_L for the following load resistors: 1.0 k, 1.5 k, 2.0 k, 2.5 k, and 3.0 kΩ? Compare your results with those of problem 12.9.

12.18 Determine the Norton equivalent DC circuit as seen between the terminals a and b in the circuit shown in Figure P12.4. What are the terminal voltage V_{ab} and the load current I_L for the following load resistors: 2.0 k, 3.0 k, and 4.0 kΩ? (*Hint:* When determining the Norton resistance you may want to try a Δ–Y transformation.) Compare your results with those of problem 12.10.

12.19 Determine the Norton equivalent DC circuit as seen between the terminals a and b in the circuit shown in Figure P12.5. What are the terminal voltage V_{ab} and the load current I_L for the following load resistors: 1.0 k, 1.5 k, 2.0 k, 2.5 k, and 3.0 kΩ? Compare your results with those for problem 12.11. Are they the same? Can you explain why or why not?

12.20 Determine the Norton equivalent DC circuit as seen between the terminals a and b in the circuit shown in Figure P12.6. What are the terminal voltage, V_{ab}, and the load current, I_L, for the following load resistors: 2.0 k, 3.0 k, and 4.0 kΩ (*Hint:* When determining the Norton resistance you may want to try a Δ–Y transformation.) Compare your results with those for problem 12.12. Are they the same? Can you explain why or why not?

The following questions will demonstrate that you are able to convert between Thévenin and Norton equivalent DC circuits using source conversions.

12.21 How would you go about converting a Thévenin equivalent DC circuit to a Norton equivalent DC circuit?

12.22 Explain the relationship between a Thévenin equivalent DC circuit and a practical voltage source; and between a Norton equivalent DC circuit and a practical current source.

12.23 Convert the Thévenin equivalent DC circuits that you determined in problems 12.7, 12.9, and 12.11 to the Norton equivalent DC circuits. Are they the same circuits that you found in problems 12.15, 12.17, and 12.19? Can you explain why or why not?

12.24 Convert the Thévenin equivalent DC circuits that you determined in problems 12.8, 12.10, and 12.12 to the Norton equivalent DC circuits. Are they the same circuits that you found in problems 12.16, 12.18, and 12.20? Can you explain why or why not?

The following questions will demonstrate that you are able to explain what a Thévenin equivalent circuit is and how to determine the Thévenin equivalent circuit for a given AC circuit.

12.25 Briefly state what a Thévenin equivalent AC circuit is.

12.26 Clearly describe the procedure that is used to determine a Thévenin equivalent AC circuit.

12.27 Determine the Thévenin equivalent AC circuit as seen by the 500 Ω resistor between terminals a and b in the circuit shown in Figure P12.7. What is the voltage $\tilde{V}_{ab}$ for the following load resistors: 400, 500, and 600 Ω?

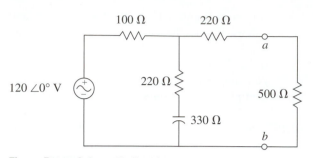

Figure P12.7 Schematic Drawing for Problems 12.27 and 12.35

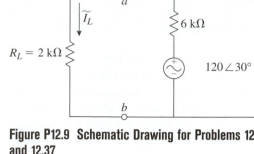

Figure P12.9 Schematic Drawing for Problems 12.29 and 12.37

12.30 Determine the Thévenin equivalent AC circuit as seen between terminals *a* and *b* in the circuit shown in Figure P12.10. What are the terminal voltage, $\tilde{V}_{ab}$, and the load current, $\tilde{I}_L$, for the following load resistors: 2.0 k, 3.0 k, and 4.0 kΩ? (*Hint:* When determining the Norton impedance you may want to try a Δ–Y transformation.)

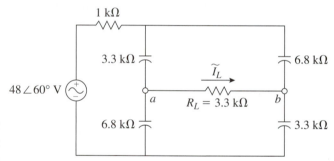

Figure P12.10 Schematic Drawing for Problems 12.30 and 12.38

12.28 Determine the Thévenin equivalent AC circuit as seen between terminals *a* and *b* in the circuit shown in Figure P12.8. What is the terminal voltage $\tilde{V}_{ab}$ for the following load resistors: 300, 500, and 700 Ω?

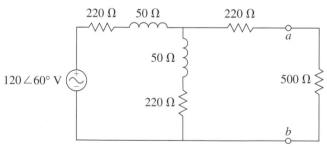

Figure P12.8 Schematic Drawing for Problems 12.28 and 12.36

12.29 Determine the Thévenin equivalent AC circuit as seen between terminals *a* and *b* in the circuit shown in Figure P12.9. What are the terminal voltage $\tilde{V}_{ab}$ and the load current $\tilde{I}_L$ for the following load resistors: 1.0 k, 1.5 k, 2.0 k, 2.5 k, and 3.0 kΩ?

12.31 Determine the Thévenin equivalent AC circuit as seen between terminals *a* and *b* in the circuit shown in Figure P12.11. What are the terminal voltage $\tilde{V}_{ab}$ and the load current $\tilde{I}_L$ for the following load resistors: 1.0 k, 1.5 k, 2.0 k, 2.5 k, and 3.0 kΩ? Compare your results with those for problem 12.29. Are they the same? Can you explain why or why not?

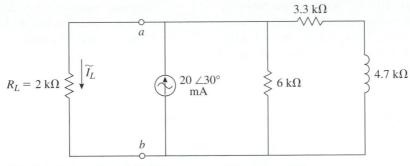

Figure P12.11 Schematic Drawing for Problems 12.31 and 12.39

12.32 Determine the Thévenin equivalent AC circuit as seen between terminals a and b in the circuit shown in Figure P12.12. What are the terminal voltage $\tilde{V}_{ab}$ and the load current $\tilde{I}_L$ for the following load resistors: 2.0 k, 3.0 k, and 4.0 kΩ?

(*Hint:* When determining the Norton impedance you may want to try a Δ–Y transformation.) Compare your results with those for problem 12.30. Are they the same? Can you explain why or why not?

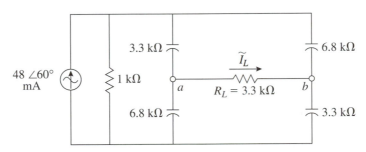

Figure P12.12 Schematic Drawing for Problems 12.32 and 12.40

The following questions will demonstrate that you are able to explain what a Norton equivalent circuit is and how to determine the Norton equivalent circuit for a given AC circuit.

12.33 Briefly state what a Norton equivalent AC circuit is.

12.34 Clearly describe the procedure that is used to determine a Norton equivalent AC circuit.

12.35 Determine the Norton equivalent AC circuit as seen by the 500 Ω resistor between terminals a and b in the circuit shown in Figure P12.7. What is the terminal voltage $\tilde{V}_{ab}$ for the following load resistors: 300, 400, 500, 600, and 700 Ω? Compare your results with those of problem 12.27.

12.36 Determine the Norton equivalent AC circuit as seen between the terminals a and b in the circuit shown in Figure P12.8. What is the terminal voltage $\tilde{V}_{ab}$ for the following load resistors: 300, 400, 500, 600, and 700 Ω? Compare your results with those of problem 12.28.

12.37 Determine the Norton equivalent AC circuit as seen between the terminals a and b in the circuit shown in Figure P12.9. What are the terminal voltage $\tilde{V}_{ab}$ and the load current $\tilde{I}_L$ for the following load resistors: 1.0 k, 1.5 k, 2.0 k, 2.5 k, and

3.0 kΩ? Compare your results with those of problem 12.29.

12.38 Determine the Norton equivalent AC circuit as seen between the terminals a and b in the circuit shown in Figure P12.10. What are the terminal voltage $\tilde{V}_{ab}$ and the load current $\tilde{I}_L$ for the following load resistors: 2.0 k, 3.0 k, and 4.0 kΩ? (*Hint:* When determining the Norton impedance you may want to try a Δ–Y transformation.) Compare your results with those of problem 12.30.

12.39 Determine the Norton equivalent AC circuit as seen between the terminals a and b in the circuit shown in Figure P12.11. What are the terminal voltage $\tilde{V}_{ab}$ and the load current $\tilde{I}_L$ for the following load resistors: 1.0 k, 1.5 k, 2.0 k, 2.5 k, and 3.0 kΩ? Compare your results with those for problem 12.31. Are they the same? Can you explain why or why not?

12.40 Determine the Norton equivalent AC circuit as seen between the terminals a and b in the circuit shown in Figure P12.12. What are the terminal voltage $\tilde{V}_{ab}$ and the load current $\tilde{I}_L$ for the following load resistors: 2.0 k, 3.0 k, and 4.0 kΩ? (*Hint:* When determining the Norton impedance you may want to try a Δ–Y transformation.) Compare your results with those for problem 12.32. Are they the same? Can you explain why or why not?

The following questions will demonstrate that you are able to convert between Thévenin and Norton equivalent AC circuits. You will also demonstrate that you are able to solve for voltages and currents for a given load using the equivalent circuit.

12.41 How would you go about converting a Thévenin equivalent AC circuit to a Norton equivalent AC circuit?

12.42 Explain the relationship between a Thévenin equivalent AC circuit and a practical voltage source; and between a Norton equivalent AC circuit and a practical current source.

12.43 Convert the Thévenin equivalent AC circuits that you determined in problems 12.27, 12.29, and 12.31 to the Norton equivalent AC circuits. Are they the same circuits that you found in problems 12.35, 12.37, and 12.39? Can you explain why or why not?

12.44 Convert the Thévenin equivalent AC circuits that you determined in problems 12.28, 12.30, and 12.32 to the Norton equivalent AC circuits. Are they the same circuits that you found in problems 12.36, 12.38, and 12.40? Can you explain why or why not?

The following questions will also demonstrate that you are able to determine the load resistance required for maximum power transfer from a DC source.

12.45 For the circuit shown in Figure P12.13, answer the following:
   a. What is the value of the load resistor R_L that, if connected between terminals *a* and *b*, would result in maximum power being delivered by the 50 V DC source?
   b. What is the maximum power that is delivered?
   c. Use a circuit simulation software program to verify your answer.

   b. What is the maximum power that is delivered?
   c. What is the relationship between Figures P12.13 and P12.14?
   d. Use a circuit simulation software program to verify your answer.

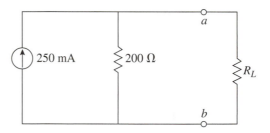

Figure P12.14 Schematic Drawing for Problem 12.46

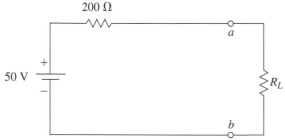

Figure P12.13 Schematic Drawing for Problem 12.45

12.46 For the circuit shown in Figure P12.14, answer the following:
   a. What is the value of the load resistor R_L that, if connected between terminals *a* and *b*, would result in maximum power being delivered by the 250 mA DC source?

12.47 For the circuits shown in Figures P12.1, P12.3, and P12.5, answer the following:
   a. What is the value of the load resistor R_L that, if connected between terminals *a* and *b*, would result in maximum power being delivered to the load?
   b. What is the maximum power that is delivered?
   c. Use a circuit simulation software program and simulate the original circuit to verify your answer.

12.48 For the circuits shown in Figures P12.2, P12.4, and P12.6, answer the following:
  a. What is the value of the load resistor R_L that, if connected between terminals a and b, would result in maximum power being delivered to the load?
  b. What is the maximum power that is delivered?
  c. Use a circuit simulation software program and simulate the original circuit to verify your answer.

The following questions will demonstrate that you are able to determine the load resistance or impedance required for maximum power transfer from an AC source.

12.49 For the circuit shown in Figure P12.15, answer the following:
  a. What is the value of the load impedance, $\tilde{Z}_L$ that, if connected between terminals a and b, would result in maximum power being delivered by the AC source if the load is purely resistive?
  b. What is the value of the load impedance $\tilde{Z}_L$ that, if connected between terminals a and b, would result in maximum power being delivered by the AC source if the load is complex?
  c. What is the maximum power that is delivered to a purely resistive load?
  d. What is the maximum power that is delivered to a complex load?
  e. Use a circuit simulation software program to verify your results.

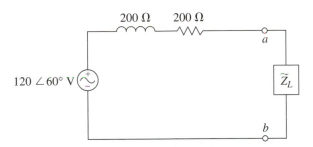

Figure P12.15 Schematic Drawing for Problem 12.49

12.50 For the circuit shown in Figure P12.16, answer the following:
  a. What is the value of the load impedance $\tilde{Z}_L$ that, if connected between terminals a and b, would result in maximum power being delivered by the AC source if the load is purely resistive?
  b. What is the value of the load impedance $\tilde{Z}_L$ that, if connected between terminals a and b, would result in maximum power being delivered by the AC source if the load is complex?
  c. What is the maximum power that is delivered to a purely resistive load?
  d. What is the maximum power that is delivered to a complex load?
  e. What is the relationship between Figures P12.15 and P12.16?

  f. Use a circuit simulation software program to verify your results.

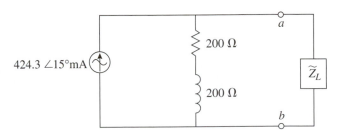

Figure P12.16 Schematic Drawing for Problem 12.50

12.51 For the circuits shown in Figures P12.7, P12.9, and P12.11, answer the following:
  a. What is the value of the load impedance $\tilde{Z}_L$ that, if connected between terminals a and b, would result in maximum power being delivered by the AC source if the load is purely resistive?
  b. What is the value of the load impedance $\tilde{Z}_L$ that, if connected between terminals a and b, would result in maximum power being delivered by the AC source if the load is complex?
  c. What is the maximum power that is delivered to a purely resistive load?
  d. Use a circuit simulation software program to verify your results.

12.52 For the circuits shown in Figures P12.8, P12.10, and P12.12, answer the following:
  a. What is the value of the load impedance $\tilde{Z}_L$ that, if connected between terminals a and b, would result in maximum power being delivered by the AC source if the load is purely resistive?
  b. What is the value of the load impedance $\tilde{Z}_L$ that, if connected between terminals a and b, would result in maximum power being delivered by the AC source if the load is complex?
  c. What is the maximum power that is delivered to a purely resistive load?
  d. Use a circuit simulation software program to verify your results.

TRANSFORMERS AND MUTUAL INDUCTORS

As a result of successfully completing this chapter, you should be able to:

1. Describe how a voltage is induced into a coil from the changing magnetic field of another coil that is driven by an AC source.
2. State the assumptions for an ideal transformer.
3. Explain how the ideal transformer transforms voltage levels.
4. Calculate voltage, current, and impedance levels in circuits that contain an ideal transformer or a practical transformer with wire losses.
5. Explain and apply the dot convention.
6. Describe the differences between a mutual inductor and an ideal transformer.
7. Calculate voltage, current, and impedance levels in circuits that contain a mutual inductor.
8. Describe series-aiding and series-opposing mutual inductors.
9. State the assumptions under which a mutual inductor may be considered an ideal transformer.

13.1 ANOTHER LOOK AT FARADAY'S LAW: TWO COILS

The ideal transformer and the mutual inductor are components that have both an input and an output set of terminals. They use magnetic fields to connect the input to output and, simultaneously, the output to the input. The transformer and the mutual inductor must do something beneficial to the signal as it passes through; otherwise, they would not be useful electric components. Two coils will be examined to illustrate the basic operation of the transformer and the mutual inductor. If one coil ("coil 1") has an AC source connected to it, as shown in Figure 13.1, an AC magnetic field is generated. What happens if a second coil ("coil 2") is placed into the region of the AC magnetic field that is generated by coil 1? An AC voltage will be generated in coil 2. Why? Faraday's law! A time-changing magnetic field will induce a voltage in the conductor.

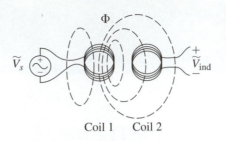

Figure 13.1 Two Coils with Magnetic Flux Linkage

Coil 1   Coil 2

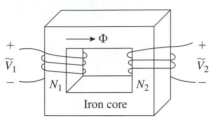

Figure 13.2 Ideal Transformer Sketch

Iron core

Figure 13.3 Picture of an Iron-Core Transformer Photo courtesy of J.W. Miller

In this and subsequent discussions, the coils are assumed to be oriented so that some of the magnetic flux goes through both coils—recall from Chapter 6 that the magnetic flux must "cut" across the conductors (or vice versa) for voltage to be induced.

Thus, there are two inductors in this scenario. A voltage is induced in coil (inductor) 1, as it would be for any inductor. The important new idea here is that a voltage is induced in coil 2 due to the time-changing magnetic field of coil 1. The terminals of coil 1 are the "input" and the terminals of coil 2 are the "output" referred to in the preceding paragraph. Now, one can ask how this situation can become useful. What if coil 2 has more wire turns than coil 1? Then there will be a longer length of conductor in coil 2 immersed in the time-changing magnetic field relative to coil 1. This longer length of conductor will thus have more voltage induced into it. In other words, the voltage at the output will be larger than the input voltage—the magnitude of the voltage has been *transformed*.

The ability to change voltage levels is the key benefit and usefulness of transformers and mutual inductors. The ability to change current and impedance levels with these components will also be examined in the next two sections. In particular, how the ideal transformer works, its input–output relationships, and examples of circuits containing ideal transformers will be covered. This coverage will be repeated for mutual inductors. The difference between the ideal transformer, a practical transformer, and the mutual inductor will be established in the subsequent discussion.

13.2 The Ideal Transformer (And Practical Ones, Too)

A transformer is a magnetic device that utilizes a magnetic core, usually iron or steel, and two coils. See the sketch in Figure 13.2 and the picture in Figure 13.3. The term *ideal transformer* is used because it is assumed that all the magnetic flux, Φ, is confined to the magnetic core (often called an "iron core" even though the core many not be pure iron) and

Figure 13.4 V_{ind} per Turn and Total Coil Voltage

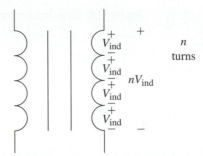

goes through both coils. Furthermore, losses in the magnetic core and in the coils are assumed to be zero (hence this component must be ideal).

A close examination of the magnetic flux will help in understanding the operation of the ideal transformer. Because of the iron core, all of the magnetic flux will go through (link) both coils. Thus the magnetic flux through coil 1 equals the magnetic flux through coil 2 ($\Phi_1 = \Phi_2$). But by Faraday's law, the voltage induced into each coil is proportional to the rate of change of the magnetic flux, i.e., how fast the magnetic flux is changing. Here is where an important realization occurs. How does the rate of change of magnetic flux in coil 1 compare with that in coil 2? *It must be the same.* This is because the magnetic flux linking both coils is the same, and if there is a fluctuation in that magnetic flux, then it must be the same for *both* coils.

The second important realization concerns the actual amount of voltage induced into each coil. Each turn will have a voltage induced into it by the changing magnetic flux. By KVL, the total voltage of one coil must be the sum of the voltages for each turn in that coil because the turns are in series. Figure 13.4 shows the schematic symbol of the ideal transformer. Hence the total voltage induced into each coil must be NV_{ind}, where V_{ind} is the voltage induced into *one* turn, N is the number of turns, and the turns are assumed to be identical. But the voltage induced into each turn is directly proportional ($\propto$) to the magnetic flux, Φ, again because the rate of change of flux is the same for both coils in an ideal transformer. Thus,

$$V_{ind} \propto \Phi \text{ for each turn,}$$

$$V_1 \propto N_1\Phi \text{ for coil 1, } \to V_1 = k N_1 \Phi, \text{ and}$$

$$V_2 \propto N_2 \Phi \text{ for coil 2} \to V_2 = k N_2 \Phi,$$

where k is a proportionality constant that is the same for both coils because the transformer core is the same for both coils (in a later course, the proportionality will be formally replaced with an equality using calculus).

In order to make this understanding of transformer operation useful for circuit analysis and predictions, take the ratio of the two coil voltages:

$$\frac{V_1}{V_2} = \frac{kN_1\Phi}{kN_2\Phi} = \frac{N_1}{N_2} \tag{13.1}$$

The proportionality constant and the flux are the same for both the numerator and denominator, and hence cancel in the ratio. Thus a very useful equation that relates the input/output voltage ratio to the coil turns ratio is obtained:

$$\frac{V_1}{V_2} = \frac{N_1}{N_2} \tag{13.2}$$

where only voltage magnitudes have been indicated. This will be generalized to phasors later in this section after the concepts have been firmly established.

The schematic symbol for an ideal transformer (or its practical equivalent, an iron-core transformer) and the *defined* voltage polarities and current directions are shown in Figure 13.5. The input side is called the *primary* and the output side is called the

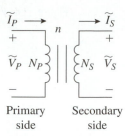

Figure 13.5 Ideal Transformer Schematic Symbol

secondary. The two parallel lines between the coils indicate a magnetic core. The lower-case *n* denotes the *turns ratio,* and it is the ratio of the number of primary turns to the number of secondary turns:

$$n \equiv \frac{N_P}{N_S} = \frac{V_P}{V_S} \tag{13.3}$$

Beware! There are some subtle implications in Equation (13.3):

■ The turns ratio is normally used to relate the primary and secondary voltage *magnitudes.* In an ideal transformer, *n* could be used to relate the voltage phasors, but this is only approximately true for a practical transformer with losses.

■ Some books and papers use *n* for V_S/V_P, instead of V_P/V_S, or use the symbol *a* instead of *n*.

■ Distinguish between V_S (or I_S) for the source voltage and for the voltage on the secondary side of the transformer. Similarly, I_P is the primary current, not necessarily a peak current (although the primary current may be expressed as either a peak or an RMS value). As before, you are responsible for knowing which variable is which, and for labeling them uniquely to prevent confusion.

If the transformer is a *step-down* transformer, $n > 1$. Which is higher, the primary or the secondary voltage? What is being stepped down? Conversely, if $n < 1$, it is a *step-up* transformer, and there is a fraction of a turn in the primary for every turn in the secondary (more turns in the secondary than in the primary). The voltage is being stepped up from the primary side to the secondary side of the transformer.

AC circuit techniques are used along with Equation (13.3) to analyze circuits that contain a transformer, as illustrated in the following example.

Example 13.2.1

Determine the voltage and current magnitudes in the secondary of the circuit in Figure 13.6.

Given:

$$V_P = V_{\text{in}} = 120.0 \ V_{\text{RMS}}$$
$$N_P = 2000$$
$$N_S = 100$$
$$Z_L = R_L = 1 \ \text{k}\Omega \ (\text{magnitude of the load impedance})$$

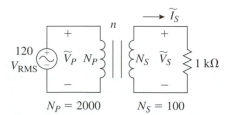

$N_P = 2000$ $N_S = 100$

Figure 13.6 Transformer Circuit for Example 13.2.1

Desired:

$$V_S$$
$$I_S$$

Strategy:

$$n = \frac{N_P}{N_S}$$

$$V_S = \frac{V_P}{n}$$

$$I_S = \frac{V_S}{Z_L}$$

Solution: _____

$$n = \frac{N_P}{N_S} = \frac{2000}{100} = 20$$

$$V_S = \frac{V_P}{n} = \frac{120.0}{20} = 6.000 = 6.00\ V_{\text{RMS}}$$

$$I_S = \frac{V_S}{Z_L} = \frac{6.000}{1\ k\Omega} = 6.00\ mA_{\text{RMS}}$$

The primary voltage of 120 V was transformed to 6 V in the secondary.

_____

Can the secondary current be determined without knowing the load impedance? Yes, if the primary current is known. In fact there is a relationship between the primary and secondary currents similar to that of the voltages. The key is in recognizing that there is no real power loss in an ideal transformer:

$$P_{\text{in}} = P_{\text{out}} \tag{13.4}$$

$$V_P I_P = V_S I_S \tag{13.5}$$

$$n = \frac{V_P}{V_S} = \frac{\dfrac{P_{\text{in}}}{I_P}}{\dfrac{P_{\text{out}}}{I_S}} = \frac{I_S}{I_P} \tag{13.6}$$

Thus the turns ratio n establishes both the primary to secondary voltage ratio and current ratio in an ideal transformer.

As mentioned earlier, the ideal transformer relationships can be generalized to include phasors, but only for the ideal transformer case. The ideal transformer does *not* change the phase relationship of a signal; hence the turns-ratio relationship of Equation (13.3) is generalized to:

$$n = \frac{\tilde{V}_P}{\tilde{V}_S} = \frac{\tilde{I}_S}{\tilde{I}_P} \tag{13.7}$$

Thus the variables in Figure 13.5 have tildes over them. Furthermore, because the phase relationship is maintained, the complex power must be the same on both sides of the ideal transformer; hence the power relationship of Equations (13.4) and (13.5) are generalized to:

$$\tilde{S}_P = \tilde{S}_S \tag{13.8}$$

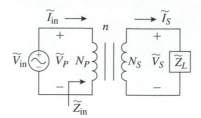

Figure 13.7 Transformer Circuit Impedances

$$\boxed{\tilde{V}_P \tilde{I}_P^* = \tilde{V}_S \tilde{I}_S^*} \tag{13.9}$$

where RMS phasors are used in Equation (13.9.)

If both the voltage and current relationships between the primary and the secondary are known, then can a corresponding impedance relationship be found? Determine the input impedance, as designated in Figure 13.7:

$$\tilde{Z}_{\text{in}} = \frac{\tilde{V}_{\text{in}}}{\tilde{I}_{\text{in}}} = \frac{\tilde{V}_P}{\tilde{I}_P} = \frac{n\tilde{V}_S}{\dfrac{\tilde{I}_S}{n}} = n^2 \frac{\tilde{V}_S}{\tilde{I}_S} \tag{13.10}$$

But $\tilde{V}_S = \tilde{V}_L$ and $\tilde{I}_S = \tilde{I}_L$; thus, $\tilde{Z}_{\text{in}} = n^2 \dfrac{\tilde{V}_L}{\tilde{I}_L} = n^2 \tilde{Z}_L$

$$\boxed{\tilde{Z}_{\text{in}} = n^2 \tilde{Z}_L} \tag{13.11}$$

Thus, the transformer also transforms impedance levels. This is a tremendously useful input–output relationship for ideal transformers, as illustrated in the following example.

Example 13.2.2

For the transformer circuit shown in Figure 13.8, determine (a) the primary current, (b) the secondary voltage and current, (c) the real and complex input powers, and (d) the input impedance.

Given:

$$\tilde{V}_P = 120\angle 0° \ V_{\text{RMS}}$$
$$N_P = 1000$$
$$N_S = 60$$
$$\tilde{Z}_L = 5 - j13 \ \Omega$$

Desired:

$$\tilde{I}_P, \ \tilde{V}_S, \ \tilde{I}_S, \ \tilde{Z}_{\text{in}}, \ \tilde{S}_{\text{in}}, \ P_{\text{in}}$$

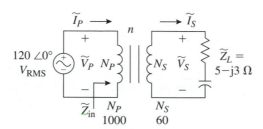

Figure 13.8 Circuit for Example 13.2.2

Strategy:

$$n = \frac{N_P}{N_S}$$

$$\tilde{Z}_{in} = n^2\tilde{Z}_L$$

$$\tilde{I}_P = \frac{\tilde{V}_P}{\tilde{Z}_{in}}$$

$$\tilde{V}_S = \frac{\tilde{V}_P}{n}$$

$$\tilde{I}_S = \frac{\tilde{V}_S}{\tilde{Z}_L}$$

$$\tilde{S}_{in} = \tilde{V}_P\tilde{I}_P^*$$

$$P_{in} = Re(\tilde{S}_{in})$$

Solution: _____

$$n = \frac{N_P}{N_S} = \frac{1000}{60} = 16.667$$

$$\tilde{Z}_{in} = n^2\tilde{Z}_L = 277.778(5 - j3) = 1619.7\angle -30.964° \ \Omega$$

$$= 1388.9 - j833.33 = 1620\angle -31.0° \ \Omega = 1390 - j833 \ \Omega$$

$$\tilde{I}_P = \frac{\tilde{V}_P}{\tilde{Z}_{in}} = \frac{120\angle 0°}{1619.7\angle -30.964°} = 74.087\angle +30.964° \text{mA}_{RMS} = 74.1\angle +31.0° \text{ mA}_{RMS}$$

$$\tilde{V}_S = \frac{\tilde{V}_P}{n} = \frac{120\angle 0°}{16.667} = 7.2000\angle 0° \ V_{RMS} = 7.20\angle 0.0° \ V_{RMS}$$

$$\tilde{I}_S = \frac{\tilde{V}_S}{\tilde{Z}_L} = \frac{7.200}{5 - j3} = 1.2348\angle +30.964° \ A_{RMS} = 1.23\angle +31.0° \ A_{RMS}$$

$$\tilde{S}_{in} = \tilde{V}_P\tilde{I}_P^* = (120\angle 0°)(74.087 \times 10^{-3}\angle +30.964°)^*$$

$$= 8.8905\angle -30.964° \ VA = 8.89\angle -31.0° \ VA$$

$$P_{in} = Re(\tilde{S}) = 7.6235 \ W = 7.62 \ W$$

$$\text{check: } n = \frac{\tilde{I}_S}{\tilde{I}_P} = \frac{1.2348\angle +30.964°}{0.074087\angle +30.964°} = 16.667 \quad ok$$

The voltage has been transformed from 120 V to 7.2 V, and the current has been transformed from 74.1 mA to 1.23 A, both from the primary side to the secondary side of the transformer. Furthermore, the load impedance of $5 - j3 \ \Omega$ has been transformed to $1390 - j833 \ \Omega$ on the primary side. The transformer has transformed the voltage, current, and impedance levels in the circuit.

_____

The ideal transformer had important underlying assumptions: all of the magnetic flux linked both coils and the losses in the magnetic core and primary and secondary conductors were zero. A practical transformer violates these ideal assumptions, but generally the assumptions are nearly true. The flux linkage is generally only a few percent from ideal. In a simple model of a practical transformer (there are more complicated models too), the losses can be accounted for by adding resistances to the ideal transformer model, as shown in Figure 13.9. These additional resistances are *inside* the transformer terminals. The ideal transformer voltage–current and impedance relationships are valid only for the ideal coils.

One must perform additional calculations to determine the voltages at the transformer terminals, as illustrated in Example 13.2.3.

Example 13.2.3

Consider the transformer circuit shown in Figure 13.8, but with nonzero conductor losses, as modeled in Figure 13.9. Determine (*a*) the primary current, (*b*) the secondary voltage and current, (*c*) the real and complex input powers, and (*d*) the input impedance. Assume at least three significant digits for all numbers.

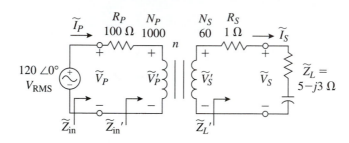

Figure 13.9 Circuit for Example 13.2.3

Given:

$$\tilde{V}_P = 120\angle 0° \ V_{RMS}$$
$$N_P = 1000$$
$$N_S = 60$$
$$R_P = 100 \ \Omega$$
$$R_S = 1 \ \Omega$$
$$\tilde{Z}_L = 5 - j3 \ \Omega$$

Desired:

$$\tilde{I}_P, \ \tilde{V}_S, \ \tilde{I}_S, \ \tilde{Z}_{in}, \ \tilde{S}_{in}, \ P_{in}$$

Strategy:

$$n = \frac{N_P}{N_S}$$
$$\tilde{Z}'_L = R_S + \tilde{Z}_L$$
$$\tilde{Z}'_{in} = n^2 \ \tilde{Z}'_L$$
$$\tilde{Z}_{in} = R_P + \tilde{Z}'_{in}$$
$$\tilde{I}_P = \frac{\tilde{V}_P}{\tilde{Z}_{in}}$$

Note that $\tilde{V}_S \neq \dfrac{\tilde{V}_P}{n}$; instead, $\tilde{V}'_S = \dfrac{\tilde{V}'_P}{n}$, but neither primed quantity is known. Instead, try the current turns-ratio relationship:

$$\tilde{I}_S = n\tilde{I}_P$$
$$\tilde{V}_S = \tilde{I}_S\tilde{Z}_L$$
$$\tilde{S}_{in} = \tilde{V}_P\tilde{I}_P^*$$
$$P_{in} = \text{Re}(\tilde{S}_{in})$$

Solution: _____

$$n = \frac{N_P}{N_S} = \frac{1000}{60} = 16.667$$

$$\tilde{Z}'_L = R_S + \tilde{Z}'_L = 1 + (5 - j3) = 6 - j3 \ \Omega$$

$$\tilde{Z}'_{in} = n^2 \tilde{Z}'_L = (16.667)^2 (6 - j3) = 1863.5 \angle -26.565° \ \Omega$$

$$\tilde{Z}_{in} = R_P + \tilde{Z}'_{in} = 100 + 1863.5 \angle -26.565°$$
$$= 1953.4 \angle -25.253° \ \Omega = 1950 \angle -25.3° \ \Omega$$

$$\tilde{I}_P = \frac{\tilde{V}_P}{\tilde{Z}_{in}} = \frac{120 \angle 0°}{1953.4 \angle -25.253°}$$
$$= 61.431 \ mA \angle +25.253° = 61.4 \angle +25.3° \ mA_{rms}$$

$$\tilde{I}_S = n \tilde{I}_P = (16.667)(61.431 \ mA \angle +25.253°)$$
$$= 1.0239 \angle +25.253° = 1.02 \angle +25.3° \ A_{RMS}$$

$$\tilde{V}_S = \tilde{I}_S \tilde{Z}_L = (1.0239 \angle +25.253°)(5 - j3)$$
$$= 5.9701 \angle -5.711° \ V = 5.97 \angle -5.7° V$$

$$\tilde{S}_{in} = \tilde{V}_P \tilde{I}_P^* = (120 \angle 0°)(61.431 \ mA \angle +25.253°)^*$$
$$= 7.3717 \angle -25.253° \ VA = 7.37 \angle -25.3° \ VA$$

$$P_{in} = Re(\tilde{S}) = 6.6672 \ W = 6.67 \ W$$

check: KVL: $\tilde{V}'_P = \tilde{V}_P - \tilde{I}_P R_P = 120 \angle 0° - (61.431 \ mA \angle +25.253°) (100)$
$$= 114.47 \angle -1.312°V$$

KVL: $\tilde{V}'_S = \tilde{V}_S + \tilde{I}_S R_S = 5.9701 \angle -5.711° + (1.0239 \angle +25.253°)(1)$
$$= 6.8683 \angle -1.312° \ V$$

$$n = \frac{\tilde{V}'_P}{\tilde{V}'_S} = \frac{114.47 \angle -1.312°}{6.8683 \angle -1.312°} = 16.666 \cong 16.7 \ ok$$

In a transformer the wire of each coil has two ends. How does one know which end is positive for a given known polarity on the other coil, especially on a schematic? The *dot convention* has been established to address this issue. This convention (an agreement on how notation will be used) is the same for both the ideal transformer and the mutual inductor, and it provides a way to determine voltage polarities in these magnetically coupled components. See Figure 13.10. The convention is as follows. If the *current* is entering the dot from the external circuit on one side of the transformer, the *voltage polarity* is positive at the dot on the other side of the transformer. This is especially useful in circuit analysis with more than one transformer in the circuit. Note that the defined direction of $\tilde{I}_2$ is *reversed* relative to that used previously for $\tilde{I}_S$. In the previous equations, the quantity $(-\tilde{I}_2)$ must be used if the direction for $\tilde{I}_2$ is utilized.

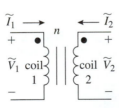

Figure 13.10 The Dot Convention

Information Research Exercise 13.2.1   (library and/or web)

Investigate the following practical transformers and their uses:

a. power
b. audio
c. isolation
d. autotransformer

13.3 GENERALIZATION OF THE TRANSFORMER: THE MUTUAL INDUCTOR

The ideal transformer has important underlying assumptions: all of the magnetic flux links both coils, and the losses in the magnetic core and the primary and secondary conductors are zero. A mutual inductor violates the first assumption significantly. A considerable percentage of the magnetic flux does not link both coils. How could this happen? Consider a core that is nonmagnetic, perhaps a plastic mold. Then there is no strong magnetic material to guide the magnetic flux, and only a fraction of the total magnetic flux will link the second coil. Why would anyone ever use this arrangement? At high frequencies, such as in the upper kilohertz and the megahertz ranges, the magnetic field losses in iron and steel become prohibitive. Ferrites, which are iron-oxide compounds, have lower losses at higher frequencies than iron cores, but the losses are still significant. Hence, to reduce core losses at high frequencies, nonmagnetic cores are often used. Sometimes, no core is desired, such as when the primary and secondary coils are on items that move relative to each other. Magnetic coupling occurs through the air in these cases. Hence, the mutual inductor is an important case of magnetic coupling, and it will be covered in this section. A picture of a few mutual inductors is shown in Figure 13.11.

The discussion in this section is divided into the following parts:

■ Development of the voltage–current relationships for the mutual inductor

■ Analysis of circuits that contain a mutual inductor

■ Relationship between the ideal transformer and the mutual inductor

13.3.1 Development of the voltage–current relationships for the mutual inductor

The purpose of this section is to develop the voltage–current relationships for the mutual inductor so that they can be used in the analysis of circuits that contain a mutual inductor. The mutual inductor is the general case of two coupled coils. The ideal transformer, with complete coupling of the magnetic flux from one coil to the other, and vice versa, is a special case of the mutual inductor. As with the ideal transformer, the mutual inductor has an "input" and an "output." In order to develop the voltage–current relationships for the mutual inductor, consider the two coupled coils in Figure 13.12. The coils are close enough so that part of the flux of one coil links to the other coil, and vice versa. The *M* stands for *mutual inductance,* and it is explained later in the context of the mutual inductor operation. The explanation that follows is separated into parts that may appear sequential, but all these actions are occurring simultaneously—keep this in mind as the discussion progresses.

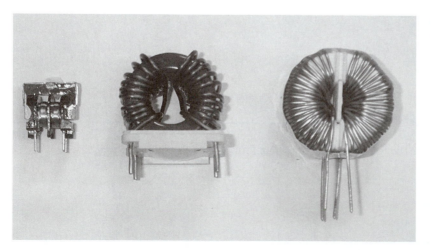

Figure 13.11 A Few Mutual Inductors
Photo courtesy of J. W. Miller

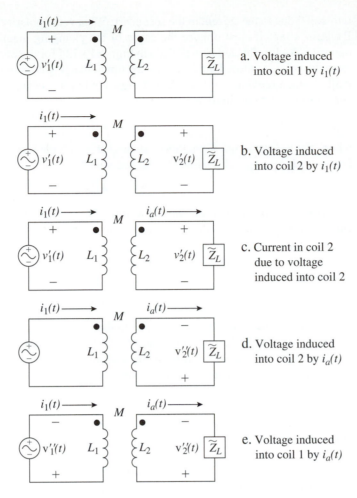

Figure 13.12 Mutual Inductor Operation

a. Voltage induced into coil 1 by $i_1(t)$

b. Voltage induced into coil 2 by $i_1(t)$

c. Current in coil 2 due to voltage induced into coil 2

d. Voltage induced into coil 2 by $i_a(t)$

e. Voltage induced into coil 1 by $i_a(t)$

Consider the voltage of the AC source in Figure 13.12a during the first quarter of the sinusoidal cycle. The voltage is increasing. Hence, $v_1'(t)$ is increasing (lowercase variables are also being used because this is a time-domain explanation). The primary current, $i_1(t)$, is also increasing, but it lags $v_1'(t)$ due to the inductance of coil 1. It would lag by 90° if only coil 1 were present. However, some of the magnetic flux of coil 1 links to coil 2, and a voltage is induced into coil 2 due to the changing magnetic field of coil 1, as shown in Figure 13.12b. Another inductance can be used to predict this second action. The phasor voltage–current relationship for coil 1 by itself is:

$$\tilde{V}_1' = +j\omega L_1 \tilde{I}_1 \tag{13.12}$$

A corresponding phasor relationship can be stated for the voltage in coil 2:

$$\tilde{V}_2' = +j\omega M \tilde{I}_1 \tag{13.13}$$

In other words the time-changing current in coil 1 creates a time-changing magnetic field that induces a voltage into coil 2 (by Faraday's law) with the dot side positive per the dot convention, as shown in Figure 13.12b. The inductance, M, is not the inductance of coil 1 or 2 but is the inductance between the two coils that results in the coupling between them. Hence it is called the *mutual inductance* because it is mutual, i.e., a property that depends on both coils (and the spacing and orientation between them).

The coupling of one coil to another through the magnetic field is now established and the voltage–current relationship involves mutual inductance. However, what if there is a load attached to the secondary, as per Figure 13.12c? The voltage induced into coil 2 will cause a current to flow on the secondary side. This current, $i_a(t)$, which is in a direction opposite the defined positive direction of $i_2(t)$ (per the dot convention—see Figure 13.10), will flow through coil 2. But coil 2 has inductance, also. Hence a voltage will be induced into coil 2 due to the current in the secondary. What is the polarity of this induced voltage? By Lenz's law, it must oppose the cause of the changing magnetic field; hence it will oppose $v_2'(t)$. It is labeled here as $v_2''(t)$ in Figure 13.12d for the purpose of distinguishing it from the voltage induced into coil 2 by the time-changing current in coil 1 [$v_2'(t)$], but there really is not a *separate* voltage in coil 2. Again, all of this occurs simultaneously. The voltage–current relationship for this action is:

$$\tilde{V}_2'' = -j\omega L_2 \tilde{I}_a \tag{13.14}$$

where the negative sign is included because this voltage has a polarity opposite to $\tilde{V}_2'$. However, $\tilde{I}_2 = -\tilde{I}_a$, so Equation (13.14) becomes:

$$\tilde{V}_2'' = +j\omega L_2 \tilde{I}_2 \tag{13.15}$$

This is the point where the mutual nature of two coupled coils becomes apparent: the current in the secondary induces a voltage into the primary. Again, the mutual inductance expresses the relationship between the current in the secondary, the changing magnetic field that it creates, and the voltage induced into the primary from this time-changing magnetic field:

$$\tilde{V}_1'' = -j\omega M \tilde{I}_a \tag{13.16}$$

The voltage induced into the primary due to the secondary current $i_a(t)$ is labeled $v_1''(t)$ in Figure 13.12e. It has the opposite polarity relative to $v_1'(t)$, by the dot convention because $i_a(t)$ leaves, not enters, the dot on the secondary side. Thus, $v_1''(t)$ will be negative. However, if $\tilde{I}_2 = -\tilde{I}_a$ is inserted into Equation (13.16), then:

$$\tilde{V}_1'' = +j\omega M \tilde{I}_2 \tag{13.17}$$

We are coming "full circle" in the mutual inductor voltage–current relationship development. The primary affects the secondary, and the secondary affects the primary. Let us put it all together on each side. On the primary side, the voltage $v_1(t)$ is made up of 1) the voltage due to the inductance of coil 1 [Equation (13.12)] and 2) the voltage induced into coil 1 due to the current in coil 2 [Equation (13.17)]:

$$\tilde{V}_1 = +\tilde{V}_1' + \tilde{V}_1'' \tag{13.18}$$

$$\tilde{V}_1 = +j\omega L_1 \tilde{I}_1 + j\omega M \tilde{I}_2 \tag{13.19}$$

Similarly, on the secondary side, the voltage $v_2(t)$ is made up of 1) the voltage due to the inductance of coil 2 [Equation (13.15)] and 2) the voltage induced into coil 2 due to the time-changing current in coil 1 [Equation (13.13)]:

$$\tilde{V}_2 = +\tilde{V}_2' + \tilde{V}_2'' \tag{13.20}$$

$$\tilde{V}_2 = +j\omega M \tilde{I}_1 + j\omega L_2 \tilde{I}_2 \tag{13.21}$$

Hence, the fundamental voltage–current relationships for the mutual inductor are:

$$\boxed{\begin{aligned} \tilde{V}_1 &= +j\omega L_1 \tilde{I}_1 + j\omega M \tilde{I}_2 \\ \tilde{V}_2 &= +j\omega M \tilde{I}_1 + j\omega L_2 \tilde{I}_2 \end{aligned}} \tag{13.22}$$

Notice that the voltage–current relationships for the mutual inductor are more involved than those for the ideal transformer because voltages and currents from both the primary and the secondary sides appear in both equations.

As in the ideal transformer, the impedance levels change from one side of the mutual inductor to the other. The variables are labeled on the mutual inductor in Figure 13.13. The development of the input impedance relationship follows. The reader should label the reason for each step.

$$\tilde{V}_1 = +j\omega L_1 \tilde{I}_1 + j\omega M \tilde{I}_2 \tag{13.23}$$

$$\tilde{V}_2 = +j\omega M \tilde{I}_1 + j\omega L_2 \tilde{I}_2 \tag{13.24}$$

Figure 13.13 Mutual Inductor with Variables Labeled

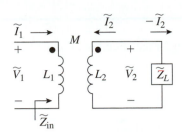

$$\tilde{Z}_L = \frac{\tilde{V}_2}{-\tilde{I}_2} \tag{13.25}$$

$$-\tilde{Z}_L\tilde{I}_2 = +j\omega M\tilde{I}_1 + j\omega L_2\tilde{I}_2 \tag{13.26}$$

$$-\tilde{I}_2(+j\omega L_2 + \tilde{Z}_L) = +j\omega M\tilde{I}_1 \tag{13.27}$$

$$\tilde{I}_2 = \frac{-j\omega M\tilde{I}_1}{+j\omega L_2 + \tilde{Z}_L} \tag{13.28}$$

$$\tilde{V}_1 = +j\omega L_1\tilde{I}_1 + j\omega M\frac{-j\omega M\tilde{I}_1}{+j\omega L_2 + \tilde{Z}_L} = +j\omega L_1\tilde{I}_1 + \frac{(\omega M)^2\tilde{I}_1}{+j\omega L_2 + \tilde{Z}_L} \tag{13.29}$$

$$\boxed{\tilde{Z}_{in} = \frac{\tilde{V}_1}{\tilde{I}_1} = +j\omega L_1 + \frac{(\omega M)^2}{+j\omega L_2 + \tilde{Z}_L}} \tag{13.30}$$

Thus, the input impedance of a mutual inductor depends on the coil inductances, the mutual inductance, the load impedance, and the frequency. Note that the load impedance is in the denominator of Equation (13.30). Sometimes the mutual inductor is referred to as an "impedance inverter," especially when the magnitude of the load impedance is much greater than the reactance of coil 1.

The voltage–current and the impedance relationships have been developed and are ready for use to analyze circuits that contain a mutual inductor. Before proceeding with the circuit analysis, one other variable is defined in the context of mutual inductors. It is called the *coupling coefficient* and is given the symbol k:

$$\boxed{k \equiv \frac{M}{\sqrt{L_1L_2}}} \tag{13.31}$$

If the coupling coefficient is 1, all the magnetic flux from one coil links into the other and vice versa. If none of the flux from one coil links into the other and vice versa, then the coupling coefficient is zero. If $0 < k < 1$, then only part of the magnetic flux from one coil links into the other and vice versa. This relationship will also be used in relating the ideal transformer and mutual inductor later in this section.

13.3.2 Analysis of circuits that contain a mutual inductor

The purpose of this section is to investigate how the voltage–current relationships for the mutual inductor are used in circuit analysis. Notice that there are two equations in four unknowns in Equation (13.22). Knowledge of the source voltage or current, either directly or indirectly, provides information on one of the unknowns. The load provides another piece of information because it relates a voltage to a current on the secondary side of the mutual inductor. Hence, that leaves two equations in two unknowns—a solvable situation. Another approach is to use the input impedance relation for a mutual inductor [Equation (13.30)]. These generic strategies are illustrated in the following example.

—— **Example 13.3.1** _____

Determine (*a*) $v_2(t)$ and (*b*) $i_1(t)$ for the circuit in Figure 13.14.

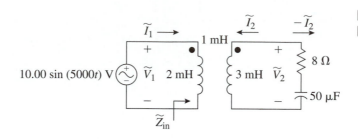

Figure 13.14 Circuit for Example 13.3.1

Given:

$$L_1 = 2.00 \text{ mH}$$
$$L_2 = 3.00 \text{ mH}$$
$$M = 1.00 \text{ mH}$$
$$\tilde{V}_S = 10.00 \angle 0° \text{ V}_{\text{peak}}$$
$$\omega = 5.000 \text{ krad/s}$$
$$\text{load: } R = 8.00 \text{ } \Omega$$
$$C = 50.0 \text{ } \mu F$$

Desired:

a. $v_2(t)$
b. $i_1(t)$

Strategy:

a. Start with the mutual inductor *i*–*v* relationships.
   Solve the second equation for $\tilde{I}_1$.
   Substitute into the first equation to eliminate $\tilde{I}_1$.
   Use Ohm's law for the load to eliminate $\tilde{I}_2$.
   Solve for $\tilde{V}_2$, insert numbers, and convert to time domain.
b. One could use a strategy similar to (*a*), or the following:
   Determine $\tilde{Z}_{\text{in}}$.
   $\tilde{I}_1 = \dfrac{\tilde{V}_S}{\tilde{Z}_{\text{in}}}$; convert back to the time domain.

Solution: _____

a. Start with the second mutual inductor equation of Equation (13.22) and solve for $\tilde{I}_1$:

$$\tilde{V}_2 = +j\omega M\tilde{I}_1 + j\omega L_2\tilde{I}_2$$
$$+j\omega M\tilde{I}_1 = +\tilde{V}_2 - j\omega L_2\tilde{I}_2$$
$$\tilde{I}_1 = \frac{+\tilde{V}_2 - j\omega L_2\tilde{I}_2}{+j\omega M}$$

Then, use the first equation of Equation (13.22):

$$\tilde{V}_1 = +j\omega L_1 \tilde{I}_1 + j\omega M \tilde{I}_2$$

and substitute for $\tilde{I}_1$:

$$\tilde{V}_1 = +j\omega L_1 \left(\frac{+\tilde{V}_2 - j\omega L_2 \tilde{I}_2}{+j\omega M} \right) + j\omega M \tilde{I}_2$$

Eliminate $\tilde{I}_2$ by substituting Ohm's law for the load:

$$\tilde{I}_2 = -\frac{\tilde{V}_2}{\tilde{Z}_L}$$

$$\tilde{V}_1 = +j\omega L_1 \left(\frac{+\tilde{V}_2 - j\omega L_2 \left(\dfrac{\tilde{V}_2}{-Z_L} \right)}{+j\omega M} \right) + j\omega M \left(\frac{\tilde{V}_2}{-\tilde{Z}_L} \right)$$

Solve for $\tilde{V}_2$:

$$\tilde{V}_1 = +\tilde{V}_2 \left[+j\omega L_1 \left(\frac{1 + j\omega L_2 \dfrac{1}{\tilde{Z}_L}}{+j\omega M} \right) - j\omega M \frac{1}{\tilde{Z}_L} \right]$$

$$\tilde{V}_1 = +\tilde{V}_2 \left[\frac{+j\omega L_1}{+j\omega M} + \frac{+j\omega L_1\, j\omega L_2}{+j\omega M \tilde{Z}_L} - \frac{j\omega M}{\tilde{Z}_L} \right]$$

$$= +\tilde{V}_2 \left[+\frac{L_1}{M} + j\omega \left(\frac{L_1 L_2}{+M\tilde{Z}_L} - \frac{M}{\tilde{Z}_L} \right) \right]$$

$$= +\tilde{V}_2 \left[+\frac{L_1}{M} + j\omega \left(\frac{L_1 L_2 - M^2}{+M\tilde{Z}_L} \right) \right]$$

$$\tilde{V}_2 = \frac{\tilde{V}_1}{+\dfrac{L_1}{M} + j\omega \left(\dfrac{L_1 L_2 - M^2}{+M\tilde{Z}_L} \right)}$$

$$\tilde{Z}_L = R - jX = 8 - j\frac{1}{5000\,(50\mu F)} = 8 - j4\ \Omega$$

$$\tilde{V}_2 = \frac{10\angle 0°}{+\dfrac{2\ \text{mH}}{1\ \text{mH}} + j5000 \left(\dfrac{2\ \text{mH} \cdot 3\ \text{mH} - (1\ \text{mH})^2}{1\ \text{mH}\,(8 - j4)} \right)} = 3.8313\angle -73.301°\ V_{peak}$$

$$v_2(t) = 3.83 \sin(5000t - 73.3°)\text{V}$$

b. $$\tilde{Z}_m = +j\omega L_1 + \frac{(\omega M)^2}{+j\omega L_2 + \tilde{Z}_L}$$

$$\tilde{Z}_{in} = +j5000\,(2\ \text{mH}) + \frac{[5000\,(1\ \text{mH})]^2}{+j5000\,(3\ \text{mH}) + (8 - j4)}$$

$$= 8.5819\angle +82.763°\ \Omega = 1.0811 + j8.5135\ \Omega$$

$$\tilde{I}_1 = \frac{\tilde{V}_S}{\tilde{Z}_{in}} = \frac{10\angle 0°}{8.5819\angle +82.763°} = 1.1653\angle -82.763°\ A_{peak}$$

$$i_1(t) = 1.17 \sin(5000t - 82.8°)\ \text{A}$$

Note the change in the impedance level: $8 - j4\ \Omega$ at the load to $1.1 + j8.5\ \Omega$ at the input. Clearly, the mutual inductor does not scale the impedance levels (versus the ideal transformer).

Thus both strategies for circuit analysis with mutual inductors are useful. As is evident from Example 13.3.1, an algebraic strategy is required to obtain a path to the solution. It is

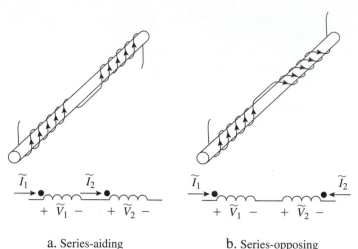

Figure 13.15 Mutual Inductors Connected in Series

a. Series-aiding

b. Series-opposing

alright to hit a dead end. Just back up in your process and keep in mind which variables are needed and which variables can be eliminated. Then manipulate your algebraic strategy toward these ends.

Another application of mutual inductors involves connecting the coils in series, as shown in Figure 13.15. There are two ways to connect the coils: such that the magnetic fields support each other, as shown in Figure 13.15a, and such that they oppose each other, as shown in Figure 13.15b. Note carefully the relative arrangement of the coils using the dot convention. In the former case, the magnetic field from one coil adds to that of the second coil, and the same for the second coil to the first coil. This arrangement is called *series-aiding*. In the latter case, the magnetic field from each coil opposes that of the other coil. This arrangement is called *series-opposing*. A short development of the total inductance of two series-aiding coils follows.

$$\tilde{V}_T = \tilde{V}_1 + \tilde{V}_2$$

$$= +j\omega L_1 \tilde{I}_1 + j\omega M \tilde{I}_2 + j\omega M \tilde{I}_1 + j\omega L_2 \tilde{I}_2 \tag{13.32}$$

But $\tilde{I}_1 = \tilde{I}_2$; hence:

$$\tilde{V}_T = +j\omega L_1 \tilde{I}_1 + j\omega M \tilde{I}_1 + j\omega M \tilde{I}_1 + j\omega L_2 \tilde{I}_1$$

$$= +j\omega \tilde{I}_1 (L_1 + M + M + L_2) \tag{13.33}$$

$$= +j\omega \tilde{I}_1 (L_T)$$

The total equivalent inductance of two coils in series aiding is:

$$\boxed{L_T = L_1 + L_2 + 2M} \tag{13.34}$$

The aiding nature of the mutual inductance is indicated in Equation (13.34). The factor of two comes from each coil aiding the other to develop twice the mutual inductance effect. One can see that an adjustable inductance could be constructed in this manner, or that two fixed inductors could be physically arranged to obtain a nonstandard inductance value. Another ramification is that if inductors are too close they couple to each other, and inductances may change from the desired values. Thus one must position inductors to prevent coupling if this scenario occurs.

If one of the coil connections is reversed, as shown in Figure 13.15b, then the mutual inductance will subtract. The reader should develop Equation (13.35) using $\tilde{I}_1 = -\tilde{I}_2$. For two mutual inductors in series-opposing:

$$\boxed{L_T = L_1 + L_2 - 2M} \tag{13.35}$$

13.3.3 Relationship between the ideal transformer and the mutual inductor

The purpose of this section is to establish the conditions when the ideal transformer model may be used in place of the mutual inductor model. This task will be accomplished through the derivation of the ideal transformer voltage–current relationships from the mutual inductor voltage–current relationships. Start with the mutual inductor relationships:

$$\tilde{V}_1 = j\omega L_1 \tilde{I}_1 + j\omega M \tilde{I}_2$$
$$\tilde{V}_2 = j\omega M \tilde{I}_1 + j\omega L_2 \tilde{I}_2 \tag{13.36}$$

Multiply the first equation in Equation (13.36) by L_2, multiply the second equation by M, and subtract the second equation from the first equation:

$$\tilde{V}_1 L_2 = j\omega L_1 L_2 \tilde{I}_1 + j\omega M L_2 \tilde{I}_2$$
$$\underline{-[V_2 M = j\omega M^2 \tilde{I}_1 + j\omega M L_2 \tilde{I}_2]}$$
$$\tilde{V}_1 L_2 - \tilde{V}_2 M = j\omega(L_1 L_2 - M^2)\tilde{I}_1 \tag{13.37}$$

In an ideal transformer, all the magnetic flux links both coils; hence, k, the coupling coefficient is unity:

$$M^2 = k^2 L_1 L_2 = L_2 L_2 \tag{13.38}$$

$$M^2 - L_1 L_2 = 0 \tag{13.39}$$

$$M = \sqrt{L_1 L_2} \tag{13.40}$$

Insert Equation (13.39) into the right side of Equation (13.37):

$$\tilde{V}_1 L_2 - \tilde{V}_2 M = 0 \tag{13.41}$$

Insert Equation (13.40) into Equation (13.41) and solve for V_1/V_2:

$$\tilde{V}_1 L_2 - \tilde{V}_2 \sqrt{L_1 L_2} = 0 \tag{13.42}$$

$$\frac{\tilde{V}_1}{\tilde{V}_2} = \frac{\sqrt{L_1 L_2}}{L_2} = \sqrt{\frac{L_1}{L_2}} = n \tag{13.43}$$

where n is the turns ratio for the ideal transformer: N_1/N_2. One can show that N_1/N_2 results if the inductance of the coils are inserted into Equation (13.43). Recall from Chapter 6 that the inductance of a coil is:

$$L = \frac{\mu N^2 A}{\ell} \tag{13.44}$$

Substitute Equation (13.44) into Equation (13.43) for each inductance:

$$\frac{\tilde{V}_1}{\tilde{V}_2} = \sqrt{\frac{L_1}{L_2}} = \sqrt{\frac{\dfrac{\mu_1 N_1^2 A_1}{\ell_1}}{\dfrac{\mu_2 N_2^2 A_2}{\ell_2}}} \tag{13.45}$$

The two coils are wound on the same core. Hence, the permeability μ, cross-sectional area A, and length ℓ of the core are the same for both coils. Only the number of turns are different:

$$\frac{\tilde{V}_1}{\tilde{V}_2} = \sqrt{\frac{L_1}{L_2}} = \sqrt{\frac{N_1^2}{N_2^2}} = \frac{N_1}{N_2} = n \tag{13.46}$$

Thus, the assumption of complete magnetic flux linkage (unity coupling coefficient) results in the well-known turns-ratio relationship for the ideal transformer.

One can derive the ideal transformer input impedance from the mutual inductor input impedance:

$$\tilde{Z}_{\text{in}} = j\omega L_1 + \frac{(\omega M)^2}{\tilde{Z}_L + j\omega L_2}$$

$$= \frac{j\omega L_1(\tilde{Z}_L + j\omega L_2) + (\omega M)^2}{\tilde{Z}_L + j\omega L_2} \tag{13.47}$$

$$= \frac{j\omega L_1 \tilde{Z}_L + \omega^2(M^2 - L_1 L_2)}{\tilde{Z}_L + j\omega L_2}$$

Use Equation (13.39) for the $k = 1$ (complete magnetic flux coupling) assumption:

$$\tilde{Z}_{\text{in}} = \frac{j\omega L_1 \tilde{Z}_L}{\tilde{Z}_L + j\omega L_2} \tag{13.48}$$

If the coil reactances ωL are significantly greater than any circuit impedances, such as Z_L here, then $\omega L \gg Z_L$ and Equation (13.48) simplifies to:

$$\tilde{Z}_{\text{in}} = \frac{j\omega L_1 \tilde{Z}_L}{j\omega L_2} = \frac{L_1 \tilde{Z}_L}{L_2} = n^2 \tilde{Z}_L \tag{13.49}$$

Thus, the ideal transformer impedance transformation results when the following assumptions are made on a mutual inductor:

■ *There is complete flux linkage between the coils.*

■ *Coil reactances are much greater than other circuit impedances.*

When these assumptions are valid, the ideal transformer relationships can be used instead of the mutual inductor relationships. Practically, these assumptions are essentially met in iron-core transformers at low frequencies; hence the use of the ideal transformer model for iron-core transformers is valid, and the circuit performance predications are accurate compared to measurements.

The ideal transformer current transformation relationship follows directly from the voltage and impedance transformation relations:

$$\tilde{Z}_{\text{in}} = n^2 \tilde{Z}_L = n^2 \frac{\tilde{V}_L}{\tilde{I}_L} = n^2 \frac{\tilde{V}_2}{-\tilde{I}_2} = n \frac{\tilde{V}_1}{-\tilde{I}_2} \tag{13.50}$$

$$\tilde{Z}_{\text{in}} = \frac{\tilde{V}_1}{\tilde{I}_1} = n \frac{\tilde{V}_1}{-\tilde{I}_2} \tag{13.51}$$

$$\frac{1}{\tilde{I}_1} = n \frac{1}{-\tilde{I}_2} \tag{13.52}$$

$$n = \frac{-\tilde{I}_2}{\tilde{I}_1} \tag{13.53}$$

CHAPTER REVIEW

13.1 Another Look at Faraday's Law: Two Coils
 • A voltage is induced into a coil from the changing magnetic field of another coil that is driven by an AC source. Faraday's law is the underlying reason.
 • There are two classifications of coupled coils: transformers and mutual inductors.
 • Transformers and mutual inductors have both an input and an output set of terminals.

13.2 The Ideal Transformer (And Practical Ones, Too)
 • The entire flux of one coil links into the other coil, and vice versa, in an ideal transformer. The ideal transformer has no losses either.
 • The ideal transformer transforms voltage, current, and impedance levels with relationships that depend on the turns ratio.
 • Practical transformers have losses and incomplete flux linkage. The flux linkage, however, is still in the upper-90s percent.

- The dot convention indicates induced voltage polarity on a schematic based on current direction into or out of the dot (as viewed from the circuit external to the transformer or mutual inductor).

13.3 Generalization of the Transformer: The Mutual Inductor
- The mutual inductor has partial flux linkage between the coils. The core is often nonmagnetic or ferrite, not iron.
- The inductances of each coil and the mutual inductance are the key quantities that characterize a mutual inductor. The coupling coefficient relates these three inductances and gives a zero-to-one scale on the degree of flux linkage.

- A set of two equations and four unknowns describes the voltage–current relationships for the mutual inductor. Other circuit information, such as the source voltage and the load impedance, are required to complete the analysis of circuits that contain mutual inductors.
- The input impedance to a mutual inductor displays an impedance inversion behavior.
- The ideal transformer is a special case of the mutual inductor. The ideal transformer relationships result when the following assumptions are made on a mutual inductor:
  1. There is complete flux linkage between both coils.
  2. Coil reactances are much greater than other circuit impedances.

HOMEWORK PROBLEMS

Proper completion of the following questions will demonstrate that you can describe the interaction between the two coils forming a transformer.

13.1 Describe the physical phenomenon that causes a coil to induce a voltage across a nearby second coil. What is the source of the voltage in the second coil?

13.2 Why does an increase in the number of turns of a coil change the induced voltage across that coil?

13.3 If the output voltage of an ideal transformer has a larger magnitude than the input voltage, does that increase the power available at the output of the transformer relative to the input? Explain.

13.4 If the number of primary turns is 100 and the desired voltage ratio for a transformer is 50, what is the needed number of turns on the secondary coil?

13.5 What is the secondary voltage of a transformer with a turns ratio of 20 and a primary voltage of 100 V?

13.6 Describe the meaning of "step-up" and "step-down" when applied to transformers.

Proper completion of the following questions will demonstrate your ability to determine the voltages, the currents, and the impedances in transformer circuits.

13.7 Determine the voltage and the current in the secondary for the transformer circuit shown in Figure P13.1.

13.8 If the current in the secondary of the transformer is 2A, as shown for the circuit in Figure P13.2, (a) calculate the required turns ratio to achieve that current, (b) calculate the power dissipated in the resistor, and (c) calculate the power supplied by the voltage source.

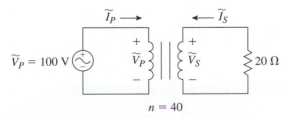

$\widetilde{I}_P \longrightarrow$ $\longleftarrow \widetilde{I}_S$

$\widetilde{V}_P = 100$ V $+ \widetilde{V}_P -$ $+ \widetilde{V}_S -$ $20 \, \Omega$

$n = 40$

Figure P13.1 Transformer Circuit of Problem 13.7

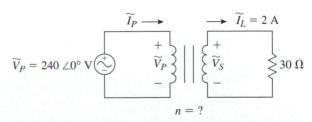

$\widetilde{I}_P \longrightarrow$ $\longrightarrow \widetilde{I}_L = 2$ A

$\widetilde{V}_P = 240 \angle 0° $ V $+ \widetilde{V}_P -$ $+ \widetilde{V}_S -$ $30 \, \Omega$

$n = ?$

Figure P13.2 Transformer Circuit of Problem 13.8

13.9 Calculate the input impedance at the primary of the transformer for the circuit of Figure P13.3.

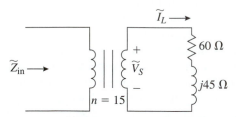

Figure P13.3 Transformer Circuit of Problem 13.9

13.10 Calculate the input impedance at the primary of the transformer for the circuit of Figure P13.4, where $N_P = 2000$ and $N_S = 40$ turns.

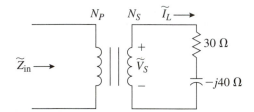

Figure P13.4 Transformer Circuit of Problem 13.10

13.11 For the transformer circuit of Figure P13.5, determine (*a*) input impedance as seen by the source, (*b*) the primary current, (*c*) the secondary voltage and current, (*d*) the complex input power, and (*e*) the power dissipated in the resistor. How does the power dissipated in the resistor relate to the complex input power? The transformer turns are $N_P = 1600$ and $N_S = 80$ turns. Assume the source voltage is RMS.

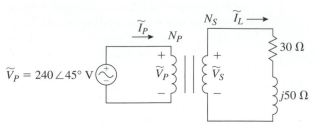

Figure P13.5 Transformer Circuit of Problem 13.11

13.12 For the transformer circuit of Figure P13.6, determine (*a*) input impedance as seen by the source, (*b*) the primary current, (*c*) the secondary voltage and current, (*d*) the complex input power, and (*e*) the power dissipated in the resistor. How does the power dissipated in the resistor relate to the complex input power? The transformer turns are $N_P = 2400$ and $N_S = 60$ turns. Assume the source voltage is RMS.

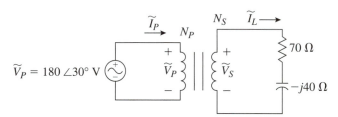

Figure P13.6 Transformer Circuit of Problem 13.12

13.13 For the transformer circuit of Figure P13.7, determine the (*a*) input impedance seen by the source, (*b*) complex power of the load impedance, and (*c*) the power dissipated in the load resistor.

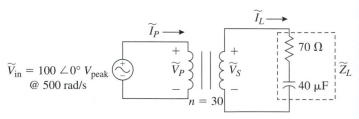

Figure P13.7 Transformer Circuit of Problem 13.13

13.14 For the transformer circuit of Figure P13.8, determine the (*a*) input impedance seen by the source, (*b*) complex power of the load impedance, and (*c*) the power dissipated in the load resistor.

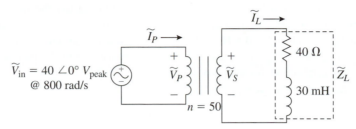

Figure P13.8 Transformer Circuit of Problem 13.14

Proper completion of the following questions will demonstrate your ability to determine the voltages, the currents, and the impedances at the input and the output of a transformer with internal losses.

13.15 For the nonideal transformer circuit of Figure P13.9, determine the (*a*) input impedance seen by the source, (*b*) complex power of the load impedance, and (*c*) the power dissipated in the
load resistor and in the two resistors that model the transformer losses. (*d*) Compare your answers to (*a*) and (*b*) with those of problem 13.13.

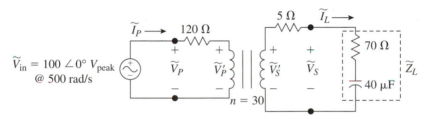

Figure P13.9 Transformer Circuit of Problem 13.15

13.16 For the nonideal transformer circuit of Figure P13.10, determine the (*a*) input impedance seen by the source, (*b*) complex power of the load impedance, and (*c*) the power dissipated in the
load resistor and in the two resistors that model the transformer losses. (*d*) Compare your answers to (*a*) and (*b*) with those of problem 13.14.

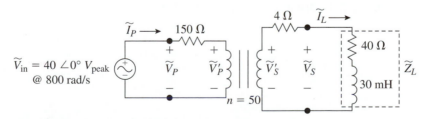

Figure P13.10 Transfomer Circuit of Problem 13.16

13.17 What purpose does the "dot" convention serve for a transformer?

Proper completion of the following questions will demonstrate your understanding of the more general interaction of two coils using the concept of mutual inductance and your ability to determine the voltages and currents at the input and the output of a mutual inductor.

13.18 What is mutual inductance? How does it differ from the inductance of a single coil?

13.19 Calculate the coupling coefficient for two coils with inductances of $L_1 = 6$ H, $L_2 = 0.5$ H, and $M = 1.2$ H.

13.20 Calculate the coupling coefficient for two coils with inductances of $L_1 = 4.5$ mH, $L_2 = 2.8$ mH, and $M = 2.8$ mH.

13.21 For the mutual inductor circuit of Figure P13.11, calculate (*a*) the input impedance seen by the source, (*b*) the current and the voltage of the load impedance, and (*c*) the current through the primary side of the mutual inductor.

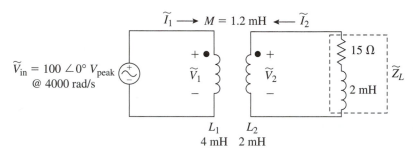

Figure P13.11 Mutual Inductor Circuit of Problem 13.21

13.22 For the mutual inductor circuit of Figure P13.12, calculate (*a*) the input impedance seen by the source, (*b*) the current and the voltage of the load impedance, and (*c*) the current through the primary side of the mutual inductor.

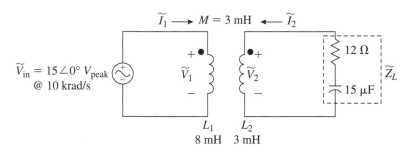

Figure P13.12 Mutual Inductor Circuit of Problem 13.22

13.23 Calculate the total inductance of two inductors in a series-aiding configuration if L_1 = 12 mH, L_2 = 6 mH, and M = 4.5 mH.

13.24 Calculate the total inductance of two inductors in a series-opposing configuration if L_1 = 20 mH, L_2 = 30 mH, and M = 14 mH.

13.25 For the mutual inductor circuit of Figure P13.13, determine which model of the transformer should be used (ideal transformer or mutual inductance model). Then calculate (a) the input impedance seen by the source, (b) the current and the voltage of the load impedance, and (c) the current through the primary side.

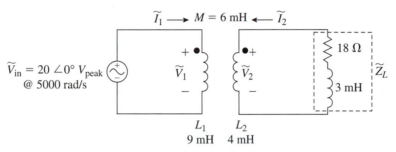

Figure P13.13 Mutual Inductor Circuit of Problem 13.25

13.26 For the mutual inductor circuit of Figure P13.14, determine which model of the transformer that should be used (ideal transformer or mutual inductance model). Then calculate (a) the input impedance seen by the source, (b) the current and the voltage of the load impedance, and (c) the current through the primary side.

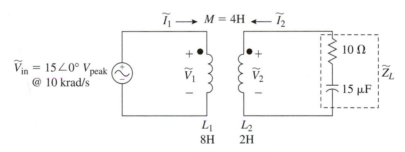

Figure P13.14 Mutual Inductor Circuit of Problem 13.26

THREE-PHASE CIRCUITS AND POWER[1]

Learning Objectives

Chapter Outline

14.1 WHAT IS THREE-PHASE? WHY IS THREE-PHASE USED?

Three-phase is the name given to a power transmission system that consists of three signals with a special phase relationship with respect to each other. The "three" indicates that three AC sinusoidal steady-state signals of the same frequency and peak (and rms) amplitude are present, and the "phase" indicates that the signals have different phases. In the case of three-phase, the signals are 120° apart, as shown in Figure 14.1.[2] The horizontal axis grid increment in Figure 14.1 is 90°. Notice that the phase angle of the third source, $-240°$, is equivalent to $+120°$. This sequence of sinusoids can be generated by a single generator that has provisions to output three sinusoidal signals or can

---

[1] The authors gratefully acknowledge the advice and review given by Dr. Glenn Wrate, Milwaukee School of Engineering, in the writing of this chapter.

[2] The system described here is technically a *balanced* three-phase circuit. There are *unbalanced* three-phase systems, where voltages and currents may not have the same magnitude and/or phase. This topic will be left for a subsequent power-related course.

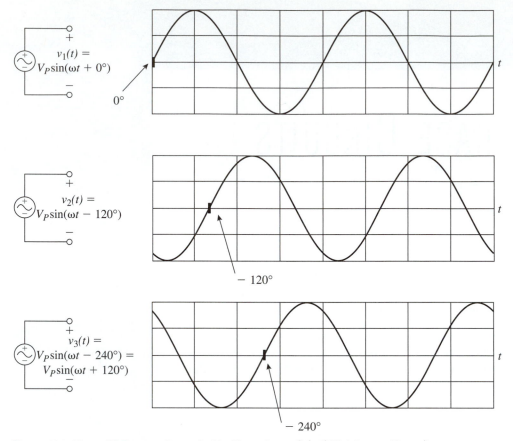

Figure 14.1 Three AC Sources Separated in Phase by 120° (90°/division on X-axes)

be generated by three synchronized separate sources. The question of Why is three-phase used? will be addressed next. Then the actual connection of the three sources in two configurations will be examined and applied to three-phase circuit analysis in Section 14.2.

A primary motivation for the use of three-phase originates in the instantaneous power $p(t)$. Recall that instantaneous power is really the power as a function of time (Section 4.3). The instantaneous power for the three sources shown in Figure 14.1 will now be determined. Assume each source is connected to a separate, identical load. The instantaneous power is:

$$p(t) = v_1(t)i_1(t) + v_2(t)i_2(t) + v_3(t)i_3(t) \tag{14.1}$$

Assume each source is connected to a purely resistive load. The voltage and the current for each phase will be in phase. The AC sinusoidal steady-state expressions for each source are inserted into Equation (14.1):

$$p(t) = V_P \sin(\omega t)I_P \sin(\omega t) + V_P \sin(\omega t - 120°)I_P \sin(\omega t - 120°)$$
$$+ V_P \sin(\omega t + 120°)I_P \sin(\omega t + 120°) \tag{14.2}$$

Note that the phase angles of each source are 120° apart. Simplify Equation (14.2):

$$p(t) = V_P I_P \sin^2(\omega t) + V_P I_P \sin^2(\omega t - 120°) + V_P I_P \sin^2(\omega t + 120°) \tag{14.3}$$

Factor out the $V_P I_P$ and use the trigonometric identity:

$$\sin^2(x) = \frac{1}{2} - \frac{1}{2}\cos(2x) \tag{14.4}$$

$$p(t) = V_P I_P \left[\frac{1}{2} - \frac{1}{2}\cos(2\omega t) + \frac{1}{2} - \frac{1}{2}\cos(2\omega t - 240°) + \frac{1}{2} - \frac{1}{2}\cos(2\omega t + 240°) \right] \tag{14.5}$$

Separate the constant terms from the cosine terms:

$$p(t) = \frac{3V_PI_P}{2} - \frac{3V_PI_P}{2}[\cos(2\omega t) + \cos(2\omega t - 240°) + \cos(2\omega t + 240°)] \qquad (14.6)$$

Note that $+240° = -120°$ and $-240° = +120°$:

$$p(t) = \frac{3V_PI_P}{2} - \frac{3V_PI_P}{2}[\cos(2\omega t) + \cos(2\omega t + 120°) + \cos(2\omega t - 120°)] \qquad (14.7)$$

Recall that a similar approach to $p(t)$ was performed in Section 4.3, but the multiplication was performed graphically. For each signal, the "pulsating" cosine term averages over time to zero, leaving the average power as:

$$P_{\text{ave}} = \frac{3V_PI_P}{2} \qquad (14.8)$$

However, the three 120°-spaced cosine terms in Equation (14.7) are special and have a profound effect on $p(t)$. Use the following trigonometric identities on the last two cosine terms in Equation (14.7) respectively:

$$\cos(a + b) = \cos(a)\cos(b) - \sin(a)\sin(b) \qquad (14.9)$$

$$\cos(a - b) = \cos(a)\cos(b) + \sin(a)\sin(b) \qquad (14.10)$$

$$p(t) = \frac{3V_PI_P}{2} - \frac{3V_PI_P}{2}[\cos(2\omega t) + \cos(2\omega t)\cos(120°) - \sin(2\omega t)\sin(120°)$$
$$+ \cos(2\omega t)\cos(120°) + \sin(2\omega t)\sin(120°)] \qquad (14.11)$$

The sine terms cancel, leaving:

$$p(t) = \frac{3V_PI_P}{2} - \frac{3V_PI_P}{2}[\cos(2\omega t) + \cos(2\omega t)\cos(120°) + \cos(2\omega t)\cos(120°)] \qquad (14.12)$$

The $\cos(120°) = -\frac{1}{2}$:

$$p(t) = \frac{3V_PI_P}{2} - \frac{3V_PI_P}{2}\left[\cos(2\omega t) - \frac{1}{2}\cos(2\omega t) - \frac{1}{2}\cos(2\omega t)\right] \qquad (14.13)$$

The cosine terms sum to zero; hence, the instantaneous power is:

$$\boxed{p(t) = \frac{3V_PI_P}{2} = 3V_{\text{RMS}}I_{\text{RMS}} = P_{\text{ave}}} \qquad (14.14)$$

What is the significance of this result? *The instantaneous power is constant.* The average power and the instantaneous power are identical. The power with three AC sinusoidal steady-state signals that are 120° apart does not pulsate. It is constant. This fact can be qualitatively visualized if the three $p(t)$ waveforms are summed, as shown in Figure 14.2.

Thus, a primary reason for the use of three-phase is that the instantaneous power is constant. This constancy of instantaneous power in three-phase can be a tremendous

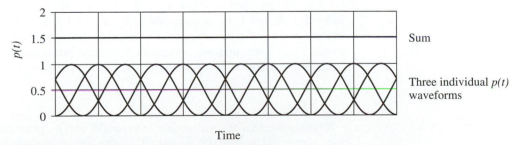

Figure 14.2 Summation of the Three p(t) Waveforms

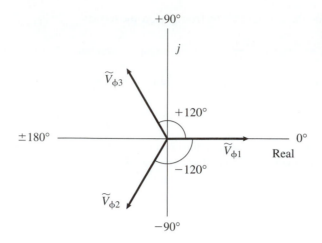

advantage over the case of the single AC source. In large motors and industrial machinery, the pulsation of power can cause vibrations that disrupt smooth operation of the equipment because the torque on the motors is not constant. In some cases the vibrations can even shorten the operational life of the equipment. There are also other reasons to use three-phase that will become apparent later in this chapter.

Another advantage of three-phase is that the total power is three times that of the power of one source, as shown in Equation (14.14). Thus, three-phase equipment has three times the power capability relative to a single-source, everything else being equal. Single-source circuits are sometimes called *single-phase* to help distinguish them from three-phase circuits.

How are three sources connected to form a three-phase circuit? If three voltage sources are connected in series, then the equivalent source is just a source that is the phasor sum of the individual source voltages. The phasors are 120° apart, as shown in Figure 14.3, so the phasor sum is zero—not a good use of sources. Similarly if three current sources are connected in parallel, then the equivalent source is just a source that is the phasor sum of the individual source currents. Again, the sum is zero and this configuration of sources would not be effective.[3] There must be other ways to connect the sources that also deliver power to loads in some "separate" manner, which was stipulated in the p(t) development. These configurations and the circuit analysis for them are covered in the next section.

14.2 THREE-PHASE CIRCUITS: CONFIGURATIONS, CONVERSIONS, ANALYSIS

Note: Effective (RMS) voltage and current values will be assumed in this section.

There are two configurations used in three-phase to connect three sources (or loads) together, as shown in Figure 14.4. The sources are connected in a "triangle" in Figure 14.4a. This configuration is called *delta* because it resembles the uppercase Greek letter Δ. In the other configuration, the sources are connected in a Y, as shown in Figure 14.4b. This configuration is called *wye,* which is the noun that describes something shaped like the letter Y.[4] The terminals of either the delta or wye configurations are called *nodes* and are labeled *A, B,* and *C* for sources and *a, b,* and *c* for loads. The wye configuration has an internal node at the junction of the wye called the *neutral point N* (for sources) or *n* (for loads). Both configurations are prevalent in three-phase systems. Each configuration has advantages and disadvantages that may be examined in a subsequent power-related course

---

[3] If three voltage sources were connected in parallel with the phase relationship of three-phase signals, what would happen? (Do *not* make this connection—you would be in peril!)
[4] *Webster's Deluxe Unabridged Dictionary*

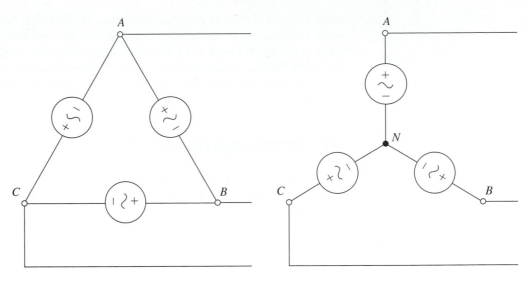

a. Delta configuration   b. Wye configuration

Figure 14.4 Three-Phase Configurations

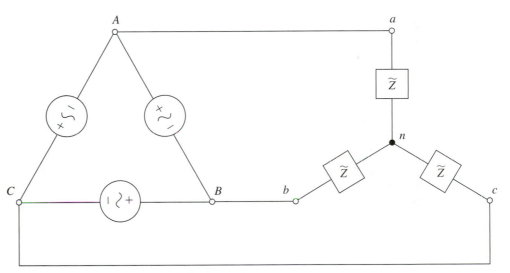

Figure 14.5 An Example of a Three-Phase Circuit

in your curriculum. The existence and use of both configurations is assumed in this introduction to three-phase discussion.

There is an amazing consequence of these three-phase configurations. Can you identify this consequence from the three-phase circuit shown in Figure 14.5? If three sources
and loads were connected as separate circuits, how many wire lines would be required? Six
lines would be required: two for each circuit. How many lines are required in the three-
phase circuit shown in Figure 14.5? Only three lines are required, and the resultant savings
in cost and resources is truly significant. The need for only three lines in three-phase systems is another primary reason for the use of three-phase.

*The transmission of electrical energy over a three-phase system requires only
three lines.*

How are three-phase circuits analyzed? The same circuit analysis principles as for single-phase AC circuits apply. Hence superposition is a valid strategy. However, the analysis
of three-phase circuits can be significantly simplified because of the identical RMS voltages in each source and load. First, two terms are defined to assist in the subsequent discussion. The term *phase* is used to refer to each branch, whether a source or a load, in

either the delta or the wye configuration.[5] Thus, there can be phase voltages and phase currents. The term *line* is used in reference to the voltages that exist between the lines and the currents through the lines in a three-phase circuit. The lines are referred to as *Aa, Bb,* and *Cc,* where the first letter is on the source connection to the line and the second letter is on the load connection to the line. Be sure to observe terminal locations carefully on each schematic.

14.2.1 Delta Configuration Analysis

In the delta configuration, the phase voltages and the line voltages are identical because they are "in parallel"—they are one and the same, as shown in Figure 14.6 for the case of a three-phase load.[6] A notation is needed to distinguish quantities in three-phase circuits. The phasor voltages in a given phase will be designated in general with a two-part subscript. The first letter will be either *L,* for line voltages, or ϕ for phase voltages. If no other letters are present in this subscript, then the line or phase voltage is being referred to generically. The next two letters are the nodes between which the voltage exists: *ab, bc,* or *ca,* where the first of these letters indicates the node for the positive side of the voltage and the second letter indicates the negative side of the voltage for a load. For a source, the notation would be *AB, BC,* and *CA.* Hence, for the delta configuration in Figure 14.6:

$$\tilde{V}_{\phi ab} = \tilde{V}_{Lab} = V_{\phi}\angle 0° \qquad (14.15)$$

$$\tilde{V}_{\phi bc} = \tilde{V}_{Lbc} = V_{\phi}\angle -120° \qquad (14.16)$$

$$\tilde{V}_{\phi ca} = \tilde{V}_{Lca} = V_{\phi}\angle +120° \qquad (14.17)$$

where V_{ϕ} is the RMS magnitude of the phase voltage. The *ab* phase voltage is arbitrarily assumed to be the reference (any phase could have been the reference). The voltages are labeled in Figure 14.6 for a three-phase load in the delta configuration and the phasor diagram for the voltages is shown in Figure 14.7 for a resistive load. Thus, in general, for the delta configuration:

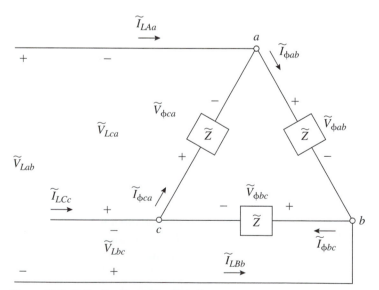

Figure 14.6 Delta Configuration Notation

---

[5] In the power utility industry, *phase* always refers to the wye configuration, not the delta configuration.

[6] This is the reason why the term *phase voltage* is not used with the delta configuration in the power utility industry—it is reserved for the wye configuration. We use phase voltage with delta in this chapter to assist in the explanation of the principles.

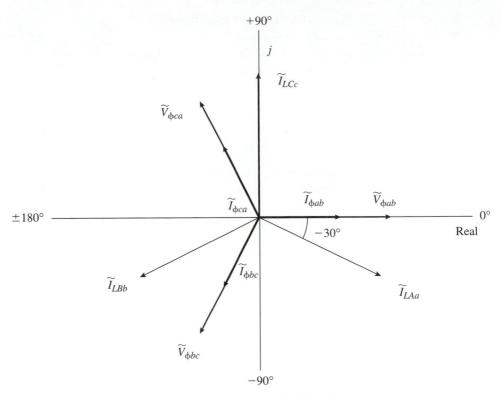

Figure 14.7 Delta Configuration Phasor Diagram for a Resistive Load

$$\boxed{\tilde{V}_\phi = \tilde{V}_L} \text{ (delta)} \qquad (14.18)$$

The phase currents and the line currents are not identical because the currents split at the three nodes in a delta configuration. Again, a notation is needed to distinguish the currents in three-phase circuits. The phasor currents in a given phase will be designated in general with a two-part subscript. The first letter will be either L, for line currents, or ϕ for phase currents. If no other letters are present in this subscript, then the line or phase current is being referred to generically. The next part of the subscript is labeled as follows:

- For phase currents in a load, the next two letters are the nodes through which the current flows: *ab, bc,* or *ca,* where the first of these letters indicates the node that the current enters the phase (branch) and the second letter indicates the node that the current leaves the phase.

- For phase currents in a source, the next two letters are the nodes through which the current flows: *AB, BC,* or *CA,* where the first of these letters indicates the node that the current leaves the phase and the second letter indicates the node that the current enters the phase. Note that the current directions in the source are opposite those in the load.

- For line currents, the first letter indicates the source terminal end of the line (where the current enters the line) and the second letter indicates the load end of the line (where the current leaves the line).

The various currents are labeled for a three-phase load in the delta configuration in Figure 14.6. Note the phase shift between the line currents and the phase currents in Figure 14.7. Why does this phase shift occur?

Consider the relationship between the line and phase currents in a delta configuration. Apply KCL to one node, say node *a*:

$$+\tilde{I}_{LAa} - \tilde{I}_{\phi ab} + \tilde{I}_{\phi ca} = 0 \qquad (14.19)$$

$$+\tilde{I}_{LAa} = +\tilde{I}_{\phi ab} - \tilde{I}_{\phi ca} \qquad (14.20)$$

Ohm's law is applied to each phase:

$$\tilde{I}_{LAa} = +\frac{\tilde{V}_{\phi ab}}{\tilde{Z}} - \frac{\tilde{V}_{\phi ca}}{\tilde{Z}} \tag{14.21}$$

$$\tilde{I}_{LAa} = \frac{V_\phi \angle 0°}{Z \angle \theta} - \frac{V_\phi \angle +120°}{Z \angle \theta} = \frac{V_\phi (1 \angle 0° - 1 \angle +120°)}{Z \angle \theta} \tag{14.22}$$

$$\tilde{I}_{La} = \frac{V_\phi (1.73205\ldots) \angle -30°}{Z \angle \theta} \tag{14.23}$$

If one were to perform the complex number subtraction in parentheses in Equation (14.22) geometrically, the magnitude in Equation (14.23) would be exactly equal to $\sqrt{3}$. Hence,

$$\tilde{I}_{LAa} = \frac{\sqrt{3}\, V_\phi}{Z \angle \theta} \angle -30° \tag{14.24}$$

The magnitude of the phase voltage divided by the magnitude of the phase impedance in Equation (14.24) is just the magnitude of the phase current. Hence one can generalize the relationship between the magnitudes of the phase and line currents in the delta configuration:

$$\boxed{I_L = \sqrt{3}\, I_\phi}\ \text{(delta)} \tag{14.25}$$

The phase shift must be approached with caution. If the angles in Equation (14.24) are simplified:

$$\tilde{I}_{LAa} = \frac{\sqrt{3}\, V_\phi}{Z} \angle(-30° - \theta) \tag{14.26}$$

The magnitude of the phase voltage divided by the magnitude of the phase impedance is the magnitude of the phase current:

$$\tilde{I}_{LAa} = \sqrt{3}\, I_\phi \angle(-30° - \theta) \tag{14.27}$$

The $-\theta$ in Equation (14.27) is just the phase shift between the voltage and the current due to the reactance of the load. It can be incorporated into the phase current because θ is the angle of the phase current relative to the angle of the phase voltage. Thus, Equation (14.27) becomes:

$$\tilde{I}_{LAa} = \sqrt{3}\, (I_\phi \angle -\theta) \angle -30° = \sqrt{3}\, \tilde{I}_\phi \angle -30° \tag{14.28}$$

Thus the line current in line Aa lags the phase current in phase ab by 30°. The phase and line current phasors are shown in Figure 14.7 for the case of a resistive load. The phase voltages and phase currents are in phase, and the line currents lag the phase currents by 30°.

What if the phase shift of the bc phase voltage were $+120°$ and the phase shift of the ca phase voltage were $-120°$?[7] Then the line current in line Aa would lead the phase current in phase ab by 30°. Instead of trying to remember all these phase relationships, if phases are needed, just draw the phasor diagram and the phase relationships become apparent. In general the relationship between the line and phase currents in the delta configuration is:

$$\boxed{\tilde{I}_L = \sqrt{3}\, \tilde{I}_\phi \angle \pm 30°}\ \text{(delta)} \tag{14.29}$$

where the plus-or-minus depends on the phase relationship between the phase voltages.

---

[7] This scenario is referred to as being of the opposite *phase sequence* relative to the original scenario. It corresponds to the opposite direction of rotation of a three-phase generator, for example.

14.2.2 Wye Configuration Analysis

The analysis of phase voltages and currents in the wye configuration is analogous to the analysis of the delta configuration. In the wye configuration, the phase currents and the line currents are identical because they are "in series"—they are one and the same, as shown in Figure 14.8 for the case of a three-phase load. A notation is again needed to distinguish quantities in three-phase circuits, and it follows the same format as for the delta configuration. The phasor currents in a given phase will be designated in general with a two-part subscript. The first letter will be either L, for line currents, or ϕ, for phase currents. If no other letters are present in this subscript, then the line or the phase current is being referred to generically. The next letter(s) refers to the node through which the current flows:

- For phase currents in a load, the next two letters are the nodes through which the current flows: *an, bn,* or *cn,* where the first of these letters indicates the node that the current enters the phase (branch), and the second letter indicates the node that the current leaves the phase. The junction of the wye, labeled *n* in Figure 14.8, is called the *neutral point.*

- For phase currents in a source, the next two letters are the nodes through which the current flows: *AN, BN,* or *CN,* where the first of these letters indicates the node that the current leaves the phase, and the second letter indicates the node that the current enters the phase. The junction of the wye in a source is labeled *N.*

- For line currents, the first letter indicates the source terminal end of the line (where the current enters the line), and the second letter indicates the load end of the line (where the current leaves the line). This notation is no different than that for the line connections to the delta configuration.

Hence for the three-phase load in the wye configuration in Figure 14.8:

$$\tilde{I}_{\phi an} = \tilde{I}_{LAa} \tag{14.30}$$

$$\tilde{I}_{\phi bn} = \tilde{I}_{LBb} \tag{14.31}$$

$$\tilde{I}_{\phi cn} = \tilde{I}_{LCc} \tag{14.32}$$

Thus, in general, for the wye configuration:

$$\boxed{\tilde{I}_{\phi} = \tilde{I}_{L}} \;(\text{wye}) \tag{14.33}$$

For voltages in the wye configuration, the first letter will be either L, for line voltages, or ϕ, for phase voltages. If no other letters are present in this subscript, then the line or the

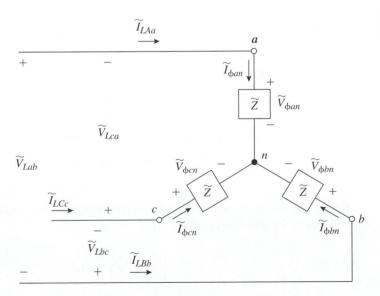

Figure 14.8 Wye Configuration Notation

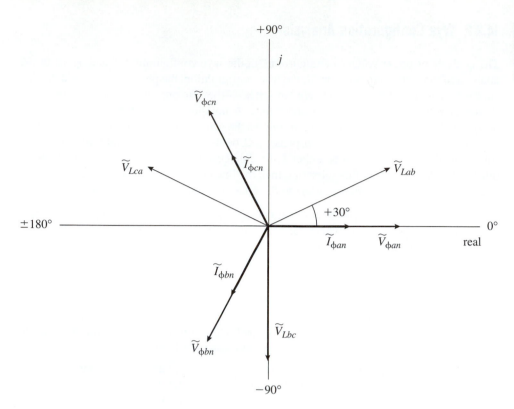

Figure 14.9 Wye Configuration Phasor Diagram for a Resistive Circuit

phase voltage is being referred to generically. For phase voltages in the load, the next two letters are the nodes between which the voltage exists: *an, bn,* or *cn,* where the first of these letters indicates the node for the positive side of the phase voltage, and the second letter indicates the negative side of the phase voltage, which is at the junction of the wye, labeled *n* in Figure 14.8. For the source, the notation would be *AN, BN,* and *CN.* Hence for the wye configuration in Figure 14.8, the phase voltages are:

$$\tilde{V}_{\phi an} = V_\phi \angle 0° \tag{14.34}$$

$$\tilde{V}_{\phi bn} = V_\phi \angle -120° \tag{14.35}$$

$$\tilde{V}_{\phi cn} = V_\phi \angle +120° \tag{14.36}$$

where V_ϕ is the RMS magnitude of the phase voltage. The *an* phase voltage is arbitrarily assumed to be the reference (any phase could have been the reference). The voltages are labeled in Figure 14.8 and the phasor diagram for the voltages is shown in Figure 14.9 for a resistive load.

The phase voltages and the line voltages are not identical because the line voltage consists of the phasor sum of two phase voltages. The various line voltages are labeled in Figure 14.8. Consider the relationship between the line and phase voltages in a wye configuration. Apply KVL to one line voltage, say, between nodes *a* and *b:*

$$+\tilde{V}_{Lab} - \tilde{V}_{\phi an} + \tilde{V}_{\phi bn} = 0 \tag{14.37}$$

$$+\tilde{V}_{Lab} = +\tilde{V}_{\phi an} - \tilde{V}_{\phi bn} \tag{14.38}$$

The phasor voltage in each phase is expressed in polar form:

$$+\tilde{V}_{Lab} = +V_\phi \angle 0° - V_\phi \angle -120° = V_\phi \left(1\angle 0° - 1\angle -120°\right) \tag{14.39}$$

$$+\tilde{V}_{Lab} = V_\phi \sqrt{3} \angle +30° \tag{14.40}$$

Hence one can generalize the relationship between the magnitudes of the phase and line voltages in the wye configuration:

$$\boxed{V_L = \sqrt{3}\, V_\phi} \; (\text{wye}) \tag{14.41}$$

The phase shift must again be approached with caution. In this setup the line voltage across lines *ab* leads the phase voltage in phase *an* by 30° because the phase shift of phase voltage *an* is 0°. However, if the phase shift of the *bn* phase voltage were +120° and the phase shift of the *cn* phase voltage were −120°,[8] then the line voltage across lines *ab* would lag the phase voltage in phase *an* by 30°. Again, if phases are needed, just draw the phasor diagram and the phase relationships become apparent. In general the relationship between the line and phase voltages in the wye configuration is:

$$\boxed{\tilde{V}_L = \sqrt{3}\,\tilde{V}_\phi\angle\pm30°}\;\text{(wye)} \tag{14.42}$$

where the plus-or-minus depends on the phase relationship between the phase voltages. The phase and the line voltage phasors are shown in Figure 14.9 for the case of a resistive load. The phase voltages and the phase currents are in phase, and the line voltages lead the phase voltages by 30°.

14.2.3 Complex Power in Three-Phase Circuits

Now the power in terms of phase and line voltages and currents can be determined. In either the delta or the wye configuration, the total complex power in a three-phase load with complex impedances is three times the complex power in one phase:

$$\boxed{\tilde{S}_T = 3\tilde{S}_\phi = 3\tilde{V}_\phi\tilde{I}_\phi^*}\;\text{(delta or wye)} \tag{14.43}$$

The phasors in Equation (14.43) can be expressed in polar form:

$$\tilde{S}_T = 3\tilde{V}_\phi\tilde{I}_\phi^* = 3V_\phi\angle\theta_{V\phi}\,(I_\phi\angle\theta_{I\phi}) \tag{14.44}$$

where the phase subscript on the angles is used to distinguish them from line phase angles. The line voltage and the phase voltage are equal in the delta configuration. Substitute in the line-phase current magnitude relationship in Equation (14.25) into Equation (14.44)

$$\tilde{S}_T = 3V_L\angle\theta_{V\phi}\left(\frac{I_L}{\sqrt{3}}\angle\theta_{I\phi}\right)^* = 3V_L\angle\theta_{V\phi}\left(\frac{I_L}{\sqrt{3}}\angle-\theta_{I\phi}\right) \tag{14.45}$$

Simplify using the rules of complex number multiplication:

$$\tilde{S}_T = \sqrt{3}V_LI_L\angle(\theta_{V\phi}-\theta_{I\phi}) \tag{14.46}$$

The phase angle is recognized as the power factor angle for the impedance in any one of the phases of the load:

$$\theta = \theta_{V\phi} - \theta_{I\phi} \tag{14.47}$$

Thus the total complex power of a load in the delta configuration is expressed in terms of the magnitudes of the line voltages and currents and the power factor angle of the load in any one of the phases as:

$$\tilde{S}_T = \sqrt{3}V_LI_L\angle\theta \tag{14.48}$$

For the wye configuration, the line and phase currents are equal. Substitute in the line-phase current relationship in Equation (14.33) and the line-phase voltage magnitude relationship in Equation (14.41) into:

$$\tilde{S}_T = 3\tilde{V}_\phi\tilde{I}_\phi^* = 3V_\phi\angle\theta_{V\phi}(I_\phi\angle\theta_{I\phi})^* \tag{14.49}$$

$$\tilde{S}_T = 3\frac{V_L}{\sqrt{3}}\angle\theta_{V\phi}\,(I_L\angle\theta_{I\phi})^* = \sqrt{3}V_L\angle\theta_{V\phi}\,(I_L\angle-\theta_{I\phi}) \tag{14.50}$$

---

[8] Again, this scenario is referred to as being of the opposite *phase sequence* relative to the original scenario.

Simplify using the rules of complex-number multiplication:

$$\tilde{S}_T = \sqrt{3}V_L I_L \angle (\theta_{V\phi} - \theta_{I\phi}) \tag{14.51}$$

The phase angle is recognized as the power factor angle for the impedance in any one of the phases of the load. Thus,

$$\boxed{\tilde{S}_T = \sqrt{3}V_L I_L \angle \theta} \text{ (wye and delta configurations)} \tag{14.52}$$

where θ is the power factor angle of the load. This result is identical to Equation (14.48). Hence the total complex power in a three-phase balanced circuit is determined from either the phase voltages and the currents using Equation (14.43) or from the line voltage and current magnitudes and the power factor angle of the load in any one of the phases using Equation (14.52). The complex power calculation techniques in Chapter 10 are applicable.

Can the total complex power be determined strictly from the *phasor* line voltage and current? Yes. The wye configuration will be used to demonstrate the result. Solve the line-phase voltage relationship in Equation (14.42) for the line voltage:

$$\tilde{V}_L = \sqrt{3}\,\tilde{V}_\phi \angle \pm 30° \rightarrow \tilde{V}_\phi = \frac{\tilde{V}_L}{\sqrt{3}\,\angle \pm 30°} = \frac{\tilde{V}_L}{\sqrt{3}} \angle \mp 30° \tag{14.53}$$

Substitute this result and the line-phase current relationship in Equation (14.33) into:

$$\tilde{S}_T = 3\tilde{V}_\phi \tilde{I}_\phi^* \tag{14.54}$$

$$\tilde{S}_T = 3\left(\frac{\tilde{V}_L}{\sqrt{3}} \angle \mp 30\right)\tilde{I}_L^* \tag{14.55}$$

$$\boxed{\tilde{S}_T = \sqrt{3}\,\tilde{V}_L \tilde{I}_L^* \,(1 \angle \mp 30)} \tag{14.56}$$

where the plus-or-minus depends on the phase relationship between the phase voltages. A similar $\pm 30°$ would have resulted if the delta configuration had been used in this development because the line and phase currents are $\pm 30°$ out of phase. Thus, the total complex power can be determined using line voltages and currents, but an additional angle of $\pm 30°$ must be incorporated into the complex power calculation.[9] Again, the line voltage is not directly across each phase in the wye configuration, and the line current is not the total current through each phase in the delta configuration. Hence this "extra $\pm 30°$" in Equation (14.56) accounts for the phase shift between the line and phase voltages in the wye or the line and phase currents in the delta.

14.2.4 Three-Phase Circuit Analysis

All of the developments in this section suggest how to approach the circuit analysis of three-phase balanced circuits: *one may analyze the three-phase circuit using only one phase and/or line.* The results for one phase/line are applicable to the other phases/lines. The phase angle may need to be adjusted for any given quantity (120° apart), but the magnitudes are identical. (If the three-phase circuit is *not* balanced, then this circuit analysis approach is invalid. This is a topic for future courses.)

How are three-phase circuits analyzed for voltage, current, and power values? The strategy is straightforward. If the source and the load are both wye (or both delta), analyze the three-phase circuit as is. If the source is in a delta configuration and the load is in a wye

---

[9] Power in a three-phase circuit is not generally measured in this manner. The interested reader should investigate the topic of the "two wattmeter method."

configuration, or vice versa, convert the load or the source to match the configuration of the other, and then analyze the circuit. Source voltages are converted using the line-phase voltage relationship in Equation (14.41) (or Equation (14.42) if phases are needed). To convert the load configuration, use the relationship between impedances in balanced delta and wye configurations from Section 11.4b:

$$\tilde{Z}_\Delta = 3\tilde{Z}_Y \tag{14.57}$$

Then, one can convert resultant voltages and currents as needed to express them in the original delta or wye configuration for the source or load that was converted. The following examples illustrate the circuit analysis techniques that were described here.

Example 14.2.1

Determine (*a*) the line voltages, (*b*) the line currents, (*c*) the phase currents in the load, and (*d*) the total load complex power for the circuit shown in Figure 14.10.

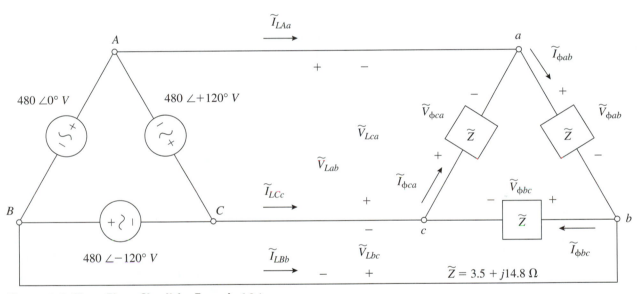

Figure 14.10 Three-Phase Circuit for Example 4.2.1

Given: source phase voltages per Figure 14.10

           load impedance in each phase: $\tilde{Z} = 3.5 + j14.8 \ \Omega$

Desired: a. $\tilde{V}_{Lab}, \tilde{V}_{Lbc}, \tilde{V}_{Lca}$

           b. $\tilde{I}_{LAa}, \tilde{I}_{LBb}, \tilde{I}_{LCc}$

           c. $\tilde{I}_{\phi ab}, \tilde{I}_{\phi bc}, \tilde{I}_{\phi ca}$

           d. $\tilde{S}_T$

Strategy: a. delta source-delta load $\rightarrow \tilde{V}_\phi = \tilde{V}_L$

           b. $\tilde{I}_\phi = \dfrac{\tilde{V}_\phi}{\tilde{Z}}$

           c. $\tilde{I}_L = \sqrt{3}\, \tilde{I}_\phi \angle \pm 30°$ (must determine the sign, too)

           d. $\tilde{S}_T = 3\tilde{V}_\phi \tilde{I}_\phi^*$

Solution:

a. $\tilde{V}_{Lab} = \tilde{V}_{\phi ab} = 480\angle 0°$ V (RMS voltages and currents)
The other line voltages have identical magnitudes; adjust the phase shifts $\pm 120°$:

$$\tilde{V}_{Lbc} = \tilde{V}_{\phi bc} = 480\angle -120° \text{ V}$$

$$\tilde{V}_{Lca} = \tilde{V}_{\phi ca} = 480\angle +120° \text{ V}$$

b. $\tilde{I}_{\phi ab} = \dfrac{\tilde{V}_{\phi ab}}{\tilde{Z}} = \dfrac{480\angle 0°}{3.5 + j14.8} = 31.562\angle -76.695° \text{ A} = 31.6\angle -76.7° \text{ A}$

The other phase currents have identical magnitudes and phase shifts $\pm 120°$ relative to this phase current:

$$\tilde{I}_{\phi bc} = 31.6\angle(-76.7° - 120°) = 31.6\angle -196.7° = 31.6\angle +163.3° \text{ A}$$

$$\tilde{I}_{\phi ca} = 31.6\angle(-76.7° + 120°) = 31.6\angle +43.3° \text{ A}$$

c.   First it must be determined whether the line currents lead or lag the phase currents. The most straightforward manner to determine this is to sketch a phasor diagram, as shown in Figure 14.11. At node a,

$$+\tilde{I}_{La} = +\tilde{I}_{\phi ab} - \tilde{I}_{\phi ca}$$

Draw $\tilde{I}_{\phi ab}$, $\tilde{I}_{\phi ca}$, and $-\tilde{I}_{\phi ca}$ on the phasor diagram.
Draw an estimate sketch for $+\tilde{I}_{La} = +\tilde{I}_{\phi ab} + (-\tilde{I}_{\phi ca})$.
It is seen from the sketch in Figure 14.11 that the line current lags the phase current, so $-30°$ must be used in $\tilde{I}_L = \sqrt{3}\,\tilde{I}_\phi \angle \pm 30°$.

$$\tilde{I}_{LAa} = \sqrt{3}\,\tilde{I}_{\phi ab}\angle -30° = \sqrt{3}\,(31.562\angle -76.695°)\angle -30°$$

$$= 54.667\angle -106.695° = 54.7\angle -106.7° \text{ A}$$

The other line currents have identical magnitudes and phase shifts $\pm 120°$ relative to this line current:

$$\tilde{I}_{LBb} = 54.7\angle -226.7° = 54.7\angle +133.3° \text{ A}$$

$$\tilde{I}_{LCc} = 54.7\angle +13.3° \text{ A}$$

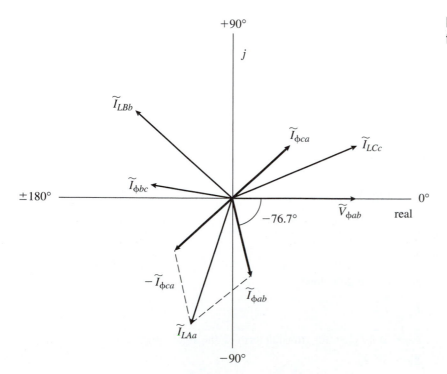

Figure 14.11 Phasor Diagram Sketch for Example 14.2.1

d.
$$\tilde{S}_T = 3\tilde{V}_{\phi ab}\tilde{I}_{\phi ab}^* = 3(480\angle 0°)(31.562\angle -76.695°)^*$$

$$= 3(480\angle 0°)(31.562\angle +76.695°) = 45{,}449\angle +76.695°$$

$$= 45.4\angle +76.7°\ kVA = 10.5\ kW + j44.2\ kVAR$$

Example 14.2.2

Determine (a) the line voltage magnitudes, (b) the line current magnitudes, and (c) the total load complex power for the circuit shown in Figure 14.12 by (1) converting the source to a wye configuration, and (2) converting the load to a delta configuration. Compare the answers from both approaches.

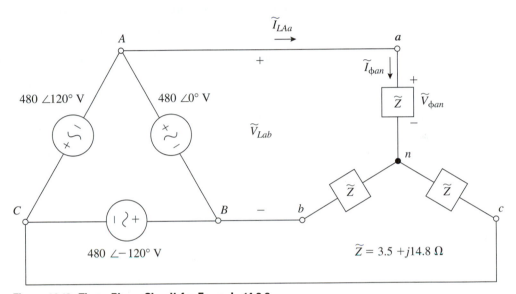

Figure 14.12 Three-Phase Circuit for Example 14.2.2

Given: source phase voltages per Figure 14.12

load impedance in each phase:

$$\tilde{Z} = 3.5 + j14.8\ \Omega$$

Desired: a. V_L

b. I_L

c. 1. S_T by source $\Delta \to Y$, $\angle \tilde{Z} = \theta$

2. S_T by load $Y \to \Delta$, $\angle \tilde{Z} = \theta$

Strategy: a. delta source $\to V_\phi = V_L$

b. source $\Delta \to Y$, $I_L = I_\phi = \dfrac{V_\phi}{Z}$

c. 1. $S_T = 3V_\phi I_\phi$ in the delta load configuration, $\angle \tilde{Z} = \theta$, $\tilde{S}_T = S_T \angle \theta$

2. load $Y \to \Delta$, $I_\phi = \dfrac{I_L}{\sqrt{3}}$, $S_T = 3V_\phi I_\phi$ in the wye load configuration, $\angle \tilde{Z} = \theta$, $\tilde{S}_T = S_T \angle \theta$

Solution:

a. $V_\phi = V_L = 480\ V$

b. The $\Delta \rightarrow Y$ conversion of the source is shown in Figure 14.13. The voltage of each phase in the wye configuration is:

$$V_\phi = \frac{V_L}{\sqrt{3}} = \frac{480}{\sqrt{3}} = 277.13 \text{ V}$$

Then V_ϕ is across each load impedance in the wye configuration (the neutral points of the source and the load are often connected, in which case they are at the same potential):

$$I_\phi = \frac{V_\phi}{Z} = \frac{277.13}{\mid 3.5 + j14.8 \mid} = \frac{277.13}{15.208} = 18.223 \text{ A}$$

The line current equals the phase current in the wye configuration:

$$I_L = I_\phi = 18.223 \text{ A}$$

c. 1. For both the source and the load in wye configurations (Figure 14.13):

$$S_T = 3V_\phi I_\phi = 3(277.13)(18.222) = 15.150 \text{ kVA}$$

The power factor angle of the load is identical to the impedance phase angle:

$$\angle \tilde{Z} = \angle(3.5 + j14.8) = 76.695°$$

Hence the complex power delivered to the load is:

$$\tilde{S}_T = S_T \angle \theta = 15.150 \angle +76.695° \text{ kVA} = 3.4865 \text{ kW} + j14.743 \text{ kVAR}$$

$$\tilde{S}_T = 15.2 \angle +76.7° \text{ kVA} = 3.49 \text{ kW} + j14.7 \text{ kVAR}$$

2. The $Y \rightarrow \Delta$ conversion of the load is shown in Figure 14.14. First convert the load impedances to the delta:

$$\tilde{Z} = \tilde{Z}_\Delta = 3\tilde{Z}_Y = 3(3.5 + j14.8) = 10.5 + j44.4 = 45.625 \angle + 76.695° \text{ } \Omega$$

Determine the magnitude of the phase current in each impedance of the load delta configuration:

$$I_\phi = \frac{V_\phi}{Z'} = \frac{480}{45.624} = 10.521 \text{ A}$$

For the source and the load in delta configurations (Figure 14.14):

$$S_T = 3V_\phi I_\phi = 3(480)(10.521) = 15.150 \text{ kVA}$$

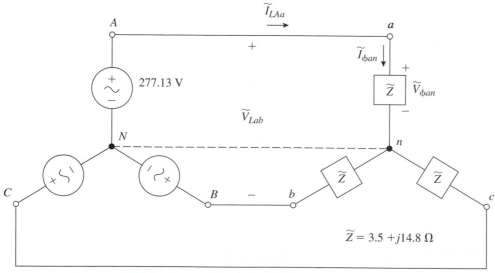

Figure 14.13 Wye Source Configuration for Example 14.2.2

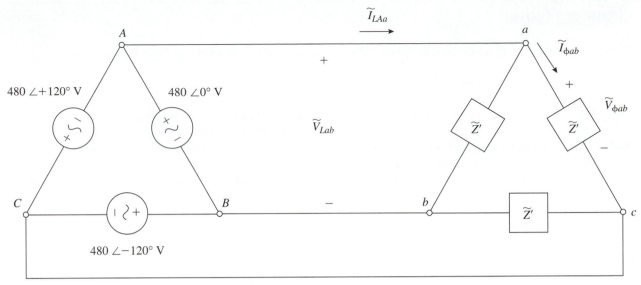

Figure 14.14 Delta Load Configuration for Example 14.2.2

The power factor angle of the load is identical to the impedance phase angle:

$$\angle \tilde{Z} = \angle(3.5 + j14.8) = 76.695°$$

Hence the complex power delivered to the load is:

$$\tilde{S}_T = S_T \angle \theta = 15.150\angle +76.695° \text{ kVA} = 3.4865 \text{ kW} + j14.743 \text{ kVAR}$$

$$\tilde{S}_T = 15.2\angle +76.7° \text{ kVA} = 3.49 \text{ kW} + j14.7 \text{ kVAR}$$

The answers to parts (1) and (2) match, as expected.

CHAPTER REVIEW

14.1 **What Is Three-Phase? Why Is Three-Phase Used?**

- Three-phase is the name given to a power transmission system with three AC sinusoidal steady-state signals of the same frequency and peak (and rms) amplitude that have different phases—the phases are 120° apart.
- Unlike single-phase circuits, the instantaneous power in a three-phase balanced circuit is constant.
- All else being equal, the total power in a three-phase circuit is three times that of a single-phase circuit.

14.2 **Three-Phase Circuits: Configurations, Conversions, Analysis**

- There are two standard configurations of three-phase loads and sources: delta (Δ) and wye (Y).
- The transmission of electrical energy over a three-phase system requires only three lines.
- The line and phase voltages are identical in a delta configuration.

- The line current is $\sqrt{3}$ times the phase current with a $\pm 30°$ phase shift in the delta configuration: $\tilde{I}_L = \sqrt{3}\,\tilde{I}_\phi \angle \pm 30°$.
- The line and phase currents are identical in a wye configuration.
- The line voltage is $\sqrt{3}$ times the phase voltage with a $\pm 30°$ phase shift in the wye configuration: $\tilde{V}_L = \sqrt{3}\,\tilde{V}_\phi \angle \pm 30°$.
- If the phase angles of voltages and currents are required, draw the phasor diagrams for the three-phase circuit to show the phase relationships of the voltages and the currents.
- The complex power in either a delta or a wye configuration can be determined from the phase voltages and currents: $\tilde{S}_T = 3\tilde{S}_\phi = 3\tilde{V}_\phi \tilde{I}_\phi^*$ or the line voltages and currents and the power factor angle of the load: $\tilde{S}_T = \sqrt{3}V_L I_L \angle \theta$.
- Delta–wye conversions are useful in the analysis of three-phase circuits with a delta source–wye load or a wye source–delta load.

HOMEWORK PROBLEMS

Completing the following problems will demonstrate that can describe three-phase and why it is used.

14.1 List at least three benefits for using three-phase power.

14.2 What is the key power result for three-phase systems?

14.3 What is the phase angle difference between each phase in a balanced three-phase circuit?

14.4 Sketch the sinusoidal voltages for each phase in a balanced three-phase circuit.

14.5 Write the sine wave equations for the waveforms in problem 14.4.

14.6 Write phasor equations for the waveforms in problem 14.4.

14.7 Sketch a three-phase wye generator.

14.8 Sketch a three-phase delta load.

14.9 Sketch a three-phase wye generator connected to a wye load.

14.10 Sketch a three-phase wye generator connected to a delta load.

Completing the following problems will demonstrate that you understand the configurations of three-phase and can perform basic analysis of three-phase circuits.

14.11 (a) Calculate the magnitude of the current supplied to each load, and (b) calculate the total complex power delivered to the load for Figure P14.1.

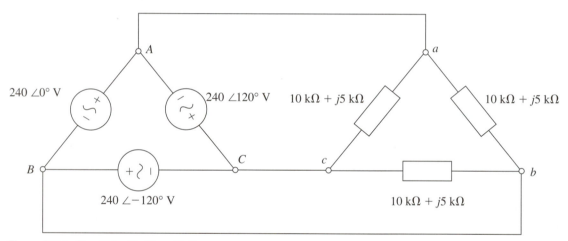

Figure P14.1 Circuit for Problem 14.11

14.12 (a) Calculate the magnitude of the current supplied to each load, and (b) calculate the total complex power delivered to the load for Figure P14.2.

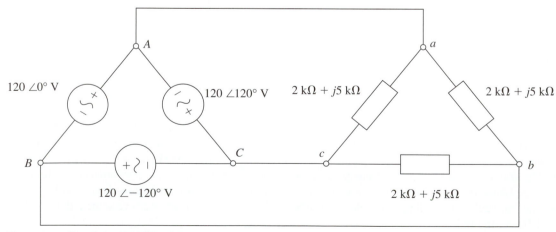

Figure 14.2 Circuit for Problem 14.12

14.13 (*a*) Calculate the magnitude of the current supplied to each load, and (*b*) calculate the total complex power delivered to the load for Figure P14.3.

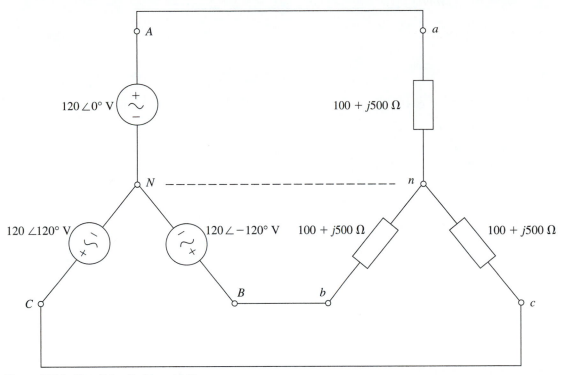

Figure P14.3 Circuit for Problem 14.13

14.14 (*a*) Calculate the magnitude of the current supplied to each load, and (*b*) calculate the total complex power delivered to the load for Figure P14.4.

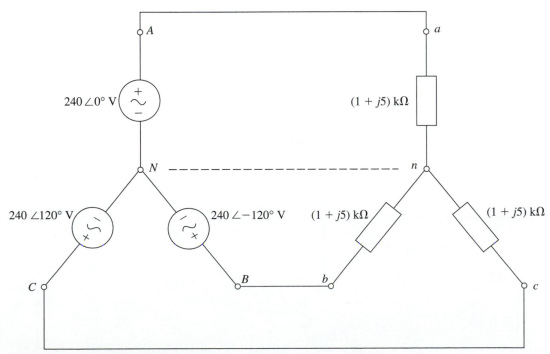

Figure P14.4 Circuit for Problem 14.14

14.15 (*a*) Calculate the magnitude of the current supplied to each load, and (*b*) calculate the total complex power delivered to the load for Figure P14.5.

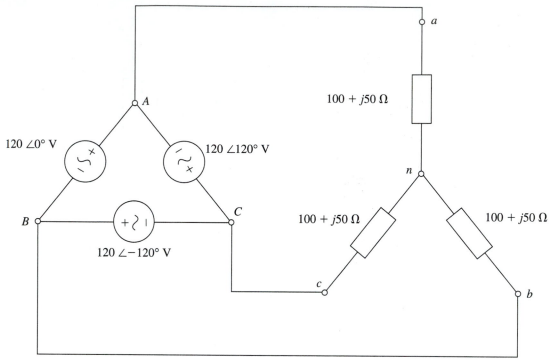

Figure 14.5 Circuit for Problem 14.15

14.16 (*a*) Calculate the magnitude of the current supplied to each load, and (*b*) calculate the total complex power delivered to the load for Figure P14.6.

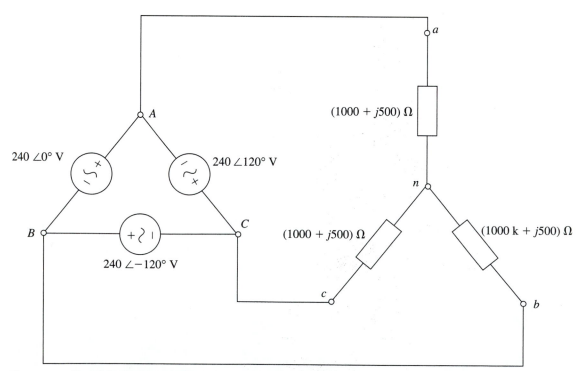

Figure 14.6 Circuit for Problem 14.16

14.17 (*a*) Calculate the magnitude of the current supplied to each load, and (*b*) calculate the total complex power delivered to the load for Figure P14.7.

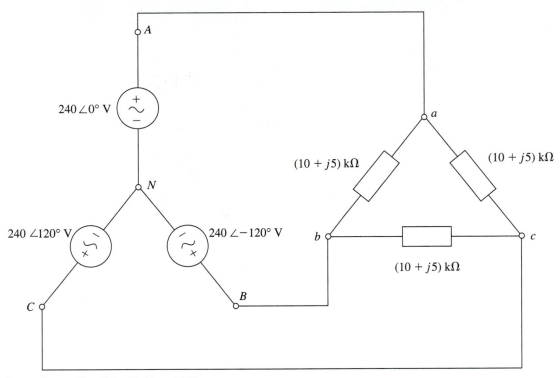

Figure P14.7 Circuit for Problem 14.17

14.18 (*a*) Calculate the magnitude of the current supplied to each load, and (*b*) calculate the total complex power delivered to the load for Figure P14.8.

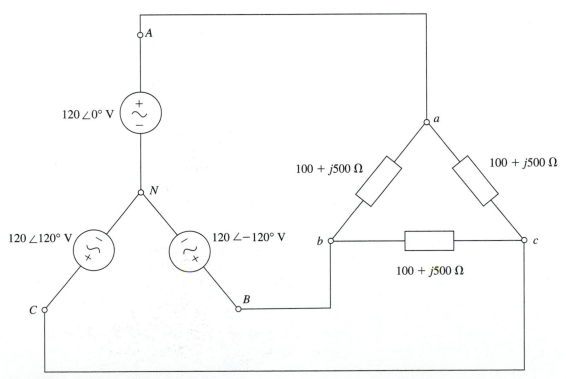

Figure P14.8 Circuit for Problem 14.18

VARIABLE-FREQUENCY CIRCUIT ANALYSIS

As a result of successfully completing this chapter, you should be able to:

1. Describe the harmonic content of a periodic signal.
2. Given the Fourier series expression for a specific periodic signal, determine the specific values of the individual terms in the series.
3. Analyze first-order circuits[1] with frequency as a variable.
4. Determine the transfer function of first-order circuits.
5. Plot the frequency response of voltage, current, impedance, or transfer function for first-order circuits.
6. Identify the four basic filter types and describe the frequency response of each.
7. Manipulate the transfer function into Bode form for first-order circuits.
8. Describe and plot the transfer function of an electric circuit on semi-logarithmic graphs.
9. Describe and calculate circuit gain in decibels (dB).
10. Draw and label the magnitude and phase Bode plots of a transfer function in Bode form.
11. Determine the Fourier series response of a first-order circuit to a periodic input signal.

Learning Objectives

You should be able to:

a. Enter complex-number formulas into a spreadsheet.
b. Create graphs with both linear and logarithmic scales in a spreadsheet.
c. Apply the laws of logarithms in algebraic expressions.
d. Manipulate algebraic and trigonometric expressions using symbols for constants as well as variables.

Mathematical Assumptions

15.1 Motivation: The Frequency Content of Periodic Signals

15.2 Analyzing a Circuit with an Unspecified Source Frequency

15.3 Transfer Function and Frequency Response

15.4 Basic Filter Types and Responses

15.5 Bode Plots—Transfer Function Setup

Chapter Outline

---

[1] First-order circuits contain either inductors *or* capacitors, not both. Circuit analysis with frequency as a variable for circuits with both inductors *and* capacitors will be introduced in Section 16.7.

15.1 MOTIVATION: THE FREQUENCY CONTENT OF PERIODIC SIGNALS

The analysis of a circuit when the source frequency is *not* specified will be examined in this chapter. This approach is useful because it would be a simple matter to insert any specific frequency of interest into the results of the analysis without having to reanalyze the entire circuit. An important benefit of this approach is that the results of the analysis could be inserted into computer programs, programmable calculators, or spreadsheets, and the calculations could then be performed automatically for hundreds of frequencies. What is the catch? Well, you will finally be able to put all those algebra skills to good use: because when the source frequency is not specified, ω is not assigned a value and must remain as a *variable* during the circuit analysis. The resulting addition, subtraction, multiplication, division, and trigonometry with variables is algebra.

The algebraic process of variable frequency circuit analysis can be applied to determine ratios of phasor voltages and phasor currents. These ratios are called steady-state *transfer functions* and are often used with electronic devices that have an output and an input. What is interesting about the transfer function is that it can be plotted as a function of frequency to see how the circuit behaves as the source frequency changes. These graphs will be called *frequency response* plots. Some interesting and extremely useful circuits, called *filters,* will be covered both to illustrate these concepts and to show how the response of a circuit can be controlled as a function of frequency. An important technique to quickly visualize the approximate frequency response plot of a circuit will be examined. These approximate plots are called *Bode plots.*

Is the ability to analyze a circuit with an unspecified source frequency worth all the algebra that will be required to perform this analysis? The answer would be no if the only signals in the world were DC and AC signals. However, the answer is actually yes because there is a large class of signals, named *periodic signals,* that can be analyzed using AC circuit analysis techniques. In fact, *any practical periodic signal can be represented as a summation of sinusoidal signals with the frequencies being integer multiples of the primary (fundamental) frequency of the signal.* Examples of several common periodic signals are given in Figure 15.1. A signal is periodic when the "pattern" repeats with the same period, just as one cycle of a sinusoidal signal repeats every period of time.

The important concepts to realize are that periodic signals contain several frequencies and that circuit behavior depends on frequency. Hence each frequency in a periodic signal will have a different response in the circuit. The significance of this chapter is that one will only have to analyze a circuit once, despite the numerous frequencies contained in a periodic signal. After the circuit analysis with frequency as a variable, each frequency is inserted separately into the expression that resulted from the analysis, usually with the aid of a spreadsheet or a computer program. Then the total response of the circuit to the periodic signal can be obtained from summing the results for all of the frequencies, again usually with the aid of a spreadsheet or a computer program.

The fundamental frequency of a periodic signal must be determined first. The fundamental frequency of any periodic signal is the frequency at which the "pattern" repeats. The period (T) is labeled on each periodic signal in Figure 15.1 and, as with sinusoidal signals, the corresponding frequency is $1/T$. The modifier "fundamental" is used because there are several frequencies contained within each periodic signal. The additional frequencies contained in periodic signals are multiples of the fundamental frequency.

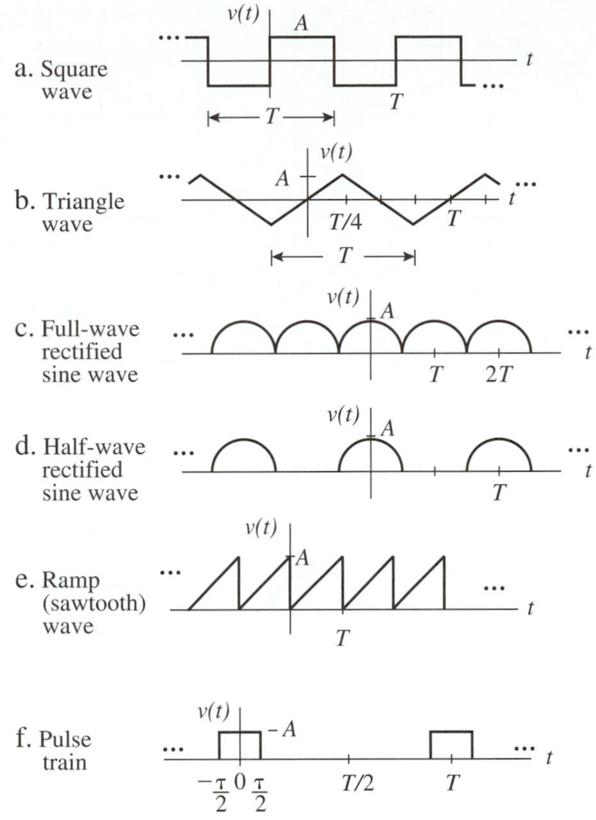

Figure 15.1 Common Periodic Signals[2]

a. Square wave

b. Triangle wave

c. Full-wave rectified sine wave

d. Half-wave rectified sine wave

e. Ramp (sawtooth) wave

f. Pulse train

Let f_o be the fundamental cyclic frequency and ω_o be the fundamental radian frequency. The term *fundamental frequency* refers to both f_o and ω_o. As with sinusoidal signals, you must determine which one is being used from the context in which it is used.

The multiples of the fundamental frequency are nf_o, where n is an integer (1, 2, 3, 4, ...). When $n = 1$, the frequency is the fundamental frequency. When $n \geq 2$, the frequencies are called *harmonics*. Thus, the sinusoid of frequency $2f_o$ is called the second harmonic, the sinusoid of frequency $3f_o$ is called the third harmonic, and so on.

How would these sinusoids of harmonic frequencies be combined to form a periodic signal? An example of the graphical summation of harmonics is shown in Figure 15.2. What wave is this summation already approaching as the number of harmonics that are included in the summation increases?

Thus a summation of sinusoids that have harmonically related frequencies results in a periodic signal. The study of an advanced topic, Fourier series,[3] would show that if the amplitudes and/or phases of the harmonics are changed, different periodic signals result. This is how the periodic signals, such as those in Figure 15.1, are different from each other. Again, the concept is that periodic signals contain harmonically related frequencies; hence one must analyze the circuit at each of those frequencies.

Is there any way to simplify and automate the circuit analysis when there are signals of several different frequencies? Yes, if the impedances of the circuit were written with frequency as a variable, then one would analyze the circuit only once. It would be straightforward to insert the amplitude and the frequency for each frequency into the resultant general expression of the phasor circuit analysis. The expression could be copied in a spreadsheet to automate the calculations. Furthermore, the circuit response could be plotted as a function of frequency (frequency response plot). The benefits of being able to analyze a circuit with an unspecified frequency, i.e., with frequency as a variable, make the algebraic circuit

---

[2] The terms *wave* and *wave form* and often used interchangeably with "signal" when the signal is periodic.

[3] See Hayt, Kemmerly, and Durbin, *Engineering Circuit Analysis,* 6th ed., chapter 18, for example.

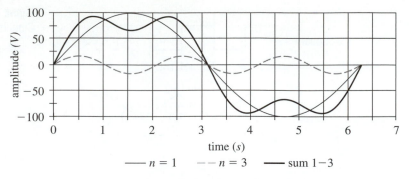

a. 1st, 3rd, and sum of harmonics

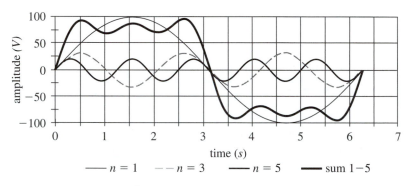

b. 1st, 3rd, 5th, and sum of harmonics

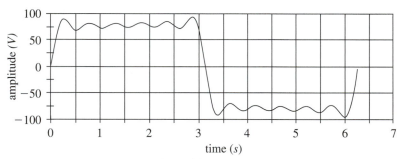

c. Sum of first 11 odd harmonics

Figure 15.2 Adding Harmonics to Produce a Square Wave

analysis worthwhile. The analysis methods and plots are covered in detail in the following sections of this chapter.

Note: The mathematics underlying the Fourier series representation of periodic signals is examined in the remainder of this section. For those who wish to proceed directly to variable frequency circuit analysis, go to Section 15.2.

An infinite series, called the Fourier series, is used to mathematically represent periodic signals. The Fourier series is a summation of an infinite number of sinusoids. The general expression for the Fourier series is:

$$v(t) = B_o + \sum_{n=1}^{\infty} B_n \sin(n\omega_o t + \theta_n) \qquad \textbf{(15.1)}$$

where: B_o = average DC voltage (or current) magnitude,

B_n = peak amplitude of the sinusoid of frequency nf_o,

θ_n = phase of the sinusoid of frequency nf_o,

ω_o = fundamental frequency, and

$n = 1, 2, 3, \ldots$

Practically, only a finite number of terms are necessary because the amplitude of the higher harmonics (terms with large n) are negligible. As an example, the Fourier series for the square wave shown in Figure 15.1 is:

$$v(t) = 0 + \sum_{\substack{n=1 \\ \text{odd}}}^{\infty} \frac{4A}{n\pi} \sin(n\omega_o t) = \sum_{n=1,3,5,\ldots}^{\infty} \frac{4A}{n\pi} \sin(n\omega_o t) \qquad \textbf{(15.2)}$$

where: $B_o = 0$ (the average DC voltage is zero),

$B_n = \dfrac{4A}{n\pi}$ (the peak amplitude of frequency nf_o; note that it depends on n),

$\theta_n = 0$ (the phase of all harmonics is zero for a square wave),

A = the peak amplitude in the square wave (see Figure 15.1), and

$n = 1, 3, 5, ..$ (the even harmonics are zero for a square wave).

Expanding the summation for a square wave [Equation (15.2)] for a few terms results in the Fourier series expression in Equation (15.3):

$$v(t) = \frac{4A}{\pi} \sin(\omega_o t) + \frac{4A}{3\pi} \sin(3\omega_o t) + \frac{4A}{5\pi} \sin(5\omega_o t) + \frac{4A}{7\pi} \sin(7\omega_o t) + \ldots \qquad \textbf{(15.3)}$$

If specific values for A and ω_o are known, then the specific amplitudes and frequencies for each term can be determined, as illustrated in Example 15.1.1.

Example 15.1.1

Write the first three terms of the Fourier series for a square wave of period 4 ms and peak amplitude 10 V.

Given:

$$A = 10 \text{ V}$$
$$T = 4 \text{ ms}$$

Desired: First three terms of the Fourier series

Strategy: Determine f_o from T.

Insert values into general Fourier series expression for a square wave.

Simplify constants.

Solution:

$$f_o = \frac{1}{T} = \frac{1}{4 \times 10^{-3}} = 250.00 \text{ Hz}$$

$$\omega_o = 2\pi f_o = 2\pi 250 = 1570.8 \text{ rad/s}$$

$$B_n = \frac{4A}{n\pi} = \frac{4(10)}{n\pi} = \frac{12.732}{n}$$

$$v(t) = \sum_{n=1,3,5,\ldots}^{\infty} \frac{12.732}{n} \sin(n1570.8t)$$

$$= \frac{12.732}{1} \sin(1570.8t) + \frac{12.732}{3} \sin(3 \times 1570.8t) + \frac{12.732}{5} \sin(5 \times 1570.8t) + \ldots$$

$$v(t) = 12.732 \sin(1570.8t) + 4.2440 \sin(4712.4t) + 2.5464 \sin(7854.0t) + \ldots$$

TABLE 15.1	periodic signal type	Fourier series expression
Fourier Series of Common Periodic Signals	square wave	$\displaystyle\sum_{n=1,3,5,\ldots}^{\infty} \frac{4A}{n\pi} \sin(n\omega_o t)$
	triangle wave	$\displaystyle\sum_{n=1,3,5,\ldots}^{\infty} \frac{8A(-1)^{(n-1)/2}}{n^2\pi^2} \sin(n\omega_o t)$
	full-wave rectified sine wave	$\displaystyle\frac{2A}{\pi} + \sum_{n=1}^{\infty} \frac{4A(-1)^{n+1}}{\pi(4n^2-1)} \cos(n\omega_o t)$
	half-wave rectified sine wave	$\displaystyle\frac{A}{\pi} + \frac{A}{2}\cos(\omega_o t) + \sum_{n=2,4,6,\ldots}^{\infty} \frac{2A(-1)^{(n/2)+1}}{\pi(n^2-1)} \cos(n\omega_o t)$
	ramp (sawtooth)	$\displaystyle\frac{A}{2} - \sum_{n=1}^{\infty} \frac{A}{n\pi} \sin(n\omega_o t)$
	pulse train	$\displaystyle\frac{A\tau}{T} + \sum_{n=1}^{\infty} \frac{2A\sin\left(\dfrac{n\pi\tau}{T}\right)}{n\pi} \cos(n\omega_o t)$

The Fourier series for a few waves are given in Table 15.1 (corresponding to the plots in Figure 15.1). Some Fourier series have the cosine instead of the sine for the function of time. Recall that a cosine wave leads the sine by 90°.

Given the Fourier series expression for a periodic wave, how does one use it to analyze the circuit response to the periodic signal? This topic is examined in Section 15.8.

15.2 Analyzing a Circuit with an Unspecified Source Frequency

In the *RC* circuit shown in Figure 15.3, what is the phasor current? If the frequency were known, the impedance of each component could be calculated. Because the frequency is intentionally not specified at this point, the impedance of each component is expressed with ω remaining as a variable:

$$\tilde{Z}_R = R = 200\ \Omega, \quad \tilde{Z}_C = -j\frac{1}{\omega C} = -j\frac{1}{\omega\, 2 \times 10^{-6}} = -j\frac{5 \times 10^5}{\omega} \tag{15.4}$$

Now the total impedance is determined:

$$\tilde{Z}_T = \tilde{Z}_R + \tilde{Z}_C = 200 - j\frac{5 \times 10^5}{\omega} \tag{15.5}$$

Note that strategy-wise we are proceeding just as if we knew the frequency. Now determine the phasor current using Ohm's law:

$$\tilde{I} = \frac{\tilde{V}_S}{\tilde{Z}_T} = \frac{10\angle 0°}{200 - j\dfrac{5 \times 10^5}{\omega}} \tag{15.6}$$

Note that the numerator in Equation (15.6) is expressed in polar form, and the denominator is expressed in rectangular form, just as one might do with complex numbers. The par-

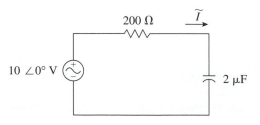

Figure 15.3 An RC Circuit with Frequency Unspecified

200 Ω $\tilde{I}$

10 ∠0° V

2 μF

	freq (Hz)	freq (rad/s)	I	|I|(A)	∠I(°)
TABLE 15.2 A Portion of Spreadsheet Calculations of I in Equation (15.6)	1.00E+01	6.28E+01	3.15627972948081E-005+1.25584380181905E-003i	1.26E-03	8.86E+01
	2.00E+01	1.26E+02	1.26012550663641E-004+2.50694004121704E-003i	2.51E-03	8.71E+01
	3.00E+01	1.88E+02	2.82637841111379E-004+3.74860079314147E-003i	3.76E-03	8.57E+01
	4.00E+01	2.51E+02	5.00267801354747E-004+4.97625584095764E-003i	5.00E-03	8.43E+01
	5.00E+01	3.14E+02	7.77293820075068E-004+6.18550768498646E-003i	6.23E-03	8.28E+01
	6.00E+01	3.77E+02	1.11169887425834E-003+7.37218212782956E-003i	7.46E-03	8.14E+01
	7.00E+01	4.40E+02	1.50109350580314E-003+8.53237326756118E-003i	8.66E-03	8.00E+01
	8.00E+01	5.03E+02	1.94275727090994E-003+9.66248196540741E-003i	9.86E-03	7.86E+01
	9.00E+01	5.65E+02	2.43368436291737E-003+1.0759247118993E-002i	1.10E-02	7.73E+01
	1.00E+02	6.28E+02	2.97063204450559E-003+1.18197693497562E-002i	1.22E-02	7.59E+01
	2.00E+02	1.26E+03	1.00849989203302E-002+2.00634678655871E-002i	2.25E-02	6.33E+01
	3.00E+02	1.88E+03	1.81221907236837E-002+2.40353019444029E-002i	3.01E-02	5.30E+01
	4.00E+02	2.51E+03	2.51323888289836E-002+2.49996494615017E-002i	3.54E-02	4.48E+01
	5.00E+02	3.14E+03	3.06136681630425E-002+2.43615830716159E-002i	3.91E-02	3.85E+01
	6.00E+02	3.77E+03	3.47279600099971E-002+2.30296937462029E-002i	4.17E-02	3.36E+01
	7.00E+02	4.40E+03	3.77903096254448E-002+2.14804092066528E-002i	4.35E-02	2.96E+01
	8.00E+02	5.03E+03	4.00844604915358E-002+1.99363750887491E-002i	4.48E-02	2.64E+01
	9.00E+02	5.65E+03	4.18252605916394E-002+1.84908249146279E-002i	4.57E-02	2.39E+01
	1.00E+03	6.28E+03	4.31661750751195E-002+1.71752753439386E-002i	4.65E-02	2.17E+01
	2.00E+03	1.26E+04	4.8096411957619E-002+9.56847713504899E-003i	4.90E-02	1.13E+01
	3.00E+03	1.88E+04	4.91356796491113E-002+6.51682191527992E-003i	4.96E-02	7.55E+00
	4.00E+03	2.51E+04	4.95101148980722E-002+4.92486219942242E-003i	4.98E-02	5.68E+00
	5.00E+03	3.14E+04	4.96853637583937E-002+3.95383562073363E-003i	4.98E-02	4.55E+00
	6.00E+03	3.77E+04	4.97810816788759E-002+3.3012105090227E-003i	4.99E-02	3.79E+00
	7.00E+03	4.40E+04	4.98389750019541E-002+2.83289972506944E-003i	4.99E-02	3.25E+00
	8.00E+03	5.03E+04	4.98766221096623E-002+2.48065967296227E-003i	4.99E-02	2.85E+00
	9.00E+03	5.65E+04	4.99024657234997E-002+2.20617335899135E-003i	5.00E-02	2.53E+00
	1.00E+04	6.28E+04	4.99209679443642E-002+1.9862922030694E-003i	5.00E-02	2.28E+00

ticular form of the complex numbers is not important, but one must be able to convert forms for use with software. Spreadsheet software, such as Microsoft Excel, requires the rectangular form of complex numbers and complex number expressions.

How is the expression in Equation (15.6) used? If ω were known, the value for ω would be inserted and the phasor current would be determined per normal AC circuit calculations. Here is where the power of computer software comes in. This expression could be entered into a spreadsheet (with complex number and functions math loaded and enabled). Then the phasor current could be determined for any and all desired frequencies. Furthermore, the phasor current magnitude and the phase could be plotted as a function of frequency. A portion of a Microsoft Excel spreadsheet is shown in Table 15.2. The cell contents for the first few cells (first column A, second column B, etc.; first row 1, second row 2, etc.) are:

A1: freq (Hz)

B1: freq (rad/s)

C1: I

D1: |I| (A)

E1: ∠I(°)

A3: 10

B3: =A3*2*PI()

C3: =IMDIV(COMPLEX(10,0),COMPLEX(200,-500000/B3))

[*Note*: This cell was entered as "=IMDIV(COMPLEX(10,0),COMPLEX (200,-5E5/B3))"]

D3: =IMABS(C3)

E3: =IMARGUMENT(C3)*180/PI()

Cells A1 through E1 are just column headings. Cells A3 through E3 are the data and the formulas for each frequency. Note that the source voltage, $10\angle 0°$, had to be expressed in rectangular form in cell C3. Also note the full precision of the complex numbers is shown in column C (3rd column).

The plots of the current magnitude as a function of frequency and the current phase (angle) as a function of frequency are shown in Figure 15.4. Is the task complete? No, the critical step of interpreting the significance of the results (analysis of results) must be performed. In Figure 15.4a, the magnitude of the current increases and then levels out as frequency increases. Why? What happens to the capacitor reactance as frequency increases? What does the total impedance of the circuit approach as frequency increases? Hence, what should the current of this circuit approach as frequency increases? In order to answer these questions, start with the effect of increasing frequency on the capacitive reactance:

$$X_C(\omega) = \left(\frac{1}{\omega C}\right) \to 0 \text{ as } \omega \text{ increases} \tag{15.7}$$

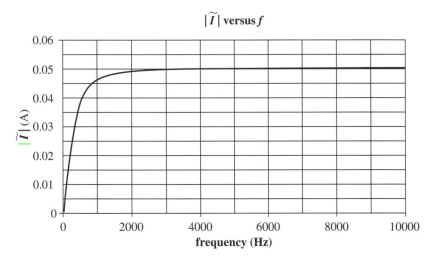

a. Current magnitude response

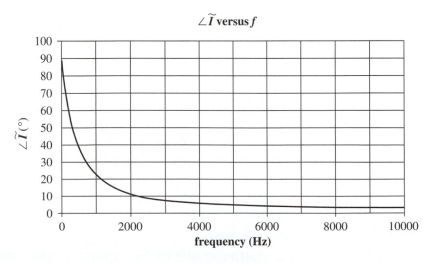

b. Phase response

Figure 15.4 Current Frequency Response [Eq. (15.6)]

The effect of the reactance on the impedance of the circuit is:

$$\tilde{Z}(\omega) = R - j\left(\frac{1}{\omega C}\right) \to R \text{ as } \omega \text{ increases} \tag{15.8}$$

Hence the effect on the circuit current as frequency increases is:

$$\tilde{I} = \frac{\tilde{V}_s}{\tilde{Z}} \to \frac{10\ \angle 0°}{R} = 50 \text{ mA} \tag{15.9}$$

In Figure 15.4b, the angle of the current approaches $+90°$ as frequency approaches zero and approaches $0°$ at high frequencies. At low frequencies, what is the reactance of the capacitor relative to that of the resistance? It is much larger. Hence the impedance of the capacitor dominates the impedance of the resistor, most of the source voltage is across the capacitor, the current is primarily determined by the reactance of the capacitor, and consequently, the current leads the voltage by almost $90°$ as frequency approaches zero. You should formulate the corresponding reasoning for the phase approaching $0°$ at high frequencies.

Notice how unwieldy writing out the transfer function expressions with numbers inserted for the constants could be if the circuit is more complicated than a *RC* series circuit. This issue can be addressed in a better way: do not write the values for *R, L,* and *C* in a given circuit, but just use symbols (*R, L,* or *C*). However, this requires that you *distinguish between symbols that represent constants,* such as *R, L,* and *C, and symbols that represent variables,* such as *f* and ω. Then, when the algebra is done, the numbers may be substituted in for the symbols that represent constants. This approach has the added benefit that once the circuit analysis has been performed, it is usable for all component values as well as for all frequencies. The previous development for the phasor current will be repeated using this approach:

$$\tilde{Z}_R = R, \quad \tilde{Z}_C = -j\frac{1}{\omega C} \tag{15.10}$$

$$\tilde{Z}_T = \tilde{Z}_R + \tilde{Z}_C = R - j\frac{1}{\omega C} \tag{15.11}$$

$$\tilde{I} = \frac{\tilde{V}_S}{\tilde{Z}_T} = \frac{V_S\ \angle 0°}{R - j\dfrac{1}{\omega C}} \tag{15.12}$$

If one substitutes the numerical values for *R* and *C* of the previous example, the phasor current will be:

$$\tilde{I} = \frac{\tilde{V}_S}{\tilde{Z}_T} = \frac{10\ \angle 0°}{200 - j\dfrac{5 \times 10^5}{\omega}} \tag{15.13}$$

which is identical to the result in Equation (15.6).

From this point forth, symbols for the component values will be utilized except when the numbers are needed to perform calculations.

Why should we learn these algebraic techniques when there are circuit simulation programs that automatically generate frequency response plots? The answer is implied above: you may not have component values to work with. Then the analysis must proceed with symbols for constants, as illustrated above. There are at least two benefits to this approach:

■ The analysis of a given circuit is performed in general. Then the same results of the analysis can be used in different applications just by changing the component values. The results can also be inserted into various software programs.

■ As one analyzes a circuit in general and then applies the results to specific circuits, one learns to visualize circuit performance from the mathematics. This approach is useful in several areas of circuit design, where one must specify the circuit performance and then generate the circuit(s) to accomplish the desired circuit performance.

15.3 TRANSFER FUNCTION AND FREQUENCY RESPONSE

As in the previous section, note that any change in the magnitude of the source voltage results in a proportionate change in the magnitude of the current. One must recalculate each time a new source voltage is used. If both sides of Equation (15.12) were divided by the phasor source voltage, then the "output" quantity, current, is expressed relative to the "input" quantity, source voltage (in practice, the input and output quantities are determined based on the purpose of the circuit). The *ratio* of the one phasor variable to another phasor variable would be the result, the ratio of current to source voltage in this case:

$$\frac{\tilde{I}}{\tilde{V}_S} = \frac{1}{R - j\dfrac{1}{\omega C}} \tag{15.14}$$

This expression has the distinct feature of no voltage or current on the right-hand side, i.e., the ratio of two phasor variables *depends only on the circuit (and frequency)*. Such a ratio of phasor variables is called a *transfer function* and the most common symbol for it is $H(j\omega)$. In general, a transfer function can be a voltage ratio, a current ratio, a voltage-to-current ratio, or a current-to-voltage ratio. The corresponding mathematical expressions would be:

$$H(j\omega) = \frac{\tilde{V}_a}{\tilde{V}_b} \tag{15.15}$$

$$H(j\omega) = \frac{\tilde{I}_a}{\tilde{I}_b} \tag{15.16}$$

$$H(j\omega) = \frac{\tilde{V}_a}{\tilde{I}_b} \tag{15.17}$$

$$H(j\omega) = \frac{\tilde{I}_a}{\tilde{V}_b} \tag{15.18}$$

where the *a* and *b* subscripts refer to voltages and currents at different points in the circuit. This is why it is called a transfer function: the "transfer" from one voltage or current in the circuit to another voltage or current in the circuit. However, if not explicitly defined, the transfer function is normally understood to be the ratio of output voltage to input voltage.

Once $H(j\omega)$ is obtained, one can find the output for any given input by rearranging the transfer function. For example, if $\tilde{V}_a$ is the output and $\tilde{V}_b$ is the input, then the output is found from:

$$\tilde{V}_a = H(j\omega)\tilde{V}_b \tag{15.19}$$

In the case examined in the previous section, if the output quantity is current and the input quantity is source voltage, then the transfer function is equal to Equation (15.14):

$$H(j\omega) = \frac{\tilde{I}}{\tilde{V}_S} = \frac{1}{R - j\dfrac{1}{\omega C}} \tag{15.20}$$

Again, numbers could be substituted for the purpose of calculations:

$$H(j\omega) = \frac{\tilde{I}}{\tilde{V}_S} = \frac{1}{200 - j\dfrac{5 \times 10^5}{\omega}} \tag{15.21}$$

TABLE 15.3	freq (Hz)	freq (rad/s)	*H*	|*H*|(A/V)	∠*H*(°)
A Portion of Spreadsheet Calculations of *H*(*jω*) in Equation (15.21)	1.00E+01	6.28E+01	3.15627972948081E-006+1.25584380181905E-004i	1.26E-04	8.86E+01
	2.00E+01	1.26E+02	1.26012550663641E-005+2.50694004121704E-004i	2.51E-04	8.71E+01
	3.00E+01	1.88E+02	2.82637841111379E-005+3.74860079314147E-004i	3.76E-04	8.57E+01
	4.00E+01	2.51E+02	5.00267801354747E-005+4.97625584095764E-004i	5.00E-04	8.43E+01
	5.00E+01	3.14E+02	7.77293820075068E-005+6.18550768498646E-004i	6.23E-04	8.28E+01
	6.00E+01	3.77E+02	1.11169887425834E-004+7.37218212782956E-004i	7.46E-04	8.14E+01
	7.00E+01	4.40E+02	1.50109350580314E-004+8.53237326756118E-004i	8.66E-04	8.00E+01
	8.00E+01	5.03E+02	1.94275727090994E-004+9.66248196540741E-004i	9.86E-04	7.86E+01
	9.00E+01	5.65E+02	2.43368436291737E-004+1.0759247118993E-003i	1.10E-03	7.73E+01
	1.00E+02	6.28E+02	2.97063204450559E-004+1.18197693497562E-003i	1.22E-03	7.59E+01
	2.00E+02	1.26E+03	1.00849989203302E-003+2.00634678655871E-003i	2.25E-03	6.33E+01
	3.00E+02	1.88E+03	1.81221907236837E-003+2.40353019444029E-003i	3.01E-03	5.30E+01
	4.00E+02	2.51E+03	2.51323888289836E-003+2.49996494615017E-003i	3.54E-03	4.48E+01
	5.00E+02	3.14E+03	3.06136681630425E-003+2.43615830716159E-003i	3.91E-03	3.85E+01
	6.00E+02	3.77E+03	3.47279600099971E-002+2.30296937462029E-003i	4.17E-03	3.36E+01
	7.00E+02	4.40E+03	3.77903096254448E-002+2.14804092066528E-003i	4.35E-03	2.96E+01
	8.00E+02	5.03E+03	4.00844604915358E-002+1.99363750887491E-003i	4.48E-03	2.64E+01
	9.00E+02	5.65E+03	4.18252605916394E-002+1.84908249146279E-003i	4.57E-03	2.39E+01
	1.00E+03	6.28E+03	4.31661750751195E-003+1.71752753439386E-003i	4.65E-03	2.17E+01
	2.00E+03	1.26E+04	4.8096411957619E-003+9.56847713504899E-004i	4.90E-03	1.13E+01
	3.00E+03	1.88E+04	4.91356796491113E-003+6.51682191527992E-004i	4.96E-03	7.55E+00
	4.00E+03	2.51E+04	4.95101148980722E-003+4.92486219942242E-004i	4.98E-03	5.68E+00
	5.00E+03	3.14E+04	4.96853637583937E-003+3.95383562073363E-004i	4.98E-03	4.55E+00
	6.00E+03	3.77E+04	4.97810816788759E-003+3.3012105090227E-004i	4.99E-03	3.79E+00
	7.00E+03	4.40E+04	4.98389750019541E-003+2.83289972506944E-004i	4.99E-03	3.25E+00
	8.00E+03	5.03E+04	4.98766221096623E-003+2.48065967296227E-004i	4.99E-03	2.85E+00
	9.00E+03	5.65E+04	4.99024657234997E-003+2.20617335899135E-004i	5.00E-03	2.53E+00
	1.00E+04	6.28E+04	4.99209679443642E-003+1.9862922030694E-004i	5.00E-03	2.28E+00

A fair question to ask at this time is, Why create this transfer function? If the source voltage or the circuit current is given, the other can be found at any given frequency; however, a study of Equation (15.20) reveals that there are no voltage or current variables in the transfer function expression of the circuit. This means that even though the transfer function is found from a ratio of phasor variables, *the transfer function depends only on the circuit itself, not on the applied signal.* The usefulness of the transfer function is in finding out how the circuit behaves as a function of frequency independent of the source voltage or current. This circuit behavior is called the frequency response and is covered next.

The transfer function, such as the one obtained in Equation (15.14), is useful in determining the behavior of a circuit versus frequency, i.e., the behavior of the circuit as the frequency of the source is varied. The behavior may not be obvious from the equation, so how can this behavior be visualized? Just as in Section 15.2, plot it. The values of frequency and the expressions for the magnitude of $H(j\omega)$ and the angle of $H(j\omega)$ are easily entered into a spreadsheet and the calculations automatically made. What is the independent variable? The dependent variables are the magnitude of $H(j\omega)$, $|H(j\omega)|$, and the angle of $H(j\omega)$, $\angle H(j\omega)$. A portion of a Microsoft Excel spreadsheet for Equation 15.21 is shown in Table 15.3. The cell contents for the first few cells (first column A, second column B, etc.; first row 1, second row 3) are:

A1: freq (Hz)

B1: freq (rad/s)

C1: H

D1: $|H|$ (A/V)

E1: ∠H(°)

A3: 10

B3: =A3*2*PI()

C3: =IMDIV(COMPLEX(1,0),COMPLEX(1,-500000/B3))

[Note: this cell was entered as "=IMDIV(COMPLEX(1,0),COMPLEX (1,-5E5/B3))"]

D3: =IMABS(C3)

E3: =IMARGUMENT(C3)*180/PI()

The graphs are generated in the spreadsheet program and are shown in Figure 15.5.

The magnitude of the transfer function increases as frequency increases. Why? What happens to the capacitor reactance as frequency increases? What does the denominator of the transfer function approach as frequency increases? Hence, what should the transfer function of this circuit approach as frequency increases? The answers to these questions begin with the dependence of capacitive reactance on frequency:

$$X_C(\omega) = \left(\frac{1}{\omega C}\right) \to 0 \text{ as } \omega \text{ increases} \tag{15.22}$$

Figure 15.5 Transfer Function Frequency Response [Eq. (15.21)]

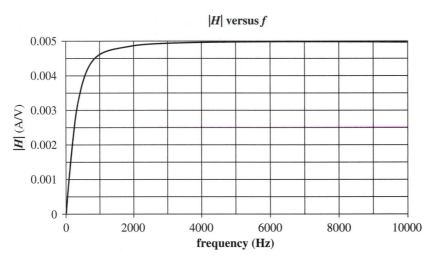

a. Magnitude response

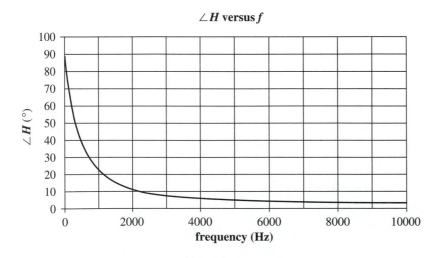

b. Phase response

The denominator of the transfer function is inspected next:

$$\text{denominator of } H(j\omega) = R - j\left(\frac{1}{\omega C}\right) \to R \text{ as } \omega \text{ increases} \tag{15.23}$$

Thus the effect on the transfer function is:

$$H(j\omega) = \frac{\tilde{I}}{\tilde{V}_S} = \frac{1}{R - j\dfrac{1}{\omega C}} \to \frac{1}{R} \text{ as } \omega \text{ increases} \tag{15.24}$$

The transfer function magnitude will approach:

$$|H(j\omega)| = \frac{I}{V_S} \to \frac{1}{R} = 0.005 \text{ A/V} \tag{15.25}$$

as shown on the magnitude graph. You should formulate the reasoning why the transfer function magnitude approaches zero as frequency approaches zero. You should also formulate the reasoning behind the phase response of this transfer function.

Important point on notation: Even though the transfer function is a complex quantity, the tilde is not normally used with it to indicate such. This is a quirk in commonly used notation. Hence $H(j\omega)$ is considered to be complex and $|H(j\omega)|$ is the magnitude of $H(j\omega)$ ($|\,|$ are called magnitude bars). The $(j\omega)$ is a reminder that the transfer function is a complex number quantity that is dependent on ω. Hence $H(j\omega)$ will be used without the tilde.

The next example has a complex algebraic expression in both the numerator and the denominator. Note again that the algebra is handled in the same manner as if the expressions were complex numbers.

____ Example 15.3.1 _____

Given the circuit in Figure 15.6, (*a*) determine the transfer function for the output-to-input phase voltage ratio, and (*b*) plot the magnitude and phase frequency responses.

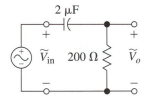

Figure 15.6 RC Circuit for Example 15.3

Given: RC circuit in Figure 15.6

Desired: $H(j\omega) = \dfrac{\tilde{V}_o}{\tilde{V}_{\text{in}}}$, $|H(j\omega)|$ versus f plot, $\angle H(j\omega)$ versus f plot

Strategy: Voltage-divider rule $\to H(j\omega)$

Insert component impedance expressions.

Simplify/clear compound fractions.

Enter equations into a spreadsheet, plot.

Solution: _____

$$\tilde{V}_o = \frac{\tilde{V}_{\text{in}}\tilde{Z}_R}{\tilde{Z}_C + \tilde{Z}_R} \tag{15.26}$$

$$H(j\omega) = \frac{\tilde{V}_o}{\tilde{V}_{\text{in}}} = \frac{\tilde{Z}_R}{\tilde{Z}_C + \tilde{Z}_R} \tag{15.27}$$

$$\tilde{Z}_R = R, \quad \tilde{Z}_C = \frac{1}{j\omega C} \tag{15.28}$$

$$H(j\omega) = \frac{\tilde{V}_o}{\tilde{V}_{\text{in}}} = \frac{R}{R + \dfrac{1}{j\omega C}} \tag{15.29}$$

Often it simplifies equation handling to clear compound fractions (this statement will become even more clear later in this chapter):

$$H(j\omega) = \frac{R}{R + \dfrac{1}{j\omega C}} \cdot \frac{j\omega C}{j\omega C} = \frac{j\omega RC}{1 + j\omega RC} \tag{15.30}$$

As before, these equations could be entered into a spreadsheet to make plots as shown in Figure 15.7. As expected, the low frequencies are blocked by the capacitor, and the high frequencies are passed. The term *blocked* has a specific meaning when used in electronics. It means that the voltage or current is only partially transferred from one point in the circuit to another point in the circuit, usually because of a voltage or current division between components. In the case of the *RC* circuit shown in Figure 15.6, most of the source voltage will drop across the capacitor at low frequencies, and only a small portion of the source voltage will drop across the resistance. Hence the capacitor has "blocked" most of the

Figure 15.7 Transfer Function Frequency Response for Example 15.3.

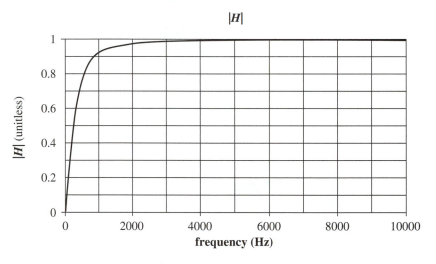

a. Magnitude response

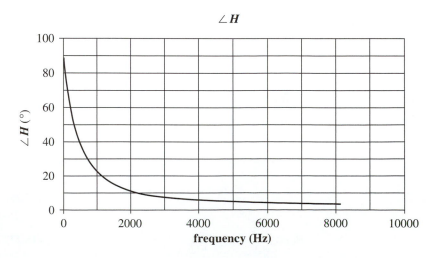

b. Phase response

source voltage at low frequencies from reaching the resistance. This wording is often shortened to "the capacitor blocks low frequencies." The term *passed* has the opposite meaning: most of the voltage (or current) from one point in the circuit reaches the other point of concern in the circuit.

The phase response is not as obvious (can you qualitatively explain it?) but is just as important as the magnitude response when total system response is predicted in later studies.

15.4 BASIC FILTER TYPES AND RESPONSES

As illustrated in the previous examples of this chapter, circuits with reactive components do *not* have a constant response versus frequency because the impedances of reactive components are frequency dependent. In general there are four common types of frequency response, as illustrated in Figure 15.8 for the ideal cases. The circuits that produce such frequency responses are called filters. Lowpass and highpass filters, which produce the frequency responses shown in Figure 15.8(a and b), will now be examined. The bandpass and bandreject frequency responses shown in Figure 15.8(c and d) are usually produced by circuits having *both* inductors and capacitors, and these are examined in Chapter 16.

The lowpass filter (LPF) response, shown in Figure 15.8a, is produced by circuits that *pass* low frequencies and *block* high frequencies. Two common lowpass filters are shown in Figure 15.9. Often the input source is not shown but is assumed when a circuit is shown and $H(j\omega)$ is the quantity of concern. However, the voltages and/or the current that make up the $H(j\omega)$ ratio are clearly labeled. In the RL circuit shown in Figure 15.9a, the series inductor passes low frequencies and blocks high frequencies. In the RC circuit shown in Figure 15.9b, the parallel capacitor "shorts" high frequencies to ground; thus the filter only passes low frequencies to the output. An example of an application of LPFs is to block out high-frequency audio noise in a voice communication channel.

The highpass filter (HPF) response, shown in Figure 15.8b, is produced by circuits that pass high frequencies and block low frequencies. Two common highpass filters are

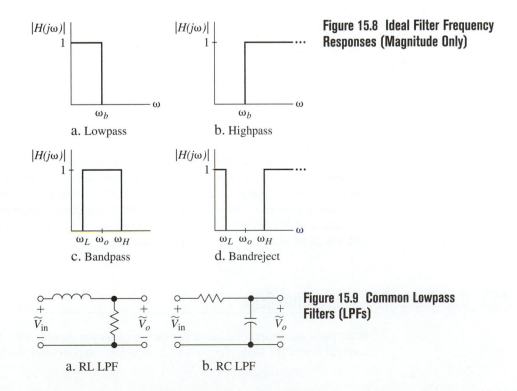

Figure 15.8 Ideal Filter Frequency Responses (Magnitude Only)

a. Lowpass

b. Highpass

c. Bandpass

d. Bandreject

Figure 15.9 Common Lowpass Filters (LPFs)

a. RL LPF

b. RC LPF

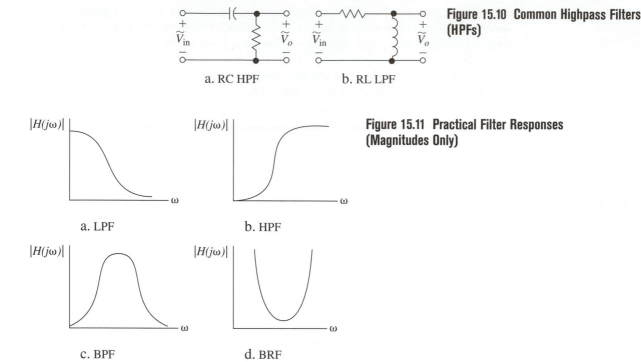

Figure 15.10 Common Highpass Filters (HPFs)

a. RC HPF   b. RL LPF

Figure 15.11 Practical Filter Responses (Magnitudes Only)

a. LPF   b. HPF

c. BPF   d. BRF

shown in Figure 15.10. In the RC circuit shown in Figure 15.10a, the series capacitor passes high frequencies and blocks low frequencies. In the RL circuit shown in Figure 15.10b, the parallel inductor "shorts" low frequencies to ground; thus the filter only passes high frequencies to the output. An example of an application of HPFs is in speaker networks where a HPF can be used to route high frequencies to the tweeter speaker. A LPF can be used to route low frequencies to the woofer speaker.

Example circuits for bandpass filters (BPFs) and bandreject filters (BRFs) will be covered in Chapter 16. A bandpass filter passes a band of frequencies and rejects the rest. It is often used in channel selector applications, such as for radio or television channels. A bandreject filter rejects a band of frequencies and passes the rest. It is often used to reject an interfering signal, such as 60 Hz, that interferes with systems. An example of an interfering signal is 60 Hz "hum" in audio products.

The frequency responses shown in Figure 15.8 are ideal. As shown by previous examples in this chapter, actual frequency responses do not "jump" instantaneously as frequency is changed, but they instead change smoothly over a given frequency range, as shown in Figure 15.11. Recall the RC circuit that was examined in the previous section. The transfer function values do not jump suddenly, but instead change in a smooth manner. The reason for this behavior is in the mathematical expression of the transfer function: a small change of frequency does not produce a huge change in the transfer function magnitude or phase.

We have already covered how to determine the transfer functions and how to plot the magnitude and phase of the frequency responses. Although straightforward, it does take some time and a spreadsheet, too. Would it not be convenient if one could quickly construct the approximate frequency response, both magnitude and phase, including quantitative values of important frequencies and magnitudes? The Bode plot is such a technique, and it is addressed in the remainder of this chapter.

15.5 BODE PLOTS—TRANSFER FUNCTION SETUP

Bode plots are an approximation to exact frequency response plots. The technique behind construction of the Bode plot allows one to quickly approximate the frequency response of

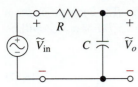

Figure 15.12 LPF Circuit for Bode Plot Illustration

the circuit. The key to this technique is *the algebraic manipulation of the transfer function into a standard form.* The technique is illustrated first for a lowpass filter. After the proper form of the transfer function is derived, logarithms and logarithmic scales are reviewed, and the decibel (dB) is introduced in Section 15.6. Then the actual Bode plot for the LPF is constructed in Section 15.7. The Bode plot for a HPF concludes the discussion.

The illustration of the Bode plot technique begins with finding the transfer function of the circuit, as before, but with additional steps: the transfer function is manipulated into polar form. This process is illustrated next for the circuit in Figure 15.12.

Analyze the circuit for the transfer function expression:

$$\tilde{Z}_R = R, \quad \tilde{Z}_C = \frac{1}{j\omega C} \tag{15.31}$$

$\tilde{Z}_C = \frac{1}{j\omega C}$ is used instead of $\tilde{Z}_C = -j\frac{1}{\omega C}$ because it is easier to clear compound fractions later.

$$\tilde{V}_o = \frac{\tilde{V}_{in}\tilde{Z}_C}{\tilde{Z}_C + \tilde{Z}_R} \tag{15.32}$$

$$H(j\omega) = \frac{\tilde{V}_o}{\tilde{V}_{in}} = \frac{\tilde{Z}_C}{\tilde{Z}_C + \tilde{Z}_R} \tag{15.33}$$

Insert the expressions for the impedances (or admittances if used):

$$H(j\omega) = \frac{\tilde{V}_o}{\tilde{V}_{in}} = \frac{\frac{1}{j\omega C}}{R + \frac{1}{j\omega C}} \tag{15.34}$$

Clear compound fractions:

$$H(j\omega) = \frac{\frac{1}{j\omega C}}{R + \frac{1}{j\omega C}} \cdot \frac{j\omega C}{j\omega C} = \frac{1}{1 + j\omega RC} \tag{15.35}$$

This is where the important additional steps in the Bode plot technique come in. *Anywhere there is a real part of a complex number, divide through by the real part to make it equal to one.* In this particular problem, the real part is already one, so this step is not required.

Manipulate all ω terms into ω/ωₐ form:

$$H(j\omega) = \frac{1}{1 + j\omega RC}$$

The next important step is to *manipulate any imaginary term into the form of a frequency variable (ω) over a constant:*

$$H(j\omega) = \frac{1}{1 + j\omega RC} = \frac{1}{1 + j\dfrac{\omega}{\left(\dfrac{1}{RC}\right)}} \tag{15.36}$$

This is the special form that will make the Bode technique so useful. The constant under ω is a very special constant that will be examined closely in Section 15.7. For now it will be called the *break frequency* and will be given the symbol ω_b.

$$H(j\omega) = \frac{1}{1 + j\dfrac{\omega}{\omega_b}} \tag{15.37}$$

where $\omega_b = 1/RC$ in this particular circuit.

Convert both the numerator and denominator into polar form:

$$H(j\omega) = \frac{1\angle 0°}{\sqrt{1^2 + \left(\dfrac{\omega}{\omega_b}\right)^2}\angle\tan^{-1}\left(\dfrac{\left(\dfrac{\omega}{\omega_b}\right)}{1}\right)} = \frac{1\angle 0°}{\sqrt{1 + \left(\dfrac{\omega}{\omega_b}\right)^2}\angle\tan^{-1}\left(\dfrac{\omega}{\omega_b}\right)} \tag{15.38}$$

The zero degree angle in the numerator has no effect on the total angle and can be ignored. Also, with the real part of complex numbers equal to one, the arc tangent expression simplifies.

Finally, separate the magnitude and phase of $H(j\omega)$ into individual equations:

$$|H(j\omega)| = \frac{1}{\sqrt{1 + \left(\dfrac{\omega}{\omega_b}\right)^2}}, \qquad \angle H(j\omega) = \angle -\tan^{-1}\left(\dfrac{\omega}{\omega_b}\right) \tag{15.39}$$

The transfer function is now in a form that will allow us to easily make a Bode plot of the transfer function. Before seeing how this is done (and why it works so well), logarithms and logarithmic scales will be reviewed and the decibel (dB) will be introduced in the next section.

15.6 LOGARITHM REVIEW, LOG SCALES AND GRAPHS, AND GAIN IN DECIBELS

15.6.1 Logarithm Review

A review of logarithms and log graphs follows. Even if you are familiar with these topics, at least skim the material. Subsequent material will be based on these concepts. Then the decibel and its use in log graphs will be established.

Logarithms are based upon the fact that every positive number can be expressed as a power of another number called the base, which is commonly 10. Any positive number y can be expressed in the form b^x, where b is a positive number except 1. In variable form, given $y = b^x$, then $x = \log_b y$, where b is the base. Examples:

$$3 = \log_2 8 \quad \text{because } 2^3 = 8$$

$$4 = \log_3 81 \quad \text{because } 3^4 = 81$$

Thus, the logarithm of a number is the exponent indicating the power to which it is necessary to raise the base in order to equal the given number. If the base is 10, the logarithm is called a *common logarithm:*

$$x = \log_{10} y, \quad x = \log y$$

No subscript after log means base 10. The reverse operation, finding the original number given the logarithm, i.e., the power of 10, is called taking the *antilog* or taking the *inverse log:*

$$y = \text{antilog}(x)$$

$$= \text{inv log}(x)$$

$$= \log^{-1}(x)$$

$$= 10^x$$

On calculators, you must hit the LOG key to take the logarithm of a number. You must use the 10^x key or the INV LOG key to take the antilog of a number. Try these examples on your calculator:

$$10^3 = 1000, \quad \log 1000 = 3$$

$$10^{-4} = 0.0001, \quad \log 0.0001 = -4$$

$$\log 500 = 2.7, \quad 10^{2.7} = 500$$

There are also several laws for mathematical operations with logarithms. These laws are stated below. Refer to a college algebra[4] book for a complete explanation of the laws of logarithms if necessary.

1. multiplication: $\log_b (MN) = \log_b M + \log_b N$
    (The log of a product equals the sum of the logs.)

2. division: $\log_b \left(\dfrac{M}{N}\right) = \log_b M - \log_b N$

    (The log of a quotient equals the log of the numerator minus the log of the denominator.)

3. exponentiation: $\log_b N^P = P \log_b N$
    (The log of a number to a power equals the product of the exponent and the log of the number.)

4. taking roots: $\log_b \sqrt[r]{N} = \left(\dfrac{1}{r}\right) \log_b N$

    (The log of the root of a number equals the log of the number divided by the root.)

Note that division is really multiplication and exponentiation combined:

$$\log_b \left(\frac{M}{N}\right) = \log_b(MN^{-1}) = \log M + \log N^{-1} = \log M + (-1) \log N = \log M - \log N$$

and that taking a root is really a special case of exponentiation:

$$\log_b \sqrt[r]{N} = \log_b N^{\frac{1}{r}} = \left(\frac{1}{r}\right) \log_b N$$

As an example of operations with logarithms, consider the multiplication of two numbers:

$$237 \cdot 452 = \text{product} = 107{,}124 = 107{,}000 \text{ (to 3 significant digits)}$$

If logarithms are used, one would add the logs and convert back:

$$\log (237 \cdot 452) = \log 237 + \log 452 = 2.375 + 2.655 = 5.030 = \log (\text{product})$$

$$\text{product} = \log^{-1} 5.030 = 10^{5.030} = \text{antilog } 5.030 = 107{,}152 = 107{,}000$$

Similarly, to divide numbers, one would subtract the logs and convert back. For example,

$$55 \div 2.4 = \frac{55}{2.4} = \text{quotient} = 22.92 = 23 \qquad \text{(to 2 significant digits)}$$

Using logarithms:

$$\log\left(\frac{55}{2.4}\right) = \log 55 - \log 2.4 = 1.360 = \log (\text{quotient})$$

$$\text{quotient} = 10^{1.360} = 22.91 = 23$$

---

[4] See Kuhfittig, *Basic Technical Mathematics with Calculus,* chapter 12, for example.

TABLE 15.4	x	$10 \log(x)$	x	$10 \log(x)$
Example Set of Numbers Showing Number Range Expansion and Compression with Logarithms (Note: Two sets of column.)	0.01	−20	1	0
	0.02	−16.9897	2	3.0103
	0.03	−15.2288	3	4.771213
	0.04	−13.9794	4	6.0206
	0.05	−13.0103	5	6.9897
	0.06	−12.2185	6	7.781513
	0.07	−11.549	7	8.45098
	0.08	−10.9691	8	9.0309
	0.09	−10.4576	9	9.542425
	0.1	−10	10	10
	0.2	−6.9897	20	13.0103
	0.3	−5.22879	30	14.77121
	0.4	−3.9794	40	16.0206
	0.5	−3.0103	50	16.9897
	0.6	−2.21849	60	17.78151
	0.7	−1.54902	70	18.45098
	0.8	−0.9691	80	19.0309
	0.9	−0.45757	90	19.54243
	1	0	100	20

These methods of mathematical operations were commonly utilized for numerical calculations before calculators became readily available. Now, logarithms are rarely used for this purpose. However, one must still know, understand, and be able to use the laws of logarithms when *variables* are involved, as will be illustrated in the next section.

What are logs used for in a practical sense? They are used to compress a large range of numbers into a manageable range. They are also used to expand a small range of numbers into a manageable range. See the data in Table 15.4, for example. The numbers from 0.01 to 100 are listed in the "x" column. The spread in the data is 99.99 (100 − 0.01). The range of data points in the "x" column of Table 15.4 are subdivided into four ranges:

$$0.01 \text{ to } 0.1 \text{ in increments of } 0.01$$

$$0.1 \text{ to } 1.0 \text{ in increments of } 0.1$$

$$1 \text{ to } 10 \text{ in increments of } 1$$

$$10 \text{ to } 100 \text{ in increments of } 10$$

Each number in the left column (x) is inserted into the equation $10 \log(x)$ and the result is placed in the "$10 \log(x)$" column (the reason for multiplying the $\log(x)$ by 10 will be discussed later in this section). What is the spread in values for $10 \log(x) +20 − (−20) = 40$? The subranges are:

$$0.01 \text{ to } 0.1 \text{ in increments of } 0.01 \Rightarrow −20 \text{ to } −10$$

$$0.1 \text{ to } 1.0 \text{ in increments of } 0.1 \Rightarrow \quad −10 \text{ to } 0$$

$$1 \text{ to } 10 \text{ in increments of } 1 \Rightarrow \quad 0 \text{ to } +10$$

$$10 \text{ to } 100 \text{ in increments of } 10 \Rightarrow \quad +10 \text{ to } +20$$

The 0.01 to 1.0 range of the numbers, i.e., the numbers that were less than or equal to one with a range of 0.99, were expanded into a range of −20 to 0, i.e., a range of 20. The 1 to 100 range of numbers, i.e., the numbers that were greater than or equal to one with a range of 99, were compressed into a range of 0 to +20, i.e., a range of 20. Hence, *logarithms compress large ranges of numbers and expand small ranges of numbers.*

A primary application for this use of logarithms in circuits is for determining the gain of a circuit with a convenient range of numbers. The gain of a circuit is related to its transfer function, which will be examined shortly. The logarithm is also used for the frequency axis in frequency response plots for circuits, as discussed next.

15.6.2 Log Scales and Graphs

Another use of logarithms is to build logarithmic scales. For example, the x-axis in a frequency response plot, i.e., frequency, is usually plotted on a logarithmic scale. The reason that log scales are used is similar to the reason logarithms are used. The numbers in Table 15.4 will be used to illustrate this problem and solution. Consider a linear scale, i.e., a scale with uniform spacing of the divisions, as shown in Figure 15.13a. This scale is common in x-y graphs. The numbers in the "x" column of Table 15.4 are plotted on the linear scale in Figure 15.13a. Notice how the numbers in the range of 0.1 – 10 are so closely spaced that the markers on the graph overlap each other and are not distinguishable. On the other hand, the numbers in the range 10–100 are spaced rather far apart, and any variations between the points would not be shown. These effects illustrate the problem with linear scales that is solved by plotting on a log scale.

A log scale is a scale where the divisions are separated logarithmically, that is, by the powers of 10. A log scale will be constructed to illustrate this concept. The same set of numbers in Table 15.4 will be replotted as follows. Take the log of each number and multiply each result by 10 (the multiplication by 10 will be justified later). The results of these calculations are listed in the "10 log(x)" column in Table 15.4. Plot these results on a linear scale, as shown in Figure 15.13b. Notice how the entire range of 0.01 to 100, a factor of 10^4, is now in a manageable range of -20 to $+20$, a range of 40.

Now, plot the original numbers on a log scale, as shown in Figure 15.13c. The locations of the points in Figures 15.13 (b and c) are identical. What can you conclude from this observation?

Taking the log of numbers and plotting them on a linear scale is equivalent to directly plotting the original numbers on a log scale.

However, the log scale has original numbers marked, not the log of the numbers as on the linear scale. Plotting directly on a log scale saves you from taking the log of the original

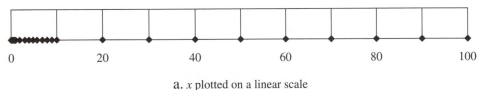

Figure 15.13 Linear and Log Scales

0   20   40   60   80   100

a. x plotted on a linear scale

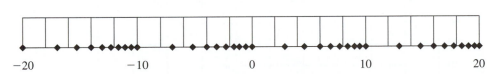

-20 -10 0   10   20

b. 10 log(x) plotted on a linear scale

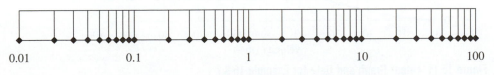

0.01   0.1   1   10   100

c. x plotted on a log scale

numbers. Note that a log scale is naturally divided into *decades* (factor of 10 ranges). Examples of decades: 0.01 to 0.1, 0.1 to 1, 1 to 10, 10 to 100, 100 to 1000, etc. (on both ends of this list).

Another important observation from Figure 15.13c is that a log scale *never* reaches zero. In fact the decades in the region between zero and one on the linear scale become an infinite number of decades on a log scale that each occupy an equal length on the scale as do the decades above one. Each decade is given equal spacing on the log scale. Thus:

The space on a log scale is the same for any decade,

whether from 0.01 to 0.1 or 10 to 100. Thus the data points in a decade of small numbers are just as distinguishable as the data points in a decade of large numbers, as illustrated in Figure 15.13c.

Now that log scales have been established, they may be used in graphs. A *log–log* graph has a log scale on both the *x* and *y* axes. The following example will illustrate its use.

▬ Example 15.6.1

Plot the kinetic energy versus velocity for the data in Table 15.5 on (*a*) a linear graph and (*b*) a log–log graph.

TABLE 15.5	velocity (m/s)	kinetic energy (J)
Data for Example 15.6.1	2	40
	7	490
	11	1210
	15	2250
	28	7840
	32	10240
	46	21160
	63	39690
	88	77440

Solution:

The data from Table 15.5 is plotted on a linear graph in Figure 15.14 and is plotted on a log–log graph in Figure 15.15. Note the curved line on the linear graph became a straight

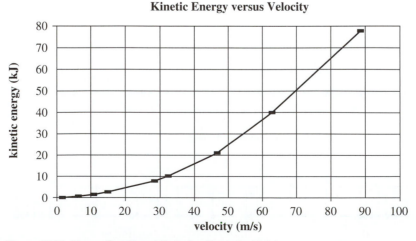

Figure 15.14 Linear Graph and Data for Example 15.6.1

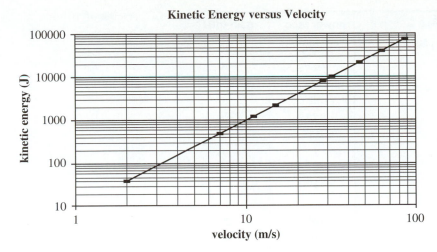

Kinetic Energy versus Velocity

Figure 15.15 Log–Log Graph and Data for Example 15.6.1

line on the log–log graph. This occurrence is not true in general but is true for the kinetic energy–velocity relationship.[5] Straight lines on log scales will also be used in Bode plots, as illustrated later in this chapter.

15.6.3 Gain in Decibels

The use of logarithms and log scales in transfer functions and frequency response plots will be covered now. Reinspect the log scale in Figure 15.13c. Notice that the *original* numbers appear on a logarithmic scale. This is convenient for frequency, where one uses prefixes such as k and M to express large numbers, but it becomes awkward for gain.

On the vertical axis of the magnitude Bode plot, the transfer function magnitude is related to a quantity called *gain*. The *power gain* of a circuit is defined as the output power over the input power:

$$G \equiv \frac{P_o}{P_{in}} \tag{15.40}$$

Gain is a unitless ratio. It is sometimes used for voltage and current ratios also, so one must know or determine from context which gain is being used. Power gain is being examined here. The specific problem with gain as a ratio is that the large spread (range) of numbers becomes awkward. For example, a given filter may have a transfer function magnitude that varies from 1 to $0.00000001 = 1 \times 10^{-8}$. To compress this range of numbers, the decibel (dB) is used. The basis for the decibel is the Bel, which is defined as:

$$G(\text{Bels}) \equiv \log\left(\frac{P_o}{P_{in}}\right) \tag{15.41}$$

where G is the power gain in Bels. Gain may be greater than one ($G = 1 = 0$ Bel), such as for amplifiers, or it may be less than one, such as for circuits with resistors, capacitors, and inductors only. Note that, fundamentally, the Bel is not a unit because the units of a power ratio cancel; instead, it indicates the log of a power ratio. Equation (15.41) tends to produce numbers that are somewhat small. For example, a power ratio of ½ is -0.3 Bel. Thus, the convention of the decibel (dB) is most commonly used:

$$G(\text{dB}) \equiv 10 \log\left(\frac{P_o}{P_{in}}\right) \tag{15.42}$$

---

[5] The interested reader should investigate *power relationships* and the related topic of *exponential relationships*.

TABLE 15.6	power ratio	gain (dB)
Power Ratios and Associated Gains in dB	0.0001	−40
	0.001	−30
	0.01	−20
	0.1	−10
	0.25	−6
	0.5	−3
	1	0
	2	3
	4	6
	10	10
	100	20
	1000	30
	10000	40

TABLE 15.7	voltage ratio	gain (dB)
Voltage Ratios and Associated Gains in dB	0.0001	−80
	0.001	−60
	0.01	−40
	0.1	−20
	0.5	−6
	1	0
	2	6
	10	20
	100	40
	1000	60
	10000	80

The multiplying factor of 10 produces numbers that are in 1s and 10s, "convenient" gain numbers with which to work (recall that this same factor was used in the log scales discussion of Figure 15.13). A short listing of power ratios and gains is given in Table 15.6.

How is the gain in decibels utilized in Bode plots? The definition of gain is extended to include all transfer function ratios. The most common ratio is the output voltage to input voltage. If the powers in Equation (15.42) are expressed in terms of voltages, then:

$$G(\text{dB}) = 10 \log \left(\frac{\frac{V_o^2}{R_o}}{\frac{V_{in}^2}{R_{in}}} \right) \tag{15.43}$$

Here is where the gain in decibels is extended to include voltage ratios. The resistances are "ignored" by assuming they have the same value so that a voltage ratio is formed:

$$G(\text{dB}) = 10 \log \left(\frac{V_o^2}{V_{in}^2} \right) = 10 \log \left(\frac{V_o}{V_{in}} \right)^2 = 20 \log \left(\frac{V_o}{V_{in}} \right) = 20 \log |H(j\omega)| \tag{15.44}$$

The overall result is:

$$G(\text{dB}) = 20 \log |H(j\omega)| \tag{15.45}$$

where it is understood that the resistances are not included, i.e., that the gain in decibels on Bode plots is 20 times the log of the transfer function ratio. Note that if H is a voltage ratio, then the voltage ratio-to-gain in decibels relationship is different than the power ratio-to-gain in decibels relationship, as illustrated in Table 15.7 (compare to Table 15.6). The numbers are still in a convenient, manageable range.

15.7 BODE PLOTS—PLOTTING FROM THE TRANSFER FUNCTION

Now that logarithms have been reviewed, we are in a position to construct Bode plots. What are Bode plots? Bode plots are *approximations* to the frequency response plots of a circuit. The approximation is that the frequency response over finite frequency ranges is approximated by straight line segments. "Plots" is used in the plural sense because there are both magnitude and phase plots. However, the use of straight lines is valid only if these plots are made on semilog graphs, where the frequency scale is logarithmic and represents the independent variable on the x axis. The linear scale is the magnitude in decibels or phase in degrees of the transfer function and represents the dependent variable on the y axis. Typical Bode plot grids are shown in Figure 15.16. The grid is called a semilog graph because only one scale, the frequency scale, is logarithmic. The magnitude and phase plots are discussed next.

A Bode plot for the magnitude of the transfer function will be *gain (dB) on a linear scale for the vertical axis versus frequency on a log scale for the horizontal axis.* Recall that taking the log of a set of numbers and plotting them on a linear scale is identical to

Figure 15.16 Typical Bode Plot Grids

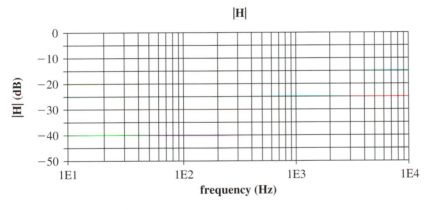

a. Bode plot grid for the magnitude of H

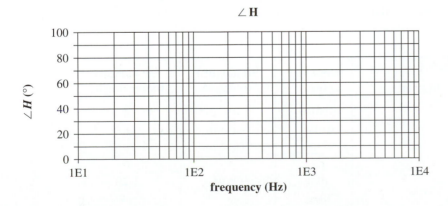

b. Bode plot grid for the phase of H

plotting the original numbers on a log scale (can you justify this? If not, reread section 15.6.). The difference is that on the log scale, the original numbers are used to label the axis. On a linear scale, the log of the numbers are used. These numbers are, in this case, the gain in decibels, which contains the log of the numbers. The vertical scale (y-axis) is linear, but again is made logarithmic when the data is in decibel form.

Frequency is plotted directly on the log scale on the horizontal axis. Thus the original frequency numbers appear on the horizontal axis, but the scale is logarithmic.

The phase of the transfer function is directly plotted on a linear scale for the vertical axis (no logs for phase). The horizontal axis is the same: frequency is plotted on the log scale.

The example in Section 15.5 will now be continued to illustrate how to make the Bode plot approximation of the frequency response of a circuit. The magnitude and the phase of the transfer function for the circuit of Section 15.5, Equation (15.39), is repeated in Equation (15.46):

$$|H(j\omega)| = \frac{1}{\sqrt{1 + \left(\dfrac{\omega}{\omega_b}\right)^2}}, \quad \angle H(j\omega) = \angle -\tan^{-1}\left(\frac{\omega}{\omega_b}\right) \tag{15.46}$$

where $\omega_b = \dfrac{1}{RC}$. Convert the magnitude of $H(j\omega)$ in Equation (15.46) into decibel form:

$$|H(j\omega)|(\text{dB}) = 20\log|H(j\omega)| = 20\log\left[\frac{1}{\sqrt{1 + \left(\dfrac{\omega}{\omega_b}\right)^2}}\right] \tag{15.47}$$

The square root in the denominator can be moved to the numerator:

$$|H(j\omega)|(\text{dB}) = 20\log|H(j\omega)| = 20\log\left\{\left[\sqrt{1 + \left(\frac{\omega}{\omega_b}\right)^2}\right]^{-1}\right\} = -20\log\sqrt{1 + \left(\frac{\omega}{\omega_b}\right)^2} \tag{15.48}$$

where the power is "brought out front" of the log by the rules of exponents inside logarithms. Thus the magnitude of the transfer function in decibels for this RC LPF is:

$$|H(j\omega)|(\text{dB}) = -20\log\sqrt{1 + \left(\frac{\omega}{\omega_b}\right)^2}, \quad \omega_b = \frac{1}{RC} \tag{15.49}$$

Examine some typical values for $|H(j\omega)|$. Insert frequencies from 0.001 to 1000 times the break frequency into Equation (15.49), as shown in Table 15.8. Note that for frequencies less than the break frequency, the magnitude of the square-root term is approximately 1 and the $|H(j\omega)|$ is 0 dB. However, for frequencies greater than the break frequency, the magnitude has a -20 dB/dec (decade) slope. The change in slope occurs at the break fre-

TABLE 15.8 $\qquad$ $\|H(j\omega)\|$ Calculations for Equation (15.49)	frequency (ω)	$\|H(j\omega)\|$(dB) (precise)	$\|H(j\omega)\|$(dB) (approximate)
	$0.001\omega_b$	-4×10^{-6}	0
	$0.01\omega_b$	-4×10^{-4}	0
	$0.1\omega_b$	-4×10^{-2}	0
	$1\omega_b$	-3.01	0
	$10\omega_b$	-20.04	-20
	$100\omega_b$	-40	-40
	$1000\omega_b$	-60	-60

quency. These observations are the key to the Bode magnitude plot. Why does the $-20\,\text{dB/}$ decade slope occur? For frequencies significantly greater than the break frequency,

$$\left(\frac{\omega}{\omega_b}\right) \gg 1 \tag{15.50}$$

Then the square root approaches:

$$\sqrt{1 + \left(\frac{\omega}{\omega_b}\right)^2} \rightarrow \sqrt{\left(\frac{\omega}{\omega_b}\right)^2} \rightarrow \left(\frac{\omega}{\omega_b}\right) \tag{15.51}$$

and the transfer function approaches:

$$|H(j\omega)|(\text{dB}) = -20\log\sqrt{1 + \left(\frac{\omega}{\omega_b}\right)^2} \rightarrow -20\log\left(\frac{\omega}{\omega_b}\right) \tag{15.52}$$

The power of ω is one. Hence every factor of 10 increase in ω increases the log by 1. This changes $|H(j\omega)|$ by an increment of 20 because of the -20 multiplying factor in front of the log. Hence, $-20\,\text{dB/decade}$ is the slope of the straight line above the break frequency on the magnitude Bode plot. The reader should reflect on how the laws of logarithms and log scales lead to such the simple yet important result of 20 dB/decade on a magnitude Bode plot.

In order to illustrate this circuit performance approximation, insert actual component values and make some plots. If $R = 10\,\text{k}\Omega$ and $C = 0.1\,\mu\text{F}$, the break frequency for this circuit is:

$$\omega_b = \frac{1}{RC} = \frac{1}{(10\,\text{k}\Omega)(0.1\,\mu)} = 1000\,\text{rad/s} \tag{15.53}$$

This ω_b is the frequency at which the magnitude response "breaks" from 0 dB. This point is the transition from one straight line to another straight line of a different slope. The actual frequency response plot (magnitude) and the Bode magnitude plot are shown together in Figure 15.17 for comparison. When ω is significantly below ω_b, the 1 under the square root in Equation (15.49) dominates and the $|H(j\omega)|$ is 0 dB. When ω is significantly above ω_b, the squared term dominates and $|H(j\omega)|$ changes at a rate of 20 dB/decade starting at ω_b (in this case, in the negative dB direction). This is the power of the Bode plot: *once the break frequency is known, the magnitude plot can be automatically plotted.* One does *not* calculate $|H(j\omega)|$ at various frequencies (such as Table 15.8). One plots $|H(j\omega)|$ at the break frequency and then plots straight line segments of the proper slopes above and below the break frequency. However, the Bode plot is an approximation, especially in the

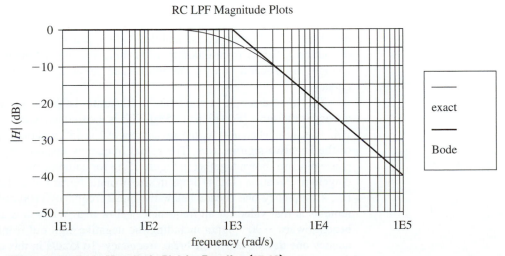

Figure 15.17 Bode Magnitude Plot for Equation (15.49)

TABLE 15.9	frequency	$\angle H(j\omega)$—precise	$\angle H(j\omega)$—approximate
$\angle H(j\omega)$ Calculations for Equation (15.54)	$0.001\ \omega_b$	$-0.06°$	$0°$
	$0.01\ \omega_b$	$-0.6°$	$0°$
	$0.1\ \omega_b$	$-5.7°$	$0°$
	$1\ \omega_b$	$-45°$	$-45°$
	$10\ \omega_b$	$-84.3°$	$-90°$
	$100\ \omega_b$	$-89.4°$	$-90°$
	$1000\ \omega_b$	$-89.9°$	$-90°$

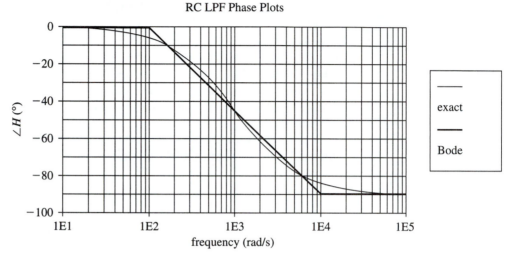

Figure 15.18 **Bode Phase Plot for Equation (15.54)**

vicinity of the break frequency. Note the 3 dB error at $\omega = \omega_b$ in the Bode approximation versus the exact calculation. Again, Bode plots are usually used for rapid estimates and checks of circuit performance. They are also used to help one think about and visualize circuit behavior when designing circuits. If the exact frequency response is needed, one should plot the exact transfer function.

The phase of the Bode plot is similarly analyzed. The phase of the transfer function from Equation (15.39) is:

$$\angle H(j\omega) = \angle -\tan^{-1}\left(\frac{\omega}{\omega_b}\right) \tag{15.54}$$

See Table 15.9 for the precise and the Bode approximate values for the angle of $H(j\omega)$ [Equation (15.54)] from 0.001 to 1000 times the break frequency. Note that the phase changes significantly only from one decade below to one decade above the break frequency.

The same example is continued to illustrate the phase performance. The break frequency is the same [Equation (15.53)]. This is the frequency at which the phase response is $-45°$ because when $\omega = \omega_b$ the arc tangent of 1 is 45°, and the total angle is $-45°$ because of the negative sign in front. The actual frequency response plot (phase) and the Bode phase plot are shown together in Figure 15.18 for comparison. When ω is below ω_b, the arc tangent in Equation (15.54) drops in size of angle toward 0° and is effectively 0° at $0.1\omega_b$, i.e., a frequency one decade below the break frequency (100 rad/s in this example), and remains at that value for all frequencies below that. When ω is above ω_b, the arc tangent heads toward $-90°$ (again including the negative sign out front) at $10\omega_b$, i.e., at a frequency one decade above the break frequency (10 krad/s in this example), and remains at that value for all frequencies above that. The power of the Bode plot is that *once the break*

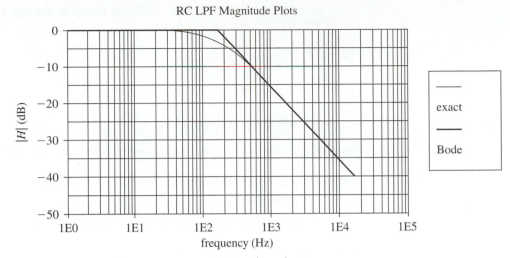

Figure 15.19 Bode Magnitude Plot for Equation (15.49)

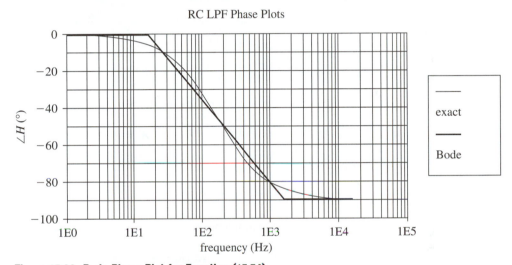

Figure 15.20 Bode Phase Plot for Equation (15.54)

frequency is known, the phase plot can be automatically plotted. Again, one does *not* calculate $\angle H(j\omega)$ at various frequencies (such as Table 15.9). One plots $\angle H(j\omega)$ at the break frequency, at one tenth of the break frequency, and at ten times the break frequency. Then straight line segments are plotted in the three frequency ranges (up to $0.1\omega_b$, $0.1\omega_b$ to $10\omega_b$, and above $10\omega_b$).

The frequency in Bode plots is usually expressed in hertz, not radians per second, because AC signals are normally specified and measured in hertz. The plots of Figures 15.17 and 15.18 are repeated in Figures 15.19 and 15.20, respectively, with the frequency expressed in hertz. The frequency in hertz is directly obtained by adding another column in a spreadsheet. The break frequency is:

$$f_b = \frac{\omega_b}{2\pi} = \frac{1000}{2\pi} = 159.155 \text{ Hz} \approx 160 \text{ kHz}$$

Two more examples are given to illustrate the Bode plot procedures.

Example 15.7.1

a. Determine the transfer function and Bode plots for the circuit in Figure 15.21.
b. Determine the filter type.

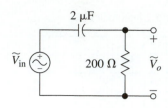

Figure 15.21 RC Circuit for Example 15.7.1

Given: RC circuit in Figure 15.21

Desired: $H(j\omega) = \dfrac{\tilde{V}_o}{\tilde{V}_{in}}$, $|H(j\omega)|$ versus f plot, $\angle H(j\omega)$ versus f plot

Strategy: Determine and manipulate the transfer function into Bode form.

Draw the Bode plots from the transfer functions.

Determine the filter type from the magnitude plot.

Solution:

Determine the transfer function:

$$\tilde{V}_o = \frac{\tilde{V}_{in}\,\tilde{Z}_R}{\tilde{Z}_C + \tilde{Z}_R}$$

$$H(j\omega) = \frac{\tilde{V}_o}{\tilde{V}_{in}} = \frac{\tilde{Z}_R}{\tilde{Z}_C + \tilde{Z}_R}$$

Insert the expressions for the impedances:

$$\tilde{Z}_R = R \qquad \tilde{Z}_C = \frac{1}{j\omega C}$$

$$H(j\omega) = \frac{\tilde{V}_o}{\tilde{V}_{in}} = \frac{R}{R + \dfrac{1}{j\omega C}}$$

Clear compound fractions:

$$H(j\omega) = \frac{R}{R + \dfrac{1}{j\omega C}} \cdot \frac{j\omega C}{j\omega C} = \frac{j\omega RC}{1 + j\omega RC}$$

(If the real part of the complex number were not equal to one, then divide through by the real part to make it so—see Example 15.7.2).

Manipulate all ω terms into ω/ω_b form:

$$H(j\omega) = \frac{j\dfrac{\omega}{\left(\dfrac{1}{RC}\right)}}{1 + j\left[\dfrac{\omega}{\left(\dfrac{1}{RC}\right)}\right]}$$

$$\omega_b = \frac{1}{RC} = \frac{1}{(200)(2\,\mu)} = 2.5 \text{ krad/s}$$

$$f_b = \frac{\omega_b}{2\pi} = 397.89 \text{ Hz} \approx 400 \text{ Hz}$$

$$H(j\omega) = \frac{j\dfrac{\omega}{\omega_b}}{1 + j\dfrac{\omega}{\omega_b}}$$

Convert both the numerator and denominator into polar form:

$$H(j\omega) = \frac{\dfrac{\omega}{\omega_b} \angle 90°}{\sqrt{1^2 + \left[\dfrac{\omega}{\omega_b}\right]^2} \angle \tan^{-1}\left[\dfrac{\omega}{\omega_b}\right]}$$

Finally, separate the magnitude and phase:

$$|H(j\omega)| = \frac{\dfrac{\omega}{\omega_b}}{\sqrt{1 + \left(\dfrac{\omega}{\omega_b}\right)^2}}, \quad \angle H(j\omega) = \angle +90° - \tan^{-1}\left(\frac{\omega}{\omega_b}\right)$$

$$|H(j\omega)|_{dB} = +20 \log\left(\frac{\omega}{\omega_b}\right) - 20 \log \sqrt{1 + \left(\frac{\omega}{\omega_b}\right)^2}$$

$$\omega_b = 2.5 \text{ krad/s}, \quad f_b \approx 400 \text{ Hz}$$

The Bode plots are sketched in Figure 15.22. With two terms in $|H|$, one must plot each term and then add them graphically.

In the magnitude plot, the numerator term gives a straight line with a +20 dB/decade slope that crosses the 0 dB axis at 400 Hz. Note that this line does not break at ω_b because the numerator of this transfer function does not include "1" in front of $j\left(\dfrac{\omega}{\omega_b}\right)$.

The denominator term is 0 dB until 400 Hz. Then it has a slope of −20 dB/decade as frequency increases. When the two traces are added for frequencies below 400 Hz, the numerator trace is added to the 0 dB of the denominator trace. Hence the total trace is identical with the numerator trace. For frequencies above 400 Hz, the numerator trace continues to increase at a slope of +20 dB/decade. The denominator decreases at a slope of −20 dB/decade. The two effects cancel when they are summed to form the total trace. Hence, above 400 Hz, the total magnitude trace is a straight horizontal line at 0 dB.

Likewise, the two traces of the phase Bode plot must be added. The numerator term is +90°; hence the trace is just a straight horizontal line at +90°. The denominator term gives a trace identical to the trace of the LPF covered previously. At frequencies of $0.1f_b = 40$ Hz or less, the arc tangent is approximately 0°. At $f = f_b = 400$ Hz, the arc tangent is 45°, but there is a negative sign because the term was in the denominator of the transfer function, so the angle of the denominator at $f = f_b$ is −45°. At frequencies of $10f_b = 4$ kHz or greater, the arc tangent is 90°, but again there is a negative, so the angle of the denominator at $f \geq 10f_b$ is −90°. The two traces are added to produce the total trace. Again, one can see how

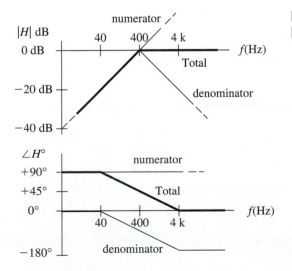

Figure 15.22 Magnitude and Phase Bode Plots for Example 15.7.1

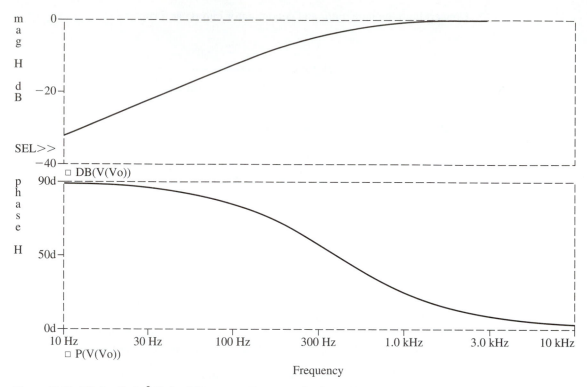

Figure 15.23 PSpice Probe[6] Plots of Frequency Response for Example 15.7.1

straightforward it is to generate a Bode plot once the logic behind the transfer function is understood.

For comparison, the results of a circuit simulation are shown in Figure 15.23. The schematic of Figure 15.21 was entered into the program along with the component values. An AC sweep was selected and the frequency range of 10 Hz to 10 kHz was entered with 101 points per decade. The magnitude and phase plots were generated using the simulation software. Notice that the Bode plots differ from the actual frequency response plots in the vicinity of "corners," i.e., at the break frequency on the magnitude plots and at $0.1f_b$ and at $10f_b$ on the phase plots. Again, Bode plots are approximations, so if a more precise plot is required, plot directly from the transfer function.

From the plots, one concludes that the circuit has more loss as the frequency is lowered. At 400 Hz and above, the circuit passes the signal with a constant response. Hence this circuit is a highpass filter (HPF).

Some transfer functions require more algebra to be manipulated into Bode form. The strategy is the same, but strategic grouping is necessary, as illustrated by the next example.

Example 15.7.2

(*a*) Determine the transfer function for the circuit in Figure 15.24. (*b*) Draw and label the Bode plots based on the transfer function. (*c*) Determine the type of filter that the circuit in Figure 15.24 is.

Given: *RC* circuit in Figure 15.24

---

[6] The Probe plot usually has a black background. The background was converted to white for readability in this text.

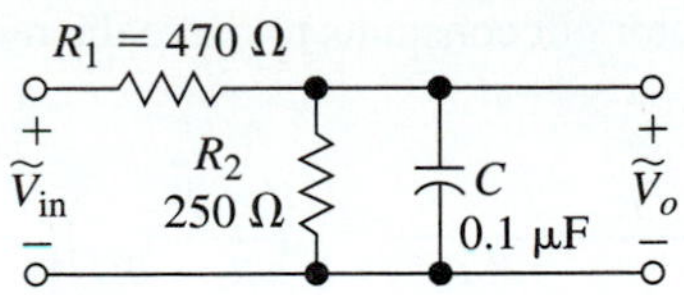

$R_1 = 470\ \Omega$

Figure 15.24 *RC* circuit for Example 15.7.2

R_2
$250\ \Omega$

C
$0.1\ \mu\text{F}$

$\tilde{V}_{in}$ $\tilde{V}_o$

Desired: $H(j\omega) = \dfrac{\tilde{V}_o}{\tilde{V}_{in}}$, $|H(j\omega)|$ versus f plot, $\angle H(j\omega)$ versus f plot, filter type

Strategy: Determine and manipulate the transfer function into Bode form.

Draw the Bode plots from the transfer functions.

Determine the filter type from the magnitude plot.

Solution:

Determine the transfer function:

$$\tilde{Z}_O = \tilde{Z}_C \parallel \tilde{Z}_{R2} = \frac{1}{\dfrac{1}{R_2} + j\omega C}$$

$$\tilde{V}_o = \frac{\tilde{V}_{in}\,\tilde{Z}_O}{\tilde{Z}_{R1} + \tilde{Z}_O}$$

$$H(j\omega) = \frac{\tilde{V}_o}{\tilde{V}_{in}} = \frac{\tilde{Z}_O}{\tilde{Z}_{R1} + \tilde{Z}_O}$$

Insert the expressions for the impedances:

$$\tilde{Z}_{R1} = R_1, \quad \tilde{Z}_O = \frac{1}{\dfrac{1}{R_2} + j\omega C}$$

$$H(j\omega) = \frac{\tilde{V}_o}{\tilde{V}_{in}} = \frac{\dfrac{1}{\dfrac{1}{R_2} + j\omega C}}{R_1 + \dfrac{1}{\dfrac{1}{R_2} + j\omega C}}$$

Clear compound fractions:

$$H(j\omega) = \frac{\dfrac{1}{\dfrac{1}{R_2} + j\omega C} \cdot \dfrac{R_2}{R_2}}{R_1 + \dfrac{1}{\dfrac{1}{R_2} + j\omega C} \cdot \dfrac{R_2}{R_2}} = \frac{\dfrac{R_2}{1 + j\omega R_2 C}}{R_1 + \dfrac{R_2}{1 + j\omega R_2 C}}$$

$$H(j\omega) = \frac{\dfrac{R_2}{1 + j\omega R_2 C}}{R_1 + \dfrac{R_2}{1 + j\omega R_2 C}} \cdot \frac{1 + j\omega R_2 C}{1 + j\omega R_2 C} = \frac{R_2}{R_1(1 + j\omega R_2 C) + R_2}$$

$$H(j\omega) = \frac{R_2}{R_1 + j\omega R_1 R_2 C + R_2} = \frac{R_2}{(R_1 + R_2) + j\omega R_1 R_2 C}$$

Either divide through or factor out constants to make the real part of complex numbers equal to 1:

$$H(j\omega) = \frac{R_2}{(R_1 + R_2)\left[1 + j\dfrac{\omega R_1 R_2 C}{R_1 + R_2}\right]} = \left(\frac{R_2}{R_1 + R_2}\right)\left(\frac{1}{1 + j\dfrac{\omega R_1 R_2 C}{R_1 + R_2}}\right)$$

Manipulate all ω terms into ω/ω_b form:

$$H(j\omega) = \left(\frac{R_2}{R_1 + R_2}\right)\frac{1}{1 + j\dfrac{\omega}{\left(\dfrac{R_1 + R_2}{R_1 R_2 C}\right)}}$$

$$\omega_b = \frac{R_1 + R_2}{R_1 R_2 C} = \frac{470 + 250}{(470)(250)(0.1\,\mu F)} = 61.28 \text{ krad/s}$$

$$f_b = 9.753 \text{ kHz} \approx 10 \text{ kHz}$$

$$H(j\omega) = \left(\frac{R_2}{R_1 + R_2}\right)\frac{1}{1 + j\dfrac{\omega}{\omega_b}}$$

Convert both the numerator and the denominator into polar form:

$$H(j\omega) = \frac{\left(\dfrac{R_2}{R_1 + R_2}\right)}{\sqrt{1^2 + \left[\dfrac{\omega}{\omega_b}\right]^2}\angle\tan^{-1}\left[\dfrac{\omega}{\omega_b}\right]}$$

Finally, separate the magnitude and phase:

$$|H(j\omega)| = \frac{\left(\dfrac{R_2}{R_1 + R_2}\right)}{\sqrt{1 + \left(\dfrac{\omega}{\omega_b}\right)^2}}, \quad \angle H(j\omega) = \angle -\tan^{-1}\left(\frac{\omega}{\omega_b}\right)$$

$$|H(j\omega)|_{dB} = +20\log\left(\frac{R_2}{R_1 + R_2}\right) - 20\log\sqrt{1 + \left(\frac{\omega}{\omega_b}\right)^2}$$

Where:

$$+20\log\left(\frac{R_2}{R_1 + R_2}\right) = +20\log\left(\frac{250}{470 + 250}\right) = -9.2 \text{ dB}$$

Then:

$$|H(j\omega)|_{dB} = -9.2 \text{ dB} - 20\log\sqrt{1 + \left(\frac{\omega}{\omega_b}\right)^2}$$

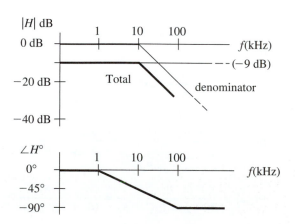

Figure 15.25 Magnitude and Phase Bode Plots for Example 15.7.2

The transfer function appears exactly like the RC LPF investigated earlier in this chapter except for the -9.2 dB term. The resistances are constants, so this term is a constant that is just an offset of some negative dB value. In fact, one can see from the circuit that at low frequencies (below the break frequency), the capacitor impedance is much larger than the resistances, so the capacitor is effectively "out of the circuit" and there is a voltage divider between the resistances. The output voltage magnitude will be less than the input voltage magnitude. Hence, at low frequencies, the magnitude response should be constant with a negative dB value, as shown in Figure 15.25. The phase term is identical with that of the LPF, so no further analysis is necessary on the phase.

The break frequency contains a familiar aspect. It appears that the equation for the total resistance of parallel resistances is embedded in the break frequency expression:

$$\omega_b = \frac{R_1 + R_2}{R_1 R_2 C} = \frac{1}{\dfrac{R_1 R_2 C}{R_1 + R_2}} = \frac{1}{\left(\dfrac{R_1 R_2}{R_1 + R_2}\right)C}$$

One can visualize this parallel resistance arrangement if an ideal voltage source is connected to the input terminals of the circuit in Figure 15.24. What is the impedance of an ideal voltage source? It is zero, i.e., a short. Hence, from the viewpoint of the capacitor, the Thévenin equivalent impedance equals the two resistances in parallel.

Thus the same Bode plot strategy works for more complicated RC and RL circuits. If both an inductor and a capacitor are present in the same circuit, then a more intricate strategy is required. This strategy is described in Section 16.7.

15.8 CIRCUIT ANALYSIS WITH PERIODIC SIGNALS

The analysis of a circuit with a periodic signal input is covered in this section. Begin the analysis by examining each individual term in a Fourier series. If only one term were given as the signal, how would one analyze the circuit? Phasor analysis of AC circuits is used. One could analyze the circuit for each significant (non-negligible amplitude) term in the Fourier series using standard AC phasor analysis. For example, the first three for the Fourier series in Example 15.1.1 are:

$$v(t) = 12.732 \sin(1570.8t) + 4.2440 \sin(4712.4t) + 2.5464 \sin(7854.0t) + \ldots$$

The first three corresponding phasors for the Fourier series in Example 15.1.1 are:

$$\tilde{V}_1 = 12.732\angle 0° \text{ @ } 1570.8 \text{ rad/s}$$
$$\tilde{V}_3 = 4.2440\angle 0° \text{ @ } 4712.4 \text{ rad/s}$$
$$\tilde{V}_5 = 2.5464\angle 0° \text{ @ } 7854.0 \text{ rad/s}$$

The phasor analysis would be labor intensive if performed with paper and calculator because each term is at a different frequency—the impedance of inductors and capacitors would have to be recalculated for each frequency. A spreadsheet or other mathematical program speeds this analysis process. Then the resultant phasor at each frequency can be returned to the time domain and the time domain terms summed to give the total resultant signal, again using a computational program such as a spreadsheet. It is critical to understand that *these phasors* **cannot** *be added directly—they are at different frequencies.*

It is important to distinguish two types of superposition. In earlier chapters, superposition was used to sum signals of the *same* frequency. The total phasor for a given voltage or current was determined, and then the total phasor expression could be converted back into the time domain. In this chapter, signals of *different* frequencies are being summed in the time domain. They *cannot* be summed as phasors. Phasors are used in the circuit analysis for *each* frequency. Then the phasor result at each frequency is converted back into the time domain. Finally, the time-domain expressions for the different frequencies are summed in the time domain. It is your responsibility to know when phasors should be summed and when time-domain expressions should be summed—determine whether the frequencies of the signals to be summed are the same or not.

If a square wave is applied to the input of an RC circuit, what is the output? The next example illustrates the process for arriving at the answer to this question. The square wave of Example 15.1.1 is used along with the RC circuit that was examined in Sections 15.5 and 15.7.

—— Example 15.8.1

A 10 V_{peak} square wave of period 4 ms is input into a series RC circuit where $R = 10\ k\Omega$ and $C = 0.1\ \mu F$. Determine the first three nonzero terms of the Fourier series output voltage across the capacitor.

Given: Circuit: RC circuit in Figure 15.12 with $R = 10\ k\Omega$ and $C = 0.1\ \mu F$

Input signal: 10 V_{peak} square wave of period 4 ms

Desired: Output waveform across C

Strategy: Use the Fourier series results from Example 15.1.1 and the transfer function results of Equation (15.37) to determine the phasors for the first three nonzero harmonics.

Determine the first three terms of the time-domain Fourier series expression for the output voltage.

Solution: _____

$$H(j\omega) = \frac{1}{1 + j\dfrac{\omega}{\omega_b}}$$

$$\omega_b = \frac{1}{RC} = 1000 \text{ rad/s}$$

$$v_{in}(t) = 12.732 \sin(1570.8t) + 4.2440 \sin(4712.4t) + 2.5464 \sin(7854.0t) + \dots$$

$$\tilde{V}_{in1} = 12.732\angle 0° \text{ @ } 1570.8 \text{ rad/s}$$

$$\tilde{V}_{in3} = 4.2440\angle 0° \text{ @ } 4712.4 \text{ rad/s}$$

$$\tilde{V}_{in5} = 2.5464\angle 0° \text{ @ } 7854.0 \text{ rad/s}$$

$$\tilde{V}_{o1} = \tilde{V}_{in1}H(j\omega) = \frac{\tilde{V}_{in1}}{1 + j\dfrac{1\omega_o}{\omega_b}} = \frac{12.732\angle 0°}{1 + j\dfrac{1570.8}{1000}} = 6.8374\ \angle -57.518° \text{ at } 1570.8 \text{ rad/s}$$

$$\tilde{V}_{o3} = \tilde{V}_{in3}H(j\omega) = \frac{\tilde{V}_{in3}}{1 + j\dfrac{3\omega_o}{\omega_b}} = \frac{4.2440\angle 0°}{1 + j\dfrac{4712.4}{1000}} = 0.88101\angle -78.109° \text{ at } 4712.4 \text{ rad/s}$$

$$\tilde{V}_{o5} = \tilde{V}_{in5}H(j\omega) = \frac{\tilde{V}_{in5}}{1 + j\dfrac{3\omega_o}{\omega_b}} = \frac{2.5464\angle 0°}{1 + j\dfrac{7854.0}{1000}} = 0.32163\angle -82.744° \text{ at } 7854.0 \text{ rad/s}$$

$$v_o(t) = 6.84\ \sin(1570.8t - 57.5°) + 0.881 \sin(4712.4t - 78.1°) + 0.322 \sin(7854.0t - 82.7°) + \dots$$

Notice that the magnitudes of the output phasors are decreasing much faster than the magnitudes of the input phasors as the harmonic number increases. Hence, one should

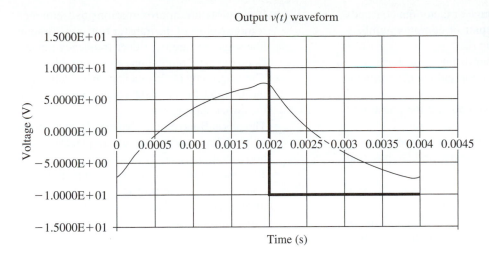

Figure 15.26 Output Waveform for Example 15.8.1 (bold–input; light trace–output)

expect that the output waveform will differ significantly from the input square-wave waveform. This scenario is illustrated in Figure 15.26, where the output waveform for the first 11 nonzero harmonics are plotted against the input square-wave signal.

CHAPTER REVIEW

15.1 Motivation: The Frequency Content of Periodic Signals
- A periodic signal consists of a summation of sinusoids. The frequencies are harmonically related.
- The Fourier series is used as a mathematical tool to represent periodic waveforms.
- Different periodic waveforms have different Fourier series.

15.2 Analyzing a Circuit with an Unspecified Source Frequency
- A circuit can be analyzed using circuit analysis techniques without specifying the frequency. Frequency is a variable.
- One must distinguish between symbols that are used for constants and symbols that are used for variables.
- The frequency response is the plots of the voltage, the current, the impedance, etc., both magnitude and phase, as a function of frequency.
- Spreadsheets are useful in plotting the frequency response of a circuit from the expression with frequency as a variable.

15.3 Transfer Function and Frequency Response
- The transfer function is the ratio of one phasor voltage or current in a circuit to another phasor voltage or current in the circuit.
- The transfer function expression depends only on the circuit components and the frequency, not on the source voltage or current.

- The output phasor voltage or current of a circuit can be determined by the input phasor voltage or current multiplied by the transfer function.
- The transfer function, magnitude and phase, can also be plotted versus frequency, and it is also called the frequency response.
- Insight into circuit performance is obtained by observing the transfer function behavior as frequency is increased or decreased. The frequency response plots help one to visualize this performance.

15.4 Basic Filter Types and Responses
- There are four fundamental filter types: lowpass (LPF), highpass (HPF), bandpass (BPF), and bandreject (BRF).
- RC and RL circuits can be constructed to form lowpass and highpass filters.
- Ideal filters have sharp corners in the frequency responses. Actual filters have smooth curves in the "corners" of the frequency responses.

15.5 Bode Plots—Transfer Function Setup
- There is a systematic approach to manipulating the transfer function into a form suitable for preparing Bode plots:
  - Analyze the circuit for the transfer function in terms of impedances and/or admittances, and insert the component expressions for impedances and/or admittances.
  - Clear compound fractions.

- Either divide through or factor out constants to make the real part of complex numbers equal to 1.
- Manipulate all ω terms into ω/ω_b form.
- Convert the numerator and the denominator into polar form.
- Separate the magnitude and the phase into separate equations.
- Take 20 log of the magnitude expression.

15.6 Logarithm Review, Log Scales and Graphs, and Gain in Decibels
- The logarithm (log) is the power of the base that equals the original number.
- There are laws for the use of logarithms in mathematical operations (multiplication, division, exponentiation, roots).
- Logarithms compress large ranges of numbers and expand ranges of small numbers.
- Taking the log of numbers and plotting them on a linear scale is equivalent to directly plotting the original numbers on a log scale.
- The space on a log scale is the same for any decade.
- Gain is power out divided by power in:

$$G = \frac{P_o}{P_{in}}$$

- The gain in decibels is defined to be 10 times the log of the gain as a power ratio:

$$G(dB) \equiv 10 \log\left(\frac{P_o}{P_{in}}\right)$$

- The gain in decibels is often used for transfer function voltage ratios:

$$G(dB) = 20 \log\left(\frac{V_o}{V_{in}}\right) = 20 \log |H(j\omega)|$$

15.7 Bode Plots—Plotting from the Transfer Function

- Bode plots are approximations to frequency response plots of the transfer function. Straight line segments are used over frequency ranges. Bode plots are plotted on semilog graphs.
- The magnitude Bode plot is gain in decibels on the vertical linear scale versus frequency on the horizontal log scale.
- The phase Bode plot is phase in degrees on the vertical linear scale versus frequency on the horizontal log scale.
- The slope of the first-order circuit response on a magnitude Bode plot is either flat or ± 20 dB/decade. The frequency where the transition between these two slopes is called the break frequency.
- The slope of the first-order circuit response on a phase Bode plot is either flat or $\pm 45°$/decade. The transition of the phase occurs from $0.1\omega_b$ and ends at $10\,\omega_b$.
- Bode plots differ from frequency response plots in the vicinity of the break frequency on the magnitude Bode plot and in the vicinities of $0.1\omega_b$ and $10\,\omega_b$ on the phase Bode plots.

15.8 Circuit Analysis with Periodic Signals
- A circuit with a periodic signal input that is represented by a Fourier series can be analyzed by using a phasor representation of each term in the Fourier series. Each phasor is valid only at the frequency of that term.
- The phasors used for a Fourier series may never be summed because they are at different frequencies. Only time domain terms may be summed in this case.
- A circuit generally changes the terms in a Fourier series in different proportions and phase shifts. Hence the output periodic signal will be different than the input periodic signal.

HOMEWORK PROBLEMS

Proper completion of the following questions will demonstrate your ability to calculate the magnitude of the frequency components comprising common periodic signals.

15.1 Calculate the coefficients of the first four nonzero terms of the Fourier series for a square wave of period 0.2 ms and peak amplitude of 20 V.

15.2 Calculate the coefficients of the first four nonzero terms of the Fourier series for a triangle wave of period 0.2 ms and peak amplitude

of 20 V. Compare the values with those of problem 15.1.

15.3 Calculate the coefficients of the first four nonzero terms of the Fourier series for a full-wave rectified sine wave of period 16.66 ms and peak amplitude of 170 V.

Your responses to the following questions will demonstrate your ability to describe transfer functions, filters, and Bode plots.

15.4 What is the significance of writing equations with ω as a variable?

15.5 What is a frequency filter? What is meant by the frequency response of a filter?

15.6 What are Bode plots? What information is displayed in the Bode plots?

15.7 Describe the particular final mathematical form of the transfer functions when used for Bode plots. Why is that form useful?

Proper completion of the following questions will demonstrate your ability to derive the transfer function for RC and RL circuits. When doing the following problems, keep in mind the process that was followed and the desired mathematical form of the end result, which specifically is aimed to clearly identify any break frequencies.

15.8 Derive the transfer function (ratio of output voltage to input voltage) for the circuit of Figure P15.1. Manipulate the expression into a proper mathematical form that clearly identifies any break frequencies and allows direct drawing of the Bode plots.

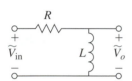

Figure P15.1 RL Circuit of Problem 15.8

15.9 Derive the transfer function (ratio of output voltage to input voltage) for the circuit of Figure P15.2. Manipulate the expression into a proper mathematical form that clearly identifies any break frequencies and allows direct drawing of the Bode plots.

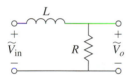

Figure P15.2 RL Circuit of Problem 15.9

15.10 Derive the transfer function (ratio of output voltage to input voltage) for the circuit of Figure P15.3. Manipulate the expression into a proper mathematical form that clearly identifies any break frequencies and allows direct drawing of the Bode plots.

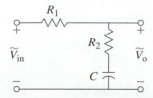

Figure P15.3 RC Circuit of Problem 15.10

15.11 Derive the transfer function (ratio of output voltage to input voltage) for the circuit of Figure P15.4. Manipulate the expression into a proper mathematical form that clearly identifies any break frequencies and allows direct drawing of the Bode plots.

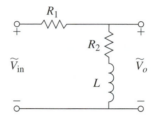

Figure P15.4 RL Circuit of Problem 15.11

15.12 Derive the transfer function (ratio of output voltage to input voltage) for the circuit of Figure P15.5. Manipulate the expression into a proper mathematical form that clearly identifies any break frequencies and allows direct drawing of the Bode plots.

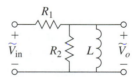

Figure P15.5 RL Circuit of Problem 15.12

15.13 Derive the transfer function (ratio of output voltage to input voltage) for the circuit of Figure P15.6. Manipulate the expression into a proper mathematical form that clearly identifies any break frequencies and allows direct drawing of the Bode plots.

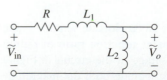

Figure P15.6 RL Circuit of Problem 15.13

15.14 Derive the transfer function (ratio of output voltage to input voltage) for the circuit of Figure P15.7. Manipulate the expression into a proper mathematical form that clearly identifies any break frequencies and allows direct drawing of the Bode plots.

15.15 Derive the transfer function (ratio of output voltage to input voltage) for the circuit of Figure P15.8. Manipulate the expression into a proper mathematical form that clearly identifies any break frequencies and allows direct drawing of the Bode plots.

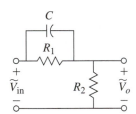

Figure P15.7 RC Circuit of Problem 15.14

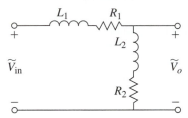

Figure P15.8 RL Circuit of Problem 15.15

Your responses to the following questions will demonstrate your ability to describe and utilize logarithms and decibels.

15.16 What are the benefits gained in converting to a log scale when plotting data with a large range of values?

15.17 What is the difference in value between the logarithms of 200 and 20? What is difference in value between the logarithms of 20 000 and 2000? Compare and explain your answers.

15.18 Use logarithms to evaluate the following:
a. 50*200*30
b. 45/(36*50)

c. $(16^3)*40$
d. $20*60/(20^4)$

15.19 What is the definition of the Bel? The definition of the decibel? What are the units of the Bel and the decibel?

15.20 Convert the following power ratios into decibels: 10, 20, 100, 200, 1.0, 0.5, 0.25, 0.125, 1000, 2000. Do you detect any pattern to the answers?

Proper completion of the following questions will demonstrate your ability to generate Bode plots from the transfer functions of RC and RL circuits.

15.21 Draw and label the Bode plots on semilog graph paper for the transfer function ratio of output voltage to input voltage of the circuit in Figure P15.9. Be sure to include both the magnitude and the phase plot. Clearly show and label the total circuit response on each plot.

and the phase plot. Clearly show and label the total circuit response on each plot.

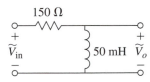

Figure P15.10 RL Circuit of Problem 15.22

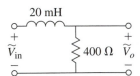

Figure P15.9 RL Circuit of Problem 15.21

15.22 Draw and label the Bode plots on semilog graph paper for the transfer function ratio of output voltage to input voltage of the circuit in Figure P15.10. Be sure to include both the magnitude

15.23 Draw and label the Bode plots on semilog graph paper for the following transfer function. Be sure to include both the magnitude and the phase plot. Clearly show and label the total circuit response on each plot.

$$H(j\omega) = \frac{120}{1 + j\dfrac{\omega}{2000}}$$

15.24 Draw and label the Bode plots on semilog graph paper for the following transfer function. Be sure to include both the magnitude and the phase plot. Clearly show and label the total circuit response on each plot.

$$H(j\omega) = \frac{800}{\sqrt{1 + \left(\frac{\omega}{5\,k}\right)^2} \angle \tan^{-1}\left(\frac{\omega}{5\,k}\right)}$$

15.25 Draw and label the Bode plots on semilog graph paper for the following transfer function. Be sure to include both the magnitude and the phase plot. Clearly show and label the total circuit response on each plot.

$$|H(j\omega)|_{dB} = +20\log\left(\frac{\omega}{100\,k}\right) - 20\log\sqrt{1 + \left(\frac{\omega}{100\,k}\right)^2}$$

$$\angle H(j\omega) = +90° - \tan^{-1}\left(\frac{\omega}{100\,k}\right)$$

15.26 Draw and label the Bode plots on semilog graph paper for the following transfer function. Be sure to include both the magnitude and the phase plot. Clearly show and label the total circuit response on each plot.

$$H(\omega) = \frac{j\frac{\omega}{100}}{\sqrt{1 + \left(\frac{\omega}{100}\right)^2} \angle \tan^{-1}\left(\frac{\omega}{100}\right)}$$

15.27 Draw and label the Bode plots on semilog graph paper for the following transfer function. Be sure to include both the magnitude and the phase plot. Clearly show and label the total circuit response on each plot.

$$H(j\omega) = \frac{j\frac{\omega}{5k}}{1 + j\frac{\omega}{5k}}$$

CHALLENGE PROBLEMS

The following problems extend the complexity of the transfer functions developed in this chapter. When doing these problems, keep in mind the process that was developed and the end result aimed for. Suggestion: *Each complex term has a magnitude and a phase value, just like the previous problems. Hence, fundamentally nothing has changed except there are more terms, not new types of terms.*

15.28 Draw and label the Bode plots on semilog graph paper for the following transfer function. Be sure to include both the magnitude and the phase plot. Clearly show and label the total circuit response on each plot.

$$H(j\omega) = \frac{\tilde{V}_{out}}{\tilde{V}_{in}} = \frac{\left(j\frac{\omega}{1000}\right)}{\left(1 + j\frac{\omega}{1000}\right)\left(1 + j\frac{\omega}{10,000}\right)}$$

15.29 Draw and label the Bode plots on semilog graph paper for the following transfer function. Be sure to include both the magnitude and the phase plot. Clearly show and label the total circuit response on each plot.

$$H(j\omega) = \frac{\tilde{V}_{out}}{\tilde{V}_{in}} = \frac{40(1 + j0.001\omega)}{(j0.005\omega)\left(1 + j\frac{\omega}{10,000}\right)}$$

Proper completion of the following questions will demonstrate your ability to determine the frequency content of a resultant periodic signal given the transfer functions of RC *and* RL *circuits.*

15.30 For each of the coefficients of the first four terms of the Fourier series for a triangle wave of problem 15.2 (period 0.2 ms and peak amplitude of 20 V), calculate the output voltage for a lowpass *RL* filter with a break-point frequency of 6 kHz. Would an *RC* lowpass filter have given a different result? *Hint:* Use the transfer function for the RL circuit that behaves like a lowpass filter. Calculate the product of the transfer function $H(j\omega)$ times the input phasor voltage for every frequency.

15.31 For each of the coefficients of the first four terms of the Fourier series for a square wave of problem 15.1 (period 0.2 ms and peak amplitude of 20 V), calculate the output voltage for a lowpass RC filter with a break-point frequency of 6 kHz (see hint in problem 15.30). Would an RL lowpass filter have given a different result?

RESONANT CIRCUITS

As a result of successfully completing this chapter, you should be able to:

1. Describe a bandpass filter.
2. Describe why both an inductor and a capacitor are utilized in bandpass filters.
3. Describe the conditions for electrical resonance.
4. Describe the mathematical strategy to develop the resonant frequency expression for a given resonant circuit.
5. Determine the resonant frequency of series, parallel, and series–parallel circuits.
6. Describe the quality factor.
7. Determine the quality factor of series, parallel, and series–parallel circuits.
8. Determine the three dB bandwidth from the resonant frequency and quality factor.
9. Decide whether a resonant circuit has a low Q or a high Q in order to determine the 3 dB determination approach.
10. Calculate the 3 dB frequencies for high Q circuits.
11. Describe Bode plots for second-order circuits including the resonant peak at the break frequency.

16.1 MOTIVATION: THE BANDPASS FILTER

Circuits with capacitors and resistances *or* inductors and resistances were covered in Chapter 15. These circuits had lowpass or highpass frequency responses. The Bode magnitude plots had a ± 20 dB/decade slope above or below the break frequency. Bandpass and bandreject filters were briefly mentioned in Section 15.4. These filters now become the focus of our attention.

The magnitude response of the transfer function for both a bandpass filter and a bandreject filter are shown again in Figure 16.1. Notice that there is a rolloff on both sides of some central frequency. The *RL* and *RC* circuits of Chapter 15 produce a rolloff on only one side of the break frequency. Hence it becomes plausible that a circuit with

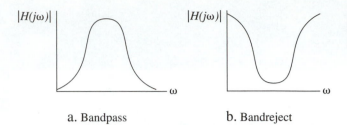

Figure 16.1 Bandpass and Bandreject Frequency Responses (Magnitude Only)

a. Bandpass   b. Bandreject

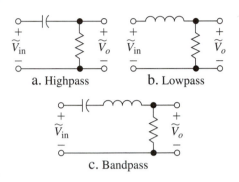

Figure 16.2 RL, RC and RLC Circuits

a. Highpass   b. Lowpass

c. Bandpass

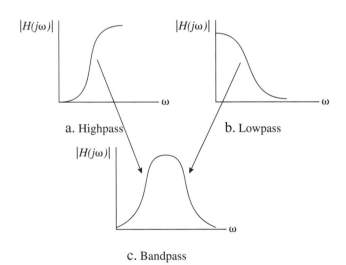

Figure 16.3 Highpass, Lowpass, and Bandpass Frequency Responses (Magnitude Only)

a. Highpass   b. Lowpass

c. Bandpass

both an inductor and a capacitor would have a rolloff on both sides of that special central frequency. Consider the *RL* and *RC* circuits shown in Figure 16.2. The *RC* highpass filter in Figure 16.2a blocks lower frequencies and passes higher frequencies, as shown in Figure 16.3a. The *RL* lowpass filter in Figure 16.2b passes lower frequencies and blocks higher frequencies, as shown in Figure 16.3b. One can qualitatively visualize that if the two circuits were combined, as shown in Figure 16.2c, that a bandpass or a bandreject frequency response would be obtained, as sketched in Figure 16.3c.

In this chapter, circuits that contain *both* inductors *and* capacitors are covered. The conditions for electrical resonance are covered in Section 16.2. It is already known that capacitive reactance decreases as frequency increases and inductive reactance increases as frequency increases. In a circuit with *both* an inductor and a capacitor, it might be expected that at some frequency, called the *resonant frequency,* the total reactance of the circuit is zero. This state of the circuit is called *resonance.* Obviously, one would like to be able to predict the resonant frequency or, alternatively, design a circuit for a desired resonant frequency. The prediction of the resonant frequency is covered in Section 16.3. This development is performed in general—examples are delayed until Section 16.6 so that all pertinent resonant circuit quantities can be determined in the context of their relationship to one another.

Once the resonant frequency is known, the next natural question to ask is, What is the frequency response of the circuit? Resonant circuits have very useful frequency responses that are characterized by bandwidth (BW) and *quality factor* (Q)[1]. Q is covered first, in Section 16.4, so that it can be related to BW. Then, the frequency response of a resonant circuit is established through the use of Q and BW in Section 16.5. A summary table and several resonant circuit examples are given in Section 16.6. The frequency response and Bode plots of resonant circuits over a wide frequency range, as opposed to the relatively narrow frequency range around resonance, are covered in Section 16.7 (optional section).

16.2 WHAT IS ELECTRICAL RESONANCE?

Resonance is defined as

$$\text{Im}[\tilde{Z}_{ckt}(\omega)] = 0 \Rightarrow X_T = 0 \tag{16.1}$$

$$\text{Im}[\tilde{Y}_{ckt}(\omega)] = 0 \Rightarrow B_T = 0 \tag{16.2}$$

i.e., the frequency at which the imaginary part of the total circuit impedance or admittance is zero when there is both an inductor and a capacitor in the circuit. The frequency at which this state occurs is called the *resonant frequency* ω_o (r/s) or f_o (Hz). *The total voltage and current for the circuit are in phase at the resonant frequency.* The power factor is unity. The circuit impedance is purely resistive at the resonant frequency. The circuit admittance is a pure conductance at the resonant frequency. Some interpretations about simple series and parallel resonance follow.

A series resonant circuit is shown in Figure 16.4. The impedance of the circuit is

$$\tilde{Z} = R + j\left(\omega L - \frac{1}{\omega C}\right) \tag{16.3}$$

At resonance, the imaginary part of the impedance must equal zero; hence:

$$\frac{1}{\omega C} = \omega L \tag{16.4}$$

i.e., the reactances have equal magnitudes; however, the inductor and the capacitor impedance phase angles are *opposite* (180° apart):

$$\tilde{Z}_L = +j\omega L = \omega L \angle + \frac{\pi}{2} \tag{16.5}$$

$$\tilde{Z}_C = -j\frac{1}{\omega C} = \frac{1}{\omega C} \angle -\frac{\pi}{2} \tag{16.6}$$

Because the inductor and the capacitor are in *series,* their impedances add and, hence, *cancel.* Thus at resonance, the total impedance is minimum, real, and equal to R. The important concept to emphasize here is that *the total impedance is real at the resonant frequency.*

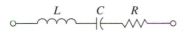

Figure 16.4 A Series Resonant Circuit

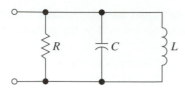

Figure 16.5 A Parallel Resonant Circuit

A parallel resonant circuit is shown in Figure 16.5. The admittance of the circuit is:

$$\tilde{Y} = G + j\left(\omega C - \frac{1}{\omega L}\right) \tag{16.7}$$

At resonance the imaginary part of the admittance must equal zero; hence,

$$\omega C = \frac{1}{\omega L} \tag{16.8}$$

i.e., the susceptances have equal magnitudes; however, the inductor and capacitor admittance phase angles are *opposite* (180° apart):

$$\tilde{Y}_C = +j\omega C = \omega C \angle +\frac{\pi}{2} \tag{16.9}$$

$$\tilde{Y}_L = -j\frac{1}{\omega L} = \frac{1}{\omega L} \angle -\frac{\pi}{2} \tag{16.10}$$

Again, the susceptances have equal magnitudes at resonance and the inductor and capacitor admittance phase angles are opposite. However, L and C are in *parallel*. The admittances of L and C add and, hence, cancel. Thus, at resonance, the total admittance is minimum, real, and equal to $G = 1/R$. Again, the concept to emphasize here is that *the total admittance is real at the resonant frequency.*

In both the series and the parallel resonant circuits, the impedances (or admittances) of the capacitor and the inductor have phase angles that are 180° apart. This implies that when one reactive element is "charging," the other is "discharging." Thus the energy is transferred back and forth between the parallel reactive elements.

16.3 DETERMINING THE RESONANT FREQUENCY

Resonance occurs in a circuit when the impedance of the circuit becomes completely real. Hence, the reactance becomes zero at a certain frequency. Note that $\omega = 0$ is *not* considered a resonant frequency. Thus, *to find the resonant frequency, the imaginary part of the total circuit impedance or admittance is found and set equal to zero.* Then one must solve for ω in terms of the R, L, and C values in the circuit. The four most important cases are considered in this section. The resonant frequency expressions are developed here, but circuit examples are delayed until Section 16.6 so that the determination of the resonant frequency can be shown in relation to the other important resonant circuit quantities.

a. **Resonant Frequency of a Series Resonant Circuit (Figure 16.4):**

The impedance of a series RLC circuit is:

$$\tilde{Z} = R + j\left(\omega L - \frac{1}{\omega C}\right) \tag{16.11}$$

At resonance, $\omega = \omega_o$, and

$$\left(\omega_o L - \frac{1}{\omega_o C}\right) = 0 \tag{16.12}$$

$$\omega_o L = \frac{1}{\omega_o C} \tag{16.13}$$

$$\omega_o = \frac{1}{\sqrt{LC}} \tag{16.14}$$

$$\tilde{Z}(\omega_o) = R \tag{16.15}$$

Thus the resonant frequency of a series resonant circuit depends on the inductor and capacitor component values, and the impedance of the circuit is purely real at resonance.

b. Resonant Frequency of a Parallel Resonant Circuit (Figure 16.5):

The admittance of a parallel RLC circuit is:

$$\tilde{Y} = G + j\left(\omega C - \frac{1}{\omega L}\right) \tag{16.16}$$

At resonance, $\omega = \omega_o$ and

$$\left(\omega_o C - \frac{1}{\omega_o L}\right) = 0 \tag{16.17}$$

$$\omega_o C = \frac{1}{\omega_o L} \tag{16.18}$$

$$\omega_o = \frac{1}{\sqrt{LC}} \tag{16.19}$$

$$\tilde{Y}(\omega_o) = G \tag{16.20}$$

Thus the resonant frequency of a parallel resonant circuit depends on the inductor and capacitor component values, and the admittance of the circuit is purely real at resonance.

Therefore, the resonant frequency of either a pure series or a pure parallel resonant circuit is:

$$\boxed{\omega_o = \frac{1}{\sqrt{LC}}, \quad f_o = \frac{1}{2\pi\sqrt{LC}}} \tag{16.21}$$

This equation also serves as a ballpark estimate on the resonant frequency of any resonant circuit, and it is one of those fundamental expressions that applies often in electronics.

c. Resonant Frequency of a Series–Parallel Resonant Circuit (Figure 16.6):

Note that in the practical inductor, the wire resistance, R_{ind}, can be significant. Start with the circuit admittance:

$$\tilde{Y} = j\omega C + \frac{1}{R_{ind} + j\omega L} \tag{16.22}$$

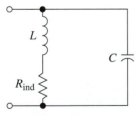

Figure 16.6 Series–Parallel Resonant Circuit

Make the denominator real by multiplying the numerator and the denominator by the complex conjugate of the denominator:

$$\tilde{Y} = +j\omega C + \frac{1}{R_{ind} + j\omega L} \cdot \left(\frac{R_{ind} - j\omega L}{R_{ind} - j\omega L}\right) \tag{16.23}$$

$$\tilde{Y} = j\omega C + \frac{R_{ind} - j\omega L}{R_{ind}^2 + \omega^2 L^2} \tag{16.24}$$

Split the second term into real and imaginary parts:

$$\tilde{Y} = +j\omega C + \frac{R_{ind}}{R_{ind}^2 + \omega^2 L^2} - j\frac{\omega L}{R_{ind}^2 + \omega^2 L^2} \tag{16.25}$$

Group the real and imaginary parts of the admittance:

$$\tilde{Y} = \frac{R_{ind}}{R_{ind}^2 + \omega^2 L^2} + j\left(\omega C - \frac{\omega L}{R_{ind}^2 + \omega^2 L^2}\right) \tag{16.26}$$

Set the imaginary part of the total admittance to zero for resonance:

$$\text{Im}(\tilde{Y}) = \omega_o C - \frac{\omega_o L}{R_{ind}^2 + \omega_o^2 L^2} = 0 \tag{16.27}$$

where $\omega = \omega_o$ at resonance. Solve for ω_o:

$$\omega_o C = \frac{\omega_o L}{R_{ind}^2 + \omega_o^2 L^2} \tag{16.28}$$

$$C = \frac{L}{R_{ind}^2 + \omega_o^2 L^2} \tag{16.29}$$

$$C(R_{ind}^2 + \omega_o^2 L^2) = L \tag{16.30}$$

$$CR_{ind}^2 + \omega_o^2 L^2 C = L \tag{16.31}$$

$$\omega_o^2 L^2 C = L - CR_{ind}^2 \tag{16.32}$$

$$\omega_o^2 = \frac{L - CR_{ind}^2}{L^2 C} = \frac{L}{L^2 C} - \frac{CR_{ind}^2}{L^2 C} \tag{16.33}$$

$$\omega_o^2 = \frac{1}{LC} - \frac{R_{ind}^2}{L^2} \tag{16.34}$$

$$\boxed{\omega_o = \sqrt{\frac{1}{LC} - \frac{R_{ind}^2}{L^2}} = \sqrt{\frac{1}{LC} - \left(\frac{R_{ind}}{L}\right)^2}} \tag{16.35}$$

Note that $\omega_o \neq 1/\sqrt{LC}$. The resistance of the inductor affects the resonant frequency in this resonant circuit configuration. The procedure illustrated in this example is general in determining the resonant frequency. Alternatively one could make a series–parallel conversion for the RL branch in this circuit and then find resonance using the parallel resonant circuit relations (see homework problem 16.14). The ω_o results should match.

The impedance of the circuit at resonance, although real, does not equal R_{ind}. One could analyze the circuit using series–parallel circuit analysis techniques to determine the impedance at resonance. Alternatively one could perform a series–parallel conversion on the inductor–resistance branch and analyze the resultant parallel circuit. (The latter approach is useful because it is also used to determine another important quantity, the quality factor, of the resonant circuit.) Note also that in the resonant condition, the *total* reactance of the circuit equaling zero does *not* imply $X_L = X_C$ in this circuit because resistances in the circuit influence the reactance calculation (think of a series–parallel conversion). Only in simple series and parallel resonant circuits does $X_L = X_C$ at ω_o.

d. **Resonant Frequency of another practical Series–Parallel Resonant Circuit (Figure 16.7):**

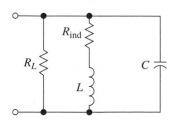

Figure 16.7 Another Practical Series–Parallel Resonant Circuit

This circuit is very similar to that in Figure 16.6 except it has a load resistance in parallel with the circuit. Note that R_L will change only the real part of the total circuit admittance:

$$\tilde{Y} = \frac{1}{R_L} + j\omega C + \frac{1}{R_{\text{ind}} + j\omega L} \qquad (16.36)$$

The additional real term in Equation (16.36) will not affect the imaginary part. Hence, the resonant frequency, which is derived from the imaginary part of the total admittance, will remain unaffected, and the resonant frequency will be the same as for the circuit in Figure 16.6, i.e., Equation (16.35). Then, what quantity will R_L affect? Think about it in the next two sections.

16.4   Q, THE QUALITY FACTOR

The quality factor is an important parameter in resonant circuits, as will become clear in Section 16.5. In this section of the chapter, the definition of Q and its relation to energy stored and power dissipated will be established. In the next section, Q will be related to frequency response. Again, the expressions for Q are developed here, but circuit examples are delayed until Section 16.6 so that the determination of Q can be shown in relation to the other important resonant circuit quantities.

Fundamentally, the quality factor Q is defined at resonance as:

$$Q \equiv (2\pi) \frac{\text{energy stored}}{\text{energy dissipated in one cycle}} = 2\pi \frac{E_s}{E_d} \qquad (16.37)$$

Q is an indication of the amount of energy stored relative to the energy loss in one cycle in the circuit. Divide the numerator and the denominator of Equation (16.37) by the period T:

$$Q = \frac{\text{energy stored } (2\pi)\left(\frac{1}{T}\right)}{\left(\frac{\text{energy dissipated in one cycle}}{T}\right)} = \frac{E_s 2\pi \left(\frac{1}{T}\right)}{\left(\frac{E_d}{T}\right)} \qquad (16.38)$$

But energy dissipated per period is power dissipated, and $2\pi/T = \omega_o$ for sinusoidal signals; hence another common expression of Q is:

$$Q = \omega_o \frac{\text{energy stored}}{\text{power dissipated}} = \omega_o \frac{E_s}{P_d} \qquad (16.39)$$

Again, note that Q indicates the amount of energy stored relative to the power losses in the circuit. The usefulness of Equation (16.39) is that the Q of a circuit can be found directly from this expression. Alternatively, the Q of a circuit can be also found by measuring the 3 dB (half-power) bandwidth of the resonant circuit (see Section 16.5).

a.   Q of a Series Resonant Circuit (Figure 16.4)

The total energy stored in a resonant circuit is the sum of the energy stored in the magnetic field of the inductor and the energy stored in the electric field of the capacitor. Because the energy alternates between the electric field of the capacitor and the magnetic field of the inductor, either can be used to determine energy stored. One can consider a moment when all of the energy is stored in the magnetic field of the inductor. The energy stored in an inductor was given in Equation 10.2 in terms of the current value. When the current through the inductor is at its peak value, then all of the energy stored in the resonant circuit is in the inductor at that moment. The current through a series circuit is the same for all components; hence for the series resonant circuit:

$$Q_s = \omega_o \frac{E_s}{P_d} = \omega_o \frac{\frac{1}{2}L I_P^2}{\frac{1}{2}I_P^2 R} \qquad (16.40)$$

where Q_s is the Q of a series resonant circuit. This expression simplifies to:

$$Q_s = \frac{\omega_o L}{R} \quad \text{(series)} \tag{16.41}$$

The quality factor of a series resonant circuit depends only on component values $\left(\text{recall } \omega_o = \dfrac{1}{\sqrt{LC}} \right)$.

b. Q of a Parallel Resonant Circuit (Figure 16.5)

Consider a moment when all the energy is stored in the electric field of the capacitor. The energy stored in a capacitor was given in Equation 10.2 in terms of the voltage value. The voltage across all components of a parallel circuit is the same; hence for the parallel resonant circuit at the peak voltage value:

$$Q_p = \omega_o \frac{E_s}{P_d} = \omega_o \frac{\frac{1}{2} C V_P^2}{\frac{1}{2} V_P^2 / R} \tag{16.42}$$

$$Q_p = \omega_o R C = \frac{\omega_o C}{G} \quad \text{where } G = \frac{1}{R} \quad \text{(parallel)} \tag{16.43}$$

The quality factor of a parallel resonant circuit depends, again, only on the component values.

c. and d. Q of a Series–Parallel Resonant Circuits (Figures 16.6 and 16.7)

The parallel resonant circuit results can be applied to the two versions of the series–parallel resonant circuits in Figures 16.6 and 16.7. For the resonant circuit in Figure 16.6, a series–parallel conversion of the RL branch results in a parallel circuit, as shown in Figure 16.8. The series-to-parallel equations were developed in Section 9.2. The parallel resistance in terms of the series resistance and reactance is:

$$R_P = \frac{R_S^2 + X_S^2}{R_S} \tag{16.44}$$

where $R_S = R_{ind}$ and $X_S = X_L$ at the resonant frequency for the series RL branch. The parallel equivalent reactance and inductance are not calculated because they are not needed in the Q calculation. Then Equation (16.43) can be used to determine the resonant circuit Q where $R = R_P$:

$$Q_p = \omega_o R C \quad \text{where} \quad R = R_P = \frac{R_{ind}^2 + X_L^2}{R_{ind}} \quad \text{(series-parallel, no } R_L\text{)} \tag{16.45}$$

The same series–parallel conversion is used for the circuit in Figure 16.7. The equivalent parallel resonant circuit is shown in Figure 16.9. The resistance R in the Q expression of Equation (16.43) is then equal to R_P in parallel with the load resistance R_L:

$$R = \frac{R_P R_L}{R_P + R_L} \tag{16.46}$$

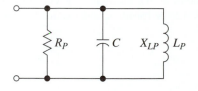

Figure 16.8 Conversion of the Series–Parallel Resonant Circuit in Figure 16.6 into an Equivalent Parallel Resonant Circuit

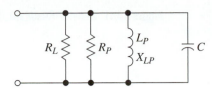

Figure 16.9 Conversion of the Series–Parallel Resonant Circuit in Figure 16.7 into an Equivalent Parallel Resonant Circuit

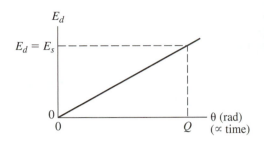

Figure 16.10 Energy Dissipated Versus Total Angle Elapsed

$$Q_P = \omega_o RC \quad \text{where} \quad R = \frac{R_P R_L}{R_P + R_L} \quad \text{and} \quad R_P = \frac{R_{\text{ind}}^2 + X_L^2}{R_{\text{ind}}} \quad \text{(series–parallel, with } R_L\text{)} \quad (16.47)$$

As previously mentioned, the *Q* can also be found from the 3 dB bandwidth and is especially found this way when determining *Q* from measurements or from computer simulation plots. This approach is examined in the next section.

An optional, alternative interpretation of *Q*:

An interesting alternative perspective of *Q* is found from examining the basic definition of *Q* from a different viewpoint. The energy dissipated in a resonant circuit will increase linearly with time because $E_d = P \cdot t$, as shown in Figure 16.10. Rearranging the basic *Q* definition:

$$Q = 2\pi \frac{E_s}{E_d} = \left(\frac{2\pi \text{ rad}}{\text{cycle}}\right)\left(\frac{(N \text{ cycles}) E_d}{E_d}\right) = 2\pi N \text{ rad} \qquad (16.48)$$

where *N* is the number of cycles it takes until the total energy dissipated in those *N* cycles equals the average energy stored, i.e., $NE_d = E_s$. Thus, *Q* is equal to the number of radians of total angle that elapse until the energy dissipated equals the energy stored in the resonant circuit. The higher the *Q* of the circuit, the larger the number of radians that must elapse before $E_s = NE_d$.

16.5 FREQUENCY RESPONSE, *Q,* AND *BW*

The discussion of frequency response, *Q,* and *BW* are discussed in general in this section. Again, circuit examples are delayed until Section 16.6 so that the determination of each quantity can be shown in relation to the other important resonant circuit quantities.

What would the power dissipated in the series resistance of a series resonant circuit be as a function of frequency? At frequencies that are low compared with the resonant frequency, the current is small due to the large reactance of the series capacitor, so $P_R = I_{\text{RMS}}^2 R$ is small. At frequencies that are high compared with the resonant frequency, the current is small due to the large reactance of the series inductor, so P_R is small again. Near resonance, the impedances of the inductor and the capacitor cancel, which results in a relatively larger current and larger P_R. A general curve for the power dissipated in the resistance of a series RLC circuit versus frequency is shown in Figure 16.11.

There are two special frequencies at which the power dissipated drops to half of the maximum (at or close to ω_o). These frequencies are called the *3 dB frequencies,* or the *half-power frequencies* and are designated by ω_L and ω_U, where ω_L is the lower 3 dB frequency and ω_U is the upper 3 dB frequency (recall that 3 dB means a factor of 2 or ½ in power). In

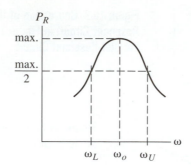

Figure 16.11 General Resonant Curve for Power Dissipated

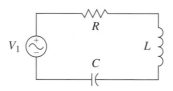

Figure 16.12 Series RLC Circuit for 3 dB Bandwidth Illustration

pure series or parallel resonant circuits (this is a qualification), these frequencies correspond to the frequencies where the voltage or the current (depending on the circuit) drops by a factor of the square root of 2 ($\sqrt{2}$). In general,

$$P_R = \frac{|\tilde{V}_R|^2}{R} = |\tilde{I}_R|^2 R \qquad (16.49)$$

If V_R or I_R in Equation (16.49) decreases by a factor of $\sqrt{2}$, then squaring the ($\sqrt{2}$) gives a factor of 2. Thus, the power drops by a factor of 2. For the voltage or the current to drop by a $\sqrt{2}$ factor, the impedance must drop or rise (again, depending on either a series or parallel resonant circuit) by a $\sqrt{2}$ factor, respectively. If one uses ω_3 to be ω_L or ω_U, one solves the following to find ω_3:

$$Z(\omega_3) = \left(\sqrt{2} \text{ or } \frac{1}{\sqrt{2}}\right) Z(\omega_o) \qquad (16.50)$$

where: $Z(\omega_3)$ is the *magnitude* of the total circuit impedance $\tilde{Z}(\omega)$ at $\omega_3 = \omega_L$ or ω_U, the $\sqrt{2}$ is chosen if $Z(\omega_o) = Z_{min}$ (such as in series resonant circuits), the $\frac{1}{\sqrt{2}}$ is chosen if $Z(\omega_o) = Z_{max}$ (such as in parallel resonant circuits), and $Z(\omega_o)$ is the magnitude of the total circuit impedance at the resonant frequency ω_o.

Note that the impedance of the resonant circuit at resonance, $\tilde{Z}(\omega_o)$, is a purely real impedance, i.e., $Z(\omega_o) = \tilde{Z}(\omega_o)$.

One solves for the frequencies that make Equation (16.50) true. The nonzero frequency solutions are the 3 dB frequencies. Finally, the 3 dB *BW* is the difference between the upper and the lower 3 dB frequencies:

$$BW = \omega_U - \omega_L \quad \text{(r/s)} \qquad (16.51)$$

$$BW = f_U - f_L \quad \text{(Hz)} \qquad (16.52)$$

A series RLC circuit will be used to illustrate the determination of the 3 dB *BW*. The circuit shown in Figure 16.12 has an impedance of

$$\tilde{Z}(\omega) = R + j\left(\omega L - \frac{1}{\omega C}\right) \qquad (16.53)$$

The magnitude of the total circuit impedance is:

$$Z(\omega) = \sqrt{R^2 + \left(\omega L - \frac{1}{\omega C}\right)^2} \qquad (16.54)$$

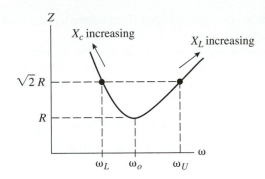

and is sketched in Figure 16.13. As expected for a series resonant circuit, the impedance is minimum at resonance and equals R.

$$\tilde{Z}(\omega_o) = R + j0 = R \tag{16.55}$$

The power in the resistance at resonance is maximum, as per Figure 16.11, and is:

$$P_R(\omega_o) = \frac{V_1^2}{R} \tag{16.56}$$

where V_1 is the RMS magnitude of the source voltage (all across R at resonance).

The condition in Equation (16.50) is used to determine the 3 dB frequencies:

$$Z(\omega_3) = \sqrt{2}\, Z(\omega_o) = \sqrt{2}\, R \tag{16.57}$$

First, the magnitude of the total impedance, Equation (16.54), at $\omega = \omega_3$, is inserted into Equation (16.57):

$$\sqrt{R^2 + \left(\omega_3 L - \frac{1}{\omega_3 C}\right)^2} = \sqrt{2}R \tag{16.58}$$

Square both sides and simplify:

$$R^2 + \left(\omega_3 L - \frac{1}{\omega_3 C}\right)^2 = 2R^2 \tag{16.59}$$

$$\left(\omega_3 L - \frac{1}{\omega_3 C}\right)^2 = R^2 \tag{16.60}$$

$$\sqrt{\left(\omega_3 L - \frac{1}{\omega_3 C}\right)^2} = \sqrt{R^2} \tag{16.61}$$

$$\left(\omega_3 L - \frac{1}{\omega_3 C}\right) = \pm R \tag{16.62}$$

Multiply through by $\omega_3 C$ to clear fractions:

$$\omega_3^2 LC - 1 = \pm \omega_3 RC \tag{16.63}$$

$$\omega_3^2 LC \pm RC\omega_3 - 1 = 0 \tag{16.64}$$

Use the quadratic equation on Equation (16.64) to solve for ω_3:

$$\omega_3 = \frac{-(\pm RC) \pm \sqrt{(RC)^2 + 4LC}}{2LC} = \mp \frac{R}{2L} \pm \sqrt{\left(\frac{R}{2L}\right)^2 + \frac{1}{LC}} \tag{16.65}$$

Because the radical is greater than $R/2L$ and because ω_3 values are positive, only the positive sign for the square root produces valid results:

$$\omega_L = -\frac{R}{2L} + \sqrt{\left(\frac{R}{2L}\right)^2 + \frac{1}{LC}} \tag{16.66}$$

$$\omega_U = +\frac{R}{2L} + \sqrt{\left(\frac{R}{2L}\right)^2 + \frac{1}{LC}} \tag{16.67}$$

Thus the 3 dB frequencies are identified. They depend on the component values, just as the resonant frequency did.

One could insert ω_L or ω_U into $\tilde{Z}(\omega)$, find I, and then calculate P_R to show $P_R(\omega_3) = \frac{1}{2}P_R(\omega_o)$, i.e., that the real power at the 3 dB frequencies is actually half of the real power at resonance. A clever alternative for this series resonant circuit arises from noting that the magnitude of the total circuit impedance at the 3 dB frequencies is a factor of $\sqrt{2}$ larger than the impedance at resonance. If $\omega L - \dfrac{1}{\omega C} = \pm R$ when $\omega = \omega_3$ in the impedance expression for the series RLC circuit, then:

$$\tilde{Z}(\omega_3) = R + j\left(\omega_3 L - \frac{1}{\omega_3 C}\right) = R \pm jR \tag{16.68}$$

$$\tilde{Z}(\omega_3) = R \pm jR = \sqrt{2}\,R\angle\pm45^2 \tag{16.69}$$

and the magnitude of the total circuit impedance at the 3 dB frequencies is a factor of $\sqrt{2}$ larger than the impedance at resonance (that equals R). Hence the current at the 3 dB frequencies becomes

$$\tilde{I}(\omega_3) = \frac{\tilde{V}_1}{\tilde{Z}(\omega_3)} = \frac{\tilde{V}_1}{\sqrt{2}\,R\angle\pm45°} \tag{16.70}$$

i.e., it decreases by a factor of $\sqrt{2}$. The real power is dissipated only in the resistance:

$$P_R(\omega_3) = |\tilde{I}|^2 R = \left|\frac{\tilde{V}_1}{\sqrt{2}R\angle\pm45°}\right|^2 R = \frac{V_1^2}{2R^2}R = \frac{V_1^2}{2R} = \frac{1}{2}P_R(\omega_o) \tag{16.71}$$

Thus the real power at the 3 dB frequency is half of the power at the resonant frequency. A similar analysis could be performed for each of the other resonant circuits. Remember that the center frequency, the 3 dB frequencies, and Q must be separately derived for *each* different resonant circuit. However, the power at the 3 dB frequencies is always half of the power at the resonant frequency.

The relationship between the *BW* and Q is simple, yet amazingly powerful. One may obtain the relationship for the series resonant circuit by subtracting ω_L in Equation (16.66) from ω_U in Equation (16.67):

$$\omega_U - \omega_L = +\frac{R}{2L} + \sqrt{\left(\frac{R}{2L}\right)^2 + \frac{1}{LC}} - \left[-\frac{R}{2L} + \sqrt{\left(\frac{R}{2L}\right)^2 + \frac{1}{LC}}\right] = +2\frac{R}{2L} = \frac{R}{L} \tag{16.72}$$

But the 3 dB bandwidth equals $\omega_U - \omega_L$:

$$BW = \omega_U - \omega_L = \frac{R}{L} \tag{16.73}$$

and $Q_s = \dfrac{\omega_o L}{R}$; hence for the series resonant circuit:

$$BW = \omega_U - \omega_L = \frac{R}{L} = \frac{Q_S}{\omega_o} \tag{16.74}$$

The Q can be solved for:

$$\boxed{Q = \frac{\omega_o}{BW(\text{rad/s})} = \frac{f_o}{BW(\text{Hz})}} \tag{16.75}$$

and this result, although illustrated for the specific case of the series resonant circuit, is very general. From Equation (16.75), Q and 3 dB *BW* are inversely proportional, as shown in Figure 16.14 for a parallel resonant circuit. As Q increases, the 3 dB *BW* decreases, and vice versa. The *BW* is usually an important design aspect. It generally sets the separation of adjacent channels of information that are separated in frequency. An example of this use is AM and FM radio. When one changes stations, the center frequency is really being

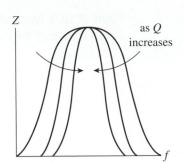

Figure 16.14 Bandwidth-*Q* Relationship for a Parallel Resonant Circuit

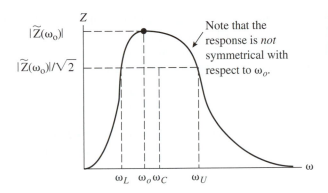

Figure 16.15 A Typical Low-*Q* Resonant Circuit Frequency Response

Note that the response is *not* symmetrical with respect to ω_o.

changed. The signals from each station must be prevented from overlapping each other in terms of frequency ranges. The bandpass filters, similar to the resonant circuits covered in this chapter, are utilized for this purpose. The bandwidth of the circuit is used to let the desired signal pass while rejecting the frequencies of signals from other stations.

Q affects the symmetry of the frequency response. For a low Q resonant circuit ($Q <$ 10 is a good rule of thumb), a typical response is shown in Figure 16.15. Note how the response is *not* symmetrical with respect to ω_o. The center frequency, which is *not* equal to the resonant frequency, is half way between the 3 dB frequencies:

$$\text{center frequency} = \omega_C = \frac{\omega_U + \omega_L}{2} \tag{16.76}$$

For the series resonant circuit case, if Equations (16.66) and (16.67) are inserted into Equation (16.76):

$$\omega_C = \frac{\omega_U + \omega_L}{2} = +\frac{R}{2L} + \sqrt{\left(\frac{R}{2L}\right)^2 + \frac{1}{LC}} + \left[-\frac{R}{2L} + \sqrt{\left(\frac{R}{2L}\right)^2 + \frac{1}{LC}}\right]$$

$$= \sqrt{\left(\frac{R}{2L}\right)^2 + \frac{1}{LC}} \tag{16.77}$$

This result is *greater than* the resonant frequency:

$$\omega_C = \sqrt{\left(\frac{R}{2L}\right)^2 + \frac{1}{LC}} > \omega_o = \sqrt{\frac{1}{LC}} \tag{16.78}$$

The reason for this lack of symmetry in the low Q case is the relative influence among the components in the resonant circuit. As another example, in a series–parallel resonant circuit, as frequency decreases, the inductive reactance becomes negligible relative to the resistance of the inductor, yet the capacitance has its full "influence" because it is in parallel with the circuit terminals. In general, the asymmetry increases as the Q decreases, as this effect becomes exaggerated in low Q circuits.

There is another consideration, too: the peak of the voltage or current response a low Q resonant response is not necessarily at the resonant frequency. The impedance (or admittance) is real at the resonant frequency, but it may not be purely real at the peak of the

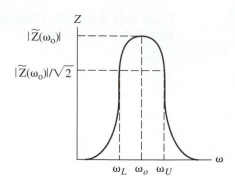

Figure 16.16 A Typical High-Q Resonant Circuit Rrequency Response

response when the Q is low. Use of circuit simulation to plot the frequency response and determine the frequency for the peak of the response is appropriate—see Example 16.6.3.

In a high-Q ($Q \geq 10$) resonant circuit, the asymmetry can usually be ignored, as sketched in Figure 16.16. Here, $\omega_o \approx \omega_c$, because in Equation (16.78):

$$\left(\frac{R}{2L}\right)^2 \ll \frac{1}{LC} \tag{16.79}$$

that is, the resistance is low enough to not change the resonant frequency appreciably. Thus the following relationships are obtained from the symmetrical frequency response:

$$\omega_U = \omega_o + \frac{BW}{2} = \omega_o + \frac{\omega_o}{2Q} \tag{16.80}$$

$$\omega_L = \omega_o - \frac{BW}{2} = \omega_o - \frac{\omega_o}{2Q} \tag{16.81}$$

In the lab, one usually measures the 3 dB BW in order to calculate the Q of the resonant circuit. However, in circuit analysis, one should determine Q or BW, whichever is easier. If ω_U and ω_L are needed and the circuit has a high Q, then Equations (16.80) and (16.81) can be utilized. However, if the circuit has a low Q, one *must* solve for ω_U and ω_L because the response is unsymmetrical about ω_o.

$$\omega_{(U,L)} \neq \omega_o \pm \frac{\omega_o}{2Q} \text{ (low-Q circuits)} \tag{16.82}$$

Although equations for the low Q 3 dB frequencies could be derived for the other resonant circuit configurations, this is often more efficiently performed using circuit simulation software. Once the component values for the desired resonant frequency, Q, and half-power bandwidth are determined, it is straightforward to enter the circuit into a circuit simulation program and determine the 3 dB frequencies from the frequency response plot. This procedure is illustrated in Example 16.6.3 in the next section.

16.6 SUMMARY TABLE AND EXAMPLES

A summary of several resonant circuit equations is given in Table 16.1. If $Q > 10$, the 3 dB frequencies are $\pm BW/2$ from the resonant frequency.

Example 16.6.1

For the circuit in Figure 16.17, determine (*a*) ω_o and f_o, (*b*) Q, (*c*) BW, (*d*) f_U and f_L, and (*e*) the magnitude of the voltage across the inductor at resonance [$V_L(\omega_o)$]. (*f*) Sketch and label the current magnitude frequency response.

Given: Resonant circuit in Figure 16.17

TABLE 16.1	Type	ω_o	Q	BW
Resonant Circuit Reference Table	Series	$\omega_o = \dfrac{1}{\sqrt{LC}}$	$Q_s = \dfrac{\omega_o L}{R} = \dfrac{1}{\omega_o RC}$	$BW\text{ (Hz)} = \dfrac{f_o}{Q_s}$ $BW\text{ (rad/s)} = \dfrac{\omega_o}{Q_s}$
	Parallel	$\omega_o = \dfrac{1}{\sqrt{LC}}$	$Q_p = \omega_o CR = \dfrac{R}{\omega_o L}$	$BW\text{ (Hz)} = \dfrac{f_o}{Q_p}$ $BW\text{ (rad/s)} = \dfrac{\omega_o}{Q_p}$
	Series–Parallel	$\omega_o = \sqrt{\dfrac{1}{LC} - \dfrac{R_{\text{ind}}^2}{L^2}}$	Convert the series $R_{\text{ind}}L$ branch to parallel R_P and X_P;	(use parallel)
	Series–Parallel With a Parallel Resistance	Same as for a series–parallel	combine parallel R's if present; determine $Q = Q_p$.	(use parallel)
	Other RLC Resonant Circuits	Find $\tilde{Z}_T$ or $\tilde{Y}_T$ Set $\text{Im}(\tilde{Z}_T$ or $\tilde{Y}_T) = 0$ Solve for ω_o	$BW = f_o/Q$ is general. Alternatively, find BW by solving for the two 3 dB frequencies.	

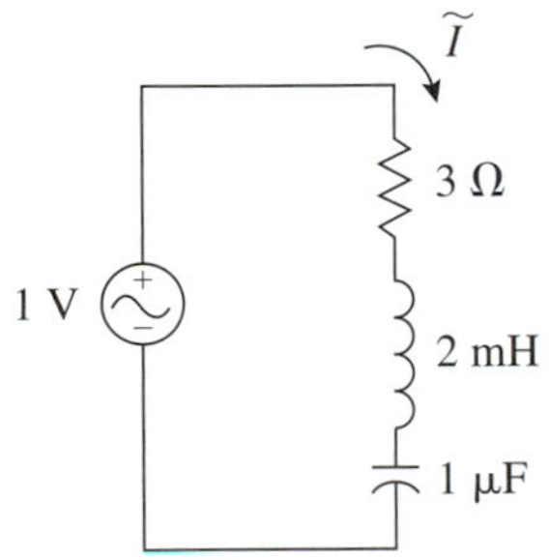

Figure 16.17 Series Resonant Circuit for Example 16.6.1

Desired: a. ω_o and f_o
　　　　　b. Q_5
　　　　　c. BW
　　　　　d. f_U and f_L
　　　　　e. $V_L(\omega_o)$
　　　　　f. I frequency response

Strategy: $\omega_o = \dfrac{1}{\sqrt{LC}}, \quad f_o = \dfrac{\omega_o}{2\pi}$

$Q_s = \dfrac{\omega_o L}{R}$

$BW = \dfrac{f_o}{Q}$

Is $Q > 10$? Use appropriate 3 dB frequency determination technique.

$$\tilde{I}(f_o) = \frac{1\angle 0°}{\tilde{Z}(\omega_o)}$$

$$\tilde{I}(f_L) = \frac{1\angle 0°}{\tilde{Z}(f_L)}, \quad I(f_L) = I(f_U)$$

$$V_L(\omega_o) = I(\omega_o)\,X_L(\omega_o)$$

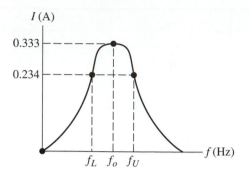

Figure 16.18 Current Response for Example 16.6.1

Solution: _____

$$\omega_o = \frac{1}{\sqrt{LC}} = 22,361 \text{ rad/s}$$

$$f_o = \frac{\omega_o}{2\pi} = 3.5588 \text{ kHz}$$

$$Q_s = \frac{\omega_o L}{R} = \frac{22,361(.002)}{3} = 14.907 > 10 \text{ (high } Q)$$

$$BW = \frac{f_o}{Q} = \frac{3.5588 \text{ kHz}}{14.907} = 238.73 \text{ Hz}$$

Because $Q > 10$:

$$f_U = f_o + \frac{BW}{2} = 3.6782 \text{ kHz} \quad (23,111 \text{ rad/s})$$

$$f_L = f_o - \frac{BW}{2} = 3.4394 \text{ kHz} \quad (21,611 \text{ rad/s})$$

$$\tilde{I}(f_o) = \frac{1\angle 0°}{\tilde{Z}(\omega_o)} = \frac{1\angle 0°}{3} = 0.33333 \text{ A}$$

$$\tilde{I}(f_L) = \frac{1\angle 0°}{\tilde{Z}(f_L)} = \frac{1\angle 0°}{3 + j21,611(.002) - j\dfrac{1}{21,611 \ (10^{-6})}} = \frac{1}{3 - j3.0507} = 0.234 \angle +45.5° \text{ A}$$

Note:
$$I(f_L) = \frac{I(f_o)}{\sqrt{2}} = \frac{0.33333}{\sqrt{2}} = 0.236$$

i.e., the current at the lower (and upper) 3 dB frequencies is a factor of $\sqrt{2}$ less than the current at the resonant frequency. See Figure 16.18.

$$V_L(\omega_o) = I(\omega_o) X_L(\omega_o) = .3333 (22,361) (.002) = 14.9 \text{ V}$$

Note that the voltage across the inductor is approximately Q times the source voltage:

$$V_L(\omega_o) \approx QV_S$$

Hence one must watch parts ratings in resonant circuits. Does this fact violate KVL? The voltage across the capacitor will also be approximately Q times the source voltage, but it will be 180° out of phase with the inductor voltage. Hence KVL is not violated.

_____ **Example 16.6.2** _____

For the circuit in Figure 16.19, determine (*a*) ω_o and f_o, (*b*) Q, (*c*) BW, (*d*) f_U and f_L, and (*e*) the magnitude of the current through the inductor at resonance [$I_L (\omega_o)$]. (*f*) Sketch and label the *V* response curve.

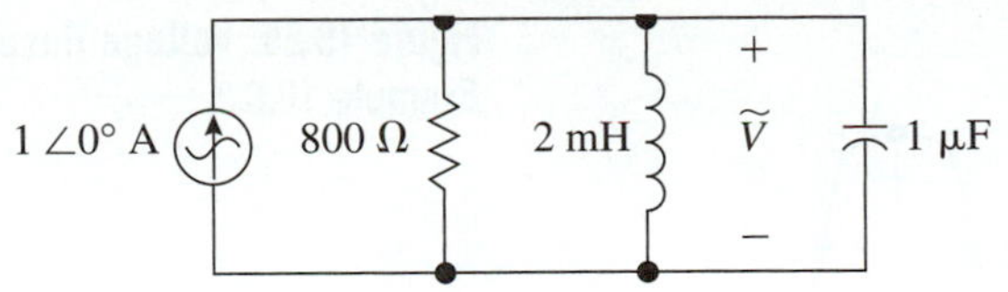

Figure 16.19 Parallel Resonant Circuit for Example 16.6.2

Given: Resonant circuit in Figure 16.19

Desired: a. ω_o and f_o

          b. Q_P

          c. BW

          d. f_U and f_L

          e. $I_L(\omega_o)$

          f. V response curve

Strategy:
$$\omega_o = \frac{1}{\sqrt{LC}}, \quad f_o = \frac{\omega_o}{2\pi}$$

$$Q_p = \frac{\omega_o C}{G} = \omega_o CR$$

$$BW = \frac{f_o}{Q_P}$$

Is $Q > 10$? Use appropriate 3 dB frequency determination technique.

$$\tilde{V}(\omega_o) = (1\angle 0°)\, \tilde{Z}_T(\omega_o)$$

$$\tilde{V}(\omega_L) = (1\angle 0°)\, \tilde{Z}_T(\omega_L) = \frac{1\angle 0°}{\tilde{Y}_T(\omega_L)}$$

$$I_L(\omega_o) = \frac{V(\omega_o)}{X_L(\omega_o)}$$

Solution: _____

$$\omega_o = \frac{1}{\sqrt{LC}} = 22{,}361 \text{ rad/s}; \quad f_o = 3.5588 \text{ kHz}$$

$$Q_p = \frac{\omega_o C}{G} = \omega_o CR = 22{,}361\,(10^{-6})800 = 17.889 > 10$$

$$BW = \frac{f_o}{Q_P} = 198.94 \text{ Hz}$$

$$f_U = f_o + \frac{BW}{2} = 3658.3 \text{ Hz } (22{,}986 \text{ rad/s})$$

$$f_L = f_o - \frac{BW}{2} = 3459.4 \text{ Hz } (21{,}736 \text{ rad/s})$$

$$\tilde{V}(\omega_o) = (1\angle 0°)(800) = 800 \text{ V}$$

$$\tilde{V}(\omega_L) = (1\angle 0°)\, \tilde{Z}_T(\omega_L) = \frac{1\angle 0°}{\tilde{Y}_T(\omega_L)} = \frac{1}{\dfrac{1}{800} - j\dfrac{1}{21{,}736\,(.002)} + j21{,}736\,(10^{-6})}$$

$$= \frac{1}{1.250 \times 10^{-3} - j1.2673 \times 10^{-3}} = \frac{1}{1.7800 \times 10^{-3}\angle -45.394°} = 562\angle +45.4°$$

$$\text{Note: } V(\omega_L) \approx \frac{V(\omega_o)}{\sqrt{2}} = \frac{800}{\sqrt{2}} = 566 \text{ V}$$

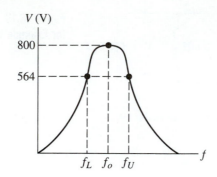

Figure 16.20 Voltage Response for Example 16.6.2

i.e., the voltage at the lower (and upper) 3 dB frequencies is a factor of $\sqrt{2}$ less than the voltage at the resonant frequency.

$$I_L(\omega_o) = \frac{V(\omega_o)}{X_L(\omega_o)} = \frac{800}{22{,}361\,(.002)} = 17.89 \text{ A}$$

Note that the current through the inductor is approximately Q times the source current:

$$I_L(\omega_o) \approx QI_s$$

Hence, one must watch parts ratings in resonant circuits. Does this fact violate KCL? The current through the capacitor will also be approximately Q times the source current, but it will be 180° out of phase with the inductor current. Hence KCL is not violated. A sketch of the frequency response is shown in Figure 16.20. It is symmetrical about the resonant frequency because $Q > 10$.

___

Example 16.6.3

For the circuit in Figure 16.21, determine (a) ω_o and f_o, (b) Q, (c) BW, and (d) f_L and f_U. (e) Sketch and label the voltage response curve.

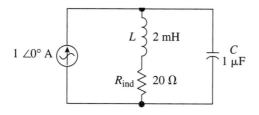

Figure 16.21 Series–Parallel Resonant Circuit for Example 16.6.3

Given: Resonant circuit in Figure 16.21

Desired: a. ω_o and f_o
          b. Q_P
          c. BW
          d. f_L and f_U
          e. the voltage response curve

Strategy:

$$\omega_o = \sqrt{\frac{1}{LC} - \frac{R_{ind}^2}{L^2}}$$

$$R_{ind} \rightarrow R_P = R:$$

$$\tilde{Z}_{branch} = R_{ind} + j\omega L$$

$$\tilde{Y} = \frac{1}{\tilde{Z}_{branch}} = G_p - jB_{Lp}$$

$$R_P = \frac{1}{G_P} = R$$

$$Q_p = \omega_o CR$$

$$BW = \frac{f_o}{Q_P}$$

Is $Q > 10$? Use appropriate 3 dB frequency determination technique.

Solution: _____

$$\omega_o = \sqrt{\frac{1}{LC} - \frac{R_{ind}^2}{L^2}} = 20{,}000 \text{ rad/s}$$

Note: $\omega_o < \dfrac{1}{\sqrt{LC}}$ (the R_{ind} in series with L changes f_o)

$$f_o = 3.1831 \text{ kHz}$$

$$R_{ind} \rightarrow R_P = R:$$

$$\tilde{Z}_{branch} = R_{ind} + j\omega L = 20 + j20k\,(0.002) = 20 + j40 \ \Omega$$

$$\tilde{Y} = \frac{1}{\tilde{Z}_{branch}} = \frac{1}{20 + j40} = 0.010000 - j0.020000 = G_P - jB_P$$

$$R_P = \frac{1}{G_P} = \frac{1}{0.01} = 100 \ \Omega$$

$$Q_p = \omega_o CR = 20 \text{ krad/s}(1 \ \mu F)(100 \ \Omega) = 2.0000$$

$$Q_P = 2.00 << 10 \rightarrow Low \ Q$$

$$BW = \frac{f_o}{Q} = 1591.5 \text{ Hz}$$

Because $Q < 10 \rightarrow$ response is not symmetrical around f_o. Simulate the circuit → frequency response curve, read off break frequencies.

The center frequency and the 3 dB frequencies are indicated on the frequency response plot in Figure 16.22. Note that the center frequency of 3.528 kHz differs from the resonant frequency of 3.183 kHz due to the low Q of this resonant circuit. The 3 dB frequencies and the BW are:

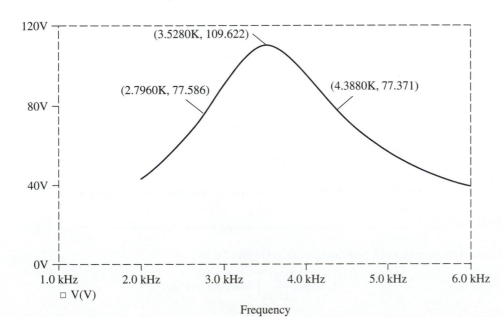

Figure 16.22 PSpice® Probe Plot for Example 16.6.3
PSpice simulation output, used with permission of Cadence Design Systems, Inc.

$$f_U = 4.3880 \text{ kHz}$$

$$f_L = 2.7960 \text{ kHz}$$

$$BW = f_U - f_L = 4.3880 - 2.7960 = 1.592 \text{ kHz}$$

which matches the BW as determined from Q. Thus in a low-Q resonant circuit, the center frequency and the resonant frequency differ, and the BW is not centered on the resonant frequency.

___ Example 16.6.4 _____

For the resonant circuit shown in Figure 6.23, determine: (a) ω_o and f_o, (b) Q, (c) f_L and f_U, and (d) the BW if the component values are:

$$R_x = 1500 \ \Omega$$

$$L = 2 \text{ mH}$$

$$R_{\text{ind}} = 2 \ \Omega$$

$$C = 1 \ \mu\text{F}$$

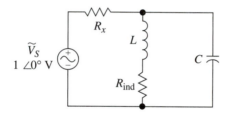

Figure 16.23 Series–Parallel Resonant Circuit for Example 16.6.4

Given: resonant circuit in Figure 16.23

Desired: a. ω_o and f_o
            b. Q_P
            c. f_L and f_U
            d. the BW

Strategy: Convert the voltage source–series resistance to a current source–parallel resistance.

$$\omega_o = \sqrt{\frac{1}{LC} - \frac{R_{\text{ind}}^2}{L^2}}$$

$$R_{\text{ind}} \rightarrow R_P$$

$$R = R_P \| R_X$$

$$Q_P = \omega_o CR$$

$$BW = \frac{f_o}{Q}$$

Is $Q > 10$? Use appropriate 3 dB frequency determination technique.

Solution: _____

Convert the source to its Norton equivalent (Figure 16.24):

$$\tilde{I}_S = \tilde{V}_S \left(\frac{1}{1500}\right) = 666.67 \ \angle 0° \ \mu\text{A}$$

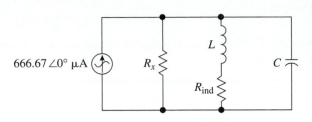

Figure 16.24 Source Conversion of Source in Figure 16.23

$666.67 \angle 0° \ \mu A$ R_x L R_{ind} C

Calculate ω_o (R_x does not influence ω_o)

$$\omega_o = \sqrt{\frac{1}{LC} - \frac{R_{\text{ind}}^2}{L^2}} = 22{,}338 \text{ rad/s}, \quad f_o = 3555.3 \text{ Hz}$$

Convert the series R_{ind}–L to a parallel R_p–X_p at ω_o (see Figure 16.25)

$$\tilde{Z} = R_{\text{ind}} + j\omega L = 2 + j22{,}338 \, (0.002) = 2 + j44.677 \ \Omega = 44.721 \angle +87.437°$$

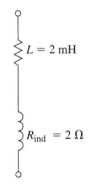
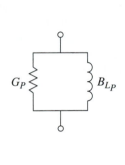

Figure 16.25 Series RL Branch Conversion to a Parallel Equivalent for Example 16.6.4

$L = 2$ mH

$R_{\text{ind}} = 2 \ \Omega$

G_P B_{Lp}

a. Series branch   b. Parallel equivalent

$$\tilde{Y} = \frac{1}{\tilde{Z}} = \frac{1}{44.721 \angle +87.437°} = 0.022361 \angle -87.437°$$

$$= 1.000 \times 10^{-3} - j2.2338 \times 10^{-2} = G_P - jB_{Lp}$$

$$R_P = \frac{1}{G_P} = 1000 \ \Omega$$

$$X_{Lp} = \frac{1}{B_{Lp}} = 44.766 \ \Omega$$

The equivalent parallel resonant circuit in Figure 16.26 is *valid only* at $\omega = \omega_o$.

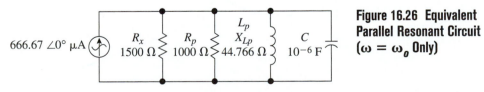

Figure 16.26 Equivalent Parallel Resonant Circuit ($\omega = \omega_o$ Only)

$666.67 \angle 0° \ \mu A$ R_x $1500 \ \Omega$ R_p $1000 \ \Omega$ $\frac{L_p}{X_{Lp}}$ $44.766 \ \Omega$ C 10^{-6} F

$$Q_p = \frac{\omega_o C}{G} = \omega_o CR = \omega_o C\left(\frac{R_p R_x}{R_p + R_x}\right) = 22{,}338 \, (10^{-6})\left(\frac{1000 \, (1500)}{1000 + 1500}\right)$$

$$Q_p = 13.403 > 10 \rightarrow \text{can use:} \quad f_L = f_o - \frac{BW}{2}, \quad f_U = f_o + \frac{BW}{2}$$

$$BW = \frac{f_o}{Q} = \frac{3555.3}{13.403} = 265.26 \text{ Hz}$$

$$f_L = 3422.6 \text{ Hz}, \quad f_U = 3687.9 \text{ Hz}$$

Perspective: The equivalent parallel resonant circuit was used to determine Q only (to see if it was > 10).

16.7 Bode Plots for Second-Order Circuits

A second-order circuit has both an inductor and a capacitor. We have examined the frequency response in the vicinity of the resonant frequency, but what if we look over a wider frequency range? A more extensive frequency response results relative to the first-order circuits in Chapter 15. The frequency response in dB for the circuit in Example 16.6.1 is plotted over a wider frequency range in Figure 16.27. The output voltage is taken across the capacitor for this plot. The resonant peak still occurs at the resonant frequency of 3.6 kHz. However, note the slope of the response at frequencies above the resonant frequency, especially away from the resonant peak. The slope is *−40 dB/decade*. Both the inductor and the capacitor are contributing to the lowpass filter response above resonance. The inductor "blocks" higher frequencies, and the capacitor "shorts" higher frequencies to ground. The capacitor contributes −20 dB/decade, and the inductor contributes −20 dB/decade. Hence, −40 dB/decade results in this scenario.

This example illustrates the nature of second-order circuits: resonant peaks and 40 dB/decade slopes. Thus one expects that the transfer function is more involved. The transfer function for the output voltage across the capacitor to the input voltage is:

$$H(j\omega) = \frac{\tilde{V}_o}{\tilde{V}_{\text{in}}} = \frac{\tilde{Z}_C}{\tilde{Z}_R + \tilde{Z}_L + \tilde{Z}_C} \qquad (16.83)$$

$$H(j\omega) = \frac{\dfrac{1}{j\omega C}}{R + j\omega L + \dfrac{1}{j\omega C}} \qquad (16.84)$$

$$H(j\omega) = \frac{\dfrac{1}{j\omega C}}{R + j\omega L + \dfrac{1}{j\omega C}} \cdot \frac{j\omega C}{j\omega C} = \frac{1}{(R)(j\omega C) + (j\omega L)(j\omega C) + 1} \qquad (16.85)$$

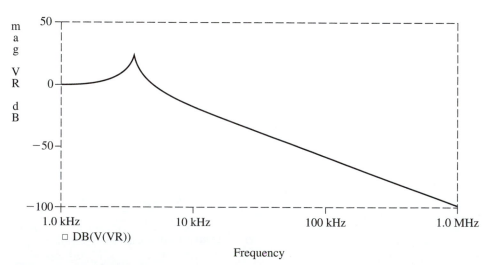

Figure 16.27 Frequency Response Plot for Example 16.6.1
PSpice® simulation output, used with permission of Cadence Design Systems, Inc.

$$H(j\omega) = \frac{1}{(1 - \omega^2 LC) + j\omega RC} \tag{16.86}$$

$$H(j\omega) = \frac{1}{\left[1 - \dfrac{\omega^2}{\left(\dfrac{1}{LC}\right)}\right] + j\omega RC} = \frac{1}{\left[1 - \dfrac{\omega^2}{\omega_b^2}\right] + j\omega RC} \tag{16.87}$$

where $\omega_b = \dfrac{1}{\sqrt{LC}}$ for the series RLC circuit. Note that Equation (16.87) differs from the $\left(1 + j\dfrac{\omega}{\omega_b}\right)$ form that was obtained for first-order circuits in Chapter 15.

Thus the frequency-squared term occurs in the transfer function of second-order circuits. When the transfer function is manipulated for preparing the Bode plots, the 20 log () is taken of the denominator (in this case). Above the break frequency, the ω^2 term will dominate, and the power of 2 will multiply with the 20, giving 40 dB/decade. This brief development and the explanation demonstrate how the 40 dB/decade slopes arise in Bode plots for second-order circuits. Detailed analyses of various second-order circuits are deferred to later courses.

CHAPTER REVIEW

16.1 Motivation: The Bandpass Filter
- Bandpass and bandreject filters require both an inductor and a capacitor in the circuit.

16.2 What Is Electrical Resonance?
- Resonance is defined as the frequency at which the impedance (or admittance) of the resonant circuit is real. Equivalently the imaginary part of the impedance (or admittance) is zero.
- At resonance, the total voltage and total current for the circuit are in phase.

16.3 Determining the Resonant Frequency
- To find the resonant frequency, the imaginary part of the total circuit impedance or admittance is found and set equal to zero, and one solves for the resonant frequency.
- The resonant frequency of series and parallel resonant circuits is:

$$\omega_o = \frac{1}{\sqrt{LC}}, \quad f_o = \frac{1}{2\pi\sqrt{LC}}$$

- The resonant frequency of series–parallel resonant circuits is:

$$\omega_o = \sqrt{\frac{1}{LC} - \frac{R_{ind}^2}{L^2}} = \sqrt{\frac{1}{LC} - \left(\frac{R_{ind}}{L}\right)^2}$$

16.4 Q, the Quality Factor
- Q is an indication of the amount of energy present (stored) relative to the energy loss in the circuit per cycle.

- The Q of a series resonant circuit is: $Q_s = \dfrac{\omega_o L}{R}$
- The Q of a parallel resonant circuit is:

$$Q_p = \omega_o RC = \frac{\omega_o C}{G}$$

- To determine the Q of a series–parallel resonant circuit, the series R_{ind}–L branch is converted to a parallel circuit, any parallel resistances are combined, and the equation for the Q of a parallel circuit is utilized.

16.5 Frequency Response, Q, and BW
- The 3 dB frequencies are defined to be where the power dissipated in the circuit resistance is half of the maximum power.
- The 3 dB bandwidth is the difference between the 3 dB frequencies:

$$BW = f_U - f_L \, (\text{Hz}) \quad \text{or} \quad BW = \omega_U - \omega_L \, (rad/s)$$

- Q is directly related to the resonant frequency and BW: $Q = \dfrac{\omega_o}{BW(\text{r/s})} = \dfrac{f_o}{BW(\text{Hz})}$
- The 3 dB frequencies are centered about the resonant frequency in high Q resonant circuits.
- The 3 dB frequencies are not centered about the resonant frequency in low Q resonant circuits.

16.6 Summary Table and Examples
- (See Table 16.1)

16.7 Bode Plots for Second-Order Circuits

- Bode plots for second-order circuits are more involved than for first-order circuits.

- In the frequency response of second-order circuits, ±40 dB/decade slopes and resonant peaks may occur.

HOMEWORK PROBLEMS

Proper completion of the following questions will demonstrate your ability to describe resonance in RLC circuits.

16.1 What defines the resonance state in an RLC circuit? Briefly explain.

16.2 For series or parallel RLC circuits, what can be said about (a) the circuit impedance at resonance and (b) the phase difference between the source voltage and the source current?

16.3 For a series resonant circuit, what is the difference in phase between the inductor and capacitor voltages at resonance? What is the phase difference at other frequencies?

16.4 For a parallel resonant circuit, what is the difference in phase between the inductor and capacitor currents at resonance? What is the phase difference at other frequencies?

Proper completion of the following questions will demonstrate your ability to calculate the resonant frequency, Q, and bandwidth for series and parallel RLC circuits.

16.5 What is the meaning of Q, the quality factor?

16.6 What is the meaning of bandwidth? How does bandwidth relate to Q?

16.7 How are bandwidth and the lower and upper 3dB frequencies related to each other? Why isn't the resonant frequency not in the center of the bandwidth in a low Q resonant circuit?

16.8 Calculate the resonant frequency in both radians/sec and hertz, Q, and the bandwidth for a series RLC circuit where $R = 4 \ \Omega$, $L = 40$ mH, and $C = 20 \ \mu$F.

16.9 Calculate the resonant frequency in both radians/sec and hertz, Q, and the lower and upper 3dB frequencies for a series RLC circuit where $R = 8 \ \Omega$, $L = 1$ mH, and $C = 5 \ \mu$F.

16.10 Calculate the resonant frequency in both radians/sec and hertz, Q, and bandwidth for a parallel RLC circuit where $R = 10$ kΩ, $L = 0.1$ mH, and $C = 20$ pF.

16.11 What happens to the resonant frequency, Q, and bandwidth as the circuit resistance is increased for a series resonant circuit? Recalculate the resonant frequency, Q, and the bandwidth for problem 16.8 after doubling the value of R. Is the effect the same for a parallel circuit when the parallel value of R is doubled?

Proper completion of the following questions will demonstrate your ability to calculate the resonant frequency and impedance for series–parallel RLC circuits and series–parallel with a parallel resistor RLC circuits.

16.12 Calculate the resonant frequency for the RLC circuit shown in Figure P16.1. By what percent is the actual resonant frequency shifted in comparison to the approximation where only L and C determine the resonant frequency?

is the actual resonant frequency shifted in comparison to the approximation where only L and C determine the resonant frequency?

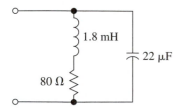

Figure P16.1 RLC Circuit of Problem 16.12

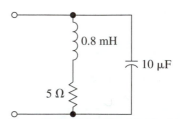

Figure P16.2 RLC Circuit of Problem 16.13

16.13 Calculate the resonant frequency for the RLC circuit shown in Figure P16.2. By what percent

16.14 In Section 16.3 of the text, the equation for the resonant frequency of a series–parallel circuit was developed using an expression of admittance. Instead, derive the equation for ω_o by first converting the RL series branch into parallel branches and then use the same formulation as was used for the parallel resonant circuit.

16.15 Calculate the total impedance at resonance for the circuit of Figure P16.3, given that $\omega_o = 1.424$ Mrad/s for the circuit.

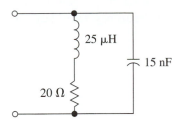

Figure P16.3 RLC Circuit of Problem 16.15

16.16 Calculate the total impedance at resonance for the circuit of Figure P16.4, given that $\omega_o = 22.22$ Mrad/s for the circuit.

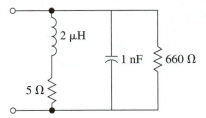

Figure P16.4 RLC Circuit of Problem 16.16

Proper completion of the following questions will demonstrate your ability to calculate resonant frequency, Q, bandwith, and 3 dB frequencies, and sketch the frequency response for series and parallel/RLC circuits.

16.17 For the series RLC circuit of Figure P16.5 calculate (a) ω_0, (b) Q, (c) ω_L and ω_U, (d) BW, and (e) the percent difference between the center frequency of the bandwidth and ω_0, (f) total impedance at ω_0, ω_L, and ω_U, and (g) sketch the response of the magnitude of I versus frequency. Label key values.

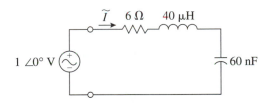

Figure P16.5 Series RLC Circuit of Problem 16.17

16.18 For the series RLC circuit of Figure P16.6 calculate (a) ω_0, (b) Q, (c) ω_L and ω_U, (d) BW, and (e) the percent difference between the center frequency of the bandwidth and ω_0, (f) total impedance at ω_0, ω_L, and ω_U, (g) the circuit current at resonance, and (h) sketch the response of the magnitude of I versus frequency. Label key values. Compare your answers for ω_0, Q, and BW with those of problem 16.17.

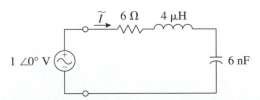

Figure P16.6 Series RLC Circuit of Problem 16.18

16.19 For the parallel RLC circuit of Figure P16.7 calculate (a) ω_0, (b) Q, (c) ω_L and ω_U, (d) BW, and (e) total impedance at ω_0, ω_L, and ω_U, (f) the circuit voltage at resonance, and (g) sketch the response of the magnitude of V versus frequency. Label key values.

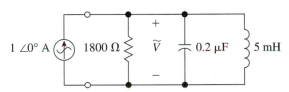

Figure P16.7 Parallel RLC Circuit of Problem 16.19

16.20 For the parallel RLC circuit of Figure P16.8 calculate (a) ω_0, (b) Q, (c) ω_L and ω_U, (d) BW, and (e) total impedance at ω_0, ω_L, and ω_U, (f) the circuit voltage at resonance, and (g) sketch the response of the magnitude of V versus frequency. Label key values.

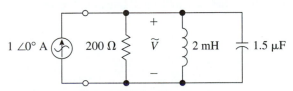

Figure P16.8 Parallel RLC Circuit of Problem 16.20

For the following problems the values of ω_L and ω_U may or may not be well centered about ω_0, depending on the value of Q. As an optional exercise, it is suggested that ω_L and ω_U be calculated using computer simulations for circuit values of Q < 10.

16.21 For the series–parallel RLC circuit of Figure P16.9 calculate (*a*) ω_0, (*b*) *Q*, (*c*) *BW*, (*d*) the total impedance at ω_0, and (*e*) the circuit voltage at resonance, and (*f*) sketch the response of the magnitude of *V* versus frequency. Label key values.

16.23 For the series–parallel RLC circuit of Figure P16.11 calculate (*a*) ω_0, (*b*) *Q*, (*c*) *BW*, (*d*) the total impedance at ω_0, and (*e*) the circuit voltage at resonance, and (*f*) sketch the response of the magnitude of *V* versus frequency. Label key values.

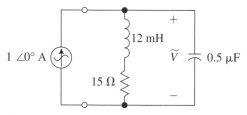

Figure P16.9 Series-Parallel RLC Circuit of Problem 16.21

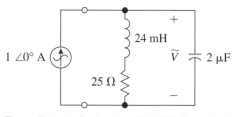

Figure P16.11 Series-Parallel RLC Circuit of Problem 16.23

16.22 The series–parallel RLC circuit of Figure P16.10 is the same as for problem 16.21, except a parallel resistor has been added. For Figure 16.10 determine the effect of the added resistor by again calculating (*a*) ω_0, (*b*) *Q*, (*c*) *BW*, (*d*) the total impedance at ω_0, and (*e*) the circuit voltage at resonance, and (*f*) sketch the response of the magnitude of *V* versus frequency. Label key values. Compare your results to those of problem 16.21.

16.24 The series–parallel RLC circuit of Figure P16.12 is the same as for problem 16.23, except a parallel resistor has been added. For Figure 16.12 determine the effect of the added resistor by again calculating (*a*) ω_0, (*b*) *Q*, (*c*) *BW*, (*d*) the total impedance at ω_0, and (*e*) the circuit voltage at resonance, and (*f*) sketch the response of the magnitude of *V* versus frequency, label key values. Compare your results to those of problem 16.23.

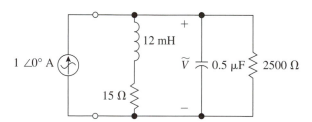

Figure P16.10 Series-Parallel RLC Circuit of Problem 16.22

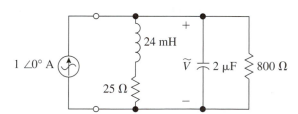

Figure P16.12 Series-Parallel RLC Circuit of Problem 16.24

BIBLIOGRAPHY

Feynman, Leighton, Sands, *The Feynman Lectures on Physics, Commemorative Issue,* Addison Wesley, Harlow, England, 1989.

Hayt and Buck, *Engineering Electromagnetics,* 6th ed., McGraw-Hill, New York, 2002.

Hayt, Kemmerly, and Durbin, *Engineering Circuit Analysis,* 6th ed., McGraw-Hill, New York, 2002.

Irwin, Kearns, *Introduction to Electrical Engineering,* Prentice Hall, Englewood Cliffs, New Jersey, 1995.

Kuhfittig, *Basic Technical Mathematics with Calculus,* 2nd ed., Brooks Cole, Pacific Grove, California, 1989.

Nilsson and Riedel, *Electric Circuits,* 6th ed., Prentice Hall, Upper Saddle River, New Jersey, 2001.

Answers to Selected Odd-Numbered Homework Problems

Chapter 1

1.1
   a. electrical to mechanical
   b. electrical to chemical
   c. electrical to light
   d. electrical to acoustic
1.3   coffee maker, mixer, toaster
1.5   A fan transforms electrical energy to mechanical energy and is used for cooling. A television transforms electrical energy into sound and light for entertainment. An oven transforms electrical energy to heat for cooking.
1.7   The most common form of nonelectrical energy used to produce electrical energy is fossil fuel. This is used to generate steam to turn turbines for generating energy.
1.9
   a. $1000 = 1 \times 10^3$
   b. 1.0×10^3
   c. 999
   d. 999.0
   e. 0.043
   f. 0.0432
   g. 7.38
   h. 7.37
1.11
   a. 385.0
   b. -53.541×10^{-3}
   c. 3.33
   d. 1.13
   e. -0.003×10^{-3}
   f. 339.41×10^3
   g. 27.68
   h. 7.083×10^{-6}
1.13
   a. 1.094
   b. 5.598×10^3
   c. 414
   d. 200 (80 has one significant digit)
   e. 230.00
   f. 0.40
   g. 341
   h. 5.2

1.15
   a. range 0.037, maximum value 0.954, minimum value 0.916
   b. range 0.53, maximum value 265.3, minimum value 264.7
   c. range 0.14, maximum value 1.51, minimum value 1.37
1.17
   a. 70 to 74 s
   b. 70.5 to 73.5 s
1.19
   a. accuracy 3.0% range, 0.30 s
   b. accuracy 0.67% range, 0.3 s
   c. accuracy 2.0% range, 0.30 s
1.21
   a. $85
   b. two significant digits
1.23   86.9 mF

Chapter 2

2.1   7.02 MJ
2.3   13.2 mW
2.5 $1.69
2.7   13.0 kJ
2.9   43.0 kJ
2.11
   a. 5.13 kWh
   b. $0.62
2.13   Energy is the ability to do work; power is how fast work is done.
2.15   133 mA
2.17   13.2 C
2.19   144 kJ
2.21   208 mA
2.23   4.36 kW
2.25   1.58 A
2.27   8.13 kW
2.29
   a. $1 k\Omega$
   b. brown, black, red

2.31

	Nominal Value	Maximum Value	Minimum Value	Range
A	2.2 kΩ	2.42 kΩ	1.98 kΩ	440 Ω
B	10 kΩ	10.5 kΩ	9.5 kΩ	1 kΩ
C	2 Ω	2.1 Ω	1.9 Ω	0.2 Ω
D	37 kΩ	40.7 kΩ	33.3 kΩ	7.4 kΩ
E	580 kΩ	609 kΩ	551 kΩ	58 kΩ
F	4.3 kΩ	5.16 kΩ	3.44 kΩ	1.72 kΩ
G	10 Ω	11 Ω	9 Ω	2 Ω
H	1.4 Ω	1.47 Ω	1.33 Ω	0.14 Ω

2.33   43.5 V
2.35   240 A
2.37
    a. 27.6 W
    b. 0.037 hp
2.39   190. W
2.41   1.56 mW
2.43   100 W
2.45   54.3 mΩ
2.47   6.15 ft
2.49   0.873 ton if 10,500 BTU/lb and 0.591 ton if 15,500 BTU/lb. Both assume 100% efficiency.

Chapter 3

3.1
    a. 12 kΩ
    b. 3.75 V
    c. 7.88 mW
3.3
    a. 1.29 mA
    b. 13.9 kΩ
    c. 7.38 V
    d. 7.38 V
3.5
    R1 = 0.6 Ω
    R2 = 1.4 Ω
    R3 = 2.5 Ω

3.7   20 AWG copper
3.9
    a. 41.7 mS
    b. 24 Ω
    c. 5.00 A
    d. 600 W
3.11
    a. 3.95 mS
    b. 253 Ω
    c. 0.593 A
    d. 68.2 mA
3.13
    a. (see figure at bottom of page)
    b. microwave 8.33 A, lamps 0.833 A each, motor 1.55 A, 0.333 A each
    c. 13.2 A
    d. 20 A breaker, 12 AWG copper wire
3.15
    a. 17.0 Ω
    b. 365 mA
    c. 253 mV
3.17
    a. 2.45 kΩ
    b. 22.5 mA
    c. 4.6 mA
    d. 0.917 V
3.19 $V_3 = 9.6$ kV and $V_S = 49.1$ kV
3.21   −4.89 V
3.23   1.14 mA
3.25 $I_T = -597\ \mu A$ and $V_{ab} = 26.8$ V
3.27   An ideal voltage source delivers a constant voltage at any current level, and an ideal current source delivers a constant current at any voltage level.
3.29
    a. 15.0 mS
    b. 66.7 Ω
    c. −1.33 A
    d. +0.44 A
    e. −222 mA
    f. $I_1 - I_2 + I_3 + I_T = 0$

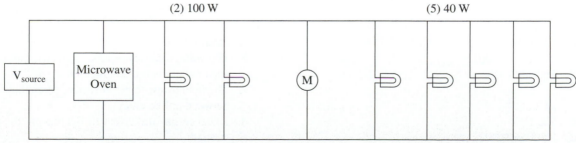

(2) 100 W   (5) 40 W

Ceiling Fan Motor

3.31

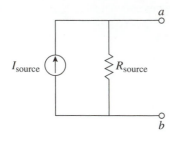

3.33

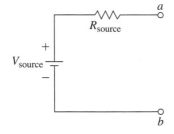

3.35 simulation
3.37 simulation
3.39 simulation
3.41

$$P_T = P_1 + P_2 + \ldots + P_N$$
$$V_T I_T = V_1 I_1 + V_2 I_2 + \ldots + V_N I_N$$
$$I_T = I_1 = I_2 = \ldots = I_N \text{ in a series circuit.}$$
$$\therefore V_T = V_1 + V_2 + \ldots + V_N$$
Q.E.D.

Chapter 4

4.1 The plot of voltage or current versus time, i.e., the plot of an electronic quantity versus time, is called a *waveform*.

4.3 Probably the most common and useful electronic signal is the alternating current sinusoidal steady-state signal, which is abbreviated AC.

4.5 AC along with some new components that will be introduced soon, the capacitor and the inductor, can be used to select a channel, such as on a radio or a television. Many electromechanical devices (motors and generators, for example) use or produce AC to operate properly and efficiently.

4.7
  a. time-independent variable
  b. angular velocity

4.9
  a. 10 V
  b. 1000 Hz
  c. 1/1000 or 0.001 s
  d. $2\pi1000$ rad/s

4.11
  a. 24 V
  b. 500 Hz
  c. 1/500 or 0.002
  d. $2\pi500$ rad/s

4.13
  a. 0.5 A
  b. 0.3536 A
  c. 0.3536 A
  d. 0.000 A

4.17
  a. 17.5 V
  b. 62.83 krad/s
  c. $-306° = +54°$
  d. $v(t) = 17.5 \sin(2\pi10kt + 54°)$
  e. $v(t) = 17.5 \cos(2\pi10kt - 36°)$

4.19
  a. 161 mA
  b. 51.4 mA
  c. 51.4 mA
  d. 321 mA
  e. 257 mA
  f. 321 μ(micro)A

4.21
  a. 12 kΩ
  b. 3.75 V
  c. 7.88 mW

4.23
  a. 1.29 mA
  b. 13.9 kΩ
  c. 7.38 V
  d. 7.38 V

4.25
  a. 72.6 V
  b. 1.96 A
  c. 1.96 A

4.27
  a. 1.12 Ω
  b. 0.91 A
  c. 3.64 V

4.29
  a. 30 V
  b. 500 mA

4.31
  a. 60 V
  b. 333 mA
4.33 simulation
4.35 *R*3 goes 5% low

Chapter 5

5.1
  a. electric field
  b. The voltage causes separated electric charges to accumulate on the conductors.
  c. The field is due to the attractive force that exists between unlike charges.
  d. Distance has the effect of decreasing the field strength.

5.3   A capacitor is an electrical component that allows for the storage of electrical energy in an electric field.

5.5   Electric energy. Energy is stored in the electric field by the work done in separating the electric charge existing on the plates of the capacitor.

5.7   The physical parameters are the plate area, the distance between the plates, and the dielectric properties of the insulating material between the plates.

5.9   1.20 nF

5.11   11.5 cm^2

5.13   120. μC

5.15   2.26 mC

5.17   2.50 kV

5.19   60.0 μF

5.21   42.7 nF

5.23   6.67 pF

5.25   150. nF

5.27   A capacitor opposes an instantaneous change in voltage because that represents an instantaneous movement of electric charge, which requires an infinite amount of work to be done in zero time. It takes time to move charge.

5.29   The resistor limits the maximum value of the current.

5.31   The current leads the voltage by 90°.

5.33   The breakdown voltage is the voltage that, when applied across the capacitor, will possibly damage the capacitor due to breakdown of the dielectric.

5.35   parallel plates, rolled-up, multilayer

5.37   The leakage resistance represents a low current path through the capacitance and is modeled as a resistor in parallel with an ideal capacitance.

Chapter 6

6.1   Eight poles will result. Each piece will be a magnet with its own north and south poles.

6.3   The inductor would resist the sudden change in current with a drastic increase in the voltage across the inductor.

6.5   magnetic field; the energy is stored in the magnetic field that is located in the space in and around the inductor.

6.7   Physical parameters are permeability of the core, number of turns, cross-sectional area, length of the coil

6.9   0.400 mWb
       2.00 μJ

6.11   231 μH
        Wheeler's equation

6.13   629 turns

6.15   283 turns

6.17   5.03 m

6.19   10.9 mH

6.21   233 mH

6.23   0.803 H

6.25   68.4 mH

6.27   The inductor opposes an instantaneous change of current because that would require an instantaneous change of magnetic flux, which induces a voltage (Faraday's law) that has a polarity, which opposes the current change (Lenz's law). The instantaneous change of magnetic flux would require an infinite amount of work to be done in zero time.

6.29   The resistor limits the value that the current through the inductor reaches, since the inductor has zero voltage across it in the DC steady-state condition.

6.31   The voltage leads the current by 90°.

6.33   Large relative permeability values of the inductor core are the only realistic way of achieving larger values of inductance.

6.35   magnetic core coil, air-core coil, toroid

Chapter 7

7.1
   a.   exponential growth

x	$f(x)$
0	25
0.5	41.21803
1	67.95705
1.5	112.0422
2	184.7264
2.5	304.5623

   b.   This is an unstable function.

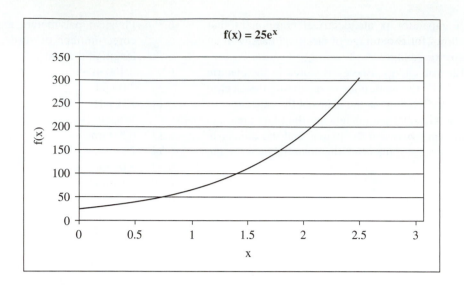

7.3

a. exponential
   buildup

x	$f(x)=6(1-e^{-x})$
0	0
1	3.79273
2	5.18780
3	5.70128
4	5.89010
5	5.95957

b. This is a stable
   function.

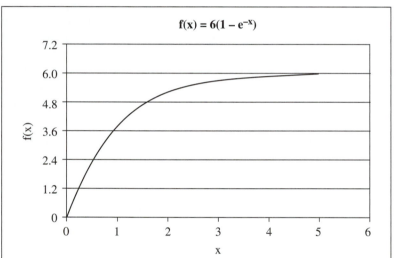

7.5

a. exponential
   decay

t	$f(t)=33e^{(-t/3.3)}$
0	33
1	24.37303159
4	9.819658535
7	3.956245384
14	0.474299319
17	0.191090605

b. This is a stable
   function

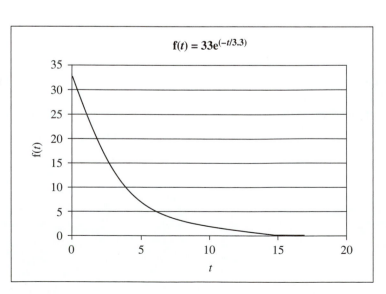

7.7

    a. exponential buildup

t	$f(t)=15(1-e^{(-t/144)})$
0	0
120	8.481026872
260	12.53425339
430	14.2427493
570	14.71357631
725	14.90237992

    b. This is a stable function.

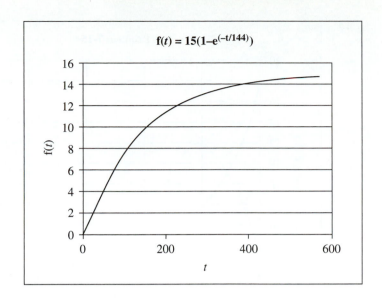

$f(t) = 15(1-e^{(-t/144)})$

7.9

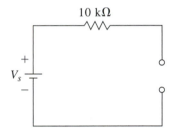

7.11

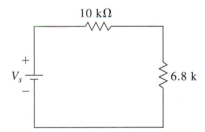

7.13

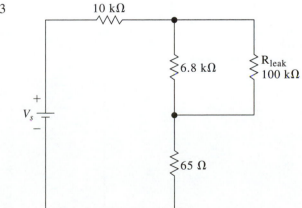

7.15

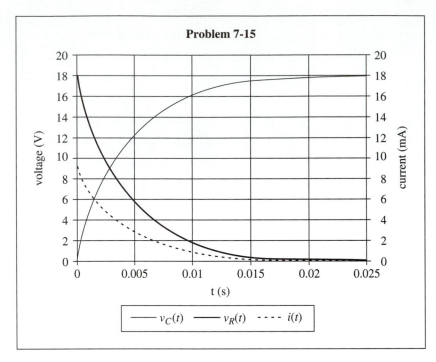

7.17

a. $v_c(t) = 12.5\ (e^{-t/99\mathrm{ms}})$

$i(t) = 4.17e^{-t/99\mathrm{ms}}$

$v_r(t) = 12.5\ (e^{-t/99\mathrm{ms}})$

b.   137ms

c.

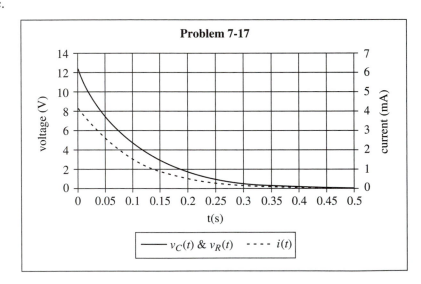

7.19

a. $v_L(t) = 25\ e^{-t/150\mathrm{ps}}$

$i(t) = 125(1 - e^{-t/150\mathrm{ps}})$

$v_R(t) = 25(1 - e^{-t/150\mathrm{ps}})$

b.   8.28μs

c.

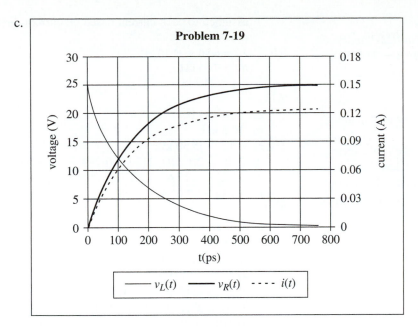

7.21

  a. $V_L(t) = -36e^{-t/0.278})$

       $i(t) = 2\,e^{-t/0.278}$

       $V_R(t) = 20\,e^{-t/0.278}$

  b. 222 ms

  c.

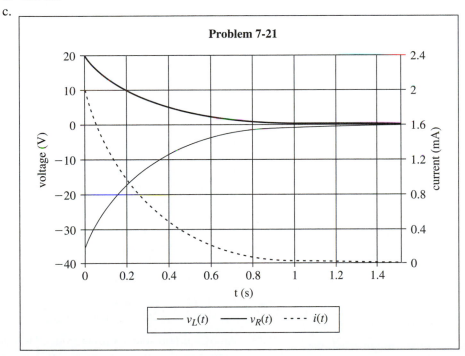

Chapter 8

8.1

  a. 20 V

  b. 62.83 krad/

  c. 0°

  d. $v(t) = 20\sin(62.83\text{ krad/s } t)$

8.3 $32.5\sin(62.83\text{krad/s } t + 27.8°)$

8.5

  a. Complex numbers are "two-dimensional" numbers: $a + jb$. They contain two pieces of information.

  b. j indicates that the imaginary part has a phase shift of $\pm 90°$ with respect to the real part.

  c. Complex numbers eliminate the need to manipulate trigonometric functions in AC circuit analysis.

8.7 $41.5 - j26.8$

8.9 $20 \sin(62.83\text{krad/s } t) + 17.5 \sin(62.83\text{krad/s } t + 60°) = 32.5 \sin(62.83\text{krad/s } t + 27.8°)$

8.11 $v(t) = 27 \sin(\omega t - 130°)$

8.13 3.62 mV

8.15

   a. A capacitor always opposes an instantaneous change of voltage.

   b. The voltage changes in response to the change of charge on the plates, which takes time.

   c. After the capacitor is fully charged, after five time constants, the capacitor acts like an open circuit.

   d. The capacitor will charge with a DC source to the level of the DC source.

   e. The current leads the voltage by 90°. The reductance is a function of frequency.

8.17   None—it only stores energy.

8.19   None—all power is dissipated.

8.21   Capacitive reactance increases as frequency decreases, because there is more time for charge to build up on the plates of the capacitor.

8.23

   a. 120 Hz

   b. leading

   c. $56.6\angle 45°$ kΩ or $40.0 + j40.0$ kΩ

   d. inductive

8.25   25.5 mA, AC sinusoidal steady-state signal

8.27

   a. $531.9\angle -90°$ Ω

   b. $1.2\angle 0°$ kΩ

   c. $1.312\angle -23.9°$ kΩ

   d. $7.62\angle 23.9°$ mA

   e. $4.05\angle -66.1°$ V

   f. $7.618\angle +23.9°$ V

8.29

   a. $62.83\angle 90°$ Ω

   b. $36\angle 0°$ Ω

   c. $72.4\angle 60.19°$ Ω

   d. $724.14\angle 90.19°$ V

   e. $360\angle 30°$ V

   f. $628.3\angle 120°$ V

Chapter 9

9.1   The total impedance of a series circuit equals the sum of the individual impedances in series.

9.3   The total current in a series circuit equals the current through each individual component in series. The current is the same throughout a series circuit.

9.5   The total admittance of a parallel circuit equals the sum of the individual admittances in parallel.

9.7   The total current in a parallel circuit equals the sum of the currents in the individual branches. In this summation, proper signs must be assigned to each current based on current direction.

9.9 $\tilde{Z}_T = 6.80 + j2.10$ kΩ
$R_P = 7.45$ kΩ
$X_{Lp} = 24.1$ kΩ

9.11

   a. $-j15.9$ kΩ

   b. 10.0 kΩ

   c. $7.17 - j5.50$ kΩ

   d. $212\angle -2.1°$ kΩ, $212 \sin(6283 + -2.1°)$ kΩ

   e. $21.2\angle -2.1°$ A

   f. $13.3\angle +87.9°$ A

9.13

   a. 157 Ω

   b. 200. Ω

   c. $124\angle +51.9°$ Ω

   d. $0.267\angle -51.9°$ A
$0.267 \sin(1571 + -51.9°)$

   e. $0.165\angle 0°$ A, $0.165 \sin(1571+)$

   f. $0.210\angle -90°$ A, $0.210 \sin(1571 + -90°)$

9.15 $\tilde{Y}_T = 1.47 + j0.417$ mS
$R_S = 629$ Ω
$X_{Cs} = 178$ Ω

9.17 $\tilde{I}_T = 32.3\angle +23.8°$ mA
$\tilde{V}_C = 12.9\angle -66.2°$ V
$\tilde{V}_R = 22.0\angle +23.8°$ V
$\tilde{V}_L = 3.23\angle +113.8°$ V
$\tilde{V}_C + \tilde{V}_R + \tilde{V}_L = 24.0\angle 0°$ V $= \tilde{V}_S$ ok

9.19 $\tilde{I}_T = 4.53\angle +28.1°$ mA
$\tilde{I}_C = 4.80\angle +90.0°$ mA
$\tilde{I}_L = 2.67\angle -90.0°$ mA
$\tilde{I}_R = 4.00\angle 0°$ mA
$\tilde{I}_C + \tilde{I}_R + \tilde{I}_L = 4.53\angle +28.1°$ mA $= \tilde{I}_T$ ok

9.21 $\tilde{I}_T = 5.91\angle -16.4°$ mA
$\tilde{I}_1 = 2.36\angle +45.0°$ mA
$\tilde{I}_2 = 5.21\angle -39.8°$ mA
$\tilde{V}_{ab} = 28.3\angle 45.0°$ V $(a+, b-)$

9.23 $\tilde{I}_T = 36.1\angle +13.8°$ mA
$\tilde{I}_1 = 26.2\angle +89.5°$ mA
$\tilde{I}_2 = 24.2\angle +39.2°$ mA
$\tilde{V}_{ab} = 7.43\angle +13.8°$ V $(a+, b-)$

9.25 $\tilde{I}_1 = 2.04\angle +94.5°$ mA
$\tilde{I}_2 = 3.43\angle -170.2°$ mA
$\tilde{V}_{ab} = 7.66\angle -22.3°$ V $(a+, b-)$

9.27 $\tilde{I}_1 = 0.142\angle -83.4°$ mA
$\tilde{V}_{ab} = 24.6\angle -165.7°$ V $(a+, b-)$

9.29   An AC current source differs from a DC current source in that it generates an AC sinusoidal steady-state current with a constant peak amplitude and a constant frequency instead of a constant DC current. Hence an AC current source must have both a magnitude and a phase angle specified instead of just the magnitude that is required for a DC current source.

9.31 $\tilde{I}_R = 58.3\angle -60.95°$ mA
$\tilde{I}_L = 87.4\angle -150.9°$ mA
$\tilde{I}_C = 192\angle +29.05°$ mA
$\tilde{V}_T = 19.2\angle -60.95°$ V

9.33 $\tilde{V}_C = 8.16\angle -90.0°\text{ V}$
        $\tilde{V}_R = 12.0\angle 0°\text{ V}$
        $\tilde{V}_L = 5.64\angle +90.0°\text{ V}$
        $\tilde{V}_T = 12.3\angle -11.9°\text{ V}$

9.35   (simulation results should match corresponding problem results)

Chapter 10

10.1   Reactive power in an inductor or a capacitor really represents stored energy, and real power in a resistance represents energy that is converted.

10.3   Positive instantaneous power is rate of energy being stored into the electric field of the capacitor or into the magnetic field of the inductor. Negative instantaneous power is the rate of energy being drawn from the electric field of the capacitor or from the magnetic field of the inductor.

10.5   pure reactive power

10.7   complex power: VA
        apparent power: VA
        real power: W
        reactive power: VAR

10.9
    a. $13.9\angle -14.9°\text{ VA}$
    b. 13.5 W
    c. 0 W
    d. 0 VAR
    e. 3.57 VAR

10.11
    a. $1.08\angle -41.4°\text{kVA}$
    b. 811 W
    c. 0 W
    d. 0 W
    e. 0 VAR
    f. 1.44 kVAR
    g. 2.15 kVAR

10.13
    a. $613\angle -71.2°\text{ VA}$
    b. 198 W
    c. 0 W
    d. 0 W
    e. 0 VAR
    f. 112 VAR
    g. 692 VAR

10.15
    a. $\tilde{S} = 9.08\angle +14.4°\text{ VA}$
    b. pf = 0.968

10.17 $X_L = 4.04\text{ k}\Omega$
        $\text{pf}_{before} = 0.967$, leading
        $\text{pf}_{after} = 1.00$

10.19 $X_L = 24.8\ \Omega$
        $\text{pf}_{before} = 0.323$, leading

        $\text{pf}_{after} = 1.00$

10.21
    a. $\text{pf}_{before} = 0.924$, lagging
    b. 27.4 Ω, in parallel

10.23
    a. 7.96 kVA
    b. 6.27 kW, 4.90 kVAR (capacitive)
    c. $\theta = -38.0°$, pf = 0.788 (leading)
    d. capacitive
    e. 2.94 Ω, in parallel

10.25 reduction in total circuit current, cost savings, smaller wire sizes

Chapter 11

11.1
    a. Mesh currents are needed to apply KVL around the mesh. If the component has two mesh currents, both must be expressed with proper signs for voltage polarities. The resulting KVL equations are a set of simultaneous equations that are solved for the mesh currents.
    b. The mesh current is the current defined to go around each window (mesh). The branch current is the current in each branch. There can be multiple branch currents in a mesh.

11.3   Mesh analysis is an organized method of applying KVL.

11.5   −3.158 V

11.7   −4.895 V

11.9   1.135 mA

11.11
    a. 13.65 mA
    b. 3.949 mA
    c. 112.82 μA
    d. 0.338 V

11.13 The node voltage is the voltage between a nontrivial node and the reference node. KCL is then applied to each nontrivial node.

11.15 A trivial node has only two components connected to them. Nontrivial nodes are those that have at least three components connected.

11.17 KCL is applied to each nontrivial node.

11.19 −3.158 V

11.21 −4.895 V

11.23 1.135 mA

11.25
    a. 13.65 mA
    b. 3.949 mA
    c. 112.82 μA
    d. 0.338 V

11.27 Mesh currents are needed to apply KVL around the loop. If the component has two mesh currents, both must be expressed with proper sign for voltage polarities. The resulting KVL equations are a set of simultaneous equations that are solved for the mesh currents. The mesh current is the current in each circuit window. The branch current is the current in each branch. There can be multiple branch currents in a mesh.

11.29 Mesh analysis is an organized method of applying KVL.

11.31
  $28.3\angle45.0°$ V
  $5.91\angle-16.4°$ mA
  $2.356\angle45.0°$ mA
  $5.21\angle-39.8°$ mA

11.33
  $36.1\angle+13.8°$ mA
  $26.2\angle89.5°$ mA
  $24.2\angle39.2°$ mA

11.35
  $5.95\angle+18.3°$ V
  $1.56\angle-57.2°$ mA
  $3.69\angle+174.2°$ mA

11.37
  $15.3\angle-29.7°$ V
  $107.3\angle-81.0°$ mA

11.39
  $28.3\angle45.0°$ V
  $5.91\angle-16.4°$ mA
  $2.356\angle45.0°$ mA
  $5.21\angle-39.8°$ mA

11.41
  $36.1\angle+13.8°$ mA
  $26.2\angle89.5°$ mA
  $24.2\angle39.2°$ mA

11.43 (same answers as 11.35)

11.45 (same answers as 11.37

11.47 300 Ω (each side of the delta)

11.49 632 mA

11.51 $335.4\angle26.6°$ Ω = $300 + j\angle50$ Ω (each side of the delta)

11.53 Use the inductive reactive value for the inductor, the inverse of the capacitance reactance for the capacitor, and ω=1, which is a frequency of 0.159155 Hz.

Chapter 12

12.1 An equivalent circuit has the same performance at the load as does the original circuit; however it is usually simpler to analyze than the original circuit. The analysis and consequent trends as R_L is varied is much easier to obtain.

12.3 The voltage across and current through R_L is the same in both the original circuit and the equivalent circuit connected to R_L.

12.5 A Thévenin equivalent DC circuit is a simplified representation of the original circuit that results in the same voltage and current performance for a load as the original circuit.

12.7 $V_{Th} = 33.4$ V
  $R_{Th} = 289$ Ω
  $V_{ab,300} = 17.5$ V
  $V_{ab,400} = 20.0$ V
  $V_{ab,500} = 21.8$ V
  $V_{ab,600} = 23.2$ V
  $V_{ab,700} = 24.3$ V

12.9 $V_{Th} = -5.01$ V
  $R_{Th} = 2.50$ kΩ
  $V_{ab,1.0k} = -1.43$ V
  $V_{ab,1.5k} = -1.88$ V
  $V_{ab,2.0k} = -2.22$ V
  $V_{ab,2.5k} = -2.50$ V
  $V_{ab,3.0k} = -2.73$ V
  $I_{L,1.0k} = -1.43$ mA
  $I_{L,1.5k} = -1.25$ mA
  $I_{L,2.0k} = -1.11$ A
  $I_{L,2.5k} = -1.00$ μA
  $I_{L,3.0k} = -910$ μA

12.11 $V_{Th} = -5.01$ V
  $R_{Th} = 2.50$ kΩ
  $V_{ab,1.0k} = -1.43$ V
  $V_{ab,1.5k} = -1.88$ V
  $V_{ab,2.0k} = -2.22$ V
  $V_{ab,2.5k} = -2.50$ V
  $V_{ab,3.0k} = -2.73$ V
  $I_{L,1.0k} = -1.43$ mA
  $I_{L,1.5k} = -1.25$ mA
  $I_{L,2.0k} = -1.11$ A
  $I_{L,2.5k} = -1.00$ μA
  $I_{L,3.0k} = -910$ μA

The answers are the same as problem 12.9 because V_{Th} and R_{Th} are the same.

12.13 The Norton equivalent circuit is another equivalent circuit that consists of a parallel current source and resistance.

12.15 $I_N = 119$ mA
  $R_N = 289$ Ω
  $V_{ab,300} = 17.5$ V
  $V_{ab,400} = 20.0$ V
  $V_{ab,500} = 21.8$ V
  $V_{ab,600} = 23.2$ V
  $V_{ab,700} = 24.3$ V

12.17 $I_N = -2.00$ mA
  $R_N = 2.50$ kΩ
  $V_{ab,1.0k} = -1.43$ V
  $V_{ab,1.5k} = -1.88$ V
  $V_{ab,2.0k} = -2.22$ V
  $V_{ab,2.5k} = -2.50$ V
  $V_{ab,3.0k} = -2.73$ V
  $I_{L,1.0k} = -1.43$ mA
  $I_{L,1.5k} = -1.25$ mA
  $I_{L,2.0k} = -1.11$ mA
  $I_{L,2.5k} = -1.00$ mA
  $I_{L,3.0k} = -910$ μA

12.19 The answers are the same as for problems 12.9 and 12.17 because the current source in this problem is equivalent to the voltage source in Figure 12.3 (perform a source conversion).

12.21 Do a source conversion on the Thévenin equivalent circuit to find the Norton equivalent circuit.

12.23 $I_N = 119$ mA, $R_N = 289$ Ω for problem 12.7;
$I_N = -2.00$ mA, $R_N = 2.50$ kΩ for problem 12.9;
$I_N = -2.00$ mA, $R_N = 2.50$ kΩ for problem 12.11.

12.25 A Thévenin equivalent AC circuit is a simplified representation of the original circuit that results in the same voltage and current performance for a load as the original circuit.

12.27 $\tilde{V}_{Th} = 103\angle-10.4°$ V
$\tilde{Z}_{Th} = (305 - j15.6)$ Ω
$\tilde{V}_{ab}(400) = 58.7\angle-9.2$
$\tilde{V}_{ab}(500) = 64.3\angle-9.3$
$\tilde{V}_{ab}(600) = 68.6\angle-9.4°$

12.29 $\tilde{V}_{Th} = 66.1 \angle 58.1°$ V
$\tilde{Z}_{Th} = (2.92 + j1.56)$ kΩ
$\tilde{V}_{ab}(1.0k) = 15.7 \angle36.4°$ V
$\tilde{V}_{ab}(1.5k) = 21.2\angle38.7°$ V
$\tilde{V}_{ab}(2.0k) = 25.7\angle40.5°$ V
$\tilde{V}_{ab}(2.5k) = 29.3 \angle42.1°$ V
$\tilde{V}_{ab}(3.0k) = 32.4\angle43.3°$ V
$\tilde{I}_L(1.0k) = 15.7\angle36.4°$ mA
$\tilde{I}_L(1.5k) = 14.1\angle38.7°$ mA
$\tilde{I}_L(2.0k) = 12.8\angle40.5°$ mA
$\tilde{I}_L(2.5k) = 11.7\angle42.1°$ mA
$\tilde{I}_L(3.0k) = 10.8\angle43.3°$ mA

12.31 $\tilde{V}_{Th} = 66.1 \angle58.1°$ volts
$\tilde{Z}_{Th} = (2.92 + j1.56)$ kΩ
$\tilde{V}_{ab}(1.0k) = 15.7\angle36.4°$ V
$\tilde{V}_{ab}(1.5k) = 21.2\angle38.7°$ V
$\tilde{V}_{ab}(2.0k) = 25.7\angle40.5°$ V
$\tilde{V}_{ab}(2.5k) = 29.3 \angle42.1°$ V
$\tilde{V}_{ab}(3.0k) = 32.4\angle43.3°$ V
$\tilde{I}_L(1.0k) = 15.7\angle36.4°$ mA
$\tilde{I}_L(1.5k) = 14.1\angle38.7°$ mA
$\tilde{I}_L(2.0k) = 12.8\angle40.5°$ mA
$\tilde{I}_L(2.5k) = 11.7\angle42.1°$ mA
$\tilde{I}_L(3.0k) = 10.8\angle43.3°$ mA

The results are the same as those for problem 12.29 because $\tilde{V}_{Th}$ and $\tilde{Z}_{Th}$ are the same for this circuit.

12.33 The Norton equivalent AC circuit is an equivalent circuit that consists of a parallel AC current source and a parallel impedance.

12.35 $\tilde{I}_N = 339\angle-7.6°$ mA
$\tilde{Z}_N = (305-j15.6)$ Ω
$V_{ab}(300 = 51.6\angle-8.9°$ V
$\tilde{V}_{ab}(400) = 58.7\angle-9.2°$ V
$\tilde{V}_{ab}(500) = 64.3\angle-9.3°$ V
$\tilde{V}_{ab}(600) = 68.6\angle-9.4°$ V
$\tilde{V}_{ab}(700) = 72.1\angle-9.5°$ V

12.37 $\tilde{I}_N = 20.0\angle30.0°$ mA
$\tilde{Z}_N = (2.92+j1.56)$ kΩ
$\tilde{V}_{ab}(1.0k) = 15.7\angle36.4°$ V
$\tilde{V}_{ab}(1.5k) = 21.2\angle38.7°$ V
$\tilde{V}_{ab}(2.0k) = 25.7\angle40.5°$ V
$\tilde{V}_{ab}(2.5k) = 29.3\angle42.1°$ V
$\tilde{V}_{ab}(3.0k) = 32.4\angle43.3°$ V
$\tilde{I}_L(1.0k) = 15.7\angle36.4°$ mA
$\tilde{I}_L(1.5k) = 14.1\angle38.7°$ mA

$\tilde{I}_L(2.0k) = 12.8\angle40.5°$ mA
$\tilde{I}_L(2.5k) = 11.7\angle42.1°$ mA
$\tilde{I}_L(3.0k) = 10.8\angle43.3°$ mA

12.39 $\tilde{I}_N = 20.0\angle30.0°$ mA
$\tilde{Z}_N = (2.92+j1.56)$ kΩ
$\tilde{V}_{ab}(1.0k) = 15.7\angle36.4°$ V
$\tilde{V}_{ab}(1.5k) = 21.2\angle38.7°$ V
$\tilde{V}_{ab}(2.0k) = 25.7\angle40.5°$ V
$\tilde{V}_{ab}(2.5k) = 29.3\angle42.1°$ V
$\tilde{V}_{ab}(3.0k) = 32.4\angle43.3°$ V
$\tilde{I}_L(1.0k) = 15.7\angle36.4°$ mA
$\tilde{I}_L(1.5k) = 14.1\angle38.7°$ mA
$\tilde{I}_L(2.0k) = 12.8\angle40.5°$ mA
$\tilde{I}_L(2.5k) = 11.7\angle42.1°$ mA
$\tilde{I}_L(3.0k) = 10.8\angle43.3°$ mA

The results are the same as those for problem 12.29 and 12.37 because $\tilde{I}_N$ and $\tilde{Z}_N$ are the same for the circuits.

12.41 A Thévenin equivalent AC circuit can be converted to a Norton equivalent AC circuit by doing a source conversion.

12.43 The circuits are the same as those found in problems 12.35, 12.37 and 12.39. The difference between the circuit in Figure P12.9 and P12.11 is that the voltage source in Figure P12.9 was converted to a current source in Figure P12.11.

12.45
   a. $R_L = 200$
   b. max power = 3.125 W

12.47
   a. 289 Ω; 2.50 kΩ; 2.50 kΩ
   b. 1.02 W; 2.505 mW; 2.505 mW

12.49
   a. 283 Ω
   b. $(200 - j200)$ Ω
   c. 14.9 W
   d. 18 W

12.51
   a. 305 Ω; 3.31 kΩ; 3.31 kΩ
   b. $(305 + j15.6)$ Ω; $(2.92 - j1.56)$ kΩ; $(2.92 - j1.56)$ kΩ
   c. 8.79 W; 0.351 W; 0.351 W

Chapter 13

13.1 The physical phenomenon is the changing magnetic flux of the first coil that also simultaneously couples the second coil. The changing flux (specifically the rate of change) is the source of the induced voltage in the second coil.

13.3 No, the best that can be achieved is for the transformer to be lossless and have the power at the output of the transformer be equal to the power at the input of the transformer.

13.5 5 V

13.7 $\tilde{V}_S = 2.50\angle0°$ V
$\tilde{I}_S = 0.125\angle180°$ A

13.9 $\tilde{Z}_{in} = 13.5 + j10.1$ kΩ

13.11

  a. $\tilde{Z}_{in} = 12.0 + j20.0$ kΩ

  b. $\tilde{I}_P = 10.3\angle -14.0°$ mA

  c. $\tilde{V}_S = 12.0\angle +45.0°$ V
    $\tilde{I}_L = 0.206\angle -14.0°$ mA

  d. $\tilde{S}_{in} = 1.27$ W $+ j2.12$ VAR

  e. $P_{RL} = 1.27$ W

  f. The power dissipated in the resistor is the real part of the complex power.

13.13

  a. $\tilde{Z}_{in} = 63.0 - j45.0$ kΩ

  b. $\tilde{S}_L = 52.6$ mW $- j37.5$ mVAR

  c. $P_{RL} = 52.6$ mW

13.15

  a. $\tilde{Z}_{in} = 67.6 - j45.0$ kΩ

  b. $\tilde{S}_L = 47.7$ mW $- j34.1$ mVAR

  c. $P_{RL} = 47.7$ mW
    $P_{120} = 90.9$ μW
    $P_5 = 3.41$ mW

13.17 The dot convention serves to establish the polarity of the voltage across a side of the transformer based on current direction on the other side.

13.19 k = 0.693

13.21

  a. $\tilde{Z}_{in} = 0.719 + j15.2$ Ω

  b. $\tilde{V}_S = 24.4\angle -16.1°$ V
    $\tilde{I}_L = 1.44\angle -44.1°$ A

  c. $\tilde{I}_P = 6.56\angle -87.3°$ A

13.23 $L_T = 27$ mH

13.25

  a. The mutual inductor model is used because the coil reactances are not much greater than the other reactances and resistances in the circuit.

  b. $\tilde{Z}_{in} = 10.5 + j24.7$ Ω

  c. $\tilde{V}_S = 13.3\angle 0°$ V
    $\tilde{I}_L = 0.569\angle -39.8°$ A

  d. $\tilde{I}_P = 0.747\angle -67.0°$ A

Chapter 14

14.1

  a. The instantaneous power is constant. The average power and the instantaneous power are identical. The power with three AC sinusoidal steady-state signals that are 120° apart does not pulsate.

  b. The total power is three times the power of one source.

  c. Only three lines are required, and the resultant savings in cost and resources are truly significant.

14.3 120°

14.5 $v_1(t) = V_p \sin(377t + 0°)$
    $v_2(t) = V_p \sin(377t - 120°)$
    $v_3(t) = V_p \sin(377t + 120°)$

14.7

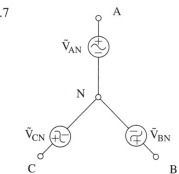

14.9

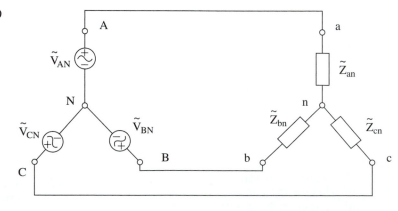

14.11 $|\tilde{I}| = 21.5$ mA

$\tilde{S}_T = 13.8$ W $+ j6.91$ VAR

14.13 $|\tilde{I}| = 235$ mA

$\tilde{S}_T = 16.6$ W $+ j83.1$ VAR

14.15 $|\tilde{I}| = 0.620$ A

$\tilde{S}_T = 115$ W $+ j57.6$ VAR

14.17 $|\tilde{I}| = 37.2$ mA

$\tilde{S}_T = 41.5$ W $+ j20.7$ VAR

Chapter 15

15.1

a. 25.5 (fundamental = first harmonic)

b. 8.49 (third harmonic)

c. 5.09 (fifth harmonic)

d. 3.64 (seventh harmonic)

15.3

a. 108.2 (DC)

b. 72.2 (fundamental = first harmonic)

c. −14.3 (second harmonic)

d. 6.18 (third harmonic)

15.5 A frequency filter is a circuit that selectively allows a range of frequencies to reach the output and rejects all remaining frequencies.

The frequency response of a filter is the magnitude and the phase of the output signal in response to the entire range of possible frequencies of the source (either voltage or current). The response is usually displayed in the form of plots of magnitude versus frequency and phase versus frequency.

15.7 The final mathematical form is arranged such that every frequency dependent component that comprises a real and an imaginary term is of the form:

$$1 + j\left(\frac{\omega}{\omega_b}\right)$$

The form is useful because it focuses on the key aspect of the break frequency and allows a direct plot of the function to be made without any actual evaluation of the function.

15.9

$$|H|_{dB} = -20 \log\sqrt{1 + \left(\frac{\omega}{\omega_b}\right)^2}$$

$$\angle H = -\tan^{-1}\left(\frac{\omega}{\omega_b}\right)$$

$$\omega_b = \frac{R}{L}$$

15.11

$$|H|_{dB} = +20 \log\left(\frac{R_2}{R_1 + R_2}\right) + 20 \log\sqrt{1 + \left(\frac{\omega}{\omega_{b1}}\right)^2} - 20 \log\sqrt{1 + \left(\frac{\omega}{\omega_{b2}}\right)^2}$$

$$\angle H = +\tan^{-1}\left(\frac{\omega}{\omega_{b1}}\right) - \tan^{-1}\left(\frac{\omega}{\omega_{b2}}\right)$$

$$\omega_{b1} = \frac{R_2}{L} \quad \omega_{b2} = \frac{R_1 + R_2}{L}$$

15.13

$$|H|_{dB} = +20 \log\left(\frac{\omega}{\omega_{b1}}\right) - 20 \log\sqrt{1 + \left(\frac{\omega}{\omega_{b2}}\right)^2}$$

$$\angle H = +90° - \tan^{-1}\left(\frac{\omega}{\omega_{b2}}\right)$$

$$\omega_{b1} = \frac{R}{L_2} \quad \omega_{b2} = \frac{R}{L_1 + L_2}$$

15.15

$$|H|_{dB} = +20 \log\left(\frac{R_2}{R_1 + R_2}\right) + 20 \log\sqrt{1 + \left(\frac{\omega}{\omega_{b1}}\right)^2} - 20 \log\sqrt{1 + \left(\frac{\omega}{\omega_{b2}}\right)^2}$$

$$\angle H = +\tan^{-1}\left(\frac{\omega}{\omega_{b1}}\right) - \tan^{-1}\left(\frac{\omega}{\omega_{b2}}\right)$$

$$\omega_{b1} = \frac{R_2}{L_2} \quad \omega_{b2} = \frac{R_1 + R_2}{L_1 + L_2}$$

15.17 2.301 and 1.301; 4.301 and 3.301; when the magnitude of any number is changed by a factor of 10, the log changes by 1.

15.19 x(Bel) $\equiv \log_{10}(x)$ where $x =$ (power1/power2), i.e., a power ratio

x(dB) $\equiv 10 \log_{10}(x)$

Both are dimensionless.

15.21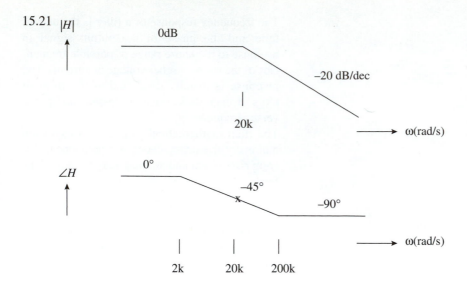

15.23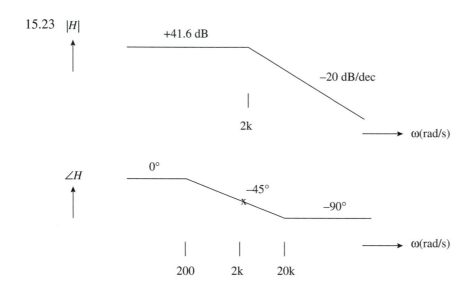

15.25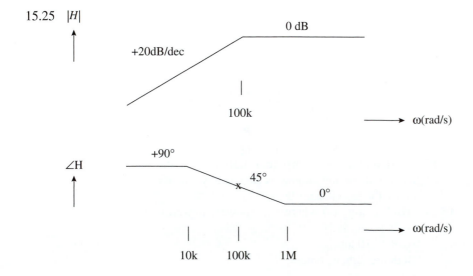

15.27 |H|

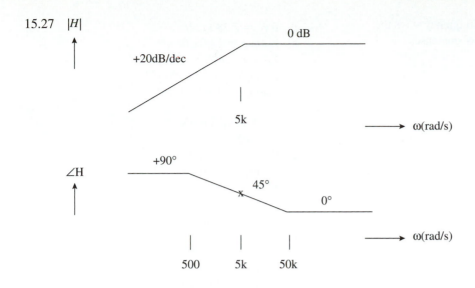

15.29 |H|

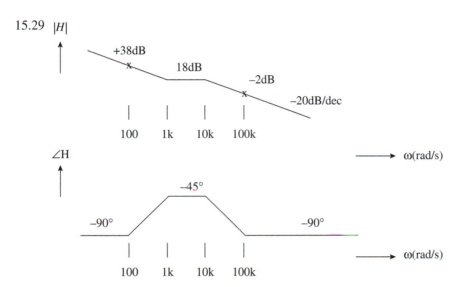

15.31

$19.6\angle-39.8°$ V (first harmonic)

$3.15\angle-68.2°$ V (third harmonic)

$1.19\angle-76.5°$ V (fifth harmonic)

$0.615\angle-80.3°$ V (seventh harmonic)

No. A RL LPF with the same break frequency will provide the same result.

Chapter 16

16.1 The imaginary part of the impedance (or admittance) is equal to zero.

16.3 The phase difference between the inductor and capacitor voltages is always 180°, regardless of whether the circuit is at resonance or not. But at resonance the voltage magnitudes are equal to each other, resulting in the inductor and capacitor impedances (or admittances) canceling each other.

16.5 Q is a measure of the amount of energy stored in a circuit relative to the energy loss per cycle in the circuit.

16.7 The bandwidth is defined as the difference in frequency between the upper and lower 3 dB frequencies. The resonant frequency is not in the center of the bandwidth because the frequency response is not symmetrical about the resonant frequency in a low Q resonant circuit.

16.9 $\omega_o = 14.1$ krad/s
$f_o = 2.25$ kHz
$Q = 1.77$
$f_L = 1.70$ kHz
$f_U = 2.98$ kHz

16.11 $\omega_o = 1.12$ krad/s
$Q = 5.59$
BW = 200. rad/s

series circuit: resonant frequency stays the same, Q halves, and BW doubles;

parallel circuit: resonant frequency stays the same, Q increases, and BW decreases

16.13 $\omega_o = 9.27$ krad/s
-17.1%

16.15 $\tilde{Z} = 834\angle 0° \, \Omega$

16.17

    a. $\omega_o = 645.5$ krad/s

    b. $Q = 4.303$

    c. $\omega_L = 575$ krad/s
       $\omega_U = 725$ krad/s

    d. BW $= 150.$krad/s

    e. 0.67%

    f. $\tilde{Z}_T(\Omega_o) = 6.00\angle 0° \, \Omega$
      $\tilde{Z}_T(\omega_L) = 6.00 - j6.00 \, \Omega$
      $\tilde{Z}_T(\omega_U) = 6.00 + j5.70 \, \Omega$

16.19

    a. $\omega_o = 31.6$ krad/s

    b. $Q = 11.4$

    c. $\omega_L = 30.2$ krad/s
      $\omega_U = 33.0$ krad/s

    d. BW $= 2.78$ krad/s

    e. $\tilde{Z}_T(\omega_o) = 1.80\angle 0° \, k\Omega$
      $\tilde{Z}_T(\omega_L) = 880 + j900 \, \Omega$
      $\tilde{Z}_T(\omega_U) = 920 - j900 \, \Omega$

    f. $\tilde{V}_T(\omega_o) = 1.80\angle 0° \, kV$

16.21

    a. $\omega_o = 12.85$ krad/s

    b. $Q = 10.3$

    c. BW $= 1.25$ krad/s

    d. $\tilde{Z}_T(\omega_o) = 1.60 \angle 0° \, k\Omega$

    e. $\tilde{V}_T(\omega_o) = 1.60 \angle 0° \, kV$

16.23

    a. $\omega_o = 4.44$ krad/s

    b. $Q = 4.27$

    c. BW $= 1.04$ krad/s

    d. $\tilde{Z}_T(\omega_o) = 480. \angle 0° \, \Omega$

    e. $\tilde{V}_T(\omega_o) = 480. \angle 0° \, V$

Index